Principles of **Vibration Analysis**

with Applications in Automotive Engineering

Other Related Publications Available from SAE International:

Vehicle Refinement
By Matthew Harrison
(Product Code: R-364)

Vehicle Vibration and Sound
By Gang Sheng
(Product Code: R-400)

Car Suspension and Handling, Fourth Edition
By Geoffrey Howard
(Product Code R-318)

Principles of **Vibration Analysis**

with Applications in Automotive Engineering

C. Q. Liu and
Ronald L. Huston

SAE International®

Warrendale, Pennsylvania
USA

SAE International®
400 Commonwealth Drive
Warrendale, PA 15096-0001 USA

E-mail: CustomerService@sae.org
Phone: 877-606-7323 (inside USA and Canada)
724-776-4970 (outside USA)
Fax: 724-776-1615

ISBN 978-0-7680-3339-7
Library of Congress Catalog Number 2010940965
SAE Order No. R-395

To purchase bulk quantities, please contact:
SAE Customer Service
E-mail: CustomerService@sae.org
Phone: 877-606-7323 (inside USA and Canada) 724-776-4970 (outside USA)
Fax: 724-776-1615

Visit the SAE Bookstore at
http://store.sae.org

Contents

Preface

Over the years there have been many outstanding books written on vibrations, including textbooks, books on analysis and design, and theoretical treatises. Why then should there be another, and, specifically, why this book?

There are several reasons: First, vibrations, as with all science and engineering subjects, is continually evolving. As new discoveries and new analyses and new experimental procedures are developed, there is an ongoing need to document these advances.

Next, with the evolution in the subject, many of the older procedures become obsolete. Traditional topical emphases become less important, and are superseded by advances in computational and measurement methods. Consequently, new topical emphases are emerging. There is a need to document these new methods for students and for practicing engineers.

Finally, it is always helpful to have a state-of-the-art treatise on fundamentals enabling future research and application.

We wrote this book attempting to satisfy each of these needs. The book summarizes basic concepts and established analytical methodologies. We then illustrate these ideas and procedures via worked-out examples. Numerous examples using Matlab are presented throughout the text. The various chapters contain problems (with answers) for readers studying the subject for the first time, and for those seeking additional expertise and/or review. In the later chapters, we illustrate application in automotive engineering with a focus upon engine balancing and engine mounting systems. Finally, the book is intended to be a quick reference for current (circa 2010) modeling and analysis methods.

The book is divided into twelve chapters, with the first of these providing an introduction to vibration terminology. In the second chapter we review the fundamentals of single-degree-of-freedom systems, and in the third chapter we consider harmonically excited motion.

The fourth chapter applies these concepts with rotors, outlining balancing methods.

In the fifth chapter we study multi-degree-of-freedom systems, and then we devote a chapter to numerical methods. In the seventh chapter we introduce the concept of transfer functions and frequency response functions together with a detailed discussion about these functions.

Chapter eight explores lumped parameter modeling and chapter nine follows with analyses of continuous systems.

Chapter ten provides an extensive exposition of engine mounting systems.

Chapter eleven summarizes experimental modal analysis methods, including the basic theory and an account of signal processing and test procedures.

In the final chapter, we conclude with some special topics, such as the frequency response function (FRF)-based substructure method, the hybrid finite element and experimental (HYFEX) method, system sensitivity analysis, and vibration absorber tuning.

Prior to writing this book we were fortunate to have studied under and been associated with academicians and engineers in industry who have significantly advanced the state of the art through their research, development, application, and exposition of both theoretical and experimental breakthroughs. We are particularly appreciative of the inspiration provided by the innovations of faculty (David L. Brown and Randall J. Allemang) and students at the Structural Dynamics Research Laboratory (SDRL) at the University of Cincinnati, and of engineers in local industries, and particularly in the automotive industry. We have relied heavily upon these sources for our exposition.

We have also, of course, relied upon many others, as cited in the references and permissions. We have attempted to acknowledge all sources, but undoubtedly have missed many which should have been cited. We apologize for any such omissions.

A number of the examples in this book have originated with the references listed in the chapters. Although we attempt to acknowledge these sources, the examples are not direct quotations. Instead they are usually done differently and often more extensively than in the quoted sources.

We appreciate the assistance of Charlotte Better in preparing the manuscript. We also appreciate the assistance of Aaron Lock in preparing the artwork for this book.

Finally, we are grateful to the Society of Automotive Engineers (SAE) for their interest and their acceptance of our work.

C. Q. Liu
Ronald L. Huston

Chapter One

Introduction

1.1 What Is Vibration?

A *vibration* is simply an oscillatory (back and forth) motion. The direction of movement reverses twice in every cycle. A motion repeating itself in equal time intervals is called a *periodic vibration*. If a periodic vibration can be represented by a simple sinusoid (a sine or cosine function), it is called a *simple harmonic vibration* or a *simple harmonic motion*. Alternatively, a vibration or oscillatory motion that is not periodic is called a *random vibration*.

If the extent of the back and forth motion (or *amplitude* of the vibration) becomes increasingly smaller, the vibration is said to be *damped* or *transient*.

The study of vibration is concerned with both the analysis of the motion itself and the analysis of the forces causing the motion.

All bodies possessing mass and elasticity are capable of vibration. Thus, most engineering machines and structures experience vibration to some degree, and their design generally requires consideration of their vibration behavior.

Vibration can be extremely useful and desired, or it can be harmful and a phenomenon to be avoided or suppressed. For example, electronically induced vibration of speaker cones produces music for entertainment. However, the vibration of an automobile affects the quality of ride; the vibration of a bridge due to an earthquake results in damage. Therefore, a thorough and precise knowledge about vibration is needed, and this forms the subject of this book.

In the study of vibration, we usually use the term *system*. A system is defined as an assemblage of components acting as a whole. Vibration of a system can be illustrated by the block diagram of Fig. 1.1-1, which includes the Input (excitation), the System (the objects we are studying), and the Output (response). Fig. 1.1-2 shows a simple impact test setup illustrating the concept of Fig. 1.1-1.

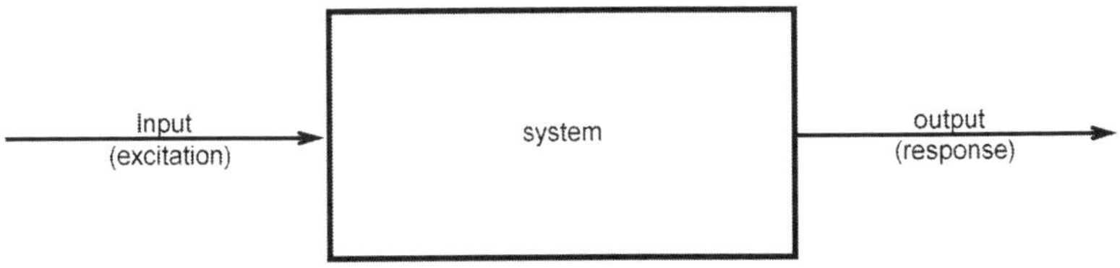

Fig. 1.1-1 A model of a vibration system.

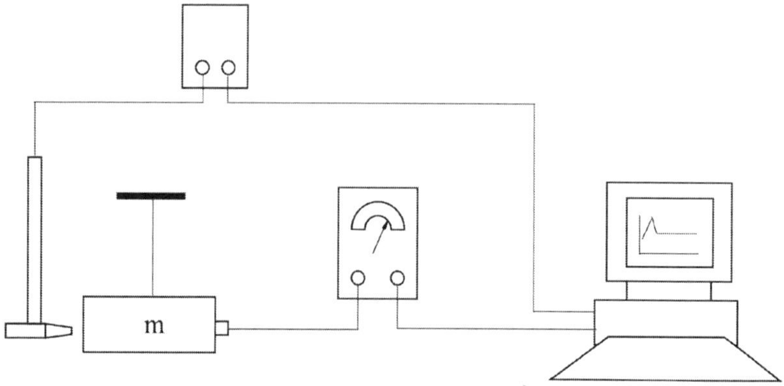

Fig. 1.1-2 Experimental impact test setup illustrating the vibration analysis of Fig. 1.1-1. **System:** mass; **Input:** by hammer; **Output:** measured by transducer.

1.2 Classification of Vibration Studies

Observe in Fig. 1.1-1 that there are three entities: the input, the system itself, and the output. Many engineering problems consist of studying the interaction of three entities, or parameters. If two of the parameters are known, the objective is to determine the third. Accordingly, we can define four types of vibration problems.

(1) Vibration Analysis

In this case, the input and the system characteristics are known but the output behavior is unknown. The problem of obtaining the response of a system when the excitation and the system characteristics are given is known as *vibration analysis*.

(2) System Design

In this case, the input and desired output are known and the objective is to design a system that exhibits the desired response. This type of problem is sometimes called *system design* or *synthesis*. It may occur that many systems are able to produce the desired response. The problem of obtaining the most satisfactory, or best, system, when all pertinent factors are taken into account, is called *optimum design*.

(3) Input Evaluation

In this case, the system characteristics and the output are known and the objective is to determine the excitation function. Here, the system is used to record the response, from which the excitation can be deduced.

(4) System Identification

In this case, the input and output are known but the system characteristics are unknown. The problem of determining the system characteristics when the excitation and the response are given is called *system identification*. Experimental modal analysis falls into this category.

1.3 Classification of Vibration

(1) Classification Based on Input

(a) Free vibration. This type of vibration occurs without a driving force. It takes place under the action of forces inherent in the system itself, or when external impressed forces are absent. The vibration is caused by an initial condition when external impressed forces are absent. For example, consider a mass suspended from a light spring as shown in Fig. 1.3-1. If the mass is displaced from its rest position and released, it will oscillate up and down.

(b) Forced vibration. This type of vibration occurs due to the excitation of external forces. Building vibrations in earthquakes and ground vibration due to an unbalanced motor are examples of forced vibration.

(c) Self-excited vibration. This type of vibration develops due to the motion itself. That is, there is a *feed-back* where more energy is put into the system than is taken out during the motion. This type of vibration in nature is different from the forced vibration. The essence of the difference can be best seen from a few examples.

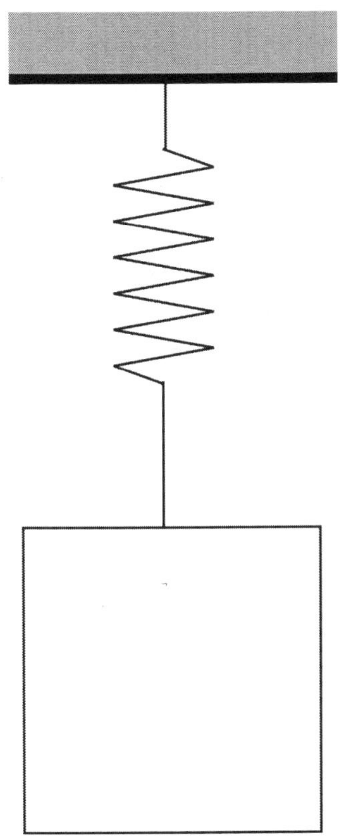

Fig. 1.3-1 A mass-spring system.

First, consider an unbalanced disk mounted on a flexible shaft supported by two bearings as shown in Fig. 1.3-2 [1.1]. The center of the disk vibrates, with the motion being maintained by the centrifugal force of the unbalance pushing the disk alternatively up and down. The amplitude increases as the rotating speed of the shaft increases, since the centrifugal force is proportional to the square of the angular velocity of the shaft. This is the reason driveline shafts of a vehicle must be balanced before being put into service, for, otherwise, the unbalanced shaft will produce serious self-excited vibration of the vehicle when driven on a highway.

Next, consider the motion of a mass-spring system on a moving belt as in Fig. 1.3-3 [1.2, 1.3, 1.4]. The belt with a horizontal section

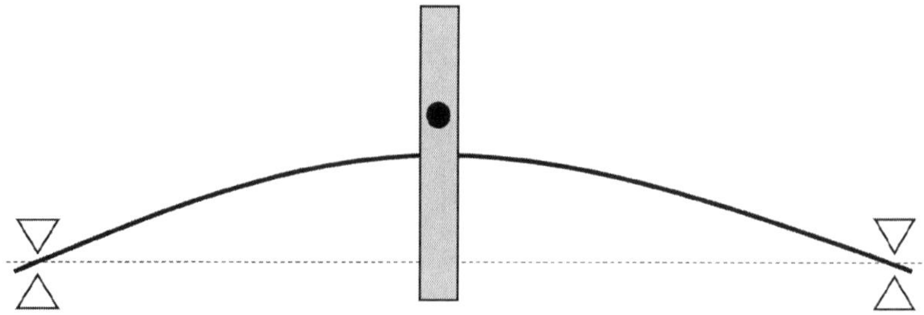

Fig. 1.3-2 An unbalanced disk mounted on a flexible shaft.

moves at constant velocity driven by a pair of pulleys. On the horizontal section there is a mass attached to a spring. This is the classical example of self-excited vibration. When the mass is moving to the left, the relative velocity between it and the belt is greater than when it is moving to the right. Thus, the coefficients of kinetic friction satisfy the relation $\mu_{kl} \leq \mu_{kr}$, where the subscripts l and r refer to the mass moving to the left and moving to the right, respectively. It is evident that the work done by the friction force while moving to the right is greater than that while moving to the left. Therefore, more energy is put into the mass-spring system than is taken out. In this type of self-excited oscillation, the amplitude continues to increase.

(2) Classification Based on Output

(a) Simple harmonic vibration. In this case, the output of the system can be described by a sine or cosine function of time. For example, for the spring-mass system shown in Fig. 1.3-4 (a), the output free vibration can be described by a sine function as

$$x(t) = A\sin(\omega t + \phi) \tag{1.3-1}$$

Fig. 1.3-4 (b) shows the time history.

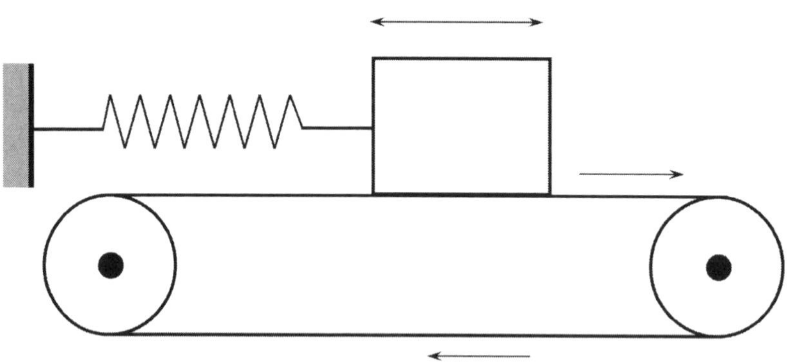

Fig. 1.3-3 A belt friction system [1.2, 1.3, 1.4].

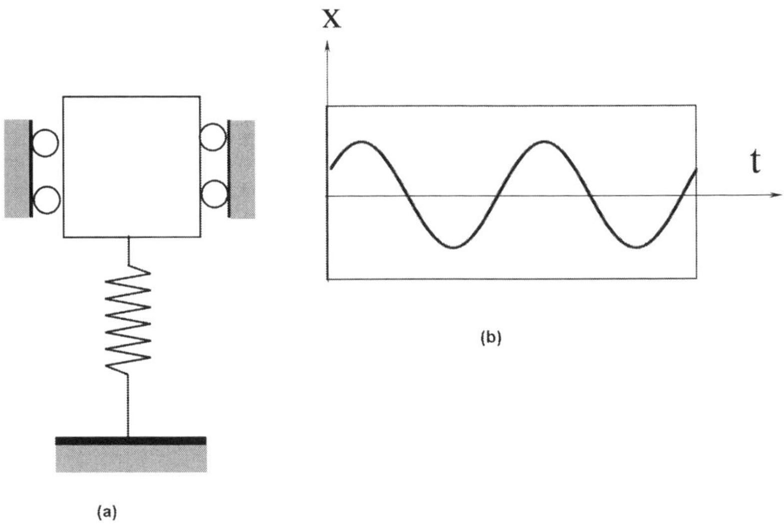

Fig. 1.3-4 Simple harmonic vibration of a spring-mass system.

(b) Periodic vibration. This type of vibration repeats itself in equal time intervals. A periodic vibration may include several frequencies simultaneously. For example, an automobile body is a multi-degree-of-freedom system, with vibration contribution at each natural frequency. Fig. 1.3-5 shows an example of a periodic signal.

(c) Transient vibration. In this case, the output of the system is not periodic, and the duration is short. Fig. 1.3-6 shows an example. An impulse or shock to a system generally produces a transient response. Due to the presence of damping,

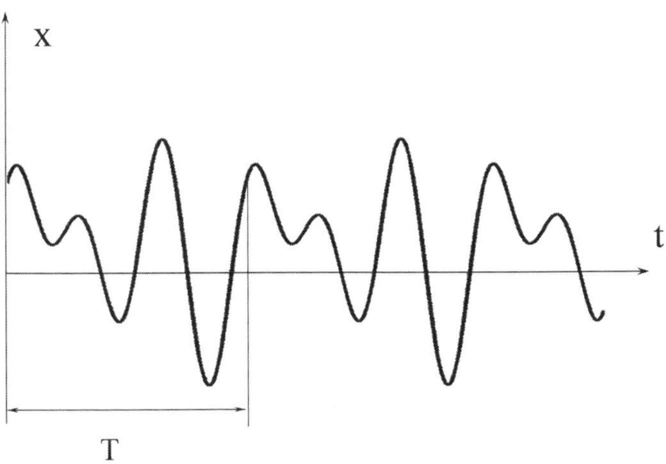

Fig. 1.3-5 Example of periodic motion.

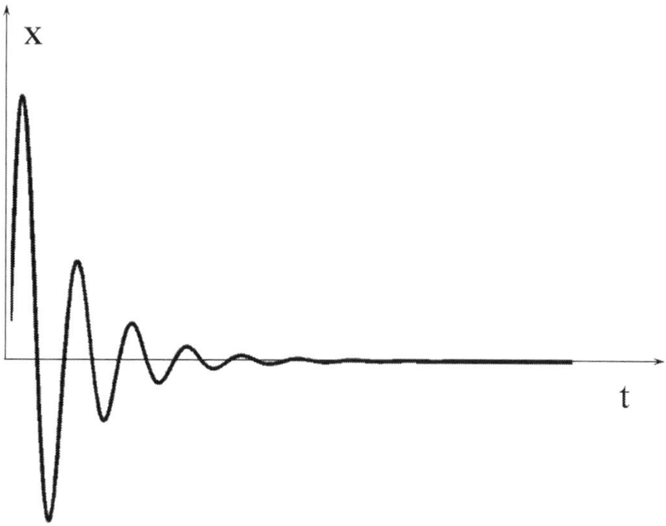

Fig. 1.3-6 Example of transient motion.

such vibrations decay and disappear after the excitation is over. A transient response can be regarded as a periodic function with an infinite period.

(d) Random vibration. In this case, the output of the system is not a deterministic function of time. That is, future instantaneous values cannot be predicted in a deterministic sense. The response can be studied only by using a statistical approach. The output of a noise generator, the heights of waves in a choppy sea, and ground motion during an earthquake are examples of this type of vibration. These phenomena all have one thing in common: the unpredictability of their instantaneous value at any future time. Fig. 1.3-7 shows a record of a random time function.

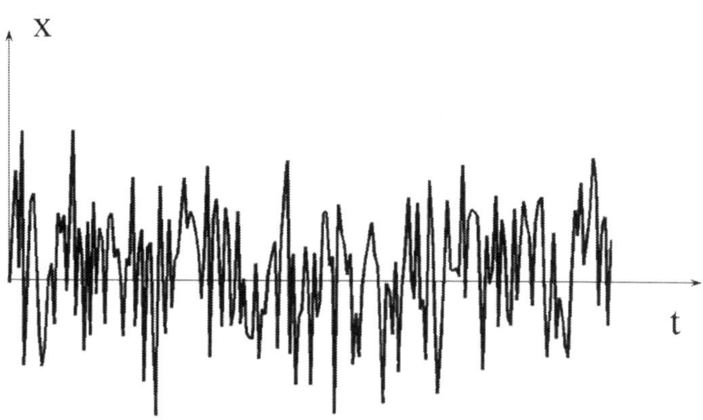

Fig. 1.3-7 A record of a random time function.

(3) Classification Based on the Degrees of Freedom of the System

(a) Single-degree-of-freedom system. The number of independent coordinates required to describe the motion of a system is called the *degrees of freedom* of the system. If the response (output) of the system can be described by one coordinate, it is called a single-degree-of-freedom system. A simple mass-spring system vibrating vertically is such an example.

(b) Multi-degree-of-freedom system. If the response (output) of the system requires two or more coordinates to describe it, the system is called a *multi-degree-of-freedom* system. Multi-degree-of-freedom systems may be modeled as arrangements of various lumped masses separated by springs and dampers. In this sense, the parameters of the system are discrete sets of finite numbers. Hence, such systems are also called *lumped-parameter systems,* or *discrete systems,* or *finite-dimensional systems.*

(c) Continuous systems. Continuous systems, also called *infinite-dimensional systems,* or *distributed-parameter systems,* have infinite degrees of freedom. The physical properties of the system are functions of spatial coordinates. The motion of the system is then described by partial differential equations.

(4) Classification Based on the Differential Equation of Motion of the System

(a) Nonlinear system. In this case, the differential equation of motion of the system includes nonlinear terms. That is, the dependent variables may be raised to powers, and cross products of the dependent variables may be present. As a consequence, the superposition principle no longer holds. Hence, the use of many usual mathematical methods are ineffective.

Consider the simple pendulum, as shown in Fig. 1.3-8. The equation of motion for this system is given by [1.5]

$$\ddot{\theta}+\frac{g}{L}\sin\theta=0 \tag{1.3-2}$$

With the expression, $\sin\theta=\theta-(1/6)\theta^3+\cdots$, Eq. 1.3-2 can be written approximately as

$$\ddot{\theta}+\frac{g}{L}\left(\theta-\frac{1}{6}\theta^3\right)=0 \tag{1.3-3}$$

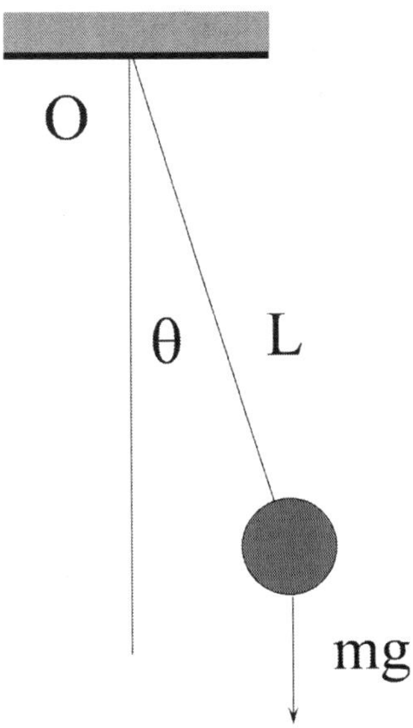

Fig. 1.3-8 A single pendulum.

This nonlinear equation is known as the free undamped *Duffing equation*, which can be written in a more general form as

$$\ddot{\theta}+\omega_0^2\theta+\alpha\theta^3=0, \tag{1.3-4}$$

where $\alpha=-\omega_0^2/6=-g/(6L)$ and $\omega_0^2=g/L$.

The following are some well-known differential equations of nonlinear vibrations:

The Duffing equation of forced vibration [1.2, 1.3, 1.6]:

$$m\ddot{x}+c\dot{x}+kx\pm\mu x^3=F\sin\omega t \tag{1.3-5}$$

The Mathieu equation [1.2, 1.7]:

$$\frac{d^2x}{d\lambda^2}+(a-2b\cos 2\lambda)x=0 \tag{1.3-6}$$

The van der Pol equation [1.2, 1.7, 1.8]:

$$\ddot{x}-\mu(1-x^2)\dot{x}+x=0 \tag{1.3-7}$$

(b) Linear system. If the motion of a system can be described by a linear differential equation with constant coefficients, it is called a *linear system.* In this case, the superposition principle holds and a large number of mathematical tools are available for analysis.

Strictly speaking, all systems are nonlinear. However, if the system obeys the *small-vibration assumption*, one can safely use the linear theory. For example, for the simple pendulum as in Fig. 1.3-8, if the angle θ remains relatively small so that $\sin\theta\approx\theta$, then the differential equation of motion becomes linear:

$$\ddot{\theta}+\frac{g}{L}\theta=0 \tag{1.3-8}$$

In using this result, of course, one must always be sure that the small vibration assumption is not violated.

1.4 Harmonic Vibration

Harmonic vibration is the simplest of the periodic vibrations. It can be illustrated by a spring-mass system as in Fig. 1.4-1. It is readily seen that the motion of the system can then be expressed as

$$x(t)=A\sin(\omega_n t+\phi), \tag{1.4-1}$$

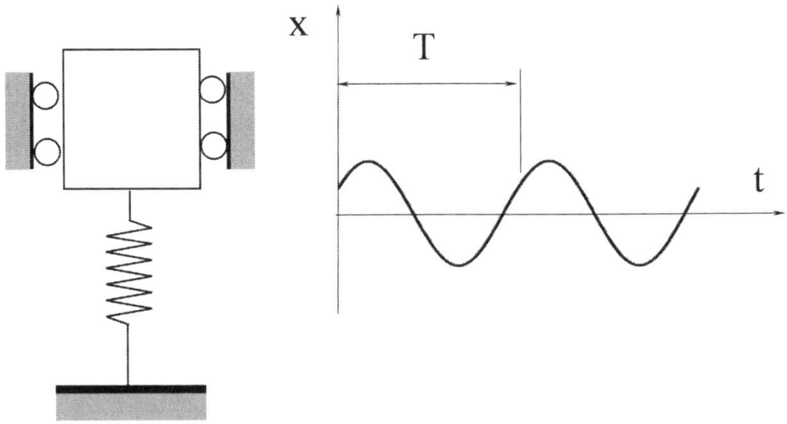

Fig. 1.4-1 Harmonic vibration of a mass-spring system.

where the constant A is the amplitude of the vibration, ω_n is the *angular natural frequency* or *circular natural frequency*, and ϕ is called the *phase angle*. It is standard to measure the time in seconds (s). The phase is measured in radians (rad) and the angular frequency is measured in radians per second (rad/s). Since the motion repeats itself in 2π radians, we have the relationship

$$\omega_n = \frac{2\pi}{T} = 2\pi f, \tag{1.4-2}$$

where T and f are, respectively, the *period* and the *frequency* of the harmonic vibration. The period T usually is measured in seconds; the frequency $f = 1/T$ usually is measured in cycles per second, also called Hertz, and simply denoted by Hz.

Successive differentiation of the displacement $x(t)$ in the form of Eq. 1.4-1 yields the velocity, $\dot{x}(t)$, given by

$$\dot{x} = A\omega_n \cos(\omega_n t + \phi) \tag{1.4-3}$$

The acceleration, $\ddot{x}(t)$, is then given by

$$\ddot{x} = -A\omega_n^2 \sin(\omega_n t + \phi) \tag{1.4-4}$$

Comparison of Eqs. 1.4-1 and 1.4-4 yields

$$\ddot{x} + \omega_n^2 x = 0 \tag{1.4-5}$$

Therefore, if the differential equation of motion of a system can be written in the form of Eq. 1.4-5, its motion is a simple harmonic vibration. The solution is given by Eq. 1.4-1.

Note that Eq. 1.4-5 is of second order. Thus, the solution (Eq. 1.4-1) includes two constants of integration. These are A and ϕ, which are determined by the initial displacement and the initial velocity, called the initial conditions. Hence, we have

$$x_0 = x(0) = A\sin(\omega_n 0 + \phi) = A\sin\phi \tag{1.4-6}$$

$$\dot{x}_0 = \dot{x}(0) = \omega_n A\cos(\omega_n 0 + \phi) = \omega_n A\cos\phi \tag{1.4-7}$$

Solving these two equations simultaneously for the two constants A and ϕ, we obtain

$$A = \frac{\sqrt{\omega_n^2 x_0^2 + \dot{x}_0^2}}{\omega_n} \quad \text{and} \quad \phi = \tan^{-1}\frac{\omega_n x_0}{\dot{x}_0} \tag{1.4-8}$$

Thus, the solution of the equation of motion for the spring-mass system is given by

$$x = \frac{\sqrt{\omega_n^2 x_0^2 + \dot{x}_0^2}}{\omega_n}\sin\left(\omega_n t + \tan^{-1}\frac{\omega_n x_0}{\dot{x}_0}\right) \tag{1.4-9}$$

1.5 Harmonic Analysis

The free vibration of a multi-degree-of-freedom system is not a simple harmonic vibration. Indeed, such vibrations contain different natural frequencies simultaneously, and result in a complex waveform that repeats periodically.

The French mathematician J. Fourier (1768–1830) showed that any periodic motion can be represented by a series of sines and cosines that are harmonically related. Based on this concept, the modern vibration analyzer can perform the so-called harmonic analysis for such complex periodic waveforms.

If $x(t)$ is a periodic function of period T, then the $x(t)$ can be represented by a *Fourier series* [1.9] as

$$x(t) = \frac{a_0}{2} + \sum_{n=1}^{\infty}\left[a_n\cos(n\omega_0 t) + b_n\sin(n\omega_0 t)\right], \tag{1.5-1}$$

where $\omega_0 = \dfrac{2\pi}{T}$ is called the fundamental frequency, and where the coefficients are given as follows:

(1) If $x(t)$ is defined over the region $-T/2<t<T/2$, then

$$a_n = \frac{\omega_0}{\pi}\int_{-T/2}^{T/2} x(t)\cos(n\omega_0 t)dt \qquad (n=0,1,2,\cdots) \tag{1.5-2a}$$

and

$$b_n = \frac{\omega_0}{\pi}\int_{-T/2}^{T/2} x(t)\sin(n\omega_0 t)dt \qquad (n=1,2,\cdots) \tag{1.5-3a}$$

(2) If $x(t)$ is defined over the region $0<t<T$, then

$$a_n = \frac{\omega_0}{\pi}\int_{0}^{T} x(t)\cos(n\omega_0 t)dt \qquad (n=0,1,2,\cdots) \tag{1.5-2b}$$

and

$$b_n = \frac{\omega_0}{\pi}\int_{0}^{T} x(t)\sin(n\omega_0 t)dt \qquad (n=1,2,\cdots) \tag{1.5-3b}$$

Note that, if $x(t)$ is an even function [that is, $x(t)=x(-t)$], then $b_n=0$, and there are no sine components. If $x(t)$ is an odd function [that is, $x(t)=-x(-t)$], then $a_n=0$, and there are no cosine components.

Next, note that Eq. 1.5-1 can be written in different forms. For example, it can be written as

$$x(t) = \frac{a_0}{2} + \sum_{n=1}^{\infty} A_n \cos(n\omega_0 t - \phi_n), \tag{1.5-4}$$

where

$$A_n = \sqrt{a_n^2 + b_n^2} = \text{Fourier spectrum; } \phi_n = \text{phase.}$$

Thus, A_n and ϕ_n (or a_n and b_n) completely define the harmonic contribution of the periodic function.

When A_n and ϕ_n are graphed against the frequency $n\omega_0$ for all n, the result is a series of discrete lines at ω_0, $2\omega_0$, etc. Such plots are called the *Fourier frequency spectrum* of the periodic time function.

Next, if we replace the sine and cosine functions in Eq. 1.5-1 using the Euler formulae

$$\cos n\omega_0 t = \frac{e^{jn\omega_0 t} + e^{-jn\omega_0 t}}{2}, \ \sin n\omega_0 t = \frac{e^{jn\omega_0 t} - e^{-jn\omega_0 t}}{2} \left(j=\sqrt{-1}\right), \tag{1.5-5}$$

and if we let

$$X_0 = \frac{a_0}{2}, \ X_n = \frac{a_n - jb_n}{2}, \ X_{-n} = \frac{a + jb_n}{2}, \tag{1.5-6}$$

then Eq. 1.5-1 can be written as

$$x(t)=\sum_{n=-\infty}^{\infty} X_n e^{jn\omega_0 t},\ \omega_0=\frac{2\pi}{T}, \tag{1.5-7}$$

where the complex Fourier coefficients are simply

$$X_n=\frac{\omega_0}{2\pi}\int_{-T/2}^{T/2} x(t)e^{-jn\omega_0 t}\,dt \tag{1.5-8}$$

Eqs. 1.5-7 and 1.5-8 are called the *periodic Fourier transform pair.*

A transient time signal is characterized by being constant for long time periods, with significant amplitude changes occurring over only a short time duration. A transient time function is not periodic. However, it can be regarded as a periodic function with an infinitely long period T. Its Fourier series representation becomes a Fourier integral representation [1.5]. Specifically, let

$$p\omega_0=\omega_p,\ (p+1)\omega_0-p\omega_0=\omega_0=\frac{2\pi}{T}=\Delta\omega_p; \tag{1.5-9}$$

then Eqs. 1.5-7 and 1.5-8 lead to

$$x(t)=\sum_{p=-\infty}^{\infty}\frac{1}{T}\left(TX_p\right)e^{j\omega_p t}=\frac{1}{2\pi}\sum_{p=-\infty}^{\infty}\left(TX_p\right)e^{j\omega_p t}\Delta\omega_p, \tag{1.5-10}$$

where

$$TX_p=\int_{-T/2}^{T/2} f(t)e^{-j\omega_p t}\,dt \tag{1.5-11}$$

Letting $T\rightarrow\infty$, and $\Delta\omega_p\rightarrow 0$, dropping the subscript p (so $\omega_p=\omega$), letting $TX_p=X(\omega)$, and replacing the summation by integration, we obtain

$$x(t)=\frac{1}{2\pi}\int_{-\infty}^{\infty} X(\omega)e^{j\omega t}\,d\omega \tag{1.5-12}$$

$$X(\omega)=\int_{-\infty}^{\infty} x(t)e^{-j\omega t}\,dt \tag{1.5-13}$$

Eqs. 1.5-12 and 1.5-13 form the *transient Fourier transform pair.* It should be noted that $X(\omega)$ is a complex function of the frequency. When $x(t)$ is a real-valued time function, $X(-\omega)$ is the complex conjugate of $X(\omega)$. Eqs. 1.5-12 and 1.5-13 are valid when $x(t)$ is bounded, defined over the region $t=-\infty$ to $t=\infty$, and satisfying

$$\int_{-\infty}^{\infty}\left|x(t)\right|dt<\infty \tag{1.5-14}$$

Fortunately, most functions describing physical phenomena satisfy these conditions.

Eqs. 1.5-12 and 1.5-13 indicate that all frequencies are present in transient time functions. Each frequency makes an infinitesimal contribution to the signal. This contribution is defined by

$$X(\omega)\frac{d\omega}{2\pi} = X(f)df, \tag{1.5-15}$$

where $d\omega$ or df are infinitesimal quantities. $X(\omega)$ is called the *Fourier spectral density*, which is a complex function, having real and imaginary parts (or amplitude and phase). When the amplitude and phase of $X(\omega)$ are graphed against the frequency ω, the result is a continuous curve.

The Fourier transform changes the domain of the function from the time domain to the frequency domain, as in Fig. 1.5-1. As an example, Fig. 1.5-2 shows some time functions and their frequency spectrum or spectral densities.

With the aid of the fast Fourier transform (FFT) [1.9], the harmonic analysis may be developed with minimal effort.

Example 1.5-1 Consider the periodic force $f(t) = -F_0$ for $0 < \omega_0 t < \pi$, and $f(t) = F_0$ for $\pi < \omega_0 t < 2\pi$, as in Fig. 1.5-3. Find the Fourier coefficients.

Solution: Observe that the force function $f(t)$ is an odd function. Thus, the values of $f(t)$ are equal and opposite at two points $\pm\omega_0 t$ at equal distance on both sides of the origin. An odd function cannot have cosine components. Hence, all the a_n coefficients are zero.

Next, the coefficients b_n are found from Eq. 1.5-3b as

$$\begin{aligned} b_n &= \frac{\omega_0}{\pi}\int_0^{\pi/\omega_0} -F_0 \sin(n\omega_0 t)dt + \frac{\omega_0}{\pi}\int_{\pi/\omega_0}^{2\pi/\omega_0} F_0 \sin(n\omega_0 t)dt \\ &= \frac{F_0\omega_0}{\pi}\left[\left.\frac{\cos(n\omega_0 t)}{n\omega_0}\right|_0^{\pi/\omega_0} - \left.\frac{\cos(n\omega_0 t)}{n\omega_0}\right|_{\pi/\omega_0}^{2\pi/\omega_0}\right] \\ &= \frac{2F_0}{\pi}\frac{\cos(n\pi)-1}{n} \end{aligned}$$

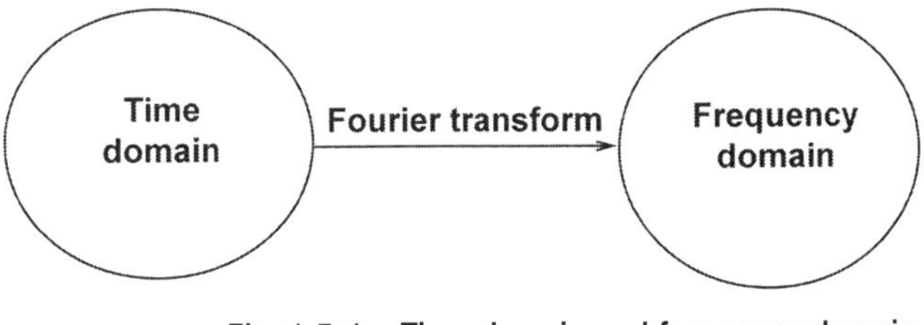

Fig. 1.5-1 Time domain and frequency domain.

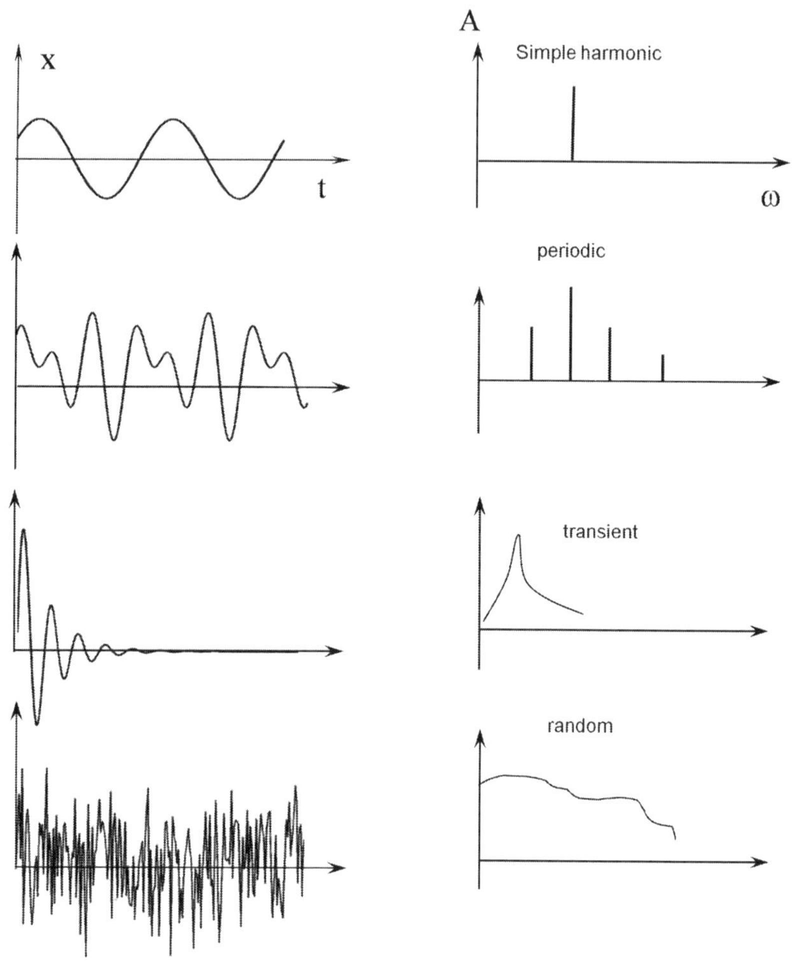

Fig. 1.5-2 Time functions and their spectrum or spectral density.

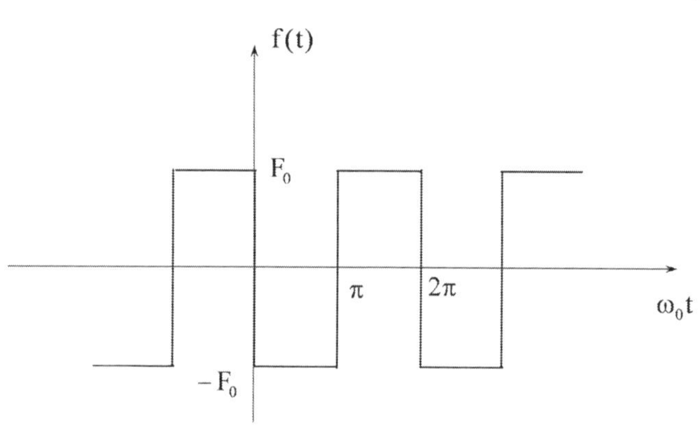

Fig. 1.5-3 Periodic force for $0 < \omega_0 t < 2\pi$.

For even orders n=2,4,6,⋯, all b_n are zero. For odd orders n=1,3,5,⋯, $\cos(n\pi)=-1$. Thus, the b_n coefficients can be expressed as

$$b_n = -\frac{4F_0}{n\pi} \ (n=1,3,5,\cdots)$$

Hence, the periodic force can be expressed as

$$f(t) = -\frac{4F_0}{\pi} \sum_{n=1,3,5,\cdots}^{\infty} \frac{1}{n} \sin(n\omega_0 t)$$

$$= -\frac{4F_0}{\pi}\left[\sin(\omega_0 t) + \frac{1}{3}\sin(3\omega_0 t) + \frac{1}{5}\sin(5\omega_0 t) + \cdots\right]$$

Example 1.5-2 Consider the periodic force during the interval $0 < \omega_0 t < 2\pi$, as shown in Fig. 1.5-4. Find the Fourier coefficients.

Solution: Note that $f(t)$ is an even function and therefore it cannot have sine components. Thus, all b_n are zero. Next, for $n=0$, we have

$$a_0 = \frac{\omega_0}{\pi}\int_0^{\pi/2\omega_0} F_0 dt - \frac{\omega_0}{\pi}\int_{\pi/2\omega_0}^{3\pi/2\omega_0} F_0 \sin\omega_0 t dt + \frac{\omega_0}{\pi}\int_{3\pi/2\omega_0}^{2\pi/\omega_0} F_0 dt = 0$$

Finally, for $n=1,2,3,\cdots$, we have

$$a_n = \frac{\omega_0}{\pi}\int_0^{\pi/2\omega_0} F_0 \cos(n\omega_0 t) dt - \frac{\omega_0}{\pi}\int_{\pi/2\omega_0}^{3\pi/2\omega_0} F_0 \cos(n\omega_0 t) dt + \frac{\omega_0}{\pi}\int_{3\pi/2\omega_0}^{2\pi/\omega_0} F_0 \cos(n\omega_0 t) dt$$

$$= \frac{2F_0}{n\pi}\left[\sin\frac{n\pi}{2} - \sin\frac{3n\pi}{2}\right]$$

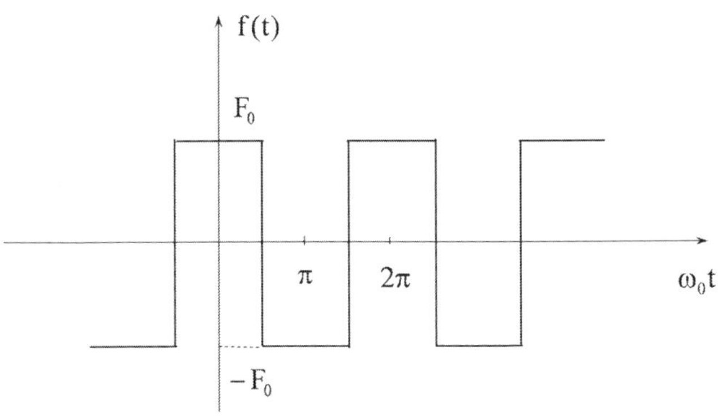

Fig. 1.5-4 Periodic force for $0 < \omega_0 t < 2\pi$.

Note that for even orders $n=2,4,6,\cdots$, all a_n are zero. For odd orders $n=1,3,5,\cdots$, $\sin(n\pi/2)=(-1)^{(n-1)/2}$ and $\sin(3n\pi/2)=-(-1)^{(n-1)/2}$. Thus, the coefficients can be expressed as

$$a_n=\frac{4F_0}{\pi}\frac{(-1)^{(n-1)/2}}{n}\quad(n=1,3,5,\cdots)$$

Hence, the periodic force function can be expressed as

$$f(t)=\frac{4F_0}{\pi}\left[\cos(\omega_0 t)-\frac{1}{3}\cos(3\omega_0 t)+\frac{1}{5}\cos(5\omega_0 t)+\cdots\right]$$

Example 1.5-3 The time function defined as

$$h(t)=\begin{cases}\dfrac{e^{-\varsigma\omega_n t}}{m\omega_d}\sin(\omega_d t),\left(t>0,\varsigma<1,\omega_d=\omega_n\sqrt{1-\varsigma^2}\right)\\ 0,(t<0)\end{cases}\tag{a}$$

is called the impulse response function. Show that its Fourier transform is given by

$$H(\omega)=\frac{1}{m\left(\omega_n^2-\omega^2+j2\varsigma\omega_n\omega\right)},\ \left(j=\sqrt{-1}\right)\tag{b}$$

Proof: From Eq. 1.5-13, we have

$$H(\omega)=\int_0^\infty h(t)e^{-j\omega t}dt=\int_0^\infty\frac{e^{-(\varsigma\omega_n+j\omega)t}}{m\omega_d}\sin(\omega_d t)dt\tag{c}$$

With the aid of the known integral

$$\int_0^\infty e^{-ax}\sin bx dx=\frac{b}{a^2+b^2}\quad(a>0),\tag{d}$$

we immediately obtain the desired result. $H(\omega)$ is called the *frequency response function*. Note that $H(-\omega)$ is the complex conjugate of $H(\omega)$. The result of this example becomes extremely significant in modal analysis.

The evaluation of the Fourier coefficients by hand is practical for only a few simple forms of $f(t)$. If $f(t)$ is taken from an actual vibration record, we will not have a mathematical expression for it. Nevertheless, with the aid of a computer and the FFT, we can numerically evaluate the Fourier coefficients.

Problems

1.1 A harmonic motion has a frequency of 30 radians per second and its maximum displacement is 0.25 mm. Determine its periodic, maximum velocity and its maximum acceleration.

Ans. $T = \pi/15$ (s), $V_{max} = 7.5$ (mm/s), $a_{max} = 225$ (mm/s^2)

1.2 A harmonic motion is expressed as $x(t) = A\sin(3t+\phi)$. The initial displacement and initial velocity are given by $x(0)$=0.25 mm, $\dot{x}(0) = 5$ mm/s. Determine the amplitude A and phase angle ϕ.

Ans. A=1.69 (mm), $\phi = 8.5°$

1.3 The equation of motion of a harmonic motion is given by $\ddot{x}+5x=0$. What is the frequency of the vibration?

Ans. $\sqrt{5}$ rad/s=0.356 Hz

1.4 Solve $m\ddot{x}+kx=0$ for m=1 kg, k=5 N/m, x_0=1 mm, $\dot{x}_0 = 0$. Plot the solution.

Ans. $x = \cos\sqrt{5}t$ (mm)

1.5 The equation of motion of a simple pendulum is written as $L\ddot{\theta}+g\theta=0$. Determine the frequency measured in radians per second and in Hz.

Ans. $\omega = \sqrt{\dfrac{g}{L}}$, $f = \dfrac{1}{2\pi}\sqrt{\dfrac{g}{L}}$

1.6 A harmonic motion can be described by $x = 3\cos\omega t - 4\sin\omega t$. Also, this can be written as $x = 5\sin(\omega t+\phi)$. Determine phase angle ϕ.

Ans. 143.13°

1.7 Determine the Fourier series for the triangular wave shown in Fig. P1.7.

Ans. $f(t) = \dfrac{1}{2} - \dfrac{4}{\pi}\sum_{n=1}^{\infty}\dfrac{1}{(2n-1)^2}\cos(2n-1)t$

1.8 Determine the Fourier series for the wave shown in Fig. P1.8.

Ans. $f(t) = \dfrac{4}{\pi^2}\sum_{n=1}^{\infty}\dfrac{(-1)^{n-1}}{(2n-1)^2}\sin(2n-1)t$

1.9 For the phase angle defined in Eq. 1.4-8, if $x_0 > 0$, $\dot{x}_0 < 0$, does the phase angle ϕ lie in the fourth quadrant or in the second quadrant?

Ans. The second quadrant.

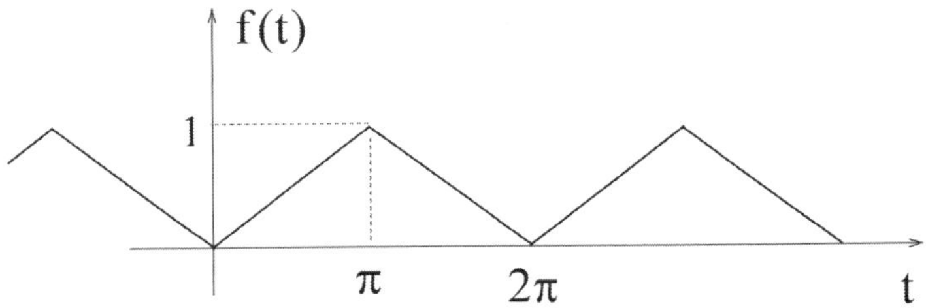

Fig. P1.7

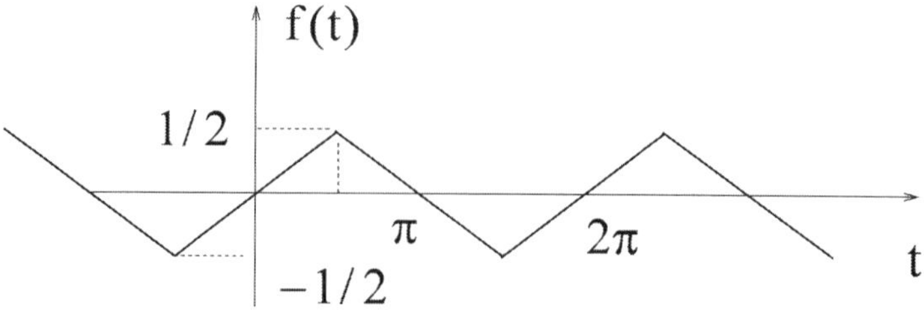

Fig. P1.8

1.10 Let the solution of Eq. 1.4-5 be written in the form

$$x = B\sin\omega_n t + C\cos\omega_n t$$

Calculate the values of B and C in terms of the initial conditions.

Ans. $B = \frac{\dot{x}_0}{\omega_n}$, $C = x_0$

References

1.1. Ehrich, F. F., *Handbook of Rotordynamics*, Kriefer Publishing Company, Malabar, FL, 2004.

1.2. Thomson, W. T., and Dahleh, M. D., *Theory of Vibration with Applications*, 5th ed., Prentice Hall, Inc., Upper Saddle River, NJ, 1998.

1.3. Tse, F. S., Morse, I. E., and Hinkle, R. T., *Mechanical Vibration, Theory and Applications*, Allyn and Bacon, Inc., Boston, 1979.

1.4. Den Hartog, J. P., *Mechanical Vibrations*, 4th ed., McGraw-Hill, New York, 1956. (Reprinted by Dover, New York, 1985.)

1.5. Meirovitch, L., *Analytical Methods in Vibrations*, Macmillan Company, New York, 1967.

1.6. Duffing, G., *Erwugene Schwingungen bei veranderlicher Eigenfrequennz*, F. Vieweg u. Sohn, Draunschweig, Germany, 1918.

1.7. McConnell, K. G., *Vibration Testing, Theory and Practice*, John Wiley & Sons, Inc., New York, 1995.

1.8. Van der Pol, B., "On Relaxation-Oscillations," *The London, Edinburgh and Dublin Philosophical Magazine and Journal of Science,* Vol. 2, 1926, pp. 978–992.

1.9. Oran Brigham, E., *The Fast Fourier Transform and Its Applications,* Prentice Hall, Inc., Upper Saddle River, NJ, 1998.

Chapter Two

Vibration of a Single-Degree-of-Freedom System

2.1 Introduction: Modeling

Modeling is a process of envisioning a physical system in a simplified form, retaining the principal characteristics but neglecting the less important features. The objective of modeling is to obtain a mathematical description (a model itself) of the simplified physical system.

In vibrations, modeling usually results in differential equations. The success of vibration modeling is measured by how well the differential equation solution describes the behavior of the physical system.

To illustrate modeling concepts, consider a relatively heavy and slightly unbalanced electric motor mounted in the middle of a steel beam, as in Fig. 2.1-1 (A). During operation, the motor's rotor experiences a rotating centrifugal force (an inertia force) due to the unbalanced mass. If this force is represented by horizontal and vertical components, the vertical component can be imagined to be producing a vertical vibration due to the elasticity of the steel beam support. This vertical vibration in turn can be modeled by the behavior of a mass-spring system as in Fig. 2.1-1 (B), where m represents

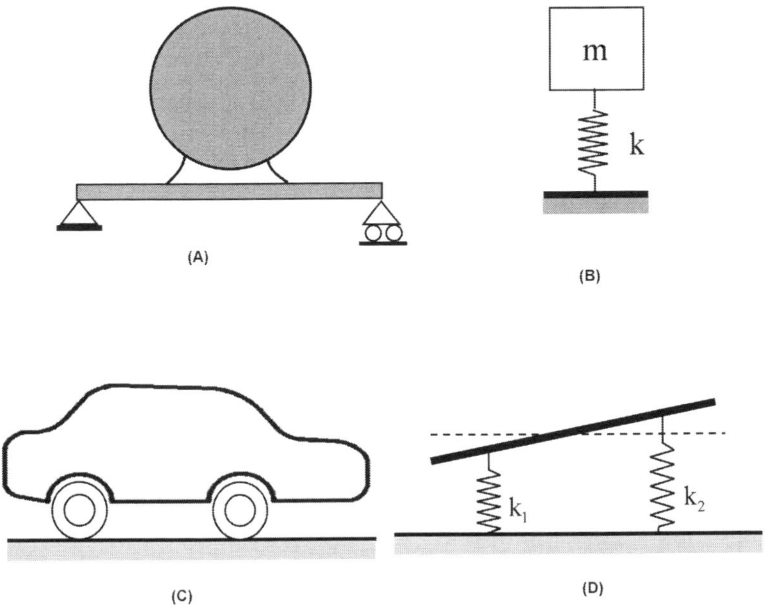

Fig. 2.1-1 Illustrations of vibration modeling.

the motor mass and k represents the beam elasticity ($k = 48EJ/L^3$, with E = Young's modulus, J = second moment of cross-section area, and L = beam length).

For a second illustration, consider an automobile body on its suspension represented by springs, as in Fig. 2.1-1 (C). If we are primarily interested in the vertical movement (or vibration) of the vehicle, we can model it as a two-degree-of-freedom system, as in Fig. 2.1-1 (D). The model is simply a rigid beam supported by two springs, as shown. For small amplitude vibration, the mass center in essence moves vertically and the beam rotates about the mass center.

The number of degrees of freedom of a vibrating system is the number of parameters (or variables) needed to define the configuration of the system.

In vibration analyses, the analysts usually model the physical system as a mass-spring system or as a collection of several mass-spring systems. Once this is done, the next step is to develop the equations of motion of the system. For simple systems, Newton's law and energy methods are commonly used. For complex systems, Lagrange's equations or Kane's equations can be used (see Sections 5.10 and 5.11, respectively).

The number of equations of motion is the same as the number of degrees of freedom of the system.

Most principles of dynamics and the resulting laws of motion can be derived from Newton's laws of motion. In this sense they are all equivalent and can be developed one from another. Newton's second law should be familiar to the reader from dynamics texts. (See, for example, [2.1, 2.2, and 2.3].)

Consider a particle with mass m acted upon by several forces. Then Newton's second law for the particle may be expressed by the equation

$$\sum \boldsymbol{F} = m\boldsymbol{a}, \tag{2.1-1}$$

where $\sum F$ represents the sum, or resultant, of all the forces acting on the particle, and $\boldsymbol{a}$ is the acceleration of the particle in an inertial (fixed), or Newtonian, reference frame.

Eq. 2.1-1 relates the forces acting on the particle and the reversed inertia force vector $m\boldsymbol{a}$.

Consider next a rigid body of mass m moving under the action of several external forces contained in a plane, as in Fig. 2.1-2. Eq. 2.1-1 is still valid for the motion of the mass center G of the body with respect to the Newtonian frame. The governing equation for the rotation motion of the body is

$$\sum \boldsymbol{T} = J\boldsymbol{\alpha}, \tag{2.1-2}$$

where $\sum \boldsymbol{T}$ is the sum of the moments of the external forces about the mass center G, $\boldsymbol{\alpha}$ is the angular acceleration of the body, and J is the moment of inertia of the body about its mass center.

Together, Eqs. 2.1-1 and 2.1-2 denote that the system of the external forces is equivalent to the system consisting of the vector $m\mathbf{a}$ attached to the mass center G and a couple with torque $J\boldsymbol{\alpha}$.

Problems involving plane motion of a rigid body are conveniently studied by drawing a *free-body diagram* similar to that of Fig. 2.1-2. Three equations of motion may then be obtained by equating the x-components, the y-components, and the moments about an arbitrary point A of the forces and vectors involved. That is,

$$\sum F_x = m\ddot{x}, \ \sum F_y = m\ddot{y}, \ \sum T_A = J_A\ddot{\theta}, \tag{2.1-3}$$

where $\ddot{x}$ and $\ddot{y}$ are acceleration components in the x and y directions, respectively; $\sum F_x$ and $\sum F_y$ are sums of external forces in the x and y directions, respectively; $\sum T_A$ represents the sum of the moments of all external forces about point A; J_A is the moment of inertia of the body about point A; and $\ddot{\theta}$ is the angular acceleration of the body.

Sometimes an energy method may be more efficient. An equation of motion is then obtained by using the principle of energy conservation. In a conservative system, the sum of the kinetic energy T and the potential energy U is constant. That is,

$$T + U = \text{constant} \tag{2.1-4}$$

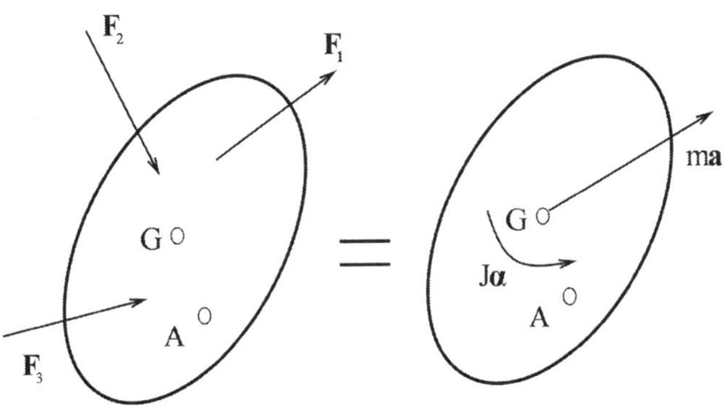

Fig. 2.1-2 Free-body diagram of a body having planar motion.

Thus, an equation of motion can be obtained by differentiating

$$\frac{d}{dt}(T+U)=0 \tag{2.1-5}$$

2.2 Free Vibration of a Single-Degree-of-Freedom System

(1) Governing Differential Equation of Motion

Consider the mass-spring system in Fig. 2.2-1. With no mass attached, the spring has a natural length L_0. When the mass m is attached and slowly released with zero velocity, the spring stretches to the static equilibrium position $L_0+\delta$, where the δ is the static deflection of the spring under the gravity force of mg. Taking the static equilibrium position as the origin of the coordinate system, and letting $x(t)$ be the displacement of the mass from the equilibrium position, the equation of motion of the system can be written as

$$m\ddot{x}=mg-k(x+\delta) \tag{2.2-1}$$

The static force balance on the mass yields the fact that

$$mg=k\delta \tag{2.2-2}$$

Thus, Eq. 2.2-1 becomes the standard equation of motion:

$$m\ddot{x}+kx=0 \tag{2.2-3}$$

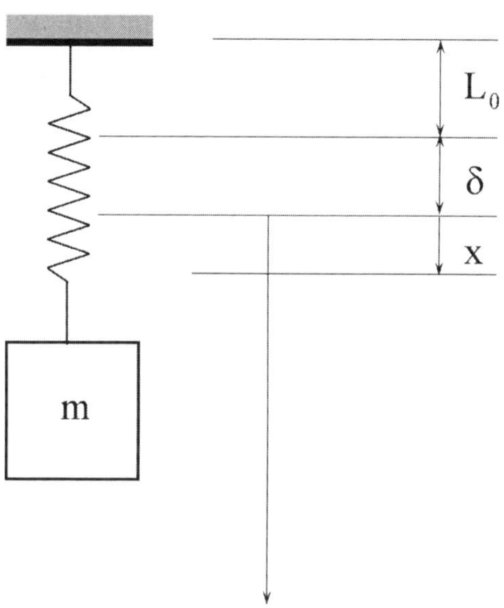

Fig. 2.2-1 A mass-spring system.

This equation my be written as

$$\ddot{x}+\omega_n^2 x=0, \tag{2.2-4}$$

where

$$\omega_n=\sqrt{\frac{k}{m}} \tag{2.2-5}$$

ω_n is called the *circular natural frequency*, measured in rad/s (radians per second).

Next, consider the shaft-disk system, as shown in Fig. 2.2-2. The shaft, or rod, exhibits a so-called restoring torque in twisting proportional to the angle of twist $\theta(t)$ of the disk.

The angle $\theta(t)$ is measured from the static equilibrium, or rest, position. The differential equation for free torsional motion of the system is

$$I\ddot{\theta}+K_{\theta}\theta=0, \qquad (2.2\text{-}6)$$

where I is the mass moment of inertia of the disk, measured in kg·m^2, and K_θ is the torsional stiffness of the shaft, measured in Nm/rad. For a uniform shaft, the torsional stiffness is given by (see the table at the end of this chapter)

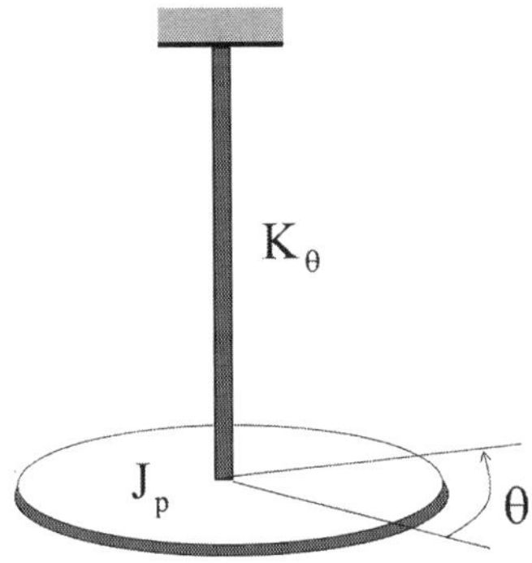

Fig. 2.2-2 A shaft-disk system.

$$K_{\theta}=\frac{GJ_{p}}{L}, \qquad (2.2\text{-}7)$$

where G is the shear modulus of rigidity of the shaft, J_p is the polar moment of inertia of the shaft, and L is the length of the shaft.

Example 2.2-1 Fig. 2.2-3 shows a system consisting of two uniform bars OA and OB with lengths L and $2L$, and masses m and $2m$, respectively. The bars are affixed to each other so that they form a right angle, as shown. The objective is to determine the natural frequency of the system.

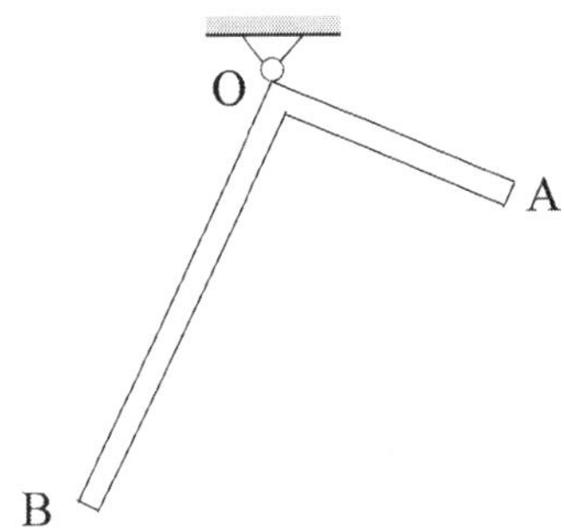

Fig. 2.2-3 A rigid right-angle bar.

Solution: Let α be the angle that OB makes with the vertical axis Oy at the static equilibrium position (see Fig. 2.2-4 (a)). Then, for equilibrium we have

$$2mgL\sin\alpha=mg(L/2)\cos\alpha \qquad \text{(a)}$$

It follows that α is determined by the expressions

$$\sin\alpha=1/\sqrt{17} \quad \text{and} \quad \cos\alpha=4/\sqrt{17} \qquad \text{(b)}$$

When the bent bar system is in static equilibrium, its mass center lies on the y-axis, a distance y_c beneath O given by

$$y_c=\frac{2mL\cos\alpha+m(L/2)\sin\alpha}{(2m+m)} \qquad \text{(c)}$$

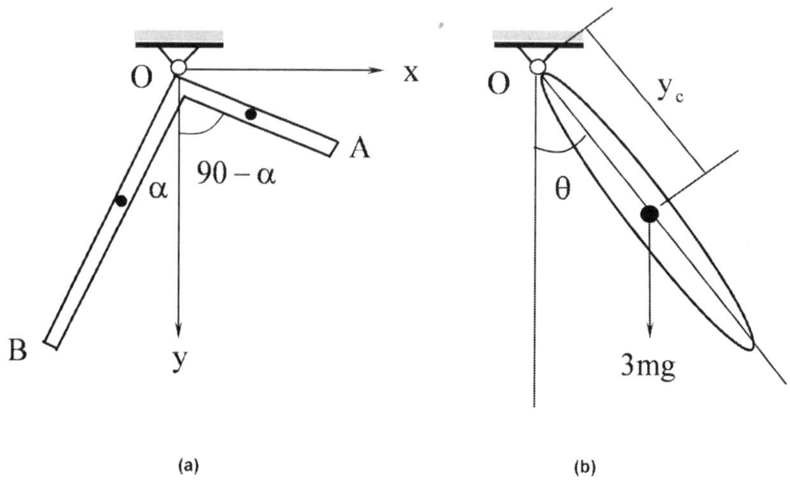

Fig. 2.2-4 Equilibrium position of the system and its equivalent system.

Substituting the results from Eq. b into Eq. c, we obtain

$$y_c = \frac{\sqrt{17}L}{6} \tag{d}$$

Next, the moment of inertia of the system about O is

$$J_o = \frac{1}{3}(2m)(2L)^2 + \frac{1}{3}m(L)^2 = 3mL^2 \tag{e}$$

Fig. 2.2-4 (b) shows a system equivalent to the original system. By taking moments about O for the equivalent system, we obtain the equation of motion as

$$J_o\ddot{\theta} = -3mg\,y_c \sin\theta \tag{f}$$

Substituting Eqs. d and e into Eq. f, and letting $\sin\theta \approx \theta$ for small vibration, we have

$$\ddot{\theta} + \frac{\sqrt{17}g}{6L}\theta = 0 \tag{g}$$

The natural frequency of oscillation is then given by

$$\omega = \sqrt{\frac{\sqrt{17}g}{6L}} \tag{h}$$

Example 2.2-2 Consider the system shown in Fig. 2.2-5 (a), consisting of an inclined light rod mounted in frictionless bearings A and B. Let a mass m be mounted at the end of another light rod with length b and rigidly affixed

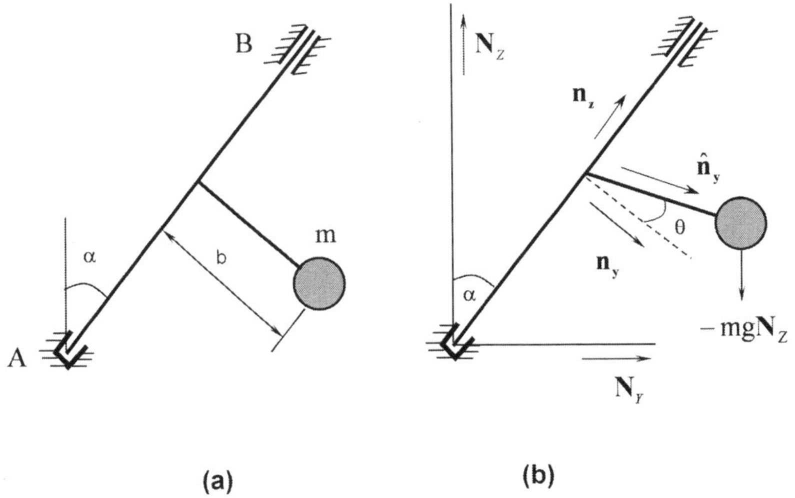

Fig. 2.2-5 A suspended mass system.

perpendicular to the inclined rod, as depicted in Fig. 2.2-5 (a). As in the previous example, the objective is to determine the natural frequency of the system.

Solution: To establish the equation of motion, we need an expression for the moment of the gravity force about the axis *AB*. To this end, we introduce unit vectors $\mathbf{N}_Y$, $\mathbf{N}_Z$, $\mathbf{n}_y$, $\hat{\mathbf{n}}_y$, and $\mathbf{n}_z$, as in Fig. 2.2-5 (b). Then the moment of the weight force about the direction n_z can be expressed as (see, for example, [2.1, 2.2, and 2.3])

$$M_{n_z} = (b\hat{\mathbf{n}}_y)\times(-mg\mathbf{N}_Z)\cdot(\mathbf{n}_z) \qquad \text{(a)}$$

The relationships between the unit vectors can be found by inspection or by using configuration graphs [2.4, 2.5] as

$$\hat{\mathbf{n}}_y = -\sin\theta\mathbf{N}_X + \cos\theta\cos\alpha\mathbf{N}_Y - \cos\theta\sin\alpha\mathbf{N}_Z \qquad \text{(b)}$$

$$\mathbf{n}_z = \sin\alpha\mathbf{N}_Y + \cos\alpha\mathbf{N}_Z \qquad \text{(c)}$$

Substituting Eqs. b and c into Eq. a, we obtain

$$M_{n_z} = -mg\,b\sin\alpha\sin\theta \qquad \text{(d)}$$

The equation of motion can be obtained by writing the moment equation about the axis *AB* as

$$mb^2\ddot{\theta} = -mgb\sin\alpha\sin\theta \qquad \text{(e)}$$

For small vibration, $\sin\theta$ is approximately equal to θ. Thus, Eq. e becomes

$$\ddot{\theta}+\frac{g\sin\alpha}{b}\theta=0 \tag{f}$$

Thus, the natural frequency of vibration of the system is given by

$$\omega_n=\sqrt{\frac{g\sin\alpha}{b}} \tag{g}$$

(2) Equivalent Spring Stiffness

If more than one spring is present in a given device, the resulting stiffness of the combined springs may be determined by considering their equivalent forces. Table 2.2-1 at the end of this chapter presents stiffnesses for various types of springs.

(a) Parallel-connected springs. The two springs attached to the mass shown in Fig. 2.2-6 are called *parallel-connected springs*. At the static equilibrium position, both springs will stretch by the same amounts due to the weight force mg of the mass. The static force balance on the mass yields

$$mg=F_1+F_2=(k_1+k_2)\delta=k_e\delta \tag{2.2-8}$$

Thus, the equivalent spring constant (stiffness) is given by

$$k_e=k_1+k_2 \tag{2.2-9}$$

Note that coaxial springs are parallel-connected springs.

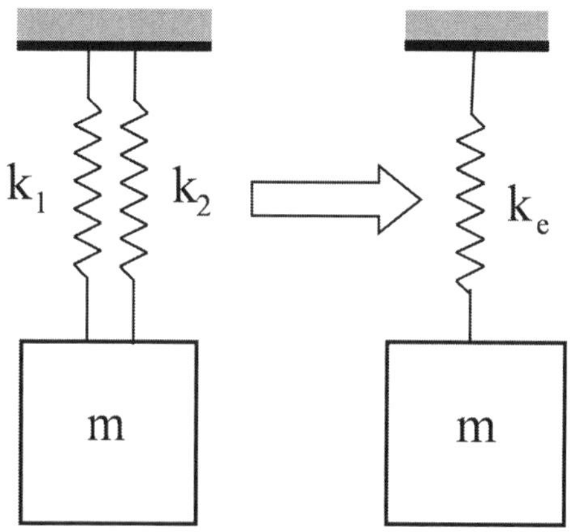

Fig. 2.2-6 Parallel-connected springs.

Example 2.2-3 Determine the natural frequency of the system shown in Fig. 2.2-7.

Solution: The springs are parallel-connected to the mass. Thus,

$$\omega_n=\sqrt{\frac{k_e}{m}}=\sqrt{\frac{k_1+k_2}{m}}$$

Example 2.2-4 Determine the natural frequency of the system as shown in Fig. 2.2-8 (A), where L = 4 m, $EJ=19.6\times10^5 N\cdot m^2$,

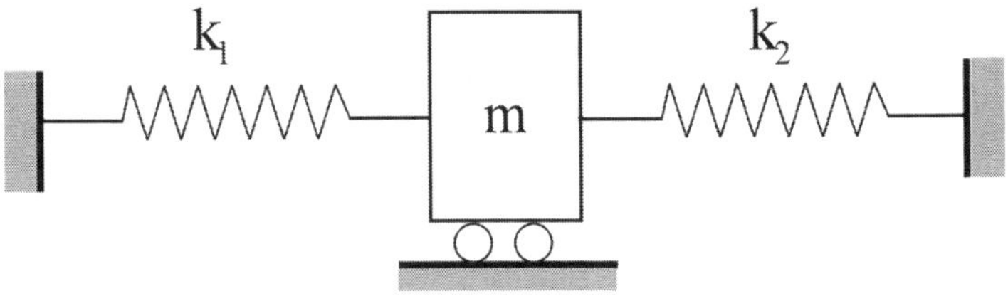

Fig. 2.2-7 System for Example 2.2-3.

$K = 4.9\times10^5$ N/m, and M = 400 kg. Assume that the beam is of negligible mass.

Solution: The equivalent spring stiffness of the beam to the mass is (see the table at the end of this chapter)

$$K_1 = \frac{48EJ}{L^3} \tag{a}$$

The springs with stiffnesses K_1 and K are parallel-connected to the mass, as in Fig. 2.2-8 (B). Hence, the equivalent spring constant for the mass is

$$k_e = K_1 + K = \frac{48\times19.6\times10^5}{4^3} + 4.9\times10^5 = 19.6\times10^5\,(\text{N/m}) \tag{b}$$

The natural frequency of the system is then given by

$$\omega_n = \sqrt{\frac{k_e}{M}} = \sqrt{\frac{19.6\times10^5}{400}} = 70 \quad (\text{rad/s})$$

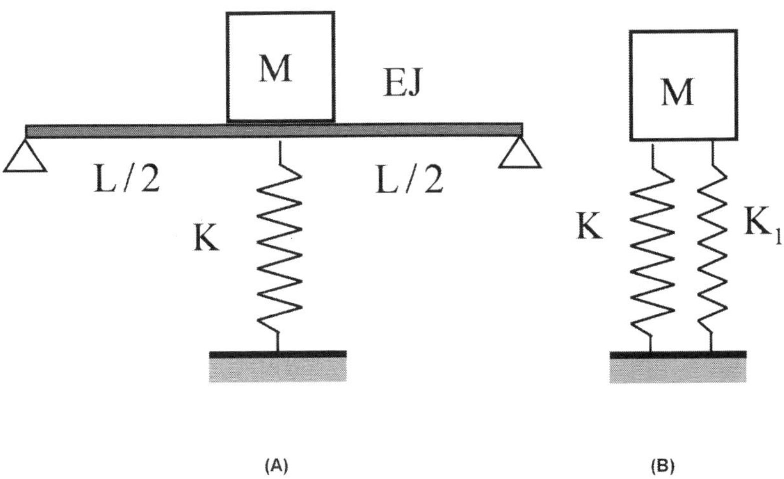

Fig. 2.2-8 System for Example 2.2-4.

or

$$f = \frac{\omega_n}{2\pi} = 11.14 \text{ (Hz)}$$

(b) Series-connected springs. The two springs attached to the mass shown in Fig. 2.2-9 are called *series-connected springs*. At the static equilibrium position, the total static displacement of the lower end of the combined spring due to the weight force mg of mass m is given by

$$\delta = \delta_1 + \delta_2 = \frac{mg}{k_1} + \frac{mg}{k_2} = mg\left(\frac{1}{k_1} + \frac{1}{k_2}\right) = mg\left(\frac{1}{k_e}\right)$$

Therefore, we have

$$\frac{1}{k_e} = \frac{1}{k_1} + \frac{1}{k_2}, \tag{2.2-10}$$

where k_e represents the equivalent spring constant and is given by

$$k_e = \frac{k_1 k_2}{k_1 + k_2} \tag{2.2-11}$$

Example 2.2-5 Consider two subway cars connected by a coupling device. Let this system be modeled as two masses connected by a spring, as in Fig. 2.2-10 (A). As before, the objective is to determine the natural frequency of vibration of the system.

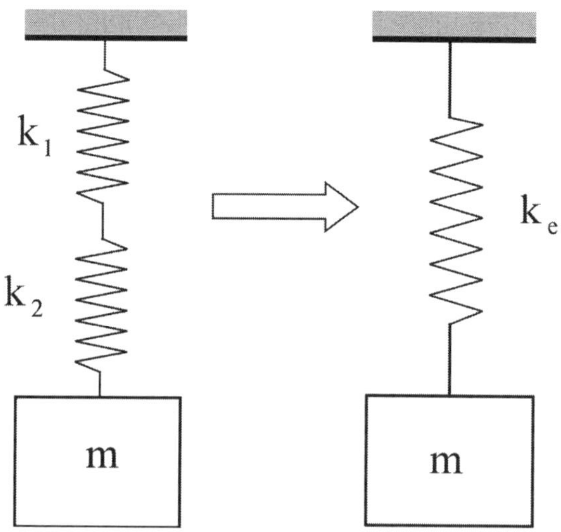

Fig. 2.2-9 Series-connected springs.

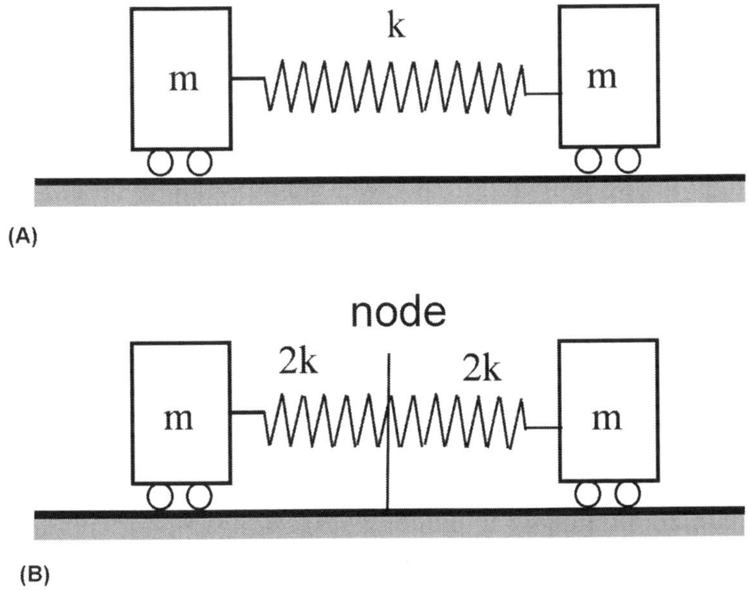

Fig. 2.2-10 For Example 2.2-5.

Solution: Although this system has two degrees of freedom, we can study it using our knowledge of the behavior of single-degree-of-freedom systems. Since there are no external horizontal forces acting on the system, the horizontal linear momentum of the system is conserved. Consequently, due to the symmetry of the system, the masses move in opposition to each other and the midpoint of the coupling spring does not move. This midpoint is sometimes called the *node*. Specifically, if the node is considered fixed, as in Fig. 2.2-10 (B), the motion of the system is unchanged. This means that the system can be split into two identical independent single-degree-of-freedom systems. The natural frequency of each system is equal to that of the original system. However, this time the stiffness of the spring for each mass is $2k$ rather than k (remember that the equivalent stiffness of the two series-connected springs is k). Thus, we obtain the natural frequency of the original system:

$$\omega_n = \sqrt{\frac{2k}{m}}$$

Example 2.2-6 Fig. 2.2-11 (A) shows a shaft with a disk at each end. The torsional stiffness of the shaft is K_θ. In the figure, I_1 and I_2 are the mass moments of inertia of the disks, respectively. Determine the natural frequency of torsional oscillation of the system.

Solution: This is also a two-degree-of-freedom system. But, as before, we can use the knowledge of single-degree-of-freedom systems to study the system.

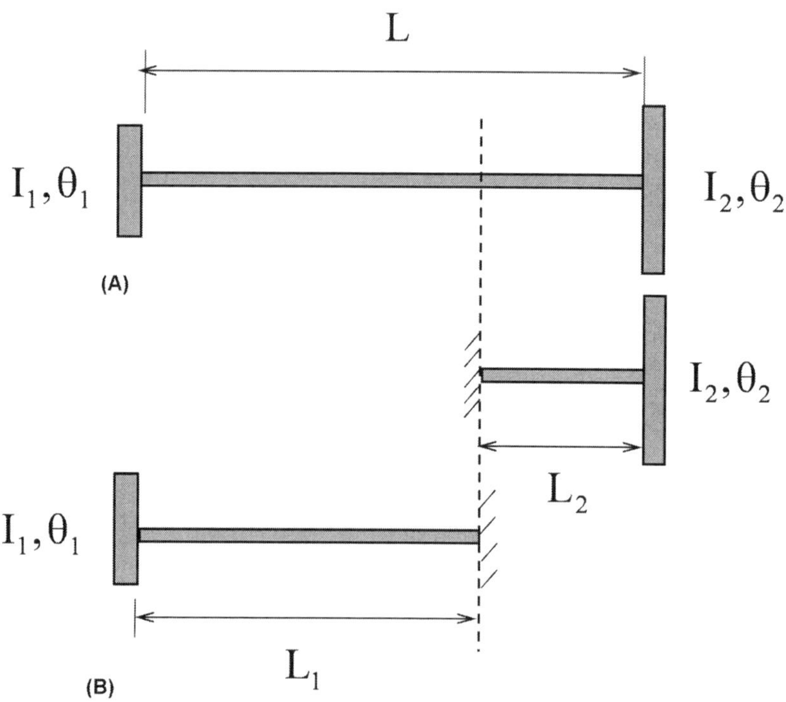

Fig. 2.2-11 A shaft-disks system.

From the principle of conservation of angular momentum, the disks are always rotating in opposition to each other. Thus, between the disks there is a single shaft cross section, which does not rotate. This cross section separates the system into two single-degree-of-freedom systems having the same natural frequency of oscillation (see Fig. 2.2-11 (B)). That is,

$$\sqrt{\frac{K_{\theta 1}}{I_1}} = \sqrt{\frac{K_{\theta 2}}{I_2}}, \qquad \text{(a)}$$

where $K_{\theta 1}$ and $K_{\theta 2}$ are the torsional stiffnesses of each single-degree-of-freedom system. Then we have

$$\frac{1}{K_{\theta 2}} = \frac{I_1}{I_2}\frac{1}{K_{\theta 1}} \qquad \text{(b)}$$

Since $K_{\theta 1}$ and $K_{\theta 2}$ are connected in series and the effective stiffness is K_θ, we have

$$\frac{1}{K_{\theta 1}} + \frac{1}{K_{\theta 2}} = \frac{1}{K_\theta} \qquad \text{(c)}$$

Substituting Eq. b into Eq. c, we obtain

$$K_{\theta 1} = \frac{I_1 + I_2}{I_2} K_\theta \tag{d}$$

Thus, the natural frequency is given by

$$\omega_n = \sqrt{\frac{K_{\theta 1}}{I_1}} = \sqrt{\frac{(I_1 + I_2) K_\theta}{I_1 I_2}} \tag{e}$$

2.3 The Energy Method

(1) The Law of Conservation of Energy

In a conservative system, the total energy is constant, and thus the differential equation of motion can be established by the principle of conservation of energy. For example, for the system shown in Fig. 2.3-1, only the spring force and the weight force do work, and the system is conservative. The kinetic energy is

$$T = \frac{1}{2} m \dot{x}^2 \tag{2.3-1}$$

Correspondingly, the potential energy is

$$U = \frac{1}{2} k \left[(x+\delta)^2 - \delta^2 \right] - mgx = \frac{1}{2} kx^2 \tag{2.3-2}$$

The total energy is constant:

$$T + U = C \tag{2.3-3}$$

That is,

$$\frac{1}{2} m \dot{x}^2 + \frac{1}{2} kx^2 = C \tag{2.3-4}$$

Taking the derivative in Eq. 2.3-4 with respect to time, we obtain

$$\ddot{x} + \frac{k}{m} x = 0 \tag{2.3-5}$$

Thus, the natural frequency is given by

$$\omega_n = \sqrt{\frac{k}{m}} \tag{2.3-6}$$

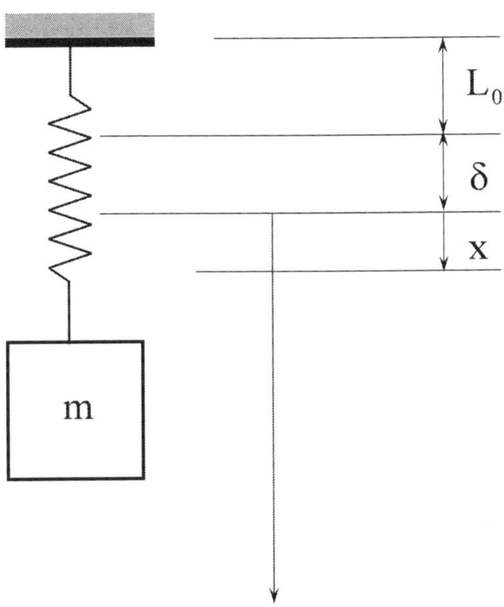

Fig. 2.3-1 A mass-spring system.

(2) The Rayleigh Method

For free harmonic vibration, the displacement and velocity can be expressed in the forms

$$x(t) = A\sin(\omega t + \phi) \tag{2.3-7}$$

$$\dot{x} = A\omega\cos(\omega t + \phi) \tag{2.3-8}$$

The kinetic energy is then given by

$$T = \frac{1}{2}m\dot{x}^2 = \frac{1}{2}mA^2\omega^2\cos^2(\omega t + \phi) \tag{2.3-9}$$

Correspondingly, the potential energy may be written as

$$U = \frac{1}{2}kx^2 = \frac{1}{2}kA^2\sin^2(\omega t + \phi) \tag{2.3-10}$$

The maximum kinetic energy will be reached at the equilibrium position (where the potential energy is zero):

$$T_{\max} = \frac{1}{2}mA^2\omega^2 \tag{2.3-11}$$

The maximum potential energy will be reached at the maximum displacement position (where the kinetic energy is zero):

$$U_{\max} = \frac{1}{2}kA^2 \tag{2.3-12}$$

According to the law of conservation of energy, we have

$$T_{\max} = U_{\max} \tag{2.3-13}$$

That is,

$$\frac{1}{2}mA^2\omega^2 = \frac{1}{2}kA^2 \tag{2.3-14}$$

We thus obtain the circular natural frequency of oscillation as

$$\omega_n = \sqrt{\frac{k}{m}} \tag{2.3-15}$$

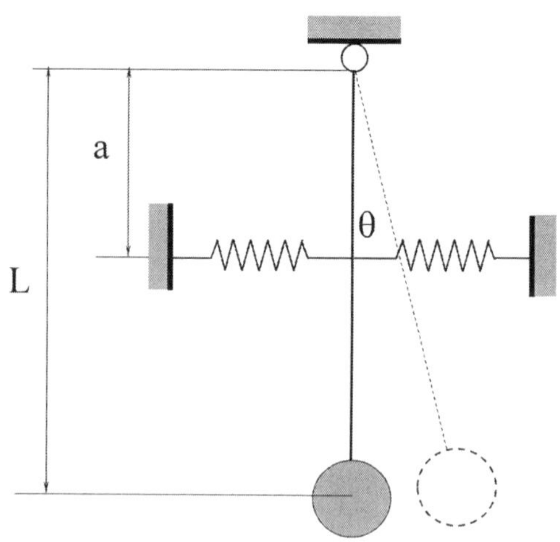

Fig. 2.3-2 A spring-restrained pendulum.

Example 2.3-1 Determine the natural frequency of the system shown in Fig. 2.3-2.

Solution (1): We derive the equation of motion by using the law of conservation of energy. The kinetic energy T and potential energy U are

$$T = \frac{1}{2} m (L\dot{\theta})^2 \tag{a}$$

and

$$U = \frac{1}{2}(2k)(a\theta)^2 + mgL(1 - \cos\theta) \tag{b}$$

Since the total energy is constant, we have

$$T + U = C \tag{c}$$

or

$$\frac{d}{dt}\left(T + U\right) = 0 \tag{d}$$

That is,

$$\ddot{\theta} + \left(\frac{2ka^2}{mL^2} + \frac{g}{L}\right)\theta = 0 \tag{e}$$

Thus, the circular natural frequency is found to be

$$\omega_n = \sqrt{\frac{2ka^2}{mL^2} + \frac{g}{L}} \tag{f}$$

Solution (2): We use the Rayleigh method. Let the motion be described as

$$\theta = \theta_{max} \sin(\omega_n t + \phi) \tag{g}$$

The maximum kinetic energy is then

$$T_{max} = \frac{1}{2} m \left(L\dot{\theta}_{max}\right)^2 = \frac{1}{2} m L^2 \theta_{max}^2 \omega_n^2 \tag{h}$$

The maximum potential energy can be expressed as

$$U_{max} = \frac{1}{2}(2k)(a\theta_{max})^2 + mgL(1 - \cos\theta_{max}) \tag{i}$$

Note that for small vibration, $1 - \cos\theta_{max} = 2\sin^2\left(\theta_{max}/2\right) \approx \frac{1}{2}\theta_{max}^2$; thus, U_{max} becomes

$$U_{max} = \left(ka^2 + \frac{1}{2} mgL\right)\theta_{max}^2 \tag{j}$$

According to the law of conservation of energy, we have

$$T_{max} = U_{max} \tag{k}$$

This leads to

$$\frac{1}{2} mL^2\theta_{max}^2 \omega_n^2 = \left(ka^2 + \frac{1}{2} mgL \right)\theta_{max}^2 \tag{l}$$

Therefore, we obtain the circular natural frequency as

$$\omega_n = \sqrt{\frac{2ka^2}{mL^2} + \frac{g}{L}} \tag{m}$$

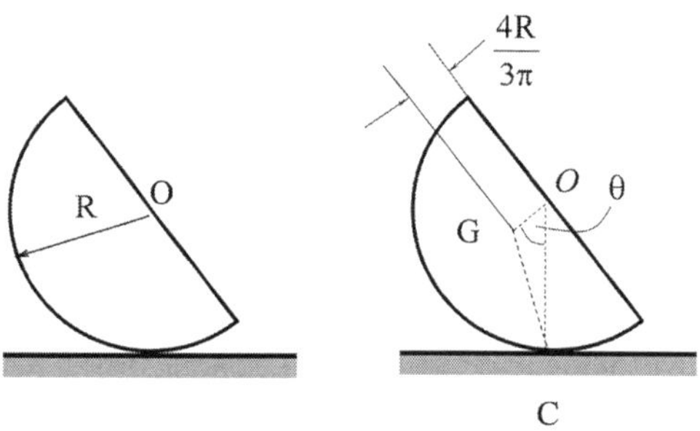

Fig. 2.3-3 A semi-cylinder rolling on a flat horizontal surface.

Example 2.3-2 Consider a semi-cylinder with mass m and radius R rolling back and forth on a flat horizontal surface, as in Fig. 2.3-3. Determine the natural frequency.

Solution: Let C be the contact point between the semi-cylinder and the flat horizontal surface, and let G be the mass center of the cylinder. Since the contact point C of the semi-cylinder has zero velocity, the velocity of G may be expressed as

$$V_G = |CG|\dot{\theta} \tag{a}$$

The distance CG can be found from the following equations:

$$|CG|^2 = R^2 + e^2 - 2Re\cos\theta, \quad \text{where} \quad e = \frac{4R}{3\pi} \tag{b}$$

Next, the moment of inertia of the semi-cylinder about G is ([2.5], p. 356)

$$J = mk_G^2, \quad \text{where} \quad k_G^2 = \frac{9\pi^2 - 32}{18\pi^2} R^2 \tag{c}$$

Thus, the kinetic energy of the semi-cylinder may be expressed as

$$T = \frac{1}{2} mV_G^2 + \frac{1}{2} J\dot{\theta}^2 \tag{d}$$

The potential energy of the semi-cylinder referred to the lowest position of the mass center G is

$$U = mge(1-\cos\theta) \tag{e}$$

Since the velocity of the contact point C of the semi-cylinder is always zero, the work done by the friction force is zero. Therefore, the total energy is constant. That is,

$$T+U=C \tag{f}$$

Hence,

$$\frac{d}{dt}\left(T+U\right)=0 \tag{g}$$

Substituting Eqs. d and e into Eq. g, we obtain

$$m\left(R^2+e^2-2Re\cos\theta\right)\dot{\theta}\ddot{\theta}+mRe\sin\theta\dot{\theta}^3+mk_G^2\dot{\theta}\ddot{\theta}+mge\sin\theta\dot{\theta}=0 \tag{h}$$

For small vibration, $\cos\theta\approx 1$, $\sin\theta\approx\theta$, and $\dot{\theta}^3\approx 0$. The above equation then becomes the familiar equation for harmonic motion:

$$\ddot{\theta}+\frac{ge}{k_G^2+(R-e)^2}\theta=0 \tag{i}$$

By inspection, the natural frequency of oscillation is

$$\omega_n=\sqrt{\frac{ge}{k_G^2+(R-e)^2}} \tag{j}$$

Example 2.3-3 Use the energy method to derive the equation of motion and then obtain the natural frequency of the pendulum of Example 2.2-2.

Solution: Let θ denote a small angle of rotation of the pendulum about the inclined axis, measured from the equilibrium position. The kinetic energy of the system is then

$$T=\frac{1}{2}m\left(b\dot{\theta}\right)^2=\frac{1}{2}\left(mb^2\right)\dot{\theta}^2 \tag{a}$$

The elevation of the mass is $b(1-\cos\theta)\sin\alpha$. The potential energy of the system is then

$$U=mgb(1-\cos\theta)\sin\alpha \tag{b}$$

The conservation of energy equation then becomes

$$\frac{1}{2}\left(mb^2\right)\dot{\theta}^2+mgb(1-\cos\theta)\sin\alpha=C \tag{c}$$

Taking the derivative with respect to time, we obtain

$$\ddot{\theta}+(g/b)\sin\alpha\sin\theta=0, \tag{d}$$

where $\sin\theta$ is approximately equal to θ for small oscillation. The natural frequency is then given by

$$\omega_n=\sqrt{\frac{g\sin\alpha}{b}} \tag{e}$$

Example 2.3-4 [2.6] A right circular cone is rolling on an inclined plane, as depicted in Fig. 2.3-4. Let the cone have generator length *L*, base radius *r*, and half central angle α. Let the plane be inclined at an angle β. Finally, let ϕ be the angle that the contacting generator of the cone makes with a fixed edge or fixed line of the inclined plane. Determine the natural frequency of oscillation of the cone.

Solution: The contacting generator of the cone with the plane is the instantaneous axis of rotation of the cone. The rolling condition is $L\dot{\phi}=r\dot{\psi}$, so that $\dot{\psi}=\dot{\phi}/\sin\alpha$, where ψ is the axial rolling angle (see Fig. 2.3-5). The angular velocity components of the cone are then

$$\omega_1=\dot{\psi}-\dot{\phi}\sin\alpha=\dot{\phi}/\sin\alpha-\dot{\phi}\sin\alpha=\dot{\phi}\cos^2\alpha/\sin\alpha \tag{a}$$

and

$$\omega_2=\dot{\phi}\cos\alpha,\ \omega_3=0 \tag{b}$$

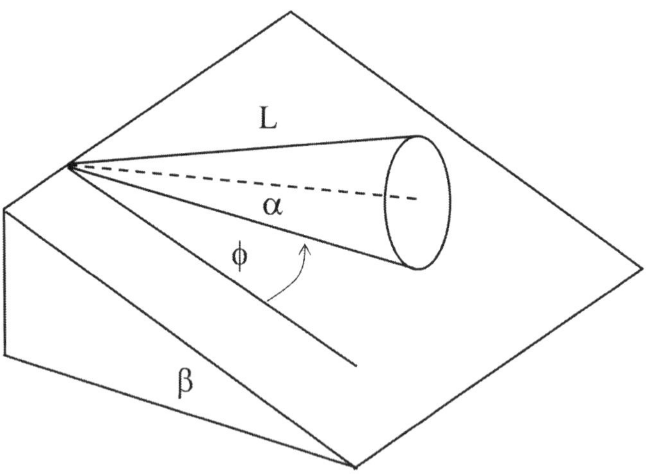

Fig. 2.3-4 A cone rolling on an inclined plane.

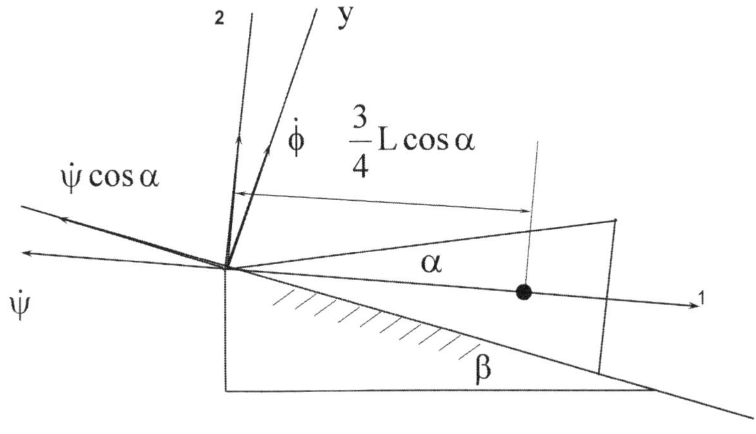

Fig. 2.3-5 Velocity components of the cone.

The velocity (or speed) of the mass center is

$$V_G = \frac{3}{4}L\cos\alpha\sin\alpha\left(\dot{\psi}\cos\alpha\right) = \frac{3}{4}L\dot{\phi}\cos^2\alpha \tag{c}$$

Next, the principal moments of inertia of the cone are ([2.5], p. 356)

$$I_{11} = \frac{3}{10}mr^2 = \frac{3}{10}mL^2\sin^2\alpha \tag{d}$$

$$I_{22} = \frac{3}{80}m\left(4r^2 + h^2\right) = \frac{3}{80}mL^2(4 - 3\cos^2\alpha) \tag{e}$$

The kinetic energy of the cone is

$$T = \frac{1}{2}mV_G^2 + \frac{1}{2}I_{11}\omega_1^2 + \frac{1}{2}I_{22}\omega_2^2 \tag{f}$$

Substituting Eqs. a to e into Eq. f, we obtain

$$T = \frac{3}{8}mL^2\dot{\phi}^2\left(\frac{1}{5} + \cos^2\alpha\right)\cos^2\alpha \tag{g}$$

Next, the potential energy relative to the lowest position of the cone is

$$U = \frac{3}{4}mgL\cos^2\alpha(1 - \cos\phi)\sin\beta, \tag{h}$$

where $(3/4)L\cos^2\alpha(1-\cos\phi)\sin\beta$ is the change in elevation of the mass center of the cone. Since the velocity of the contact generator is always zero, the work done by the friction force is zero. Therefore, the total energy is constant. That is,

$$T + U = C \tag{i}$$

Substituting Eqs. g and h into Eq. i and taking the derivative with respect to time, we obtain

$$\ddot{\phi} + \omega_n^2 \phi = 0, \tag{j}$$

where the natural frequency ω_n is given by

$$\omega_n = \sqrt{\frac{g \sin \beta}{L(1/5 + \cos^2 \alpha)}} \tag{k}$$

2.4 Effective Mass

Thus far in our evaluations of the natural frequencies in systems involving springs, we have assumed the spring to be massless. This can result in natural frequencies that are too large.

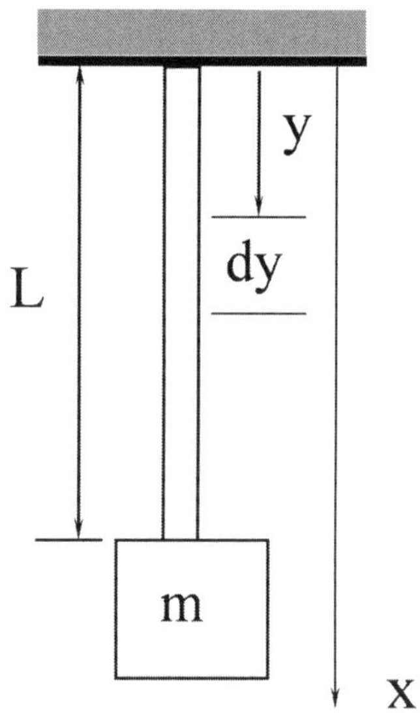

Fig. 2.4-1 A system with a massive bar-spring.

To obtain better values for the natural frequencies, we can increase the kinetic energy of the system by adding the kinetic energy of the moving spring. This requires a somewhat detailed knowledge of the moving spring.

We illustrate these concepts in the following examples.

Example 2.4-1 Determine the effect of the mass of the bar on the natural frequency of longitudinal vibration of the system shown in Fig. 2.4-1.

Solution: Consider the kinetic energy of an element dy of the bar. Let m_b be the total mass of the bar. Then $(m_b/L)dy$ is the mass of element dy. The velocity of the element is obtained by assuming that the velocity of element cross section $\dot{x}$ varies linearly over the length of the bar. Therefore, the velocity of dy is $V_{dy} = y\dot{x}/L$. Then the total kinetic energy of the bar is the sum of the kinetic energies of the elements dy, summed or integrated over the length of the bar:

$$T_{\text{spring}} = \frac{1}{2}\int_0^L \frac{m_b}{L}\left[\frac{y}{L}\dot{x}\right]^2 dy = \frac{1}{2}\left(\frac{m_b}{3}\right)\dot{x}^2 \tag{a}$$

From the form of this expression, the *effective mass* of the bar is $m_b/3$. Following the Rayleigh method, the maximum kinetic energy of the system is then

$$T_{\max} = \frac{1}{2}\left(m + \frac{m_b}{3}\right)\omega_n^2 y_m^2 \tag{b}$$

Equating this to the maximum potential energy, $(1/2)(EA/L)y_m^2$, the natural frequency of the system is seen to be

$$\omega_n = \sqrt{\frac{EA/L}{m + m_b/3}} \tag{c}$$

Similarly, we can obtain (1) the *effective mass* of a uniform cantilever beam with mass m_b for its free end lateral vibration as $(33/140)m_b$, and (2) the *effective mass* of a simply supported uniform beam with mass m_b for its middle point lateral vibration as $(17/35)m_b$. (See [2.7 and 2.8].)

2.5 Damped Free Vibration of a Single-Degree-of-Freedom System

(1) Damping Forces

Although the actual description of damping forces is difficult, ideal damping models can be assumed that will often result in satisfactory prediction of damped vibration. Of these models, viscous damping, proportional to velocity, leads to the simplest mathematical treatment.

A viscous damping force may be expressed as $F_d = c\dot{x}$, where c is a constant of proportionality, called the damping coefficient. Symbolically it is designated by a dashpot, as in Fig. 2.5-1.

(2) Differential Equation of Motion

Consider the system shown in Fig. 2.5-1. Taking the static equilibrium position as the origin of the coordinate system, the equation of motion can be written as

$$m\ddot{x} = mg - k(x + \delta) - c\dot{x} \tag{2.5-1}$$

The static force balance on the mass yields

$$mg = k\delta \tag{2.5-2}$$

Thus, the equation of motion becomes

$$m\ddot{x} + c\dot{x} + kx = 0 \tag{2.5-3}$$

Introduce the following parameters:

$$\omega_n = \sqrt{\frac{k}{m}} = \text{undamped natural frequency} \tag{2.5-4}$$

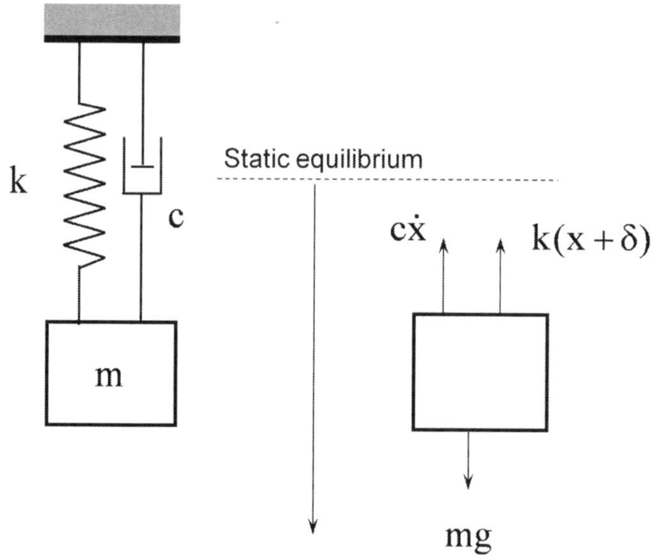

Fig. 2.5-1 Forces applied on the mass.

$$\varsigma = \frac{c}{2\sqrt{km}} = \text{damping ratio} \tag{2.5-5}$$

Eq. 2.5-3, divided through by m, in view of Eqs. 2.5-4 and 2.5-5, yields

$$\ddot{x} + 2\varsigma\omega_n\dot{x} + \omega_n^2 x = 0 \tag{2.5-6}$$

The solution of Eq. 2.5-6 is of the form $x = Ae^{\lambda t}$, where A and λ are nonzero constants to be determined. Substituting the solution form into Eq. 2.5-6, we obtain a quadratic equation in λ:

$$\lambda^2 + 2\varsigma\omega_n\lambda + \omega_n^2 = 0 \tag{2.5-7}$$

The two roots of this equation can be expressed as

$$\lambda_1 = -\varsigma\omega_n + \sqrt{\varsigma^2 - 1}\,\omega_n \tag{2.5-8a}$$

$$\lambda_2 = -\varsigma\omega_n - \sqrt{\varsigma^2 - 1}\,\omega_n \tag{2.5-8b}$$

Hence, the solution can be written as

$$x = A_1 e^{\lambda_1 t} + A_2 e^{\lambda_2 t} \tag{2.5-9}$$

(3) Discussion of the Solution

Case 1: Under-damped case: $0 < \varsigma < 1$

In this case, the discriminant of Eq. 2.5-8 is negative, resulting in a complex conjugate pair of roots. These are

$$\lambda_{1,2} = -\varsigma\omega_n \pm j\sqrt{1-\varsigma^2}\,\omega_n = -\varsigma\omega_n \pm j\omega_d, \tag{2.5-10}$$

where j is the unit imaginary number $\left(j=\sqrt{-1}\right)$ and where ω_d is defined as

$$\omega_d = \sqrt{1-\varsigma^2}\,\omega_n \tag{2.5-11}$$

ω_d is often called the *damped natural frequency.*

Recall the fundamental Euler identities

$$e^{j\theta} = \cos\theta + j\sin\theta,\ e^{-j\theta} = \cos\theta - j\sin\theta \tag{2.5-12}$$

Using these identities together with the roots of Eq. 2.5-10, the solution of Eq. 2.5-9 may be written in the form

$$\begin{aligned} x &= e^{-\varsigma\omega_n t}\left[A_1\left(\cos\omega_d t + j\sin\omega_d t\right) + A_2\left(\cos\omega_d t - j\sin\omega_d t\right)\right] \\ &= e^{-\varsigma\omega_n t}\left[\left(A_1 + A_2\right)\cos\omega_d t + j\left(A_1 - A_2\right)\sin\omega_d t\right] \end{aligned} \tag{2.5-13}$$

Since A_1 and A_2 are arbitrary constants, $\left(A_1 + A_2\right)$ and $j\left(A_1 - A_2\right)$ are also arbitrary. Therefore, for simplicity, we may write them as B_1 and B_2. Then x becomes

$$x = e^{-\varsigma\omega_n t}\left[B_1\cos\omega_d t + B_2\sin\omega_d t\right] \tag{2.5-14}$$

By defining the constant A as $\sqrt{B_1^2 + B_2^2}$ and the angle ϕ as $\tan^{-1}(B_1/B_2)$, we have $B_1 = A\sin\phi$ and $B_2 = A\cos\phi$. The solution of Eq. 2.5-14 can then be written as

$$x = Ae^{-\varsigma\omega_n t}\sin(\omega_d t + \phi), \tag{2.5-15}$$

where A (amplitude) and ϕ (phase angle) are constants to be determined from the initial displacement x_0 and the initial velocity $\dot{x}_0$.

Substituting the initial displacement x_0 into Eq. 2.5-15 and the initial velocity $\dot{x}_0$ into the derivative of Eq. 2.5-15, we obtain A and $\tan\phi$ in the forms

$$A = \sqrt{x_0^2 + \left(\frac{\dot{x}_0 + \varsigma\omega_n x_0}{\omega_d} \right)^2} \tag{2.5-16}$$

and

$$\tan\phi = \frac{x_0 \omega_d}{\dot{x}_0 + \varsigma\omega_n x_0} \tag{2.5-17}$$

Finally, the period T_d of the damped free vibration is defined as

$$T_d = \frac{2\pi}{\omega_d} = \frac{2\pi}{\sqrt{1-\varsigma^2}\,\omega_n} \tag{2.5-18}$$

Fig. 2.5-2 illustrates the importance of Eq. 2.5-15 [2.9, 2.10]. Observe that this vibration decays within the exponential envelope function $Ae^{-\varsigma\omega_n t}$, while it oscillates at the damped frequency ω_d. Both the natural frequency and the damping ratio can be obtained experimentally from this damped oscillation curve.

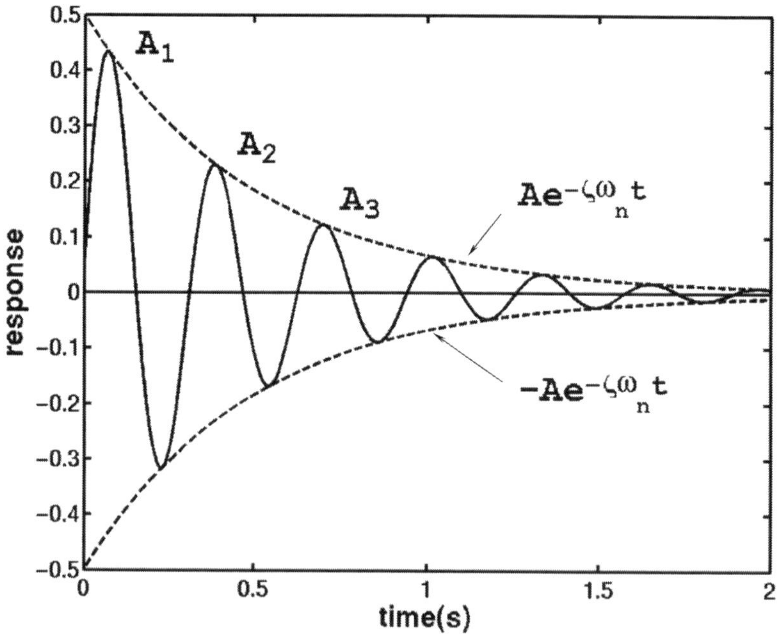

Fig. 2.5-2 Damped vibration ($0 < \varsigma < 1$) [2.9, 2.10].

The *logarithmic decrement* δ is defined as the natural logarithm of the ratio of any two successive amplitudes. That is,

$$\begin{aligned}\delta &= \ln\frac{A_1}{A_2} = \ln\frac{e^{-\varsigma\omega_n t_1}\sin(\omega_d t_1+\phi)}{e^{-\varsigma\omega_n(t_1+T_d)}\sin(\omega_d(t_1+T_d)+\phi)}\\ &= \ln\frac{e^{-\varsigma\omega_n t_1}}{e^{-\varsigma\omega_n(t_1+T_d)}} = \ln e^{\varsigma\omega_n T_d} = \varsigma\omega_n T_d\end{aligned} \tag{2.5-19}$$

Substituting the damped period $T_d = \dfrac{2\pi}{\sqrt{1-\varsigma^2}\,\omega_n}$ (see Eq. 2.5-18) into Eq. 2.5-19, the expression for the logarithmic decrement becomes

$$\delta = \frac{2\pi\varsigma}{\sqrt{1-\varsigma^2}} \approx 2\pi\varsigma \tag{2.5-20}$$

Solving this expression for the damping ratio ς yields

$$\varsigma = \frac{\delta}{\sqrt{4\pi^2+\delta^2}} \tag{2.5-21}$$

The damping ratio is the most difficult quantity to determine. Eq. 2.5-21 provides a simple method for this purpose when the value of the logarithmic decrement is known.

Observe that the ratio between two consecutive maxima is constant. Hence, the logarithmic decrement is constant, independent of both the amplitude of vibration and the time.

Case 2: Critically damped case: $\varsigma = 1.0$

In this case, the discriminant of Eq. 2.5-8 is identically zero. Consequently, we obtain double roots $\lambda_1 = \lambda_2 = -\omega_n$. The general solution of the differential equation then takes the form

$$x = (A+Bt)e^{-\omega_n t}, \tag{2.5-22}$$

where A and B are constants depending on the initial displacement and velocity. Substituting the initial displacement x_0 into Eq. 2.5-22 and the initial velocity $\dot{x}_0$ into the derivative of Eq. 2.5-22, we obtain

$$x_0 = A \tag{2.5-23}$$

$$B = x_0\omega_n + \dot{x}_0 \tag{2.5-24}$$

Hence, the solution can be written as

$$x = \left[x_0 + (x_0\omega_n + \dot{x}_0)t\right]e^{-\omega_n t} \tag{2.5-25}$$

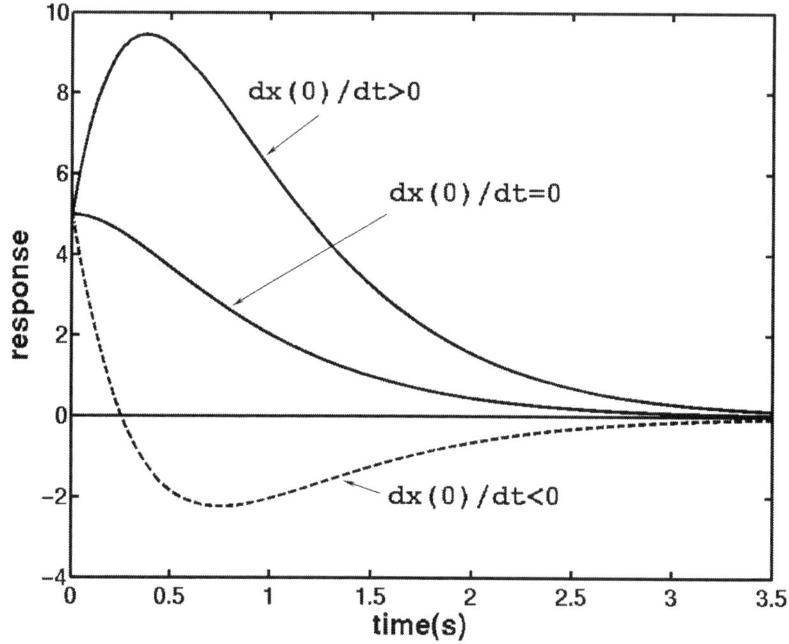

Fig. 2.5-3 Critically damped case ($\varsigma = 1.0$) [2.9, 2.10].

Eq. 2.5-25 represents *critically damped motion*. It forms a boundary between oscillatory motion and non-oscillatory motion. Fig. 2.5-3 shows three types of critically damped motion with the same initial displacement but different initial velocities [2.9].

Case 3: Over-damped case: $\varsigma > 1.0$

In this case, both roots λ_1 and λ_2 are negative. The motion is then an exponentially decreasing function with increasing time, as in Fig. 2.5-4. This motion is referred to as *aperiodic*. The general solution can be written as

$$x = Ae^{\left(-\varsigma+\sqrt{\varsigma^2-1}\right)\omega_n t} + Be^{\left(-\varsigma-\sqrt{\varsigma^2-1}\right)\omega_n t}, \tag{2.5-26}$$

where A and B are constants depending on the initial displacement x_0 and the initial velocity $\dot{x}_0$:

$$A = \frac{\dot{x}_0 + \left(\varsigma+\sqrt{\varsigma^2-1}\right)\omega_n x_0}{2\omega_n\sqrt{\varsigma^2-1}} \tag{2.5-27}$$

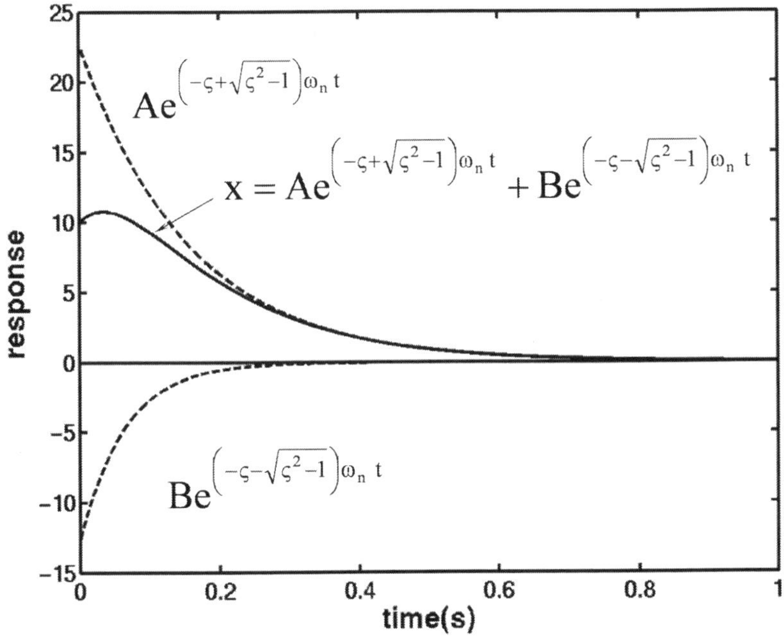

Fig. 2.5-4 Over-damped case [2.9, 2.10].

and

$$B = \frac{-\dot{x}_0 - \left(\varsigma - \sqrt{\varsigma^2 - 1}\right)\omega_n x_0}{2\omega_n \sqrt{\varsigma^2 - 1}} \tag{2.5-28}$$

Case 4: Without damping (zero damping): $\varsigma = 0$

In this case, the motion becomes free vibration. The solution can be obtained by letting $\varsigma = 0$ in Eq. 2.5-15. That is,

$$x = A\sin(\omega_n t + \phi), \tag{2.5-29}$$

where A (amplitude) and ϕ (phase angle) are the constants of integration to be determined from the initial displacement x_0 and the initial velocity $\dot{x}_0$.

Substituting the initial displacement x_0 into Eq. 2.5-29 and the initial velocity $\dot{x}_0$ into the derivative of Eq. 2.5-29, we obtain

$$A = \sqrt{x_0^2 + \left(\frac{\dot{x}_0}{\omega_n}\right)^2} \tag{2.5-30}$$

and

$$\phi = \tan^{-1}\frac{x_0\omega_n}{\dot{x}_0} \tag{2.5-31}$$

Example 2.5-1 A mass with weight 5 N suspended on a spring and a dashpot. The spring has stiffness constant 2 N/cm. After four cycles of vibration, the amplitude of the vibration is reduced to 1/12 its initial value. Determine the period and damping ratio.

Solution: Since there is vibration, this is an under-damped case. Let the amplitude at time t_1 have the form

$$A_1 = Ae^{-\varsigma\omega_n t_1} \tag{a}$$

After four cycles of vibration, the amplitude becomes

$$A_5 = Ae^{-\varsigma\omega_n(t_1+4T_d)} \tag{b}$$

Therefore, we have

$$\frac{A_1}{A_5} = e^{4\omega_n T_d} = 12 \qquad \text{or} \qquad 4\varsigma\omega_n T_d = \ln 12 \tag{c}$$

Thus, the logarithmic decrement δ is found to be

$$\delta = \varsigma\omega_n T_d = \frac{1}{4}\ln 12 = 0.621 \tag{d}$$

Since $\delta = \frac{2\pi\varsigma}{\sqrt{1-\varsigma^2}}$ (Eq. 2.5-20), we obtain the damping ratio

$$\varsigma = \frac{\delta}{\sqrt{4\pi^2+\delta^2}} = 0.0984 \tag{e}$$

Also, we can easily find the damping ratio as Eq. 2.5-20:

$$\varsigma \approx \frac{\delta}{2\pi} = 0.1 \tag{f}$$

Next, the period T_d is given by

$$T_d = \frac{2\pi}{\sqrt{1-\varsigma^2}\,\omega_n} = \frac{2\pi}{\sqrt{1-0.0984^2}}\sqrt{\frac{5/9.81}{200}} = 3.127 \ \ (\text{s}) \tag{g}$$

Table 2.2-1 Spring Stiffness [2.9] (This table is reproduced with permission of Prentice-Hall, Inc.)

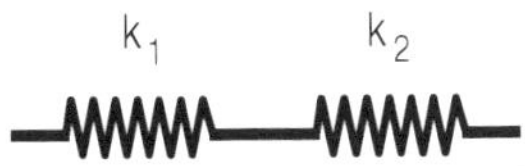	$k = \dfrac{1}{1/k_1 + 1/k_2}$
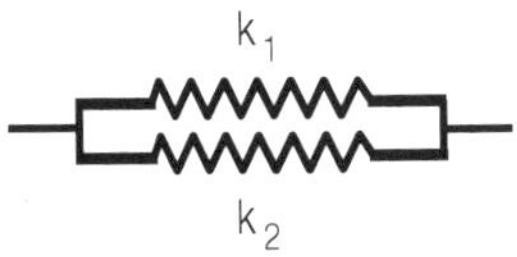	$k = k_1 = k_2$
	$k = \dfrac{EA}{L}$
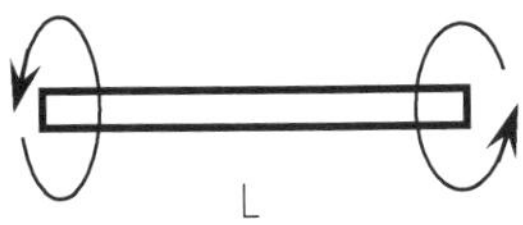	$k = \dfrac{GJ_p}{L^3}$
	$k = \dfrac{Gd^4}{64nR^3}$
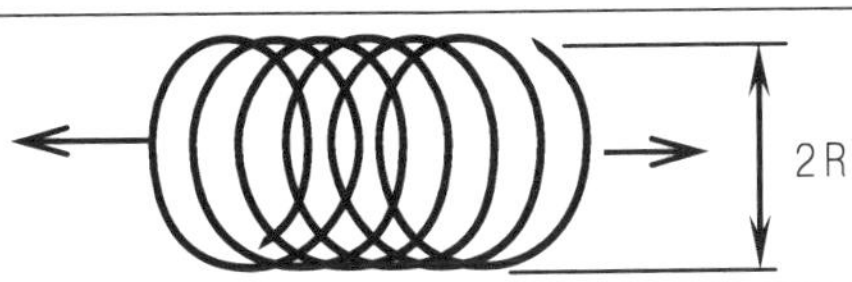	$k = \dfrac{3EI}{L^3}$
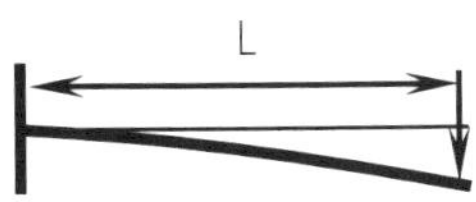	$k = \dfrac{192EI}{L^3}$
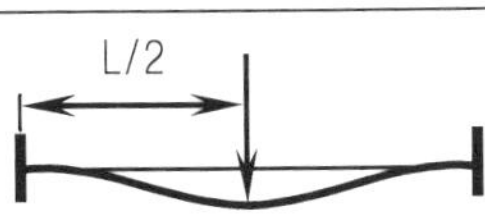	$k = \dfrac{3EJL}{(ab)^2}$
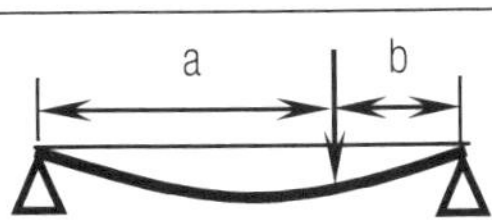	$k = \dfrac{12EJ}{L^3}$
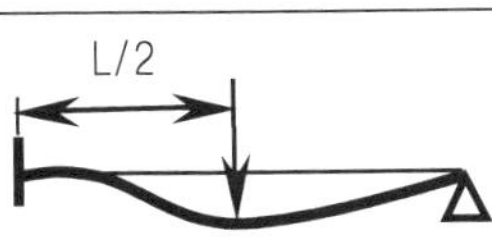	$k = \dfrac{768EJ}{7L^3}$
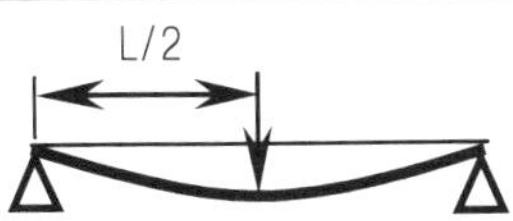	$k = \dfrac{48EJ}{L^3}$
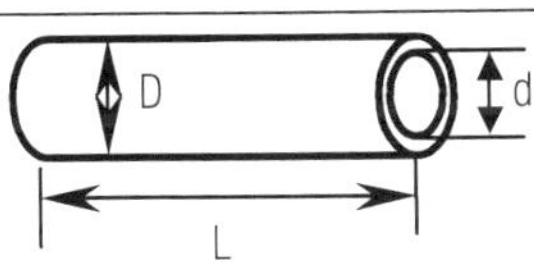	$k = \dfrac{\pi G(D^2 - d^2)}{32L}$

Problems

2.1 A spring-mass system with mass m and spring constant k_1 has a natural frequency of ω_1. If a second spring with constant k_2 is added in series with the first spring, the natural frequency becomes ω_1 / n. Determine k_2 in terms of k_1.

Ans. $k_2 = \dfrac{k_1}{n^2 - 1}$

2.2 An unknown mass M attached to the end of an unknown spring k has a natural frequency ω_1. When a known mass m is added to M, the natural frequency becomes ω_2. Determine the unknown mass M and the spring constant k.

Ans. $M = \dfrac{m\omega_2^2}{\omega_1^2 - \omega_2^2}$, $k = \dfrac{m\omega_1^2\omega_2^2}{\omega_1^2 - \omega_2^2}$

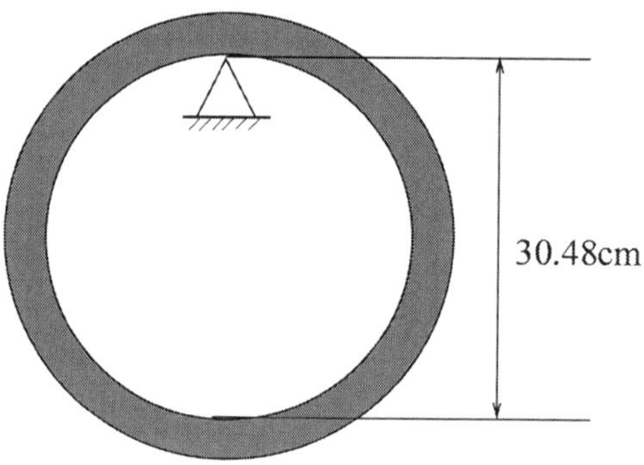

Fig. P2.3

2.3 A circular ring weighing 30 kg is allowed to swing as a pendulum about a knife-edge at its inner side, as in Fig. P2.3. If the measured period of oscillation is 1.35 seconds, determine the moment of inertia of the ring about its geometric axis.

Ans. 1.372 kg·m^2

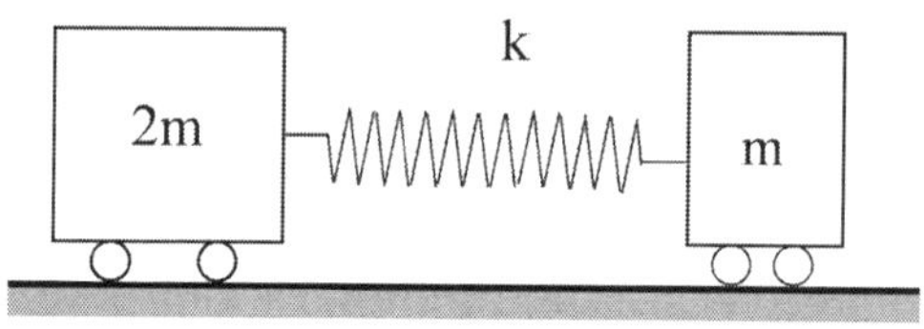

Fig. P2.4

2.4 Use the knowledge of a single-degree-of-freedom system to determine the natural frequency of the system shown in Fig. P2.4.

Ans. $\omega_n = \sqrt{\dfrac{3k}{2m}}$

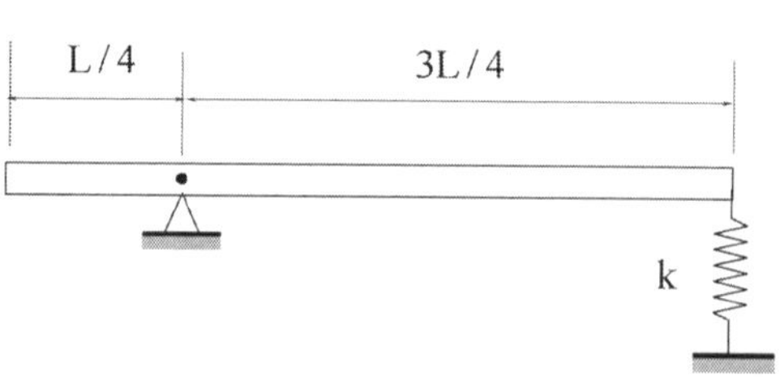

Fig. P2.5

2.5 Determine the natural frequency of the system shown in Fig. P2.5, where the mass of the uniform rod is m and the spring constant is k.

Ans. $\omega_n = 3\sqrt{\dfrac{3k}{7m}}$

2.6 Determine the natural frequency of torsional oscillation of the system shown in Fig. P2.6.

Ans. $\omega_n = \sqrt{\dfrac{3K_\theta}{2I}}$

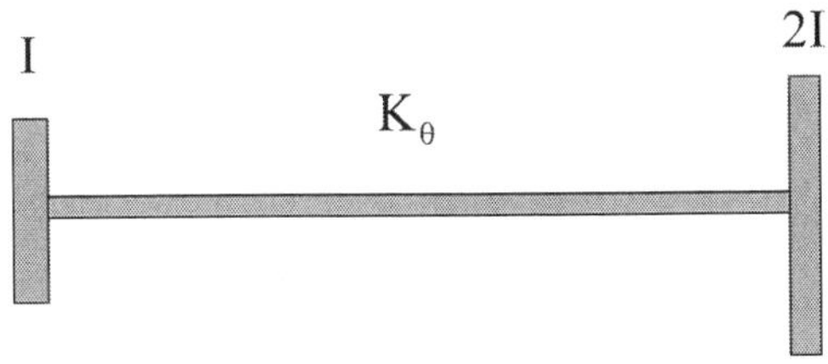

Fig. P2.6

2.7 Show that the logarithmic decrement can be expressed as

$$\delta = \frac{2\pi\varsigma}{\sqrt{1-\varsigma^2}} = \frac{1}{n}\ln\left(\frac{A_0}{A_n}\right),$$

where n is the number of cycles used, A_0 is the amplitude at time $t = t_0$, and A_n is the amplitude at time $t = t_n$.

2.8 A mass of 4 kg is attached to the end of a spring with a stiffness 5 N/cm. Determine the critical damping coefficient.

Ans. $c_r = 20\sqrt{20}$ Ns/m

2.9 A vibrating system consists of the following constants: m = 3 kg, k = 4000 N/m, c = 4 Ns/m. Determine (a) the damping ratio, (b) the damped natural frequency, and (c) the logarithmic decrement.

Ans. $\varsigma = 0.018$, $\omega_d = 36.51$ rad/s, $\delta = 0.113$

2.10 The deflection of a clamped-clamped uniform beam at point x along the axis with a load at the middle point is given by

$$y = \frac{y_m}{2}\left(1 - \cos\frac{2\pi x}{L}\right),$$

where y_m is the maximum deflection of the middle point, and L is the length of the beam. Let E, I, and m be Young's modulus, the area moment of inertia of the cross section, and the mass of the beam, respectively. Use the Rayleigh method to determine the natural frequency of transverse vibration of the beam.

Ans. $\omega_n = 22.63\sqrt{\dfrac{EI}{mL^3}}$

2.11 Find the natural frequencies for the system

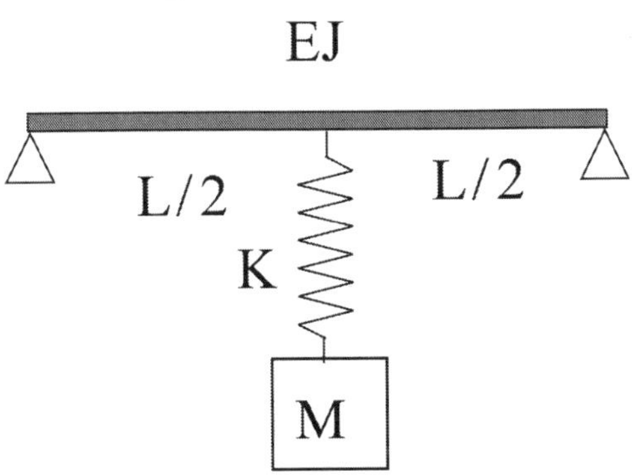

Fig. P2.11

shown in Fig. P2.11, where $L = 4$ m, $EJ = 19.6\times10^5$ N·m^2, $K = 4.9\times10^5$ N/m, and $m = 400$ kg.

Ans. $f = 4.82$ Hz

2.12 Find the natural frequency of small vibration of the system shown in Fig. P2.12 about point O, where the length of each bar is L, the mass of each bar is the same, and the angle between the two bars is 90°.

Ans. $\omega_n = 1.0299\sqrt{g/L}$

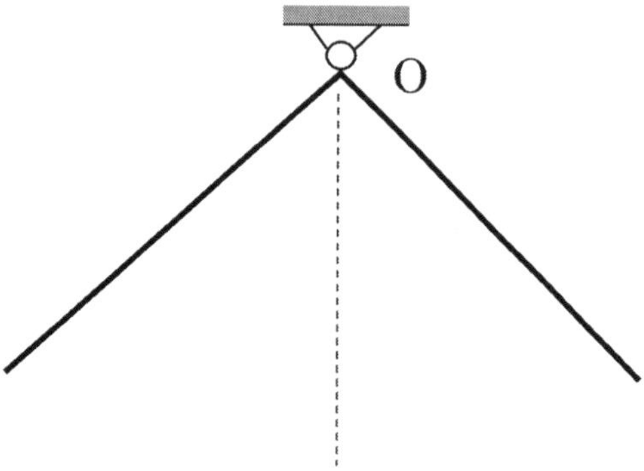

Fig. P2.12

2.13 Find the undamped and damped natural frequencies of the system shown in Fig. P2.13. Assume that the bar is uniform and its mass is m.

Ans. $\omega_n = \dfrac{a}{b}\sqrt{3k/m}$,

$\omega_d = \omega_n\sqrt{1-3b^2c^2/(4kma^2)}$

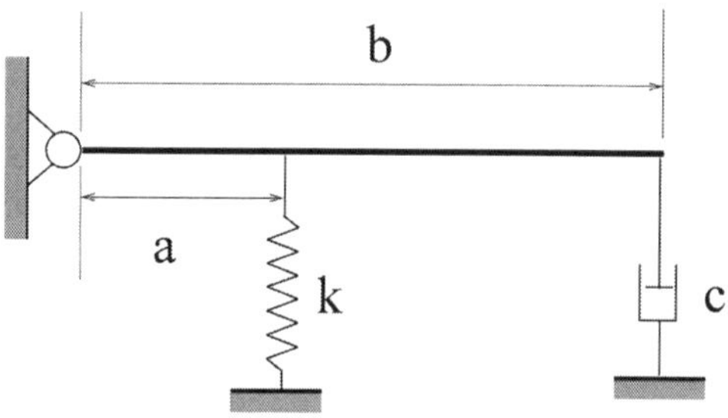

Fig. P2.13

2.14 Find the natural frequencies of the systems shown in Fig. P2.14. As a check, when $\beta = 0$, the result for the system of Fig. P2.14 (A) should be equal to that of the system of Fig. P2.14 (B).

Ans. (A) $\omega_n = \sqrt{\dfrac{ka^2 + mgb\sin\beta}{mb^2}}$, (B) $\omega_n = \dfrac{a}{b}\sqrt{\dfrac{k}{m}}$

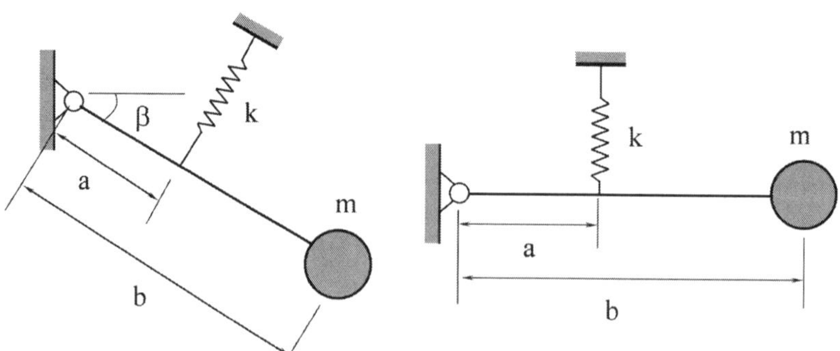

(A) (B)

Fig. P2.14

2.15 Show that the free vibratory response of a damped system can be written as

$$x = e^{-\varsigma\omega_n t}\left[x_0 \cos\sqrt{1-\varsigma^2}\,\omega_n t + \frac{\dot{x}_0 + \varsigma\omega_n x_0}{\sqrt{\varsigma^2 - 1\omega_n}} \sin\sqrt{1-\varsigma^2}\,\omega_n t\right], 0<\varsigma<1$$

$$x = e^{-\varsigma\omega_n t}\left[x_0 \cosh\sqrt{\varsigma^2 - 1\omega_n t} + \frac{\dot{x}_0 + \varsigma\omega_n x_0}{\sqrt{\varsigma^2 - 1\omega_n}} \sinh\sqrt{\varsigma^2 - 1\omega_n t}\right], \varsigma>1$$

2.16 Find the solutions of the equation $\ddot{x}+2\dot{x}+4x=0$ for the following initial conditions:

(a) $x(0)=1$ and $\dot{x}(0)=0$, (b) $x(0)=0$ and $\dot{x}(0)=3$, (c) $x(0)=1$ and $\dot{x}(0)=3$.

Ans. (a) $x=\left(2/\sqrt{3}\right)e^{-t}\sin\left(\sqrt{3}t+60°\right)$, (b) $x=\sqrt{3}e^{-t}\sin\left(\sqrt{3}t\right)$,

(c) $x=\sqrt{19/3}e^{-t}\sin\left(\sqrt{3}t+\tan^{-1}\left(\sqrt{3}/4\right)\right)$

References

2.1. Beer, F. P., and Johnston, E. R., *Vector Mechanics for Engineers*, 6th ed., McGraw-Hill, New York, 1996.

2.2. Meriam, J. L., and Kraige, L. G., *Engineering Mechanics: Dynamics*, Vol. 2, 3rd ed., John Wiley & Sons, New York, 1992.

2.3. Ginsberg, J. H., and Genin, J., *Dynamics*, 2nd ed., Wiley, New York, 1984.

2.4. Huston, R. L., *Multibody Dynamics*, Butterworth-Heinemann, Stoneham, MA, 1990.

2.5. Huston, R. L., and Liu, C. Q., *Formulas for Dynamic Analysis*, Marcel Dekker, Inc., New York, 2001.

2.6. Whittaker, E. T., *A Treatise on the Analytical Dynamics of Particles and Rigid Bodies*, 4th ed., Cambridge, London, 1937.

2.7. Tse, F. S., Morse, I. E., and Hinkle, R. T., *Mechanical Vibration, Theory and Applications*, Allyn and Bacon, Inc., Boston, 1979.

2.8. Timoshenko, S. P., Young, D. H., and Weaver, W., Jr., *Vibration Problems in Engineering*, 4th ed., John Wiley & Sons, New York, 1967.

2.9. Thomson, W. T., and Dahleh, M. D., *Theory of Vibration with Applications*, 5th ed., Prentice Hall, Inc., Upper Saddle River, NJ, 1998.

2.10. Inman, D. J., *Engineering Vibration*, Prentice Hall, Inc., Upper Saddle River, NJ, 1994.

2.11. Harris, Cyril M., ed., *Shock and Vibration Handbook*, 3rd ed., McGraw-Hill, New York, 1987.

Chapter Three

Harmonically Excited Motion

3.1 Introduction

In rotating machinery, the turning of the rotor often induces vibratory motion in surrounding structures. Such induced motion is commonly called *harmonically excited* motion. Out-of-balance tires, shaft imbalance, and loose and/or out-of-alignment bearings are also common stimulants of harmonically excited vibrations. In this chapter we consider harmonic excitation and in Chapter Four we look at imbalance in detail.

3.2 Forced Harmonic Vibration of Damped Systems

Consider the system shown in Fig. 3.2-1. It consists of a vertically vibrating block with mass m supported by a spring with stiffness k. There is a dashpot, or damper, between the block and the support which produces viscous damping with damping coefficient c. Finally, there is an oscillating force applied to the block expressed as $F_0 \sin \omega t$. This force harmonically stimulates the block, producing a harmonically excited vibration.

The system of Fig. 3.2-1 has one degree of freedom. We can obtain the equation of motion by considering a free-body diagram (FBD) of the block,

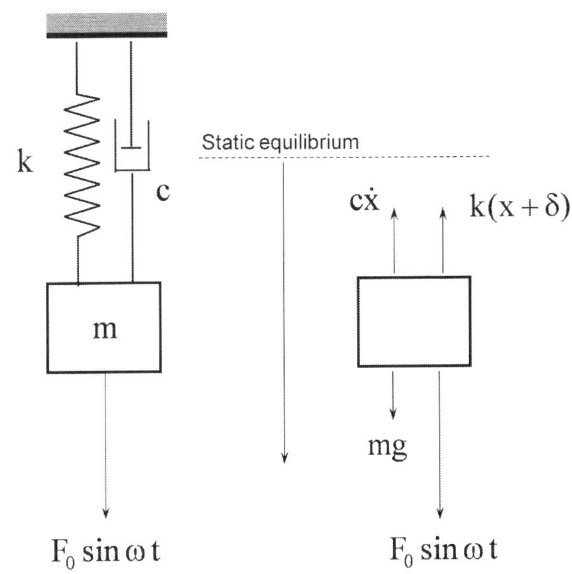

Fig. 3.2-1 Viscously damped system with harmonic excitation.

as also shown in Fig. 3.2-1. From that diagram we immediately see that the equation of motion may be written as

$$m\ddot{x} = -k(x+\delta) - c\dot{x} + mg + F_0 \sin\omega t \tag{3.2-1}$$

where δ is the downward displacement of the block during static equilibrium. Then $k\delta = mg$, and we can write Eq. 3.2-1 in the simplified form

$$m\ddot{x} + c\dot{x} + kx = F_0 \sin\omega t \tag{3.2-2}$$

We can further simplify the form of the governing equation by letting ω_n be $\sqrt{k/m}$ (the undamped natural frequency), and by letting ς be $c/2\sqrt{km}$ (the damping ratio). The equation of motion then becomes

$$\ddot{x} + 2\varsigma\omega_n\dot{x} + \omega_n^2 x = h\sin\omega t, \tag{3.2-3}$$

where $h = F_0/m$.

The solution of Eq. 3.2-3 may be written as the sum of two functions x_1 and x_2. These are (1) x_1, the *complementary function* (the general solution of the equation with the right side being zero—the *homogeneous equation*); and (2) x_2, the *particular solution* (any solution of the equation with the right hand side being $h\sin\omega t$—the *inhomogeneous equation*).

The complementary function represents a damped free vibration, as discussed in Chapter Two (see Eq. 2.5-15):

$$x_1 = Ae^{-\varsigma\omega_n t}\sin(\omega_d t + \phi) \tag{3.2-4}$$

We can obtain a particular solution of the equation by looking for it in the form of a steady-state oscillation having the same frequency as that of the excitation. Specifically, we can obtain a particular solution by assuming the form

$$x_2 = X\sin(\omega t - \psi), \tag{3.2-5}$$

where X is the amplitude of the vibration and ψ is the phase angle of displacement with respect to the excitation force.

There are two methods for finding the particular solution.

Method 1: The amplitude and phase angle can be found by substituting Eq. 3.2-5 into Eq. 3.2-3. That is,

$$-X\omega^2\sin(\omega t - \psi) + 2\varsigma\omega_n X\omega\cos(\omega t - \psi) + \omega_n^2 X\sin(\omega t - \psi) = h\sin\omega t \tag{3.2-6}$$

The right side term can be written as

$$h\sin\omega t = h\sin[(\omega t - \psi) + \psi] = h\cos\psi\sin(\omega t - \psi) + h\sin\psi\cos(\omega t - \psi) \tag{3.2-7}$$

Substituting Eq. 3.2-7 into Eq. 3.2-6, we obtain

$$\left[X(\omega_n^2-\omega^2)-h\cos\psi\right]\sin(\omega t-\psi)+\left[2\varsigma\omega_n X\omega-h\sin\psi\right]\cos(\omega t-\psi)=0 \quad (3.2\text{-}8)$$

Since Eq. 3.2-8 must hold for any and all times, the coefficients of $\sin(\omega t-\psi)$ and $\cos(\omega t-\psi)$ must vanish. That is,

$$\left[X(\omega_n^2-\omega^2)-h\cos\psi\right]=0 \quad \text{or} \quad X(\omega_n^2-\omega^2)=h\cos\psi \qquad (3.2\text{-}9)$$

and

$$\left[2\varsigma\omega_n X\omega-h\sin\psi\right]=0 \quad \text{or} \quad X2\varsigma\omega_n\omega=h\sin\psi \qquad (3.2\text{-}10)$$

From these last two equations we obtain

$$(\omega_n^2 X-\omega^2 X)^2+(2\varsigma\omega_n\omega X)^2=h^2 \qquad (3.2\text{-}11)$$

$$X=\frac{h}{\sqrt{(\omega_n^2-\omega^2)^2+(2\varsigma\omega_n\omega)^2}} \qquad (3.2\text{-}12)$$

$$\psi=\tan^{-1}\frac{2\varsigma\omega_n\omega}{\omega_n^2-\omega^2} \qquad (3.2\text{-}13)$$

Method 2: The amplitude and phase angle can be found graphically. Recall that in harmonic motion the phase angles of the velocity and acceleration precede that of the displacement by 90 and 180 degrees, respectively. Thus, the terms of Eq. 3.2-2 can be displayed graphically as in Fig. 3.2-2. This can be

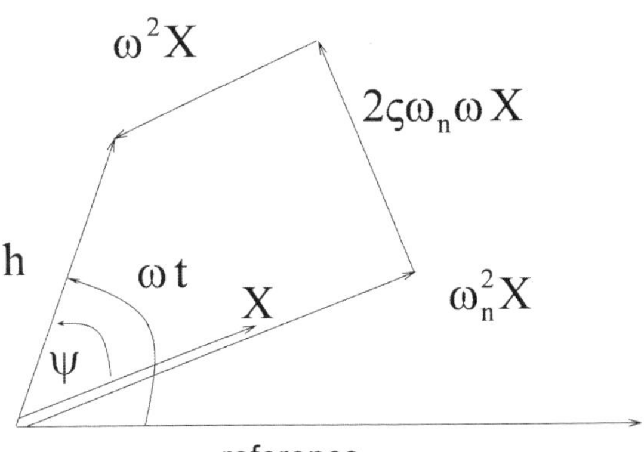

Fig. 3.2-2 Vector relationship for forced vibration with damping.

easily understood from Eq. 3.2-6, from which we obtain Eq. 3.2-11, and then Eqs. 3.2-12 and 3.2-13 as the consequences.

In summary, since the system is linear, the complete solution of Eq. 3.2-1 is the sum of the homogeneous solution, given by Eq. 3.2-4, plus the particular solution given by Eq. 3.2-5. That is,

$$x = x_1 + x_2$$

$$= Ae^{-\varsigma\omega_n t}\sin(\omega_d t+\phi)+\frac{h}{\sqrt{(\omega_n^2-\omega^2)^2+(2\varsigma\omega_n\omega)^2}}\sin\left(\omega t-\tan^{-1}\frac{2\varsigma\omega_n\omega}{\omega_n^2-\omega^2}\right) \tag{3.2-14}$$

Fig. 3.2-3 shows a graph of the solution, where the dashed line represents the first term, the homogeneous solution, which decays after a long time. The parameters used in the simulation are (1) the mass m=2 kg, (2) the spring stiffness k=3700 N/m, (3) the damping coefficient c=13 Ns/m, (4) the force amplitude F_0=65 N, (5) the excitation frequency ω=5 rad/s, (6) the initial displacement x_0=0.03 m, and (7) the initial velocity $\dot{x}_0$=1.5 m/s.

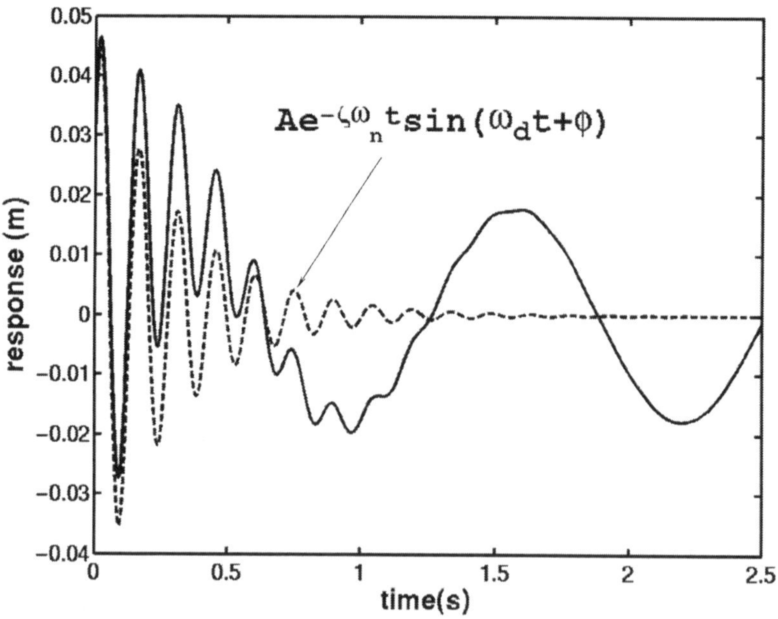

Fig. 3.2-3 Example of forced vibration.

Fig. 3.2-3 is obtained using the Matlab code:

```
clear all
k=3700; % ............................stiffness N/m
m=2; %................................mass,Kg
c=13;%............................... damping,Ns/m
F0=65;%...............................force,N
x0=0.03; % ...........................initial
                                      displacement, m
x0dot=1.5; % .........................initial velocity,
                                      m/s
omegan=sqrt(k/m); % ..................natural freq.
omega=5; % ...........................excitation freq.
myzeta=c/(2*sqrt(k*m)); % ............damping ratio
omegad=sqrt(1-myzeta^2)*omegan; % ....damped freq.
A=sqrt(x0^2+(x0dot+myzeta*omegan*x0)^2/omegad^2);
phi=atan(x0*omegad/(x0dot+myzeta*omegan*x0));
h=F0/m;
X=h/sqrt((omegan^2-omega^2)^2+(2*myzeta*omegan*omega)^2);
psi=atan(2*myzeta*omegan*omega)/(omegan^2-omega^2);
for kk=1:250
t(kk)=0.01*(kk-1);
x1(kk)=A*exp(-myzeta*omegan*t(kk))*sin(omegad*t(kk)+phi);
x2(kk)=X*sin(omega*t(kk)-psi);
x(kk)=x1(kk)+x2(kk);
end
h=plot(t,x,'k',t,x1,'--k');
set(h,'LineWidth',2);
hx=get(gca,'Xlabel');
hy=get(gca,'Ylabel');
set(hx,'String','time(s)');
set(hy,'String','response (m)');
set(hx,'FontSize',14);
set(hy,'FontSize',14);
```

Consider further the terms of Eq. 3.2-14: observe that, for large values of time t, the first term (the *homogeneous* solution) approaches zero. Thus, during the passage of time, the total solution approaches the particular solution. Therefore, the first term is sometimes called the *transient response* and the second term is called the *steady-state response*.

If the system has large damping, the transient response decays very quickly. Alternatively, if the system is lightly damped, the transient response will linger and thus it cannot be ignored.

To obtain additional insight, we introduce parameters r and X_0 as

$$r = \frac{\omega}{\omega_n} \tag{3.2-15}$$

$$X_0 = \frac{h}{\omega_n^2} = \frac{F_0}{k}, \tag{3.2-16}$$

where r expresses the ratio of the excitation frequency to the natural frequency of the system, and X_0 is the *static deflection* of the spring under a load F_0. If these are substituted into Eqs. 3.2-12 and 3.2-13, the amplitude X and the phase angle ψ can be written as

$$X = \frac{X_0}{\sqrt{(1-r^2)^2 + (2\varsigma r)^2}} \tag{3.2-17}$$

$$\psi = \tan^{-1} \frac{2\varsigma r}{1-r^2} \tag{3.2-18}$$

Dividing Eq. 3.2-17 by the static deflection of the spring X_0 yields

$$MF = \left| \frac{X}{X_0} \right| = \frac{1}{\sqrt{(1-r^2)^2 + (2\varsigma r)^2}} \tag{3.2-19}$$

MF is a dimensionless parameter providing the ratio of the maximum response amplitude to the static deflection of the spring under load F_0. MF is variously called the *magnification factor*, the *amplitude ratio*, and the *amplification factor*. MF expresses the extent to which the displacement from the harmonic excitation is a multiple of the static spring deflection.

Eqs. 3.2-18 and 3.2-19 show that the phase angle and the dimensionless amplitude are functions of only the frequency ratio and the damping ratio. Figs. 3.2-4 and 3.2-5 provide a graphical representation of these equations. The amplitude representation of Fig. 3.2-4 provides a family of curves for various damping ratios. It is readily seen that the amplitude of the forced vibration is significantly reduced by damping. Interestingly, it is seen that the maximum amplitude does *not* occur when the excitation frequency ω is equal to the natural frequency ω_n (that is, $r = 1$). Instead, the maximum amplitude occurs at the *resonance frequency* ω_r given by $\sqrt{1-2\varsigma^2}\,\omega_n$.

For forced vibration of a damped system, there are three different frequencies [3.1]:

(1) $\omega_n = \sqrt{k/m}$ = the *undamped natural frequency*

(2) $\omega_d = \sqrt{1-\varsigma^2}\,\omega_n$ = the *damped natural frequency*

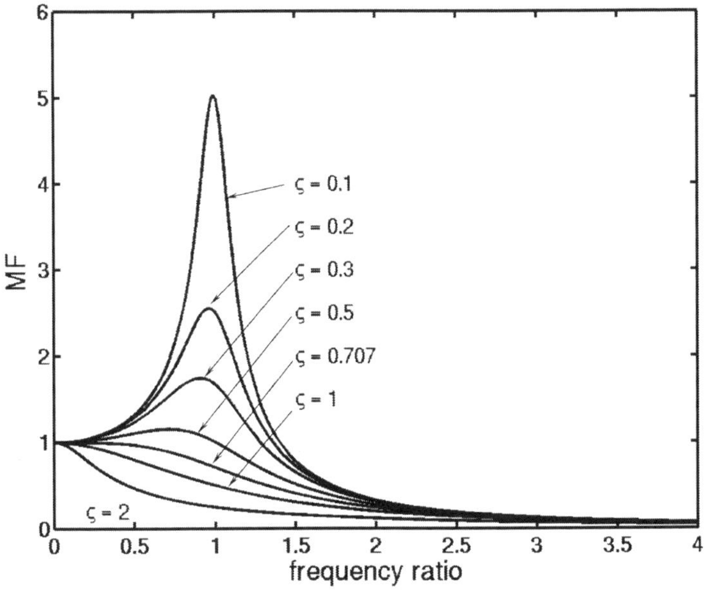

Fig. 3.2-4 Magnification factor for various degrees of damping.

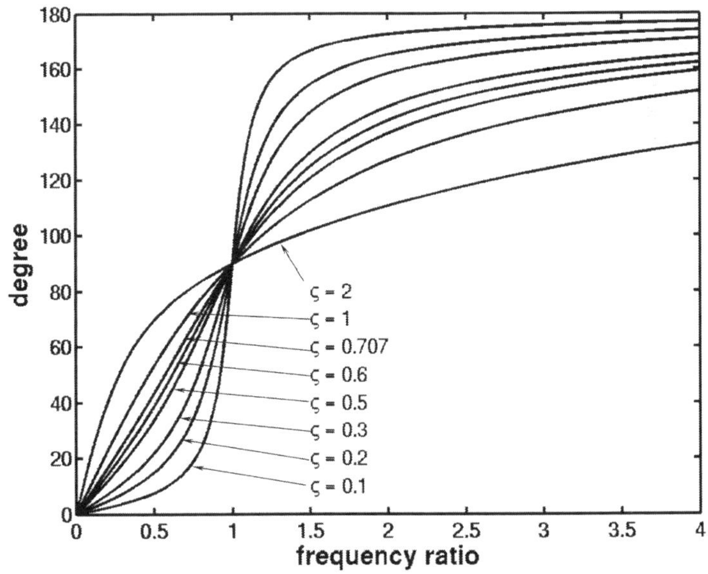

Fig. 3.2-5 The phase angle between force and displacement for various degrees of damping.

(3) $\omega_r = \sqrt{1-2\varsigma^2}\,\omega_n$ = the *resonance frequency*, for $0<\varsigma<1/\sqrt{2}$

The resonance frequency ω_r is obtained by differentiating with respect to r in Eq. 3-2-19. By setting the result equal to zero, we obtain the value $\omega_r = \sqrt{1-2\varsigma^2}\,\omega_n$.

For small damping, the undamped natural frequency ω_n, the damped natural frequency ω_d, and the resonance frequency ω_r have values close to one another. For zero damping ($\varsigma=0$), the three frequencies are the same.

Observe further that as the excitation frequency ω approaches the resonance frequency ω_r, the amplitude ratio MF approaches its maximum value $MF_{\max}$ given by

$$MF_{\max} = \frac{1}{2\varsigma\sqrt{1-\varsigma^2}} \tag{3.2-20}$$

For lightly damped systems, the resonance frequency is very close to the undamped natural frequency (that is, $\omega_r \approx \omega_n$). Thus, from Eq. 3.2-19 (with $r\approx 1$), the maximum value is approximately $1/2\varsigma$:

$$MF_{\max} \approx \frac{1}{2\varsigma} \tag{3.2-21}$$

Observe still further that the resonance frequency ω_r does not exist if the damping ratio ς is larger than $1/\sqrt{2}$, since then $\sqrt{1-2\varsigma^2}$ is imaginary. Indeed, if $\varsigma>1/\sqrt{2}$, the maximum value of the amplitude magnification factor MF occurs when the frequency ratio r is zero. As r approaches zero, the magnification factor MF approaches 1; that is, the amplitude approaches the static displacement. Finally, if the frequency ratio r becomes very large, the amplitude approaches zero.

The phase angle graph of Fig. 3.2-5 can also provide insight into the forced vibration phenomenon. As the excitation frequency ω approaches the undamped frequency ω_n (that is, $r\to 1$), the phase angle ψ passes through 90°. Eq. 3.2-18 also shows this since, as $r\to 1$, $\tan\psi\to\infty$, and thus $\psi\to\pi/2$ radians (or 90°).

Observe in Eq. 3.2-18 that since ψ is evaluated from the arctangent function, we assume that ψ lies between zero and 180 degrees (or $0\le\psi\le\pi$). From Eqs. 3.2-9 and 3.2-10 we see that the quadrant that ψ lies in is determined by the sign of $(\omega_n-\omega)$. If $(\omega_n-\omega)<0$ (or $r>1$), then $\sin\psi>0$ and $\cos\psi<0$. Thus, when the value of the tangent is negative, the phase angle ψ must be in the second quadrant rather than the fourth quadrant. When a code or calculator is used to evaluate the phase angle for the negative values of ψ, $\psi+\pi$ should be taken as the phase angle in Eq. 3.2-14.

3.3 Forced Harmonic Vibration of Undamped Systems

Consider a system with negligible damping ($\varsigma = 0$), as shown in Fig. 3.3-1. In this case, the equation of motion, Eq. 3.2-3, becomes

$$\ddot{x} + \omega_n^2 x = h \sin \omega t, \tag{3.3-1}$$

where $h = F_0/m$.

With Eq. 3.3-1 being linear in x, we may again employ the principle of superposition to obtain the general solution. That is, the general solution of Eq. 3.3-1 is the sum of the solution of the homogeneous equation (right side being zero) and a particular (that is, any) solution to the inhomogeneous equation (right side not zero). Following this procedure, we obtain the general solution of Eq. 3.3-1 as

$$x = A \sin(\omega_n t + \phi) + \frac{h}{\omega_n^2 - \omega^2} \sin \omega t, \tag{3.3-2}$$

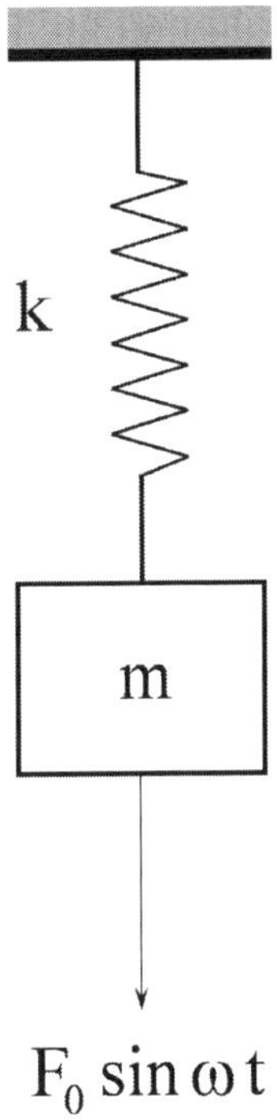

Fig. 3.3-1 Undamped spring-mass system.

where A and ϕ are constants (amplitude and phase angle) to be determined from initial conditions.

Observe that Eq. 3.3-2 may also be obtained directly from Eq. 3.2-14 by letting the damping ratio ς be zero.

Observe further that $A \sin(\omega_n t + \phi)$ may be represented as $A_1 \sin \omega_n t + A_2 \cos \omega_n t$, where A_1 and A_2 are constants. Then Eq. 3.3-2 becomes

$$x = A_1 \sin \omega_n t + A_2 \cos \omega_n t + \frac{h}{\omega_n^2 - \omega^2} \sin \omega t \tag{3.3-3}$$

If the initial values of displacement $x(0)$ and velocity $\dot{x}(0)$ are x_0 and $\dot{x}_0$, Eq. 3.3-3 yields A_1 and A_2 as

$$A_1 = \frac{\dot{x}_0}{\omega_n} - \frac{\omega}{\omega_n} \frac{h}{\omega_n^2 - \omega^2} \quad \text{and} \quad A_2 = x_0 \tag{3.3-4}$$

Substituting these values into Eq. 3.3-3 yields the response of the system as

$$x=\left(\frac{\dot{x}_0}{\omega_n}-\frac{\omega}{\omega_n}\frac{h}{\omega_n^2-\omega^2}\right)\sin\omega_n t+x_0\cos\omega_n t+\frac{h}{\omega_n^2-\omega^2}\sin\omega t \tag{3.3-5}$$

Two important phenomena occur when the excitation frequency ω becomes close to the system's natural frequency ω_n:

(1) Beating

Consider first the case where ω is nearly equal to ω_n. Let the difference be $\Delta\omega$. That is, $\Delta\omega=\omega_n-\omega$. For zero values of initial conditions x_0 and $\dot{x}_0$, the response of Eq. 3.3-5 becomes

$$x=\left(-\frac{\omega}{\omega_n}\frac{h}{\omega_n^2-\omega^2}\right)\sin\omega_n t+\frac{h}{\omega_n^2-\omega^2}\sin\omega t \tag{3.3-6}$$

Let a and b be defined as

$$a=-\frac{\omega}{\omega_n}\frac{h}{\omega_n^2-\omega^2} \quad \text{and} \quad b=\frac{h}{\omega_n^2-\omega^2} \tag{3.3-7}$$

Then the response can be expressed as [3.1]

$$\begin{aligned} x&=a\sin\omega_n t+b\sin\omega t\\ &=a\left(\sin\omega t\cos\Delta\omega t+\cos\omega t\sin\Delta\omega t\right)+b\sin\omega t,\\ &=\left(a\cos\Delta\omega t+b\right)\sin\omega t+a\sin\Delta\omega t\cos\omega t\\ &=\sqrt{\left(a\cos\Delta\omega t+b\right)^2+\left(a\sin\Delta\omega t\right)^2}\,\sin(\omega t+\psi) \end{aligned} \tag{3.3-8}$$

where the phase angle ψ can be evaluated from the expression

$$\psi=\tan^{-1}\frac{a\sin\Delta\omega t}{a\cos\Delta\omega t+b} \tag{3.3-9}$$

The amplitude can be written as

$$\sqrt{\left(a\cos\Delta\omega t+b\right)^2+\left(a\sin\Delta\omega t\right)^2}=\sqrt{a^2+b^2+2ab\cos\Delta\omega t} \tag{3.3-10}$$

The expression of Eq. 3.3-10 is seen to vary between $(a+b)$ and $(a-b)$ with a *beat frequency* $\Delta\omega$. When $\Delta\omega$ is small, the period of the beat, $T_{\text{beat}}=2\pi/\Delta\omega$, is large. Thus, the resulting motion is a rapid oscillation with a slowly varying amplitude, called a *beat*.

Fig. 3.3-2 shows the phenomenon of beats. Here $\omega_n=5$ rad/s, $\omega=0.9\omega_n$, h=2 N/kg; the dashed line is a graph of $\sqrt{a^2+b^2+2ab\cos\Delta\omega t}$. As shown in Fig. 3.3-3, when $\Delta\omega$ is very small, the beat period becomes very large. In the limit case, as $\Delta\omega$ approaches zero, the beat period approaches infinity, and, as a result, the response grows without bound.

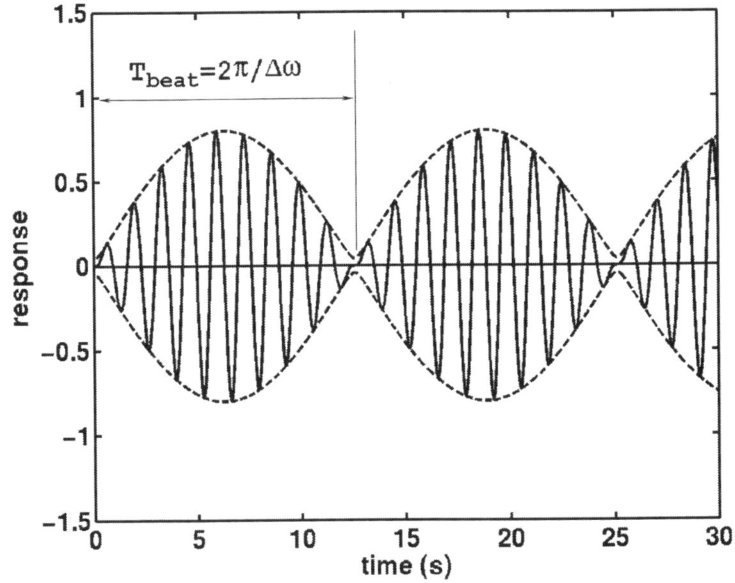

Fig. 3.3-2 Response of an undamped system of Eq. 3.3-8. Here $\omega_n = 5$ rad/s, $\omega = 0.9\omega_n$, $h = 2$ N/kg; the dashed line is the graph of $\sqrt{a^2 + b^2 + 2ab\cos\Delta\omega t}$.

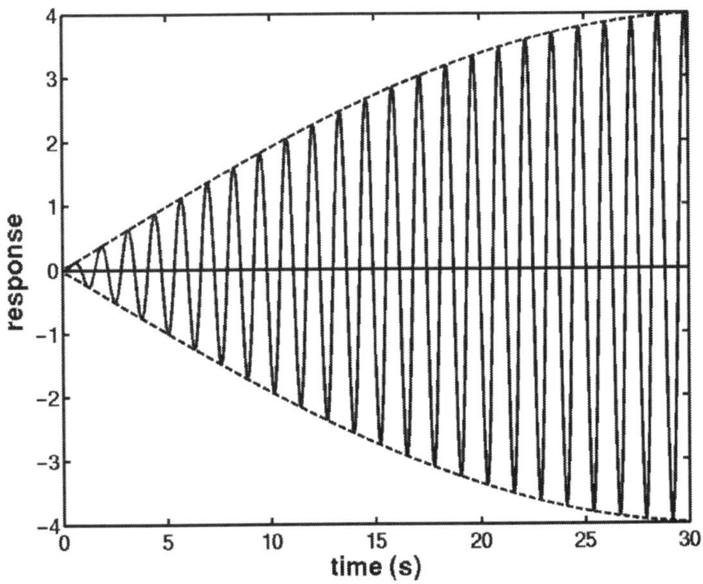

Fig. 3.3-3 Response of an undamped system of Eq. 3.3-8. Here $\omega_n = 5$ rad/s, $\omega = 0.98\omega_n$, $h = 2$ N/kg; the dashed line is the graph of $\sqrt{a^2 + b^2 + 2ab\cos\Delta\omega t}$.

(2) Resonance

Next, as the excitation frequency ω becomes exactly equal to the system's natural frequency ω_n, the solution given in Eq. 3.3-3 is not valid. In this case, the particular solution has the form

$$x_p = Bt\cos\omega t \tag{3.3-11}$$

Substitution of Eq. 3.3-11 into Eq. 3.3-1 yields

$$x_p = -\frac{h}{2\omega} t\cos\omega t \tag{3.3-12}$$

Thus the total solution, that is, the sum of the homogeneous solution and the particular solution, is now of the form ($\omega = \omega_n$):

$$x = A_1 \sin\omega_n t + A_2 \cos\omega_n t - \frac{h}{2\omega_n} t\cos\omega_n t \tag{3.3-13}$$

Let x_0 and $\dot{x}_0$ represent the initial displacement and velocity as before. Then Eq. 3.3-13 yields

$$x = \left(\frac{\dot{x}_0}{\omega_n} + \frac{h}{2\omega_n^2}\right)\sin\omega_n t + x_0 \cos\omega_n t - \frac{h}{2\omega_n} t\cos\omega_n t \tag{3.3-14}$$

Fig. 3.3-4 shows a graph of Eq. 3.3-14. It is apparent that the amplitude of vibration becomes unbounded when $\omega = \omega_n$. This important phenomenon is known as *resonance*, and the natural frequency is sometimes called the *resonance frequency*. This result, however, is primarily theoretical since all physical systems have damping.

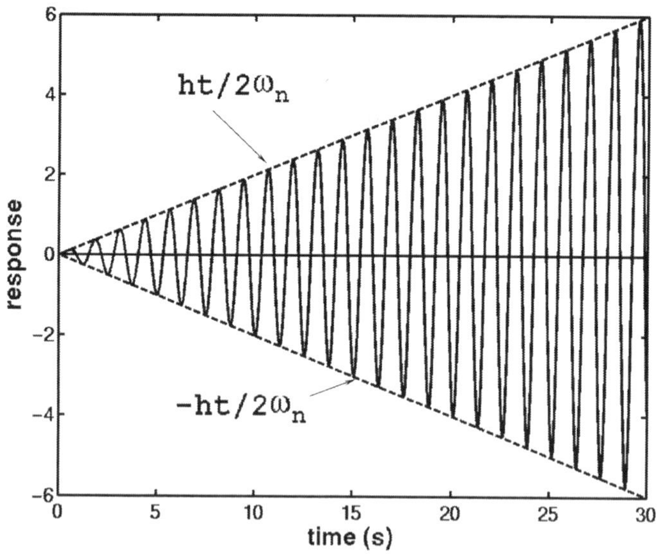

Fig. 3.3-4 Forced response of an undamped system of Eq. 3.3-14 with zero initial conditions. Here $\omega_n = 5$ rad/s, $h = 2$ N/kg.

3.4 Forced Vibration Caused by Rotating Unbalance

Unbalance in rotating machinery is a common and frequent source of vibration excitation. Perhaps the most common example is unbalanced tires on motor vehicles.

We can model unbalanced rotating machinery including unbalanced tires as a mass m at a distance e away from the axis of rotation as represented in Fig. 3.4-1, where M is the machine mass, ω is the angular speed, x measures the displacement away from equilibrium, k is a spring constant, and c is a damping coefficient. Usually m is small compared with M, and e is small compared with typical machine dimensions.

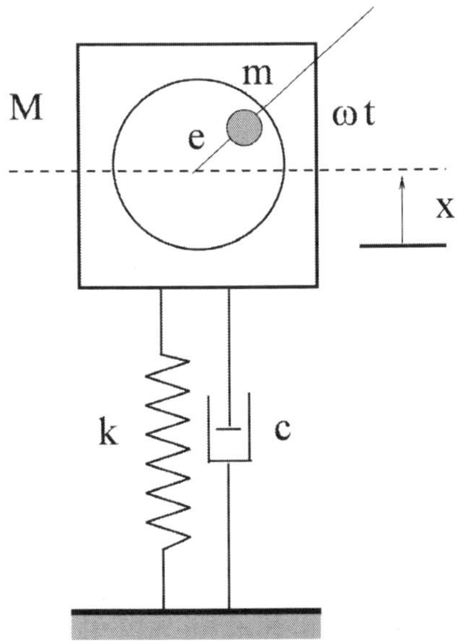

Fig. 3.4-1 Harmonic excitation due to rotating unbalance.

From Fig. 3.4-1, the equation of motion is readily seen to be

$$(M-m)\ddot{x}+m\frac{d^2}{dt^2}(x+e\sin\omega t)=-kx-c\dot{x} \qquad (3.4\text{-}1)$$

or

$$M\ddot{x}+c\dot{x}+kx=(me\omega^2)\sin\omega t \qquad (3.4\text{-}2)$$

Eq. 3.4-2 is identical to Eq. 3.2-2, where F_0 is replaced by $me\omega^2$. Introducing ω_n, h, and ς as

$$\omega_n=\sqrt{\frac{k}{M}},\ h=\frac{me\omega^2}{M},\ \varsigma=\frac{c}{2\sqrt{kM}}, \qquad (3.4\text{-}3)$$

Eq. 3.4-2 may be written as

$$\ddot{x}+2\varsigma\omega_n\dot{x}+\omega_n x=h\sin\omega t \qquad (3.4\text{-}4)$$

The solution of Eq. 3.4-4 is given by Eq. 3.2-14 and repeated here:

$$x=x_1+x_2$$
$$=Ae^{-\varsigma\omega_n t}\sin(\omega_d t+\phi)+\frac{h}{\sqrt{(\omega_n^2-\omega^2)^2+(2\varsigma\omega_n\omega)^2}}\sin\left(\omega t-\tan^{-1}\frac{2\varsigma\omega_n\omega}{\omega_n^2-\omega^2}\right), \qquad (3.4\text{-}5)$$

where A and ϕ are constants to be determined by initial conditions, and where $\omega_d = \sqrt{1-\varsigma^2}\,\omega_n$ is the damped natural frequency. Again, as discussed in Section 3.2, the phase angle is evaluated from the arctangent function, which must be evaluated assuming it lies between zero and π $(0 \le \psi \le \pi)$.

Introducing the frequency ratio r as $r = \omega / \omega_n$ and using Eq. 3.4-3, the amplitude of the particular solution can be written as

$$X = \frac{mer^2 / M}{\sqrt{(1-r^2)^2 + (2\varsigma r)^2}} \tag{3.4-6}$$

It is apparent in Eq. 3.4-6 that the amplitude of forced vibration is proportional to the quantity *me*, which is called the unbalance of the rotating part. Eq. 3.4-6 can be written in the nondimensional form

$$\frac{MX}{me} = \frac{r^2}{\sqrt{(1-r^2)^2 + (2\varsigma r)^2}} \tag{3.4-7}$$

Fig. 3.4-2 provides a graphical representation of Eq. 3.4-7 in the form of a series of curves for various damping ratios ς. The amplitude of the vibration is diminished by damping. Damping, however, is not always practical in eliminating the harmful effect of unbalance.

Another interesting property of the curves of Fig. 3.4-2 is that the maximum values do not occur at $r = 1$, but instead at a larger frequency ratio. The frequencies where the maxima occur are called *resonance*

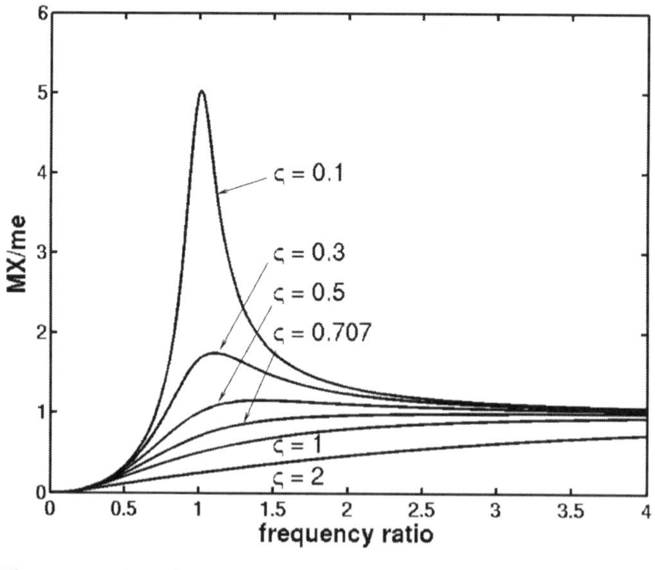

Fig. 3.4-2 Nondimensional amplitude of Eq. 3.4-7 for various damping ratios.

frequencies. The resonance frequencies can be determined by setting the derivative of (MX/me) with respect to r equal to zero. This yields

$$r_n = \frac{1}{\sqrt{1-2\varsigma^2}}, \; 0 < \varsigma < \frac{1}{\sqrt{2}} \tag{3.4-8}$$

For small values of the damping, the resonance frequency is very close to the natural frequency. Substituting this value of r_n into Eq. 3.4-7, the maximum amplitude is given by

$$\left(\frac{MX}{me}\right)_{\max} = \frac{1}{2\varsigma\sqrt{1-\varsigma^2}} \approx \frac{1}{2\varsigma} \tag{3.4-9}$$

Next, from Fig. 3.4-2, we see that as the frequency ratio r approaches zero, the amplitude of the nondimensional displacement approaches zero. As r becomes very large, the amplitude asymptotically approaches unity rather than zero for all damping ratios. Thus, the choice of damping ratio for large frequency ratios is not important. These results can be obtained by investigating the limit of (MX/me) as r approaches infinity.

Finally, for $\varsigma > 1/\sqrt{2}$, MX/me does not have a maximum value, or a peak, for any value of $r > 0$.

These observations have important implications in the design of rotating machines.

3.5 Forced Vibration Caused by Support Motion

In many cases, a dynamic system is excited through an elastic mounting. Mounting may be modeled by springs and dashpots. For example, an automobile body is excited harmonically by a road surface through tires, which can be modeled as a spring with a viscous damper, as in Fig. 3.5-1. Such systems can be modeled by considering the system to be excited by the motion of its support. This is called the *base excitation* or *support motion* problem.

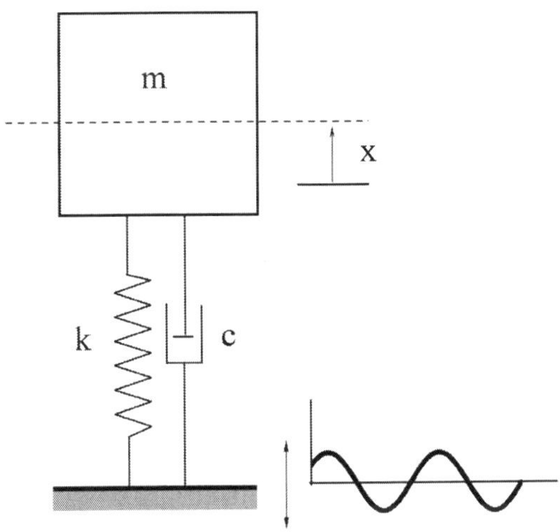

Fig. 3.5-1 System excited by support motion.

Let x_s be the displacement of the support and let x measure the displacement of the mass from the static equilibrium position. The equation of motion of the spring-mass system can then be written as

$$m\ddot{x} = -k(x - x_s) - c(\dot{x} - \dot{x}_s) \tag{3.5-1}$$

or

$$m\ddot{x} + c\dot{x} + kx = c\dot{x}_s + kx_s \tag{3.5-2}$$

Using complex algebra, we let the base motion, x_s, and the motion of the mass, x, be expressed as

$$x_s = Ae^{j\omega t} \qquad \text{and} \qquad x = Xe^{j(\omega t - \psi)} \left(j = \sqrt{-1} \right) \tag{3.5-3}$$

Then, using Euler's formula for trigonometric functions ($e^{\pm j\theta} = \cos\theta \pm j\sin\theta$), we can express the exponential functions in Eq. 3.5-3 as harmonic motions. This representation is always useful in studying vibration problems. Indeed, if the base motion is $x_s = A\sin\omega t$, then the imaginary part of x corresponds to the solution. If the base motion is $x_s = A\cos\omega t$, then the real part of the x corresponds to the solution.

Substituting the expressions of Eq. 3.5-3 into Eq. 3.5-2, we have

$$\left[(k - m\omega^2) + jc\omega \right] Xe^{-j\psi} = (k + jc\omega)A, \tag{3.5-4}$$

from which we have

$$Xe^{-j\psi} = \frac{(k + jc\omega)A}{(k - m\omega^2) + jc\omega} \tag{3.5-5}$$

The amplitude X is then given by

$$X = A\sqrt{\frac{k^2 + c^2\omega^2}{(k - m\omega^2)^2 + c^2\omega^2}} \tag{3.5-6}$$

The amplitude expression of Eq. 3.5-6 can be simplified by using the nondimensional quantities $r = \omega / \sqrt{k/m}$ and $\varsigma = c/(2\sqrt{km})$. Specifically, X takes the form

$$X = A\sqrt{\frac{1 + (2\varsigma r)^2}{(1 - r^2)^2 + (2\varsigma r)^2}} \tag{3.5-7}$$

Dividing this last expression by the amplitude A of the base motion yields

$$TR_d = \left| \frac{X}{A} \right| = \sqrt{\frac{1 + (2\varsigma r)^2}{(1 - r^2)^2 + (2\varsigma r)^2}} \tag{3.5-8}$$

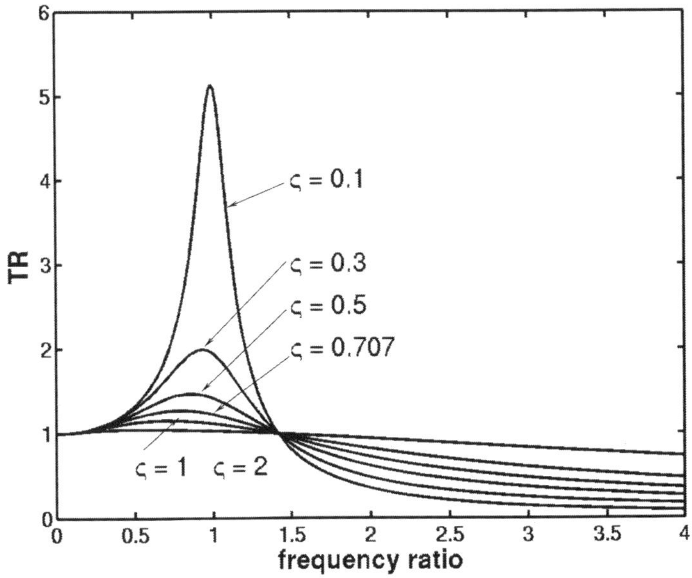

Fig. 3.5-2 Transmissibility of Eq. 3.5-8 for various degrees of damping.

TR_d expresses the ratio of the maximum response magnitude to the input displacement magnitude. This ratio is called the *displacement transmissibility.*

Fig. 3.5-2 shows the transmissibility. It is apparent in the figure that as r approaches zero, the transmissibility approaches unity, and as r becomes very large, the transmissibility asymptotically approaches zero. Interestingly, when the frequency ratio r has the value $\sqrt{2}$, the transmissibility is equal to unity for all damping ratios. Note that for $r<\sqrt{2}$, the transmissibility is greater than unity, indicating that the motion of the mass is an amplification of the base motion. Notice also that for a given $r<\sqrt{2}$, the higher damping ratio yields smaller transmissibility. However, for $r>\sqrt{2}$, the transmissibility is less than unity, indicating that the motion of the mass is smaller in amplitude than that of the base for all damping ratios, and the transmissibility increases for increasing damping ratio.

The phase angle ψ can be found by equating the real and imaginary parts in Eq. 3.5-5 to determine $\sin\psi$ and $\cos\psi$. The ratio then results in the equation determining the phase angle:

$$\tan\psi = \frac{mc\omega^2}{k^2\left(1-r^2\right)+\left(c\omega\right)^2} = \frac{2\varsigma r^3}{1-r^2+\left(2\varsigma r\right)^2} \tag{3.5-9}$$

Thus, the mass motion can be expressed as

$$x = A\sqrt{\frac{1+(2\varsigma r)^2}{(1-r^2)^2+(2\varsigma r)^2}}\exp\left[j\left(\omega t - \tan^{-1}\frac{2\varsigma r^3}{1-r^2+(2\varsigma r)^2}\right)\right] \tag{3.5-10}$$

Finally, it should be noted that the solution for the base motion $x_s = A\sin\omega t$ can be obtained by substituting this base motion into Eq. 3.5-2, and finding the particular solution. This substitution yields

$$m\ddot{x}+c\dot{x}+kx = c\omega A\cos\omega t + kA\sin\omega t \tag{3.5-11}$$

Since the equation is linear, it can be divided into two equations:

$$m\ddot{x}+c\dot{x}+kx = c\omega A\cos\omega t \tag{3.5-12}$$

and

$$m\ddot{x}+c\dot{x}+kx = kA\sin\omega t \tag{3.5-13}$$

We can obtain the particular solution by taking advantage of the linearity of the equation of motion. That is, the particular solution of Eq. 3.5-11 will be the sum of the particular solutions of Eqs. 3.5-12 and 3.5-13. Calculating these particular solutions follows directly from the analysis of Section 3.2. The same result will be obtained, but the analysis is somewhat more extensive. This is left to the reader.

3.6 Vibration Isolation

Vibration is usually an unwanted phenomenon—particularly when there are harmful and undesirable effects. Vibration can damage machinery and foundations of machinery, as well as adjacent structures. Worst of all, vibration can produce adverse health effects such as loss of hearing and nerve damage.

Not all vibration, however, is unwanted. The exceptions include pendulum clocks, cutting machines, and sound transmission devices.

Unfortunately, unwanted vibration cannot always be eliminated. In such cases, we can seek to reduce the harmful effects by isolating the vibration.

In the following paragraphs, we consider vibration isolation for two common cases: (1) a vibrating foundation causing harmful effects on a structure or machine (such as in an earthquake), and (2) a vibrating machine causing harmful effects on foundations or surrounding structures (such as occur with heavy motors and engines).

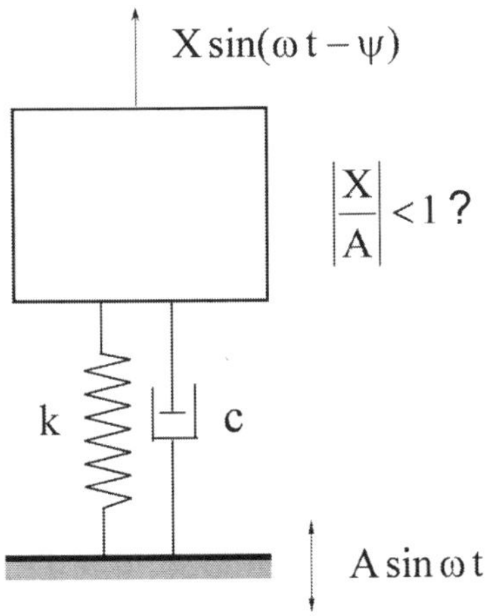

Fig. 3.6-1 Isolation of the machine from the foundation.

(1) Isolation of the Machine from the Foundation

Consider the model represented in Fig. 3.6-1. Here we seek to isolate the machine by adjusting the stiffness and the damping coefficient of the spring and damper of the model. The efficiency of the isolation is described by the *transmissibility*. As in Eq. 3.5-8, the *displacement transmissibility* is defined as

$$TR_d = \left|\frac{X}{A}\right| = \sqrt{\frac{1+(2\varsigma r)^2}{(1-r^2)^2+(2\varsigma r)^2}} \qquad (3.6\text{-}1)$$

Recall that the displacement transmissibility TR_d (or transmissibility ratio) provides the ratio of the maximum displacement of the machine to that of the foundation. The objective of the isolation provided by the spring and damper is to make TR_d less than one.

(2) Isolation of the Foundation from the Machine

The purpose of this isolation is to reduce the vibration force from the machine transmitted to the foundation. As shown in Fig. 3.6-2, let the amplitude of the vibratory force applied on the machine mass be F_0. Let the force transmitted to the foundation be F_T. The purpose of the isolation is to make the ratio of $|F_T|$ to $|F_0|$ less than one.

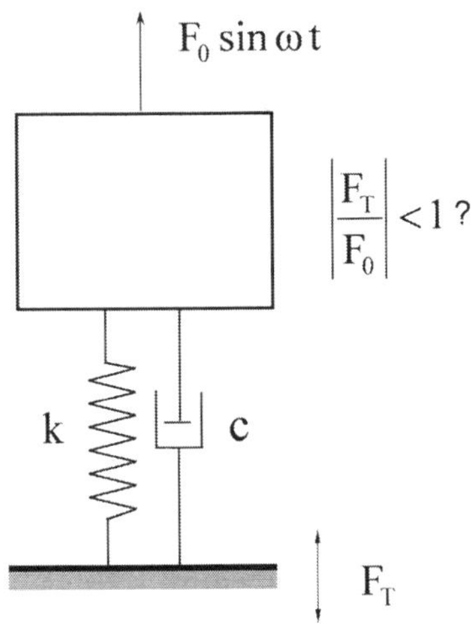

Fig. 3.6-2 Isolation of the foundation from the machine.

As shown in Fig. 3.6-2, the transmitted force through the spring and damper to the foundation is

$$F_T = \sqrt{(kX)^2+(c\omega X)^2} = kX\sqrt{1+\left(\frac{c\omega}{k}\right)^2} = kX\sqrt{1+(2\varsigma r)^2}, \qquad (3.6\text{-}2)$$

where X is the machine's vibration amplitude, which is given by Eq. 3.2-19, that is,

$$X = (MF)X_0 = \frac{X_0}{\sqrt{(1-r^2)^2 + (2\varsigma r)^2}}, \tag{3.6-3}$$

where $X_0 = (F_0/k)$ is the static deflection of the spring. Substituting Eq. 3.6-3 into Eq. 3.6-2, we obtain

$$TR_f = \left|\frac{F_T}{F_0}\right| = \sqrt{\frac{1+(2\varsigma r)^2}{(1-r^2)^2 + (2\varsigma r)^2}} \tag{3.6-4}$$

TR_f is called the *force transmissibility*, which expresses the ratio of the maximum force transmitted from the machine to the foundation to the maximum excitation force on the machine. Fig. 3.5-2 provides a typical graphical representation of TR_f.

At this point, it is helpful to make several observations and remarks.

First, a comparison of the force transmissibility Eq. 3.6-4 with the displacement transmissibility Eq. 3.6-1 indicates that they are identical. Thus, the problem of isolating a system from the motion of the foundation is identical to that of isolating disturbing forces transmitted from the system to the foundation.

Second, observe that no matter what the magnitude of damping is, vibration isolation is possible (transmissibility less than one) only for $r > \sqrt{2}$. Thus, the stiffness of the spring should be as small as possible. (Recall that r is the ratio of the applied forcing frequency ω to the natural frequency ω_n of the system, as in Eq. 3.2-15.)

Third, for $r > \sqrt{2}$, the transmissibility is less than one and vibration isolation occurs. In this isolation region, the smaller the value of the damping ratio, the smaller the transmissibility, and the better is the isolation.

Example 3.6-1 Fig. 3.6-3 depicts an isolation system. Its natural frequency is 5.5 Hz. The damping ratio of the bushing is 0.12. The amplitude of vibration of the foundation is 0.002 mm, and the maximum vibration velocity of the foundation is 0.125 mm/s. What is the amplitude of vibration of the system? What is the transmissibility?

Solution: The frequency of vibration of the foundation is

$$\omega = V / A = 0.125 / 0.002 = 62.5 \text{ rad/s} \tag{a}$$

The natural frequency of the system is

$$\omega_n = 2\pi f_n = 2\pi \times 5.5 = 34.56 \text{ rad/s} \tag{b}$$

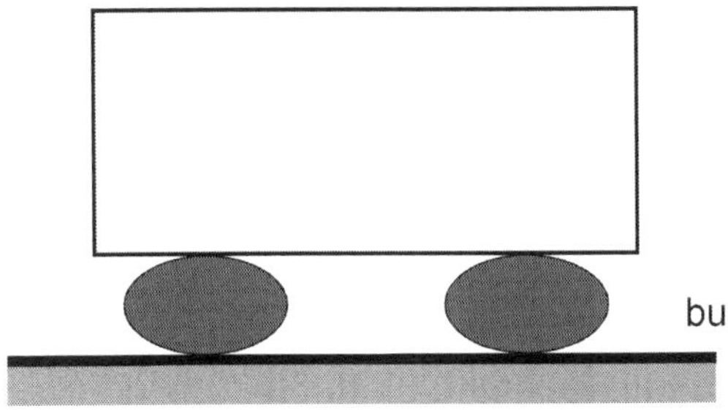

Fig. 3.6-3 For Example 3.6-1.

Hence, the frequency ratio is

$$r = \omega / \omega_n = 62.5 / 34.5 = 1.81 \tag{c}$$

From Eq. 3.6-1, the amplitude of vibration of the system is found to be

$$X = A\sqrt{\frac{1+(2\varsigma r)^2}{(1-r^2)^2+(2\varsigma r)^2}} = 0.94\times10^{-3}\ \text{mm} \tag{d}$$

The transmissibility is given by

$$TR = X / A = 0.94\times10^{-3} / 2\times10^{-3} = 0.47 \tag{e}$$

Example 3.6-2 A suspension system of a car is modeled as a spring connecting a massless wheel and a supported mass, as shown in Fig. 3.6-4. The profile of the road surface is modeled as a sinusoid function

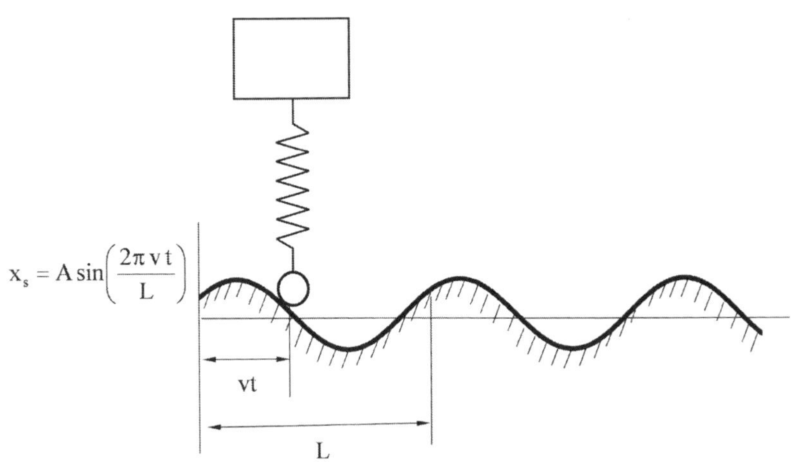

Fig. 3.6-4 A simple model of the suspension system of a car.

$x_s = A\sin\left(\frac{2\pi vt}{L}\right)$, where L is the peak-to-peak distance, and where v is the forward velocity of the car on the road. Find the vibration amplitude of the car.

Solution: We have

$$x_s = A\sin\left(\frac{2\pi vt}{L}\right),\ \omega = \frac{2\pi v}{L},\ \omega_n = \sqrt{\frac{k}{m}},\ r = \frac{\omega}{\omega_n} = \frac{2\pi v}{L}\sqrt{\frac{m}{k}},\ \varsigma = 0$$

From Eq. 3.6-1, the vibration amplitude of the car is given by

$$X = A\frac{1}{\sqrt{(1-r^2)^2}} = \frac{A}{\left|1-\frac{4\pi^2 v^2 m}{L^2 k}\right|}$$

3.7 Damping

Damping is present in all vibrating systems. Damping dissipates energy in a vibrating system. An accurate mathematical description of damping is quite complicated and not a suitable model for vibration analysis. Thus, simplified damping models have been developed that in many cases are found to be adequate in evaluating the system response. The viscous damping model is one of these. It leads to a workable analysis.

(1) Energy Dissipation Due to a Viscous Damping Force

Energy dissipation is usually determined by considering cyclic oscillation. Although the force-displacement relationships may differ, the force-displacement curves enclose an area (called the *hysteresis loop*) that is proportional to the energy lost per cycle. Thus, the energy lost per cycle due to a damping force F_d is computed as [3.2]

$$W_d = \oint F_d dx \tag{3.7-1}$$

In general, W_d is a function of both the frequency and the amplitude of the vibration.

In this section, we consider a simple case of energy dissipation: a spring-mass system with viscous damping. In the steady state, we have

$$x = X\sin(\omega t - \psi) \tag{3.7-2}$$

and

$$F_d = c\dot{x} = c\omega X\cos(\omega t - \psi) \tag{3.7-3}$$

Hence, the energy lost per cycle due to damping is

$$W_d = c\omega^2 X^2 \int_0^{2\pi/\omega} \cos^2(\omega t - \psi)dt = \pi c\omega X^2, \tag{3.7-4}$$

where X is the amplitude of vibration. Of particular interest is the energy dissipated in forced vibration at resonance. Substituting $\omega = \omega_n = \sqrt{k/m}$, and $c = 2\varsigma\sqrt{km}$ into Eq. 3.7-4, it becomes, at resonance,

$$W_d = 2\varsigma\pi kX^2, \tag{3.7-5}$$

where X is the amplitude of the vibration.

The energy dissipated per cycle can be represented graphically. From Eq. 3.7-3, the damping force can be written in terms of the displacement x as

$$F_d = \pm c\omega X\sqrt{1-\sin^2(\omega t - \psi)} = \pm c\omega\sqrt{X^2 - x^2} \tag{3.7-6}$$

This equation, in turn, can be written as

$$\left(\frac{F_d}{c\omega X}\right)^2 + \left(\frac{x}{X}\right)^2 = 1 \tag{3.7-7}$$

We recognize Eq. 3.7-7 as an equation of an ellipse, with F_d and x plotted along the vertical and horizontal axes, as in Fig. 3.7-1. The ellipse is called a *hysteresis loop*. The energy dissipated per cycle is then given by the area enclosed by the ellipse.

Next, consider a force-displacement curve for a system with viscous damping. If we add the spring force kx to F_d, we obtain

$$F = kx + F_d = kx \pm c\omega\sqrt{X^2 - x^2} \tag{3.7-8}$$

or

$$F - kx = \pm c\omega\sqrt{X^2 - x^2} \tag{3.7-9}$$

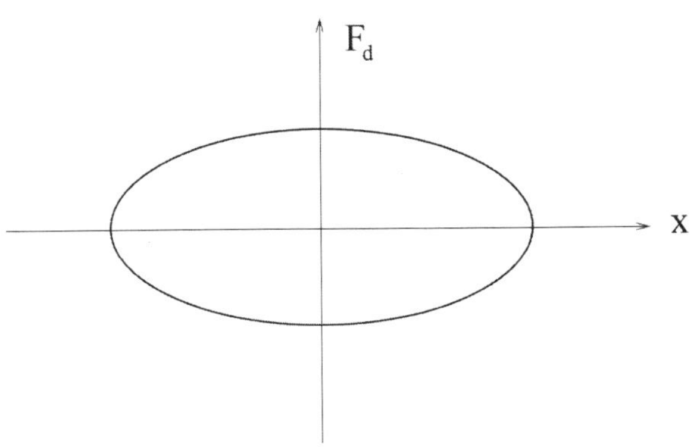

Fig. 3.7-1 Graph of damping force versus displacement defining the hysteresis loop.

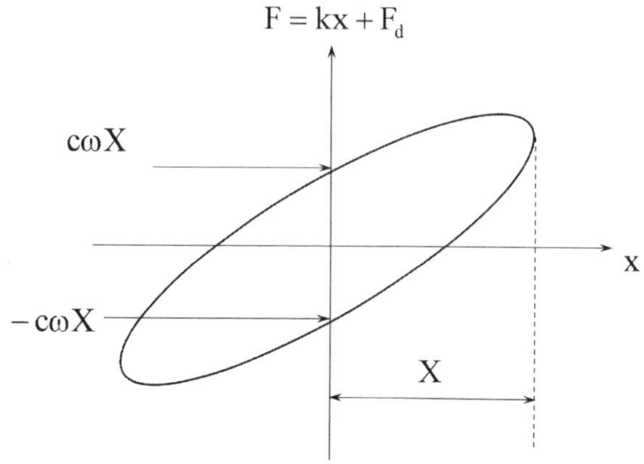

Fig. 3.7-2 Graph of total force versus displacement defining the hysteresis loop.

Squaring this expression yields

$$F^2 + \left(c^2\omega^2 + k^2\right)x^2 - \left(2k\right)xF - c^2\omega^2 X^2 = 0, \qquad (3.7\text{-}10)$$

which is the general equation for an ellipse rotated about the origin in the $F - x$ plane, as shown in Fig. 3.7-2.

(2) Structural (Hysteretic) Damping

Eq. 3.7-4 gives the energy dissipated by viscous damping in one cycle of forced vibration. This energy expression may be equated to that due to some other type of damping to obtain an *equivalent viscous damping coefficient* c_{eq} [3.3]. For example, consider *structure damping* (or *hysteretic damping*), which is attributed to internal friction in structure materials. This type of material can be tested by measuring stress (force) and strain (displacement) under controlled steady-state harmonic loading. Such tests produce *hysteresis loops* of the form shown in Fig. 3.7-3. Experience shows that for most structural materials, the energy dissipated per cycle is independent of the frequency over a wide frequency range and is proportional to the amplitude squared of the vibration. That is,

$$W_d = \alpha X^2, \qquad (3.7\text{-}11)$$

where α is a constant and X is the amplitude of vibration. Using the concept of equivalent viscous damping, we obtain from Eq. 3.7-4 the relations

$$\pi c_{eq}\omega X^2 = \alpha X^2,\ c_{eq} = \frac{\alpha}{\pi\omega} \qquad (3.7\text{-}12)$$

Therefore, the equation of motion with structural damping can be written as

$$m\ddot{x} + \left(\frac{\alpha}{\pi\omega}\right)\dot{x} + kx = F_0 \sin\omega t \qquad (3.7\text{-}13)$$

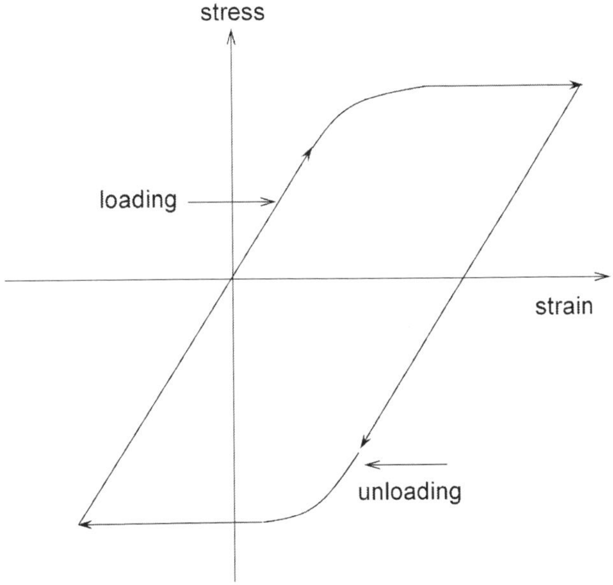

Fig. 3.7-3 Schematic of experimental stress strain curve.

(3) Complex Stiffness and Loss Factor

For harmonic vibration, $\dot{x}$ is $j\omega x$. Eq. 3.7-13 can then be written as

$$m\ddot{x}+(1+j\eta)kx=F_0 \sin\omega t, \tag{3.7-14}$$

where

$$\eta=\frac{\alpha}{\pi k} \tag{3.7-15}$$

The quantity $(1+j\eta)k$ is called the *complex stiffness*, and η is called the *structural damping loss factor*.

If we substitute $\alpha=\pi k\eta$ from Eq. 3.7-15 into Eq. 3.7-11, the energy dissipated per cycle by the structural damping is

$$W_d=\pi k\eta X^2 \tag{3.7-16}$$

Since the structural damping loss factor η is independent of the frequency, we see that ω, in Eq. 3.7-12, can be replaced by ω_n, the undamped natural frequency. Then the relationship between the structural damping loss factor η and the damping ratio ς is

$$\eta=\frac{\alpha}{\pi k}=\frac{c_{eq}\pi\omega_n}{\pi k}=\frac{2\varsigma\sqrt{km}\,\pi\sqrt{k/m}}{\pi k}=2\varsigma \tag{3.7-17}$$

(4) Sharpness of Resonance

In forced vibration, there is a quantity Q (sometimes called the *Q-factor*) related to damping that is a measure of the *sharpness* of the resonance. To develop and define this quantity, we assume viscous damping and begin our analysis with Eq. 3.2-19.

When the frequency ratio r is 1, we see from Eq. 3.2-19 that the magnification factor at resonance becomes (see Eq. 3.2-19)

$$MF_{\max} = \frac{1}{2\varsigma} \tag{3.7-18}$$

We now seek the frequency ratios on either side of resonance, where the magnification factor is $\left(1/\sqrt{2}\right) = 0.707$. These two points are called the *half-power points*. By letting the amplitude of the magnification factor be $\frac{1}{\sqrt{2}}\frac{1}{2\varsigma}$, and by squaring in Eq. 3.2-19, we have

$$\left(\frac{1}{\sqrt{2}}\right)^2\left(\frac{1}{2\varsigma}\right)^2 = \frac{1}{(1-r^2)^2+(2\varsigma r)^2} \tag{3.7-19}$$

or

$$r^4 - 2(1-2\varsigma^2)r^2 + (1-8\varsigma^2) = 0 \tag{3.7-20}$$

Solving for r^2 we obtain

$$r^2 = (1-2\varsigma^2) \pm 2\varsigma\sqrt{1-\varsigma^2} \tag{3.7-21}$$

For small damping ($\varsigma << 1$), by neglecting higher order terms of ς, we arrive at the result

$$r^2 \equiv \left(\frac{\omega}{\omega_n}\right)^2 = 1 \pm 2\varsigma \tag{3.7-22}$$

Letting the two frequencies corresponding to the roots of Eq. 3.7-20 be ω_1 and ω_2, we obtain

$$4\varsigma = \frac{\omega_2^2 - \omega_1^2}{\omega_n^2} \cong 2\left(\frac{\omega_2 - \omega_1}{\omega_n}\right) \tag{3.7-23}$$

The Q-factor is then defined as

$$Q = \frac{\omega_n}{\omega_2 - \omega_1} = \frac{1}{2\varsigma} \tag{3.7-24}$$

In summary, damping can be described by several parameters. They are

Damping ratio: ς (Eq. 2.5-5)

Logarithmic decrement: δ (Eq. 2.5-20)

Magnification factor: $MF_{\max}$ (Eq. 3.2-21)

Loss factor: η (Eq. 3.7-17)

Q-factor: Q (Eq. 3.7-24)

These parameters are related by the expression [3.4]

$$\eta = 2\varsigma = \frac{1}{MF_{\max}} = \frac{1}{Q} = \frac{\delta}{\pi} \tag{3.7-25}$$

Example 3.7-1 Determine the damping ratio ς in terms of the frequencies of the half-power points.

Solution: Let the frequencies corresponding to the half-power points be ω_1 and ω_2. Then, from Eq. 3.7-23, we have

$$\varsigma = \frac{\omega_2 - \omega_1}{2\omega_n} = \frac{\omega_2 - \omega_1}{2[(\omega_2 + \omega_1)/2]} = \frac{\omega_2 - \omega_1}{\omega_2 + \omega_1} = \frac{f_2 - f_1}{f_2 + f_1}$$

Example 3.7-2 Show that

$$\delta = \frac{\pi(f_2 - f_1)}{f_n},$$

where f_1 and f_2 are the frequencies corresponding to the half-power points.

Proof: From Eq. 3.7-25, we have

$$\delta = \pi 2\varsigma = \pi 2 \frac{f_2 - f_1}{f_2 + f_1} = \frac{\pi(f_2 - f_1)}{f_n}$$

3.8 Forced Vibration Under Periodic Excitation

When a system is excited by periodic, but not harmonic, excitation, the excitation may be expressed as a summation of harmonic functions. Indeed, any periodic function can be represented by a series of sines and cosines. Specifically, if *F(t)* is a periodic function with period *T*, then *F(t)* can be represented as (see Section 1.5)

$$F(t) = \frac{a_0}{2} + \sum_{n=1}^{\infty}\left[a_n \cos(n\omega_0 t) + b_n \sin(n\omega_0 t)\right], \tag{3.8-1}$$

where $\omega_0 = \frac{2\pi}{T}$ is called the fundamental frequency. The coefficients (a_n and b_n) can be evaluated using Eqs. 1.5-2 and 1.5-3.

Therefore, the equation of motion of a mass-spring-damper system with periodic excitation can then be written as

$$m\ddot{x}+c\dot{x}+kx=\frac{a_0}{2}+\sum_{n=1}^{n}\left[a_n\cos(n\omega_0 t)+b_n\sin(n\omega_0 t\right] \tag{3.8-2}$$

where $a_0/2$ now represents a constant force, which affects only the static equilibrium position. If we choose the static equilibrium position as the origin of the coordinate system, then this term will not be present in the equation of motion. In the following analysis, this term is then ignored.

The solution of Eq. 3.8-2 can be obtained using the superposition principle, that is,

$$x=\sum_{i=1}^{\infty}\frac{a_i\cos(i\omega_0 t-\psi_i)+b_i\sin(i\omega_0 t-\phi_i)}{k\sqrt{(1-r_i^2)^2+(2\varsigma r_i)^2}}, \tag{3.8-3}$$

where r_i is the frequency ratio defined as the ratio of $i\omega_0$ to the natural frequency ω_n, and where ψ_i and ϕ_i are phase angles. For a small damping ratio, the phase angles approach zero. Eq. 3.8-3 then becomes

$$x=\sum_{i=1}^{\infty}\frac{a_n\cos(i\omega_0 t)+b_i\sin(i\omega_0 t)}{k\left|1-r_i^2\right|} \tag{3.8-4}$$

For a forced vibration caused by periodic support motion, we may express the support movement x_s as

$$x_s=\sum_{i=1}^{\infty}[a_i\cos(i\omega_0 t)+b_i\sin(i\omega_0 t)] \tag{3.8-5}$$

Then, in view of Eq. 3.5.10, the response of the system with a small damping ratio may be expressed as

$$x=\sum_{i=1}^{\infty}\frac{a_i\cos(i\omega_0 t)+b_i\sin(i\omega_0 t)}{\left|1-r_i^2\right|} \tag{3.8-6}$$

3.9 Response to Arbitrary Excitation

Consider an arbitrary force $F(\tau)$, $0\le\tau\le t$, as represented in Fig. 3.9-1. Let F act on a mass-spring-damper system which has zero initial conditions. The equation of motion of the system is then

$$m\ddot{x}+c\dot{x}+kx=F \tag{3.9-1}$$

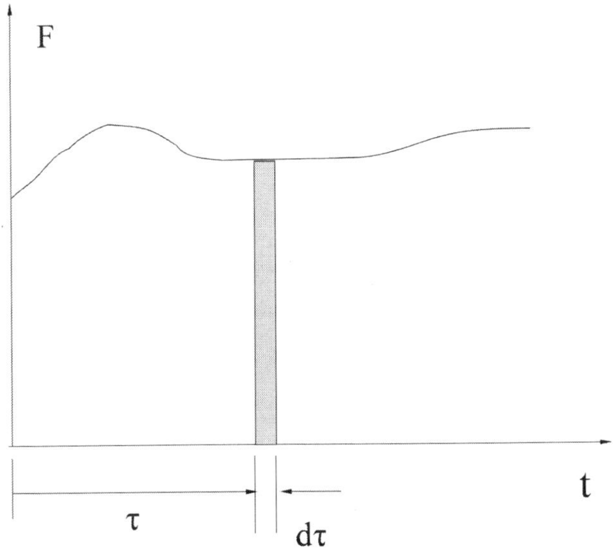

Fig. 3.9-1 Arbitrary excitation force.

Consider the arbitrary force to be a series of impulses as in Fig. 3.9-1. From $t=\tau$ to $t=\tau+d\tau$, a pulse $Fd\tau$ then acts on the mass m. For infinitesimal or very small $d\tau$, the mass will obtain a velocity increment dv as $\frac{Fd\tau}{m}$, with no displacement increment.

When $t\geq\tau$, free vibration will occur with an initial velocity $\frac{Fd\tau}{m}$ and zero initial displacement. From Eq. 2.5-15, the response of the system with initial velocity of $\dot{x}_0$ is given by

$$x=\frac{\dot{x}_0}{\omega_d}e^{-\varsigma\omega_n t}\sin(\omega_d t), \tag{3.9-2}$$

where

$$\omega_n=\sqrt{\frac{k}{m}},\ \varsigma=\frac{c}{2\sqrt{km}},\ \omega_d=\sqrt{1-\varsigma^2}\,\omega_n \tag{3.9-3}$$

When F is replaced by the Dirac delta function, the initial conditions for this unit impulse are $x_0=0$ and $\dot{x}_0=1/m$. Then Eq. 3.9-2 becomes the *impulse* (or *impulsive*) *response function*, which is usually denoted by $h(t)$, given by

$$h(t)=\frac{1}{m\omega_d}e^{-\varsigma\omega_n t}\sin(\omega_n t) \tag{3.9-4}$$

Eq. 3.9-4 shows that the impulse response function decays with an exponential envelope. This is characteristic of response to an impulse for stable linear dynamic systems. This impulse response is physically interpreted as the response to an initially imposed velocity with no initial displacement.

The impulse response function is useful for determining the response of a system to a general applied force excitation as explained in the following paragraphs.

By using Eq. 3.9-4, if $Fd\tau$ acts on the mass at $t = 0$, then the response of the system is given by

$$dx = h(t)Fd\tau = \frac{Fd\tau}{m\omega_d} e^{-\varsigma\omega_n t} \sin(\omega_d t) \tag{3.9-5}$$

If the impulse acts at $t = \tau$, then $t - \tau = 0$, so $t - \tau$ is the initial time, and the above equation should be modified as

$$dx = \frac{Fd\tau}{m\omega_d} e^{-\varsigma\omega_n (t-\tau)} \sin(\omega_d (t-\tau)) \tag{3.9-6}$$

If the excitation force $F(\tau)$ is acting from $\tau = 0$ to $\tau = t$, from the principle of superposition, the response of the system may be written as

$$x = \int_0^t \frac{F(\tau)}{m\omega_d} e^{-\varsigma\omega_n (t-\tau)} \sin(\omega_d (t-\tau)) d\tau \tag{3.9-7}$$

Eq. 3.9-7 has a number of different names, including the *convolution*, and *Duhamel's integral*, after the French mathematician J. M. C. Duhamel (1797–1872).

If an arbitrary force *F(t)* is applied to an undamped system, the response Eq. 3.9-7 becomes

$$x = \frac{1}{m\omega_n} \int_0^t F(\tau) \sin\omega_n (t-\tau) d\tau \tag{3.9-8}$$

If an arbitrary force *F(t)* is applied to an undamped system which has an initial condition other than zero, the solution must be in the form

$$x = x_0 \cos\omega_n t + \frac{\dot{x}_0}{\omega_n} \sin\omega_n t + \frac{1}{m\omega_n} \int_0^t F(\tau) \sin\omega_n (t-\tau) d\tau \tag{3.9-9}$$

Example 3.9-1 Determine the response of a single degree of freedom system, with no initial velocity, to the step (constant) excitation shown in Fig. 3.9-2.

Solution: (1) For an undamped system, let $F=F_0$, $\varsigma = 0$ in Eq. 3.9-7. We then have

$$x = \int_0^t \frac{F_0}{m\omega_n} e^{-\varsigma\omega_n (t-\tau)} \sin\omega_n (t-\tau) d\tau = \frac{F_0}{m\omega_n^2} (1 - \cos\omega_n t) = \frac{F_0}{k} (1 - \cos\omega_n t) \tag{a}$$

Fig. 3.9-2 Step function of magnitude F_0 applied at $t = 0$.

When $\omega_n t = \pi, 3\pi, \ldots$, the peak response $(2F_0/k)$ to the step excitation of magnitude F_0 is equal to twice the static deflection. Simple harmonic vibration will take place in this case.

(2) For a damped system ($\varsigma \neq 0$), Eq. 3.9-7 becomes

$$x = \frac{-F_0}{m\omega_d}\int_0^t e^{-\varsigma\omega_n(t-\tau)}\sin\omega_d(t-\tau)d(t-\tau) \tag{b}$$

By using the formula

$$\int e^{ax}\sin bx dx = \frac{e^{ax}}{a^2+b^2}\left(a\sin bx - b\cos bx\right), \tag{c}$$

we obtain

$$x = \frac{F_0}{k}\left[1 - \frac{e^{-\varsigma\omega_n t}}{\sqrt{1-\varsigma^2}}\left(\varsigma\sin\omega_d t + \sqrt{1-\varsigma^2}\cos\omega_d t\right)\right] \tag{d}$$

If the angle ψ and the associated trigonometric functions are introduced as

$$\varsigma = \sin\psi, \sqrt{1-\varsigma^2} = \cos\psi,\ \tan\psi = \frac{\varsigma}{\sqrt{1-\varsigma^2}}, \tag{e}$$

Eq. d can be written as

$$x = \frac{F_0}{k}\left[1 - \frac{e^{-\varsigma\omega_n t}}{\sqrt{1-\varsigma^2}}\cos\left(\omega_d t - \psi\right)\right] \tag{f}$$

Alternatively, we can obtain this result by simply considering the differential equation

$$\ddot{x} + 2\varsigma\omega_n\dot{x} + \omega_n^2 x = \frac{F_0}{m}, \tag{g}$$

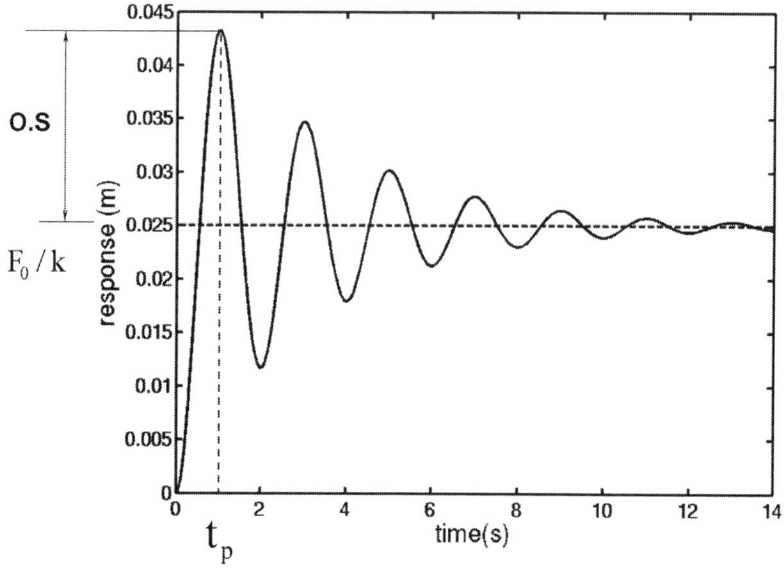

Fig. 3.9-3 Response of damped system to the step function as shown in Fig. 3.9-2. $\varsigma = 0.1$; $\omega_n = 3.16$ rad/s, $F_0 = 1000$ N, $F_0 = 25$ N, $k = 1000$ N/m.

whose solution is the sum of the solutions to the homogeneous equation and the particular solution, which, for this case, is $F_0 / m\omega_n^2$. Thus we obtain expression

$$x = Ae^{-\varsigma\omega_n t}\sin\left(\omega_d t + \phi\right) + \frac{F_0}{m\omega_n^2} \tag{h}$$

Using the initial conditions, we will obtain the same result as in Eq. f.

Examining the damped response given in Eq. f of Example 3.9-1, we observe that the response consists of the constant function F_0/k minus a decaying oscillation. For large time values, the decaying oscillation becomes negligible, and the steady state is simply F_0/k, as shown in Fig. 3.9-3.

To characterize the response to a step function, three quantities are introduced [3.5]. The first one is called the *peak time*, denoted by t_p in Fig. 3.9-3. This is the time it takes for the response to reach the largest value. The second is called the *overshoot*, denoted by O.S. in Fig. 3.9-3. O.S. is the largest value of the response *over* the steady state value. The third is called the *settling time*, which is the time it takes for the response to reach and stay within a certain percentage of the steady-state response.

To find the peak time of the response to a step force F_0, taking the derivative of the response with respect to time in Eq. f of Example 3.9-1, and setting the result equal to zero, we obtain

$$\varsigma\omega_n\cos\left(\omega_d t - \psi\right) + \sqrt{1-\varsigma^2}\,\omega_n\sin\left(\omega_d t - \psi\right) = 0 \tag{3.9-10}$$

Letting $\varsigma = \sin\psi$, $\sqrt{1-\varsigma^2} = \cos\psi$, Eq. 3.9-10 can be written as

$$\sin\psi\cos(\omega_d t - \psi) + \sin\psi\cos(\omega_d t - \psi) = 0 \tag{3.9-11}$$

Recall that $\sin\alpha\cos\beta + \cos\alpha\sin\beta \equiv \sin(\alpha+\beta)$: Eq. 3.9-11 becomes

$$\sin\omega_d t = 0 \tag{3.9-12}$$

Thus, the peak time is found to be

$$t_p = \frac{\pi}{\omega_d} = \frac{\pi}{\omega_n\sqrt{1-\varsigma^2}} \tag{3.9-13}$$

If we substitute this result into Eq. f of Example 3.9-1, the peak response is given by

$$x_{\text{max}} = \frac{F_0}{k}\left[1 + \exp\left(-\frac{\varsigma\pi}{\sqrt{1-\varsigma^2}}\right)\right] \tag{3.9-14}$$

Similarly, the overshoot is given by

$$\text{O.S.} = \frac{F_0}{k}\exp\left(-\frac{\varsigma\pi}{\sqrt{1-\varsigma^2}}\right) \tag{3.9-15}$$

Problems

3.1 A generator with 850 kg mass is mounted on the midpoint of a simply supported uniform beam. The static deflection at the midpoint of the beam is 0.25 cm. Due to damping, the amplitude of free vibration is reduced to ½ its initial value after 10 cycles of vibration. The generator is operated at 600 rpm. An unbalance mass produces a force of 2058 N. Neglecting the mass of the beam, find the amplitude of the forced vibration of the generator.

Ans. 2.663 cm

3.2 Derive the equation of motion for the system shown in Fig. P3.2 if a vertical excitation force $F = F_0 \sin\omega t$ is applied to

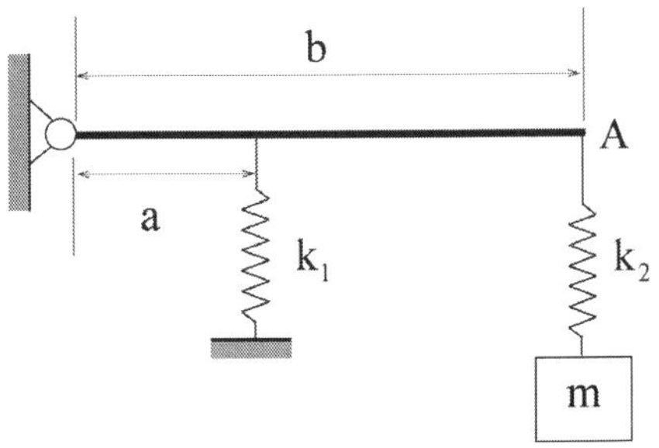

Fig. P3.2

(1) the mass m, and (2) the free end A of the bar. Assume that the bar is rigid and its mass is negligible.

Ans. (1) $m\ddot{x}+k_e x=F_0\sin\omega t$, $k_e=k_1k_2a^2/(k_1a^2+k_2b^2)$

(2) $m\ddot{x}+k_e x=F_e\sin\omega t$, $F_e=[(k_2b^2)/(k_1a^2+k_2b^2)]F_0\sin\omega t$

3.3 Show analytically that as the excitation frequency ω approaches resonance, the magnification factor approaches its maximum value $MF_{max}=\dfrac{1}{2\varsigma\sqrt{1-\varsigma^2}}$.

3.4 Show analytically that the maximum values of the curves in Fig. 3.4-2 occur at $r=1/\sqrt{1-2\varsigma^2}$.

3.5 Show analytically that when the frequency ratio r is $\sqrt{2}$, the transmissibility is equal to unity for all damping ratios.

3.6 A vertical single diesel engine with 500 kg mass is mounted on a spring with stiffness k = 200 kN/m and a damper with damping ratio $\varsigma=0.2$ (Fig. P3.6). The mass of the equivalent reciprocating parts is 10 kg and the stroke length is 200 mm. Find the amplitude of vertical vibration of the engine, the transmissibility, and the force transmitted to the foundation, if the engine is operated at 600 rpm.

Ans. X=2.2 mm; TR=0.18, F_T=706.6 N

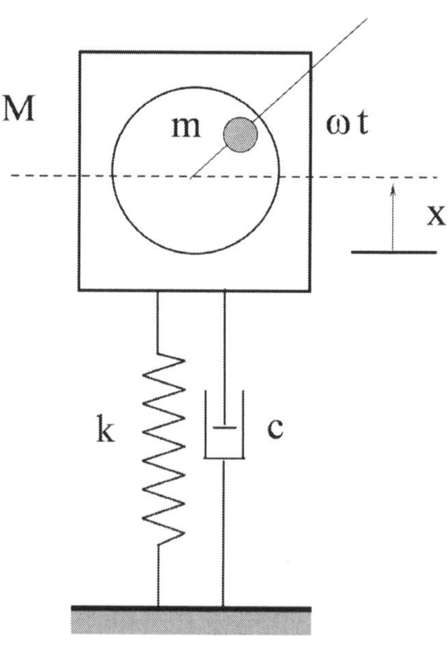

Fig. P3.6

3.7 A rotor with mass 50 kg is modeled as in Fig. P3.6. It has an unbalance of 0.05 kg·m, and it is operated at 600 rpm. It is desired that the dynamic amplitude of vibration of the rotor is to be less than 5 mm; the transmissibility is 0.1. Specify the spring and damping.

Ans. k=14.74×10^3 N/m, c=171.67 N·s/m

3.8 Derive the transmissibility equation for the system shown in Fig. P3.8.

Ans.

$$\left|\frac{F_T}{k_1X_1}\right|=\frac{\sqrt{k_2^2+c^2\omega^2}}{\sqrt{(k_1+k_2-m\omega^2)^2+c^2\omega^2}}$$

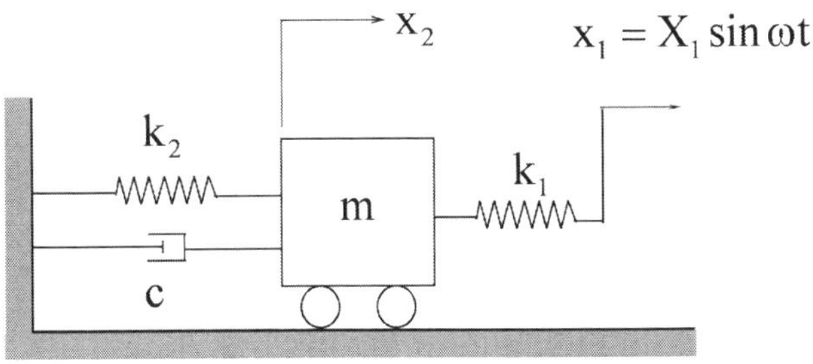

Fig. P3.8

3.9 Find the solution of forced vibration of the system shown in Fig. 3.5-1 for the base motion $x_s = A\sin\omega t$.

Ans. $$x = \frac{h}{\sqrt{(\omega_n^2-\omega^2)^2+(2\varsigma\omega_n\omega)^2}}\sin\left(\omega t - \tan^{-1}\frac{2\varsigma\omega_n\omega}{\omega_n^2-\omega^2}+\tan^{-1}\frac{c\omega}{k}\right)$$

$$h = \frac{A\sqrt{k^2+c^2\omega^2}}{m}$$

3.10 Repeat Problem 3.9 for the base acceleration $\ddot{x}_s = a\sin\omega t$.

Ans. $$x - x_s = \frac{-a}{\sqrt{(\omega_n^2-\omega^2)^2+(2\varsigma\omega_n\omega)^2}}\sin\left(\omega t - \tan^{-1}\frac{2\varsigma\omega_n\omega}{\omega_n^2-\omega^2}\right)$$

3.11 Show that for velocity squared damping ($F_d = c\dot{x}^2$), the energy dissipation per cycle is $W_d = (8/3)\omega^2 cX^3$, and the equivalent viscous damping coefficient is $c_{eq} = 8\omega cX/(3\pi)$. (Note: here c_{eq} is not a constant.)

3.12 Find the Fourier series expansion of each of the periodic functions shown in Fig. P3.12.

Ans. (1) $$F(t) = \frac{F_0}{2} - \frac{F_0}{\pi}\left(\sin\omega t + \frac{1}{2}\sin 2\omega t + \frac{1}{3}\sin 3\omega t + \cdots\right)$$

(2) $$F(t) = \frac{2F_0}{\pi}\left(\sin\omega t - \frac{1}{2}\sin 2\omega t + \frac{1}{3}\sin 3\omega t - \cdots\right)$$

3.13 A constant force F_0 is applied to an underdamped single-degree-of-freedom system at $t = 0$ with zero initial conditions. The equation of motion can be written as

$$\ddot{x} + 2\varsigma\omega_n\dot{x} + \omega_n^2 x = F_0/m$$

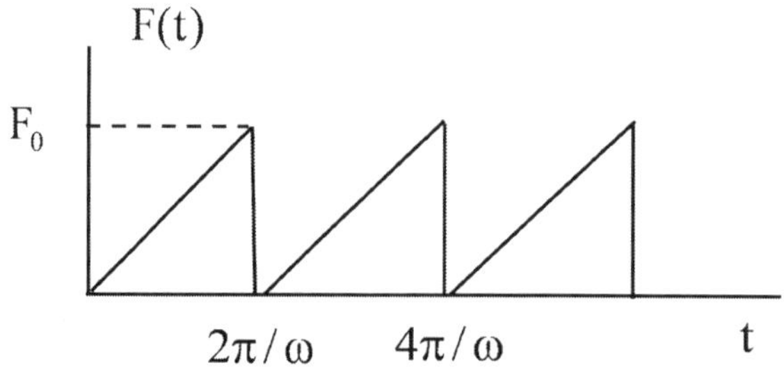

(1)

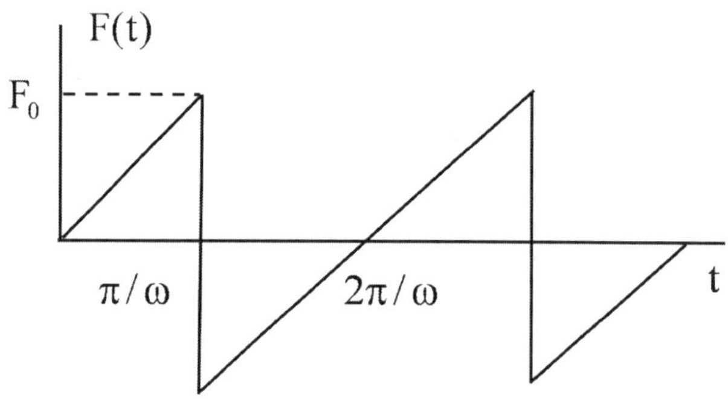

(2)

Fig. P3.12

The solution of this equation is the sum of the solution to the homogeneous equation and that of the particular solution. Show that the response of the system is given by

$$x=\frac{F_0}{k}\left[1-\frac{e^{-\varsigma\omega_n t}}{\sqrt{1-\varsigma^2}}\cos\left(\omega_d t-\psi\right)\right],\ \tan\psi=\frac{\varsigma}{\sqrt{1-\varsigma^2}}$$

3.14 Given the equation of motion and the initial conditions,

$$\ddot{x}+8\dot{x}+100x=50 \quad \text{and} \quad x(0)=1,\, \dot{x}(0)=2,$$

find the transient response of the system.

Ans. $x=0.664e^{-4t}\sin(9.165t+48.9°)$

3.15 An excitation force $F(t)=F_0\sin(\omega t+\alpha)$ is applied to the mass of a spring-mass system. Find the steady-state response of the system.

Ans. $x = \dfrac{F_0/m}{\sqrt{(\omega_n^2-\omega^2)^2+(2\varsigma\omega_n\omega)^2}}\sin(\omega t+\alpha-\psi)$

3.16 Repeat Problem 3.15 for $F(t)=F_0\cos(\omega t+\beta)$.

Ans. $x = \dfrac{F_0/m}{\sqrt{(\omega_n^2-\omega^2)^2+(2\varsigma\omega_n\omega)^2}}\cos(\omega t+\beta-\psi)$, $\tan\psi = \dfrac{2\varsigma\omega_n\omega}{\omega_n^2-\omega^2}$

3.17 An excitation force $F(t)=30\sin(10t+30°)$ is applied to the mass of a spring-mass system with mass m=15 kg, spring stiffness k=8 kN/m, and damping coefficient c=300 Ns/m. Find the solution for the initial conditions $x(0)=\dot{x}(0)=0$.

Ans. $x = e^{-10t}(0.38\cos 20.82t-1.83\sin 20.82t)+4.19\sin(10t-5.22°)$(mm)

3.18 For a damped single-degree-of-freedom system, the excitation frequency is equal to the natural frequency of the system. Find the response of the system under the initial conditions $x(0)=0$ and $\dot{x}(0)=0$.

Ans. $x = \dfrac{h}{2\varsigma\sqrt{1-\varsigma^2}\,\omega^2}e^{-\varsigma\omega t}\sin\left(\sqrt{1-\varsigma^2}\,t+\cos^{-1}\varsigma\right)-\dfrac{h}{2\varsigma\omega^2}\cos\omega t$

3.19 Find the transient response of an undamped system to the forces shown in Fig. P3.19.

Ans. (1) $x(t)=\dfrac{F_0}{m\omega_n}(1-\cos\omega_n t)$, $0\le t\le T_1$

$$x(t)=\frac{F_0}{m\omega_n}\left[1-\cos\omega_n t-\frac{t-T_1}{T_2-T_1}+\frac{\sin\omega_n(t-T_1)}{\omega_n(T_2-T_1)}\right],\ T_1\le t\le T_2$$

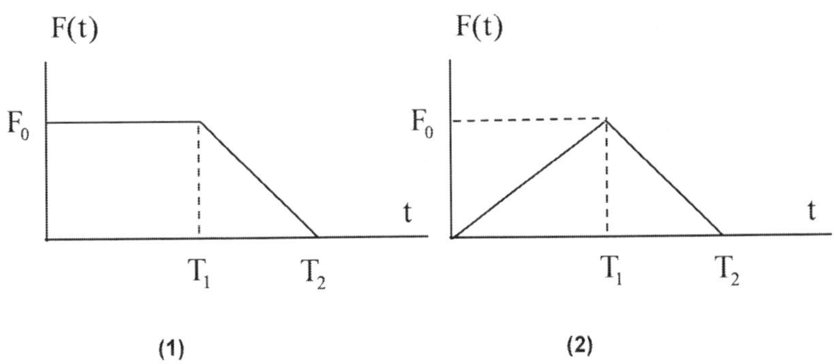

Fig. P3.19

$$x(t)=\frac{F_0}{m\omega_n}\left[-\cos\omega_n t+\frac{\sin\omega_n(t-T_1)-\sin\omega_n(t-T_2)}{\omega_n(T_2-T_1)}\right],\ t\geq T_2$$

$$(2)\ x(t)=\frac{F_0}{m\omega_n}\left(\frac{1}{T_1}-\frac{\sin\omega_n t}{\omega_n T_1}\right),\ 0\leq t\leq T_1$$

$$x(t)=\frac{F_0}{m\omega_n}\left[\frac{t}{T_1}-\frac{\sin\omega_n t}{\omega_n T_1}-\frac{T_2(t-T_1)}{T_1(T_2-T_1)}+\frac{T_2\sin\omega_n(t-T_1)}{\omega_n T_1(T_2-T_1)}\right],\ T_1\leq t\leq T_2$$

$$x(t)=\frac{F_0}{m\omega_n}\left[\frac{T_2\sin\omega_n(t-T_1)}{\omega_n T_1(T_2-T_1)}-\frac{\sin\omega_n(t-T_2)}{\omega_n(T_2-T_1)}-\frac{\sin\omega_n t}{\omega_n T_1}\right],\ t\geq T_2$$

References

3.1. Den Hartog, J. P., *Mechanical Vibrations*, 4th ed., McGraw-Hill, New York, 1956. (Reprinted by Dover, New York, 1985.)

3.2. Thomson, W. T., and Dahleh, M. D., *Theory of Vibration with Applications*, 5th ed., Prentice Hall, Inc., Upper Saddle River, NJ, 1998.

3.3. Timoshenko, S. P., Young, D. H., and Weaver, W., Jr., *Vibration Problems in Engineering*, 4th ed., John Wiley & Sons, New York, 1967.

3.4. Harris C. M., ed., *Shock and Vibration Handbook*, 3rd ed., McGraw-Hill, New York, 1987.

3.5. Inman, D. J., *Engineering Vibration*, Prentice Hall, Inc., Upper Saddle River, NJ, 1994.

Chapter Four

Balancing of Rotors

4.1 Unbalance

Unbalanced rotating equipment can produce numerous harmful effects ranging from simply unpleasant noise to destruction of the housing machine itself and even surrounding structures. Unfortunately unbalance can easily occur. All that is needed is for the rotor's mass center to be a small distance off, or away, from the axis of rotation. Alternatively, unbalance can occur if the axis of rotation does not coincide with a principal axis of inertia of the rotor.

An off-axis mass center will produce a rotating radial inertia force that in turn produces a periodic excitation of the housing structure and then reflexively on the rotor itself (see Fig. 4.1-1).

The product of the rotor mass and the off-axis (or *offset*) distance is called *unbalance* or *imbalance*.

(1) Static Unbalance

When unbalance occurs with the mass center being off the rotation axis, as can readily occur with a thin rotating disk, it can be detected by a simple static test. Specifically, the rotor assembly is placed on a pair of parallel rails, as represented in Fig. 4.1-2. The rotor assembly will then roll to a position where the offset mass center lies directly beneath the axis of rotation [4.1].

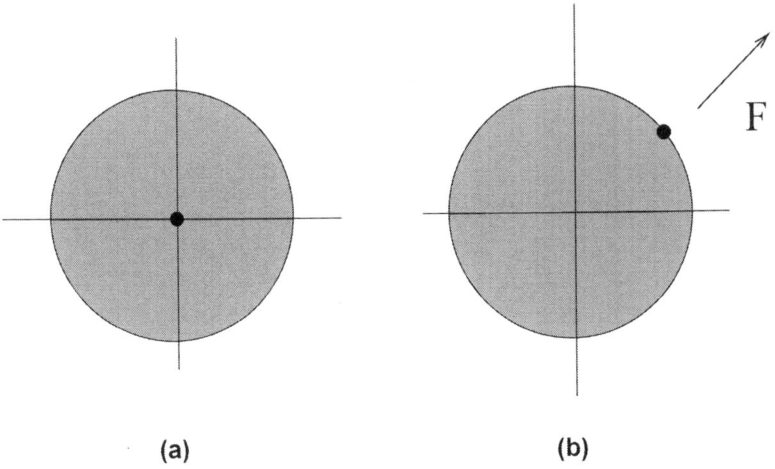

Fig. 4.1-1 A rotating shaft: (a) the mass center is on the axis of rotation; (b) the mass center is offset from the axis of rotation.

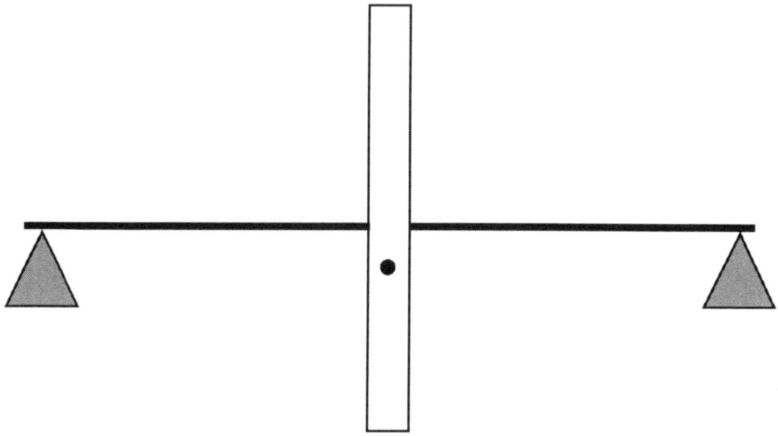

Fig. 4.1-2 Static unbalance.

(2) Dynamic Unbalance

When unbalance occurs due to the axis of rotation not coinciding with a principal inertia axis of the rotor, the unbalance cannot be detected by the simple static test above, since the mass center could lie on the rotation axis (see Fig. 4.1-3). This is called *dynamic unbalance* since it becomes apparent only when the rotor assembly is rotating [4.1].

If the rotor assembly is relatively long, it can be balanced by the addition or removal of correction weights in parallel planes perpendicular to the rotation axis [4.2].

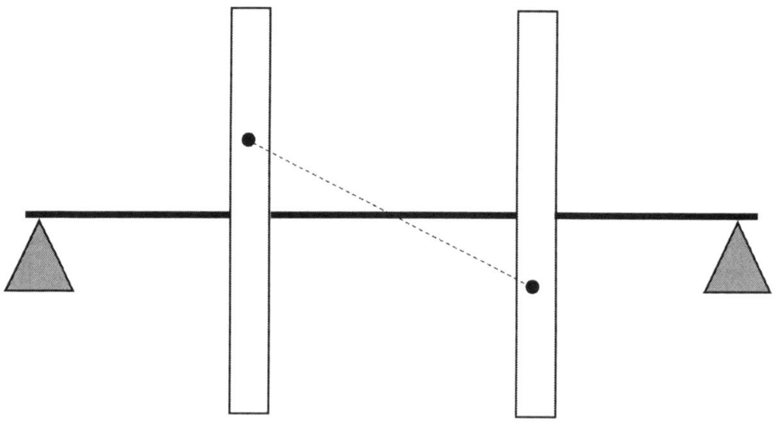

Fig. 4.1-3 Dynamic unbalance.

Example 4.1-1 Fig. 4.1-4 depicts a two-cylinder engine crankshaft. In each crank there is an imbalance of mr. What mass should be placed on the flywheels at radius r so that the crankshaft will be balanced?

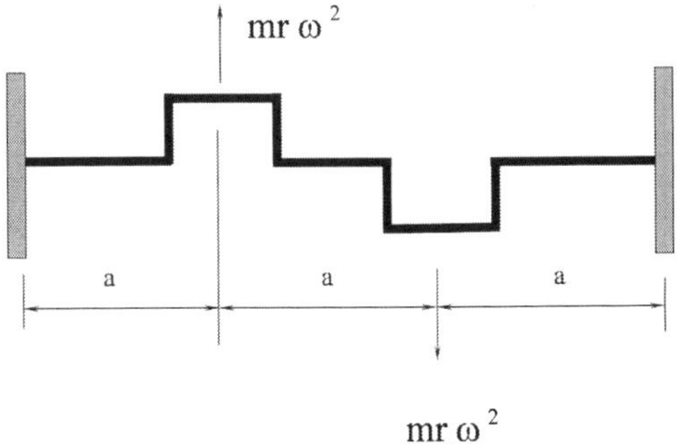

Fig. 4.1-4 An unbalanced two-cylinder engine crankshaft.

Solution: When the crankshaft is rotating with angular speed ω, the imbalance mass on each crank will produce an inertia force $mr\omega^2$. These forces produce an unbalance moment with a torque $amr\omega^2$. If we add a balance mass $(1/3)m$ on each wheel, the two balance masses will produce a moment with the same magnitude but opposite direction to the original unbalance moment. Then the crankshaft will be balanced. Fig. 4.1-5 depicts this balancing solution.

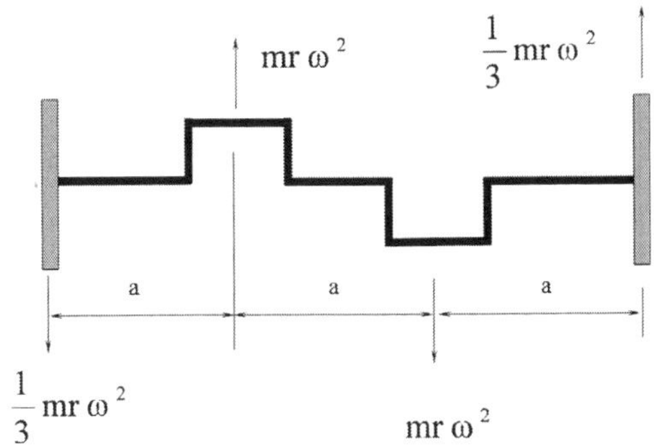

Fig. 4.1-5 A balanced two-cylinder engine crankshaft.

4.2 Whirling of Rotating Shaft

To establish some important concepts and definitions and to provide information of general utility, it is useful to consider the idealized system represented in Fig. 4.2-1 [4.1, 4.2, 4.3]. The system consists of a disk of mass m located on a light shaft supported by two bearings. The mass center G of the disk is slightly off the shaft axis.

In this model, an infinitely rigid simple support is assumed for the bearings.

The rotating shaft tends to bob out at certain speeds, and whirl takes place in a complicated manner. *Whirling* is defined as the rotation of the plane made by the bent shaft and the line of centers of the bearings. The phenomenon results from various causes such as mass unbalance, shaft flexibility, hysteresis

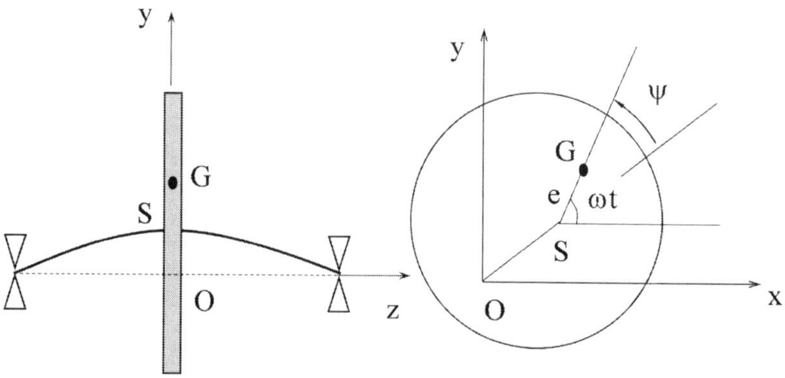

Fig. 4.2-1 Synchronous whirl of a shaft due to mass unbalance.

damping in the shaft, gyroscopic forces, fluid friction in the bearing, etc. As pointed out in Section 1.1, this motion comes under the classification of self-excited motion, in which the exciting forces including the motion are controlled by the motion itself.

In this section, we will consider the simplest case of synchronous whirl, where the whirling speed of the shaft is equal to the rotation speed of the shaft. Let e be the radial distance of the mass center G of the disk from the geometric center S of the shaft. Let O be the intersection point between the centerline of the bearings and the plane of the disk. Finally, let OS be the deflection of the shaft center.

In a synchronous whirl, O, S, and G remain fixed relative to each other, and the shaft and disk rotate at a constant speed ω.

With x_s and y_s defining the position of the shaft center S, the coordinates of the mass center G are

$$x = x_s + e\cos\omega t \qquad (4.2\text{-}1)$$

$$y = y_s + e\sin\omega t \qquad (4.2\text{-}2)$$

Assuming viscous damping proportional to the velocity of S, the equations of motion in the x and y directions are

$$m\frac{d^2}{dt^2}\left(x_s + e\cos\omega t\right) + c\dot{x}_s + kx_s = 0 \qquad (4.2\text{-}3)$$

$$m\frac{d^2}{dt^2}\left(y_s + e\sin\omega t\right) + c\dot{y}_s + ky_s = 0 \qquad (4.2\text{-}4)$$

or

$$m\ddot{x}_s + c\dot{x}_s + kx_s = me\omega^2 \cos\omega t \tag{4.2-5}$$

$$m\ddot{y}_s + c\dot{y}_s + ky_s = me\omega^2 \sin\omega t \tag{4.2-6}$$

Eqs. 4.2-5 and 4.2-6 are similar to Eq. 3.4-2. Defining the following parameters,

$$\omega_n = \sqrt{\frac{k}{m}}, \quad h = e\omega^2, \quad \varsigma = \frac{c}{2\sqrt{km}}, \tag{4.2-7}$$

we can write the steady-state solutions by inspection as

$$x_s = \frac{h}{\sqrt{(\omega_n^2 - \omega^2)^2 + (2\varsigma\omega_n\omega)^2}} \cos(\omega t - \psi) \tag{4.2-8}$$

$$y_s = \frac{h}{\sqrt{(\omega_n^2 - \omega^2)^2 + (2\varsigma\omega_n\omega)^2}} \sin(\omega t - \psi) \tag{4.2-9}$$

and

$$OS = \sqrt{x_s^2 + y_s^2} = \frac{h}{\sqrt{(\omega_n^2 - \omega^2)^2 + (2\varsigma\omega_n\omega)^2}} \tag{4.2-10}$$

$$\psi = \tan^{-1} \frac{2\varsigma\omega_n\omega}{\omega_n^2 - \omega^2} \tag{4.2-11}$$

It is evident that the line *SG=e* leads the displacement line *OS* by a phase angle ψ, which depends on the amount of damping and the rotation speed ω. When the rotation speed ω is less than the critical speed $\omega_n = \sqrt{k/m}$, or natural frequency of the shaft in lateral vibration, the phase angle lies in the first quadrant ($0 < \psi < \pi/2$). When the rotation speed ω is equal to the critical speed ω_n, the phase angle is equal to $\pi/2$, and a condition of resonance is encountered in which the amplitude is restrained only by damping. When the rotation speed ω is larger than the critical speed ω_n the phase angle lies in the second quadrant ($\pi/2 < \psi < \pi$). Table 4.2-1 lists the phase relationship of the disk-shaft system under three different speed conditions.

Table 4.2-1 Phase Relationship of Synchronous Whirl

$\omega << \omega_n$	$\omega = \omega_n$	$\omega >> \omega_n$
$0 < \psi < \pi/2$	$\psi = \pi/2$	$\pi/2 < \psi < \pi$

4.3 Experimental Balancing

In general, a long rotor, such as an automobile engine crankshaft, can be balanced experimentally. Such rotors must be spun on a balancing machine in order to detect the unbalance. Generally, if the amplitude of imbalance and the phase angle are known, correction can be obtained by removing the materials or adding the same imbalance materials to a site out of phase to the original imbalance. This processing is called balancing.

Balancing of rotating machinery may be generally classified in two areas. The first general area is balancing in the factory via a balancing machine. The second one is field, or in situ, balancing in which the rotor is balanced in place. Driveline balancing on car is such an example. Single-plane and two-plane influence coefficient methods are commonly used in industry.

(1) The Influence Coefficient Method: Single-Plane Balancing

Suppose the unbalancing mass of a rotating system is primarily confined to a single plane, as with an unbalanced disk, or an unbalanced thin rotor. Suppose further that the manifestation of the unbalance is proportional to or nearly proportional to the unbalance itself so that it may be modeled as being linear. Then we can use the *influence coefficient* method together with a couple of simple experimental measurements to detect, locate, and correct the unbalance [4.2].

The method proceeds as follows: First, since we have an essentially planar problem, it is convenient to introduce and use complex variables with $j=\sqrt{-1}$. Specifically, if $\mathbf{Z}$ is a complex variable, we can express it in the familiar forms

$$\mathbf{Z}=X+jY \quad \text{and} \quad \mathbf{Z}=Z\exp(j\phi), \tag{4.3-1}$$

where X and Y are the *horizontal and vertical components*, respectively, of $\mathbf{Z}$, and where Z and ϕ are the magnitude, or *modulus*, and phase angle, respectively, of $\mathbf{Z}$. Using Euler's formula

$$e^{j\phi}=\cos\phi+j\sin\phi \tag{4.3-2}$$

we can immediately express (X, Y) and (Z, ϕ) in terms of each other as

$$X=Z\cos\phi \qquad \text{and} \qquad Y=Z\sin\phi \tag{4.3-3}$$

and

$$Z=\sqrt{X^2+Y^2} \qquad \text{and} \qquad \phi=\tan^{-1}(Y/X) \tag{4.3-4}$$

Next, let $\mathbf{Z}$ represent the complex form of a displacement, a velocity, an acceleration, or a force; and let $\mathbf{U}$ represent the unbalance. That is, let $\mathbf{U}$ have the form

$$\mathbf{U}=m\mathbf{z}=m(x+jy), \tag{4.3-5}$$

where m is the unbalance mass and **z** is the complex position vector locating the mass relative to the axis of rotation. Then, with the system being linear, we can relate **Z** and **U** by the simple expression

$$\mathbf{Z} = \mathbf{S}\mathbf{U}, \tag{4.3-6}$$

where **S** is a complex variable known as the sensitivity or *influence coefficient*.

Third, conduct an experiment on the physical system measuring **Z** as a displacement, a velocity, an acceleration, or a force.

Fourth, let a trial (or calibration) weight be placed on the rotor, measure again the displacement, velocity, acceleration, or force, and let this measurement be called $\mathbf{Z}_t$. Then, in view of Eq. 4.3-6 and the linearity of the system, $\mathbf{Z}_t$ may be expressed as

$$\mathbf{Z}_t = \mathbf{S}(\mathbf{U}+\mathbf{U}_t), \tag{4.3-7}$$

where $\mathbf{U}_t$ is the change in the system imbalance due to the calibration weight.

Fifth, by substituting from Eq. 4.3-6 into Eq. 4.3-7, we can solve for the sensitivity **S** as

$$\mathbf{S} = \frac{\mathbf{Z}_t - \mathbf{Z}}{\mathbf{U}_t} \tag{4.3-8}$$

Finally, by substituting Eq. 4.3-8 into Eq. 4.3-6, we can obtain the original unbalance **U** of the system as

$$\mathbf{U} = \frac{\mathbf{Z}}{\mathbf{S}} = \frac{\mathbf{Z}\mathbf{U}_t}{\mathbf{Z}_t - \mathbf{Z}} \tag{4.3-9}$$

The following example illustrates the procedure.

Example 4.3-1 An unbalanced thin disk has an initial or *baseline* vibration given by

$$\mathbf{Z} = -1.563 + j1.158$$

When an unbalancing trial weight with imbalance $\mathbf{U}_t = 9e^{j0}$ is placed on the rotor, the resulting vibration $\mathbf{Z}_t$ becomes

$$\mathbf{Z}_t = 0.356 + j1.333$$

The objective is to find the sensitivity **S** and the unbalance **U** of the disk.

Solution: By using Eq. 4.3-8, we find the sensitivity **S** to be

$$\mathbf{S} = \frac{\mathbf{Z}_t - \mathbf{Z}}{\mathbf{U}_t} = 0.2132 + j0.0194$$

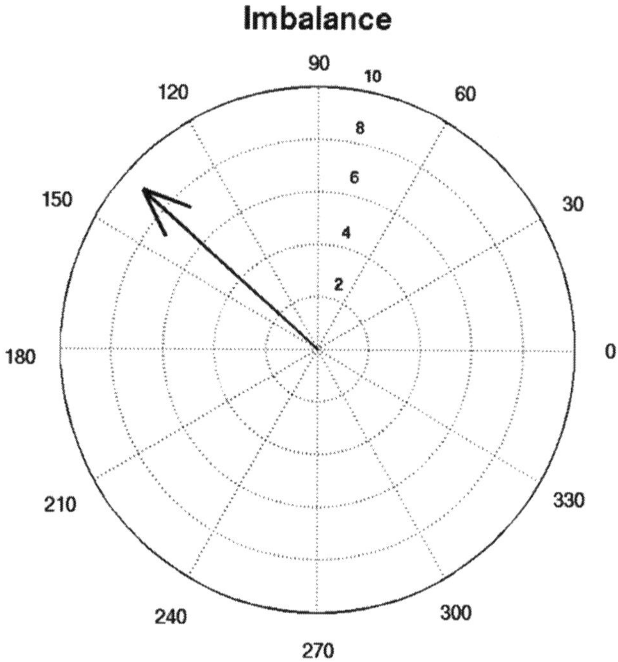

Fig. 4.3-1 Solution of Example 4.3-1.

Therefore, from Eq. 4.3-6, the imbalance **U** of the disk is given by

$$\mathrm{U} = \frac{\mathrm{Z}}{\mathrm{S}} = -6.7787 + j6.0491 = 9.085e^{j138.26^\circ}$$

(The unbalance can be easily calculated by using Matlab.) Fig. 4.3-1 shows a graphical display of the unbalance.

(2) The Influence Coefficient Method: Two-Plane Balancing

Interestingly, the foregoing method may be extended for use with systems where the imbalance occurs in more than one plane, as in dynamic imbalance.

The basic assumption of this method is that vibration measured at a particular location at a fixed angular speed is the product of a linear combination of imbalance and sensitivity, or the influence coefficients [4.2].

In this method, a trial weight is placed on a plane of the system, and the vibration responses at that plane and another plane are recorded. From these measurements, two influence coefficients are computed. Next, the trial weight is removed and placed at the second plane. The procedure is repeated to generate two more influence coefficients. This then forms a 2×2 matrix of complex influence coefficients which must be inverted for the imbalance. This

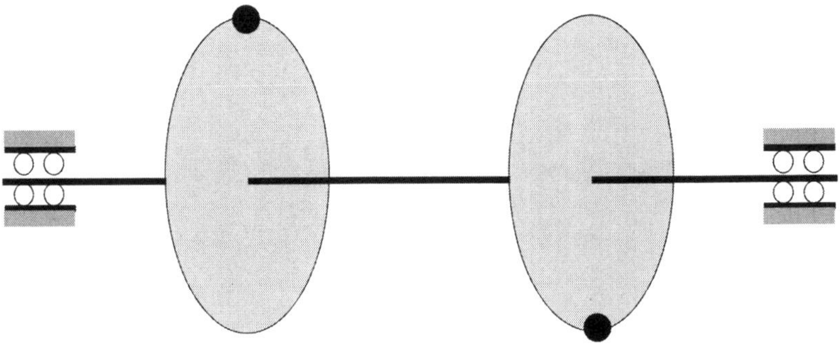

Fig. 4.3-2 Two-plane unbalance system.

method can determine the imbalance magnitudes and relative phase angles on the two planes.

Fig. 4.3-2 depicts an unbalanced system with the imbalance occurring in two planes.

Referring to Fig. 4.3-2, we can write the initial vibration readings at the two planes for the fixed angular speed ω as

$$\mathbf{Z}_1(\omega) = Z_1 e^{j\phi_1} \qquad \text{and} \qquad \mathbf{Z}_2(\omega) = Z_2 e^{j\phi_2} \tag{4.3-10}$$

The vibration is assumed to be a linear combination of the unknown unbalances $\mathbf{U}_1$ and $\mathbf{U}_2$ as follows:

$$\begin{bmatrix} \mathbf{Z}_1(\omega) \\ \mathbf{Z}_2(\omega) \end{bmatrix} = \begin{bmatrix} \mathbf{S}_{11}(\omega) & \mathbf{S}_{12}(\omega) \\ \mathbf{S}_{21}(\omega) & \mathbf{S}_{22}(\omega) \end{bmatrix} \begin{bmatrix} \mathbf{U}_1 \\ \mathbf{U}_2 \end{bmatrix} \tag{4.3-11}$$

With the placement of a trial unbalance $\mathbf{U}_{t1}$ at plane 1, new vibrations may be recorded as

$$\begin{bmatrix} \mathbf{Z}_{11}(\omega) \\ \mathbf{Z}_{21}(\omega) \end{bmatrix} = \begin{bmatrix} \mathbf{S}_{11}(\omega) & \mathbf{S}_{12}(\omega) \\ \mathbf{S}_{21}(\omega) & \mathbf{S}_{22}(\omega) \end{bmatrix} \begin{bmatrix} \mathbf{U}_1 + \mathbf{U}_{t1} \\ \mathbf{U}_2 \end{bmatrix}, \tag{4.3-12}$$

where $\mathbf{Z}_{11}$ and $\mathbf{Z}_{21}$ are the vibrations recorded at planes 1 and 2 due to the trial unbalance placed in plane 1.

The influence coefficients $\mathbf{S}_{11}$ and $\mathbf{S}_{21}$ may then be computed as

$$\mathbf{S}_{11} = \frac{\mathbf{Z}_{11} - \mathbf{Z}_1}{\mathbf{U}_{t1}}, \ \mathbf{S}_{21} = \frac{\mathbf{Z}_{21} - \mathbf{Z}_2}{\mathbf{U}_{t1}} \tag{4.3-13}$$

Next, the first trial unbalance weight is removed and a second trial unbalance weight $\mathbf{U}_{t2}$ is placed in the second plane. The resulting vibrations are

$$\begin{bmatrix} \mathbf{Z}_{12}(\omega) \\ \mathbf{Z}_{22}(\omega) \end{bmatrix} = \begin{bmatrix} \mathbf{S}_{11}(\omega) & \mathbf{S}_{12}(\omega) \\ \mathbf{S}_{21}(\omega) & \mathbf{S}_{22}(\omega) \end{bmatrix} \begin{bmatrix} \mathbf{U}_1 \\ \mathbf{U}_2 + \mathbf{U}_{t2} \end{bmatrix} \tag{4.3-14}$$

The influence coefficients $\mathbf{S}_{12}$ and $\mathbf{S}_{22}$ may then be computed as

$$\mathbf{S}_{12} = \frac{\mathbf{Z}_{12} - \mathbf{Z}_1}{\mathbf{U}_{t2}}, \ \mathbf{S}_{22} = \frac{\mathbf{Z}_{22} - \mathbf{Z}_2}{\mathbf{U}_{t2}} \tag{4.3-15}$$

The unknown imbalances of the original system at planes 1 and 2 are given by

$$\begin{bmatrix} \mathbf{U}_1 \\ \mathbf{U}_2 \end{bmatrix} = \begin{bmatrix} \mathbf{S}_{11} & \mathbf{S}_{12} \\ \mathbf{S}_{21} & \mathbf{S}_{22} \end{bmatrix}^{-1} \begin{bmatrix} \mathbf{Z}_1 \\ \mathbf{Z}_2 \end{bmatrix} \tag{4.3-16}$$

That is,

$$\mathbf{U}_1 = \frac{\mathbf{Z}_1\mathbf{S}_{22} - \mathbf{Z}_2\mathbf{S}_{12}}{\Delta} \quad \text{and} \quad \mathbf{U}_2 = \frac{\mathbf{Z}_2\mathbf{S}_{11} - \mathbf{Z}_1\mathbf{S}_{21}}{\Delta}, \tag{4.3-17}$$

where

$$\Delta = \mathbf{S}_{11}\mathbf{S}_{22} - \mathbf{S}_{12}\mathbf{S}_{21} \tag{4.3-18}$$

Example 4.3-2 For a two-plane unbalance system, the baseline responses at planes 1 and 2 are given by $\mathbf{Z}_1 = 8.6e^{j63°}$, $\mathbf{Z}_2 = 6.5e^{j206°}$. After adding the first trial unbalance $\mathbf{U}_{t1} = 10e^{j270°}$ in plane 1, the new responses are

$$\mathbf{Z}_{11} = 5.9e^{j123°} \quad \text{and} \quad \mathbf{Z}_{21} = 4.5e^{j228°}$$

Then, the first trial unbalance weight is removed and a second trial unbalance weight $\mathbf{U}_{t2} = 12e^{j180°}$ is placed in the second plane. The resulting vibrations are

$$\mathbf{Z}_{12} = 6.2e^{j36°} \quad \text{and} \quad \mathbf{Z}_{22} = 10.4e^{j162°}$$

Find the unbalance of the system.

Solution: By using Eqs. 4.3-13 and 4.3-15, we obtain

$$\mathbf{S}_{11} = 0.2714 - j0.7118; \ \mathbf{S}_{12} = -0.0926 + j0.3349$$

$$\mathbf{S}_{21} = 0.0495 + j0.2831; \ \mathbf{S}_{22} = 0.3374 - j0.5053$$

Then, from Eq. 4.3-18, we have

$$\Delta = \mathbf{S}_{11}\mathbf{S}_{22} - \mathbf{S}_{12}\mathbf{S}_{21} = -0.1687 - j0.3676$$

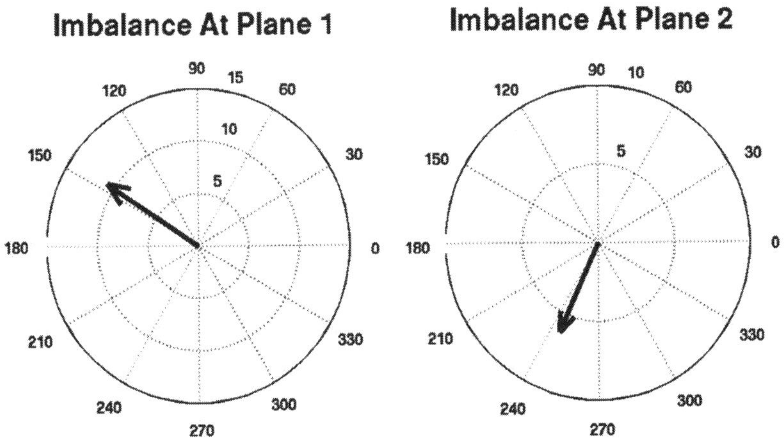

Fig. 4.3-3 Solution of Example 4.3-2 with Matlab.

By using Eqs. 4.3-17, we obtain the unbalance of the system at plane 1 and plane 2:

$$\mathbf{U}_1 = -8.9873 + j5.9238 = 10.76e^{j146^\circ}$$

$$\mathbf{U}_2 = -2.5819 - j5.6392 = 6.2e^{j245^\circ}$$

Fig. 4.3-3 shows the results.

A Matlab code for solving Example 4.3-2 is

```
% two plane balance
clear all;
% input data
Z1=8.6*(cos(63*pi/180)+i*sin(63*pi/180));
Z2=6.5*(cos(206*pi/180)+i*sin(206*pi/180));
ut1=10*(cos(270*pi/180)+i*sin(270*pi/180));
Z11=5.9*(cos(123*pi/180)+i*sin(123*pi/180));
Z21=4.5*(cos(228*pi/180)+i*sin(228*pi/180));
ut2=12*(cos(180*pi/180)+i*sin(180*pi/180));
Z12=6.2*(cos(36*pi/180)+i*sin(36*pi/180));
Z22=10.4*(cos(162*pi/180)+i*sin(162*pi/180));
% Find sensitivity matrix S:
S11=(Z11-Z1)/ut1;S21=(Z21-Z2)/ut1;
S12=(Z12-Z1)/ut2; S22=(Z22-Z2)/ut2;
% Find Imbalance of baseline:
Delta=S11*S22-S12*S21;
Imb_1=(Z1*S22-Z2*S12)/Delta
```

```
Imb_2=(Z2*S11-Z1*S21)/Delta
subplot(1,2,1), h1=compass(Imb_1,'k');
title('Imbalance At Plane 1','FontSize',15);
subplot(1,2,2), h2=compass(Imb_2,'k');
title('Imbalance At Plane 2','FontSize',15);
set(h1,'LineWidth',3);
set(h2,'LineWidth',3);
```

4.4 Inertia Forces in a Single-Cylinder Engine

Engine balancing commands the majority of attention in rotor balancing. The most challenging, and thus the most interesting, of engine balancing problems arise with multicylinder reciprocating engines [4.4–4.19].

Before looking at multicylinder problems, it is helpful to consider single-cylinder engines. The main parts of a single-cylinder engine are (1) the cylinder, (2) the piston, (3) the crank, and (4) the connecting rod. In these engines, the cylinder is usually fixed, or stationary, while the other three parts are moving: The piston and connecting rod are accelerating and decelerating, and the crank is rotating with its mass center off its axis of rotation. These movements give rise to unbalancing inertia forces, proportional to the masses of the moving parts.

Fig. 4.4-1 depicts a model of the moving parts of a single-cylinder engine with the masses of piston, connecting rod, and crank represented by point masses, or particles. In the figure, and in our subsequent analysis, we use the following notation:

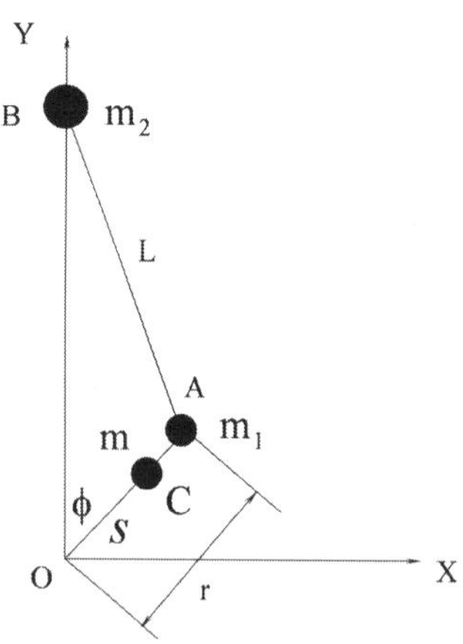

Fig. 4.4-1 A single-cylinder engine system represented by three mass particles [4.4].

r = crank radius (1/2 stroke),

L = length of connecting rod,

$s = OC$ = distance to center of gravity of crank,

e = piston pin offset,

m = mass of crank and crankpin,

m_1 = rotating mass of connecting rod, which is equal to its big-end mass,

m_2^* = small-end mass of connecting rod,

$m^* = m_1 + m$ = total mass of connecting rod,

m_2^{**} = total mass of piston, piston pin, piston pin locking device, piston rings, and oil rings,

$m_2 = m_2^* + m_2^{**}$ = reciprocating mass,

ϕ = angle of crank to cylinder axis, positive clockwise,

c = distance of the mass center of connecting rod from the crank pin A.

We will introduce other specific notation as needed.

Following Timoshenko [4.4], we replace the connecting rod by two equivalent masses m_1 and m_2^* at the crankpin A and the gudgeon pin B, respectively. These two masses are called the *big-end mass* and the *small-end mass* of the rod. The sum of these two masses must be equal to the total mass of the connecting rod, and their mass center must coincide with that of the rod. Specifically,

$$m^* = m_1 + m_2^* \qquad \text{and then} \qquad m_1 c = m_2^*(L-c)$$

From these two conditions, we obtain the relations

$$m_1 = m^*(1 - c/L) \qquad \text{and} \qquad m_2^* = m^* c/L \tag{a}$$

To the small end mass at B, we now add the masses of the piston, the piston pin, the piston pin locking devices, the piston rings, the spacer, and the oil rings; we then denote the total joint mass by m_2, which is called the *reciprocating mass*. Finally, the mass m of the crank and crankpin is assumed to be concentrated at its mass center C. With these changes, the actual system is replaced by a system of three mass particles m, m_1, and m_2, concentrated at C, A, and B, respectively.

This method of replacing the connecting rod by two concentrated masses is, strictly speaking, not precise for connecting rods of usual design [4.4, 4.7, 4.12, 4.14], since the central moment of inertia in the axial direction for the rod is different than that of the system of two concentrated masses. References [4.4] and [4.12] provide more accurate methods. Nevertheless, the inertia torque due to the connecting rod inertia is small compared with the torque of the reaction couple. Therefore, for heavy-oil engines, the above method is reasonably accurate and generally used by analysts.

The system is symmetrical in the central plane normal to the axis. Therefore, in Fig. 4.4-1, we introduce the X and Y coordinate axes, as shown. With this symmetry, the external forces acting on the system may be represented by horizontal and vertical force components $\sum F_X^e$ and $\sum F_Y^e$, and a couple with torque $\sum(F_Y^e x - F_X^e y)$ relative to the Z-axis. We are particularly interested in the forces that the operating engine exerts on its foundation (the inertia forces). Consequently, it is convenient to introduce the notation

$$F_X = -\sum F_X^e \qquad \text{and} \qquad F_Y = -\sum F_Y^e \tag{4.4-1}$$

If the inertia forces are positive, they are directed from the crank bearing outward; if they are negative, they are directed toward the crank bearing.

When the motion of the system is known, these forces can be determined by the principles of linear momentum and angular momentum. We proceed now to determine them. Noting that the speeds of points C, A, and B are $s\dot{\phi}$, $r\dot{\phi}$, and $\dot{y}$, respectively, and using the principle of linear momentum, we obtain the expressions

$$\frac{d}{dt}\left(ms\dot{\phi}\cos\phi+m_1r\dot{\phi}\cos\phi\right)=\Sigma F_X^e=-F_X \tag{4.4-2}$$

and

$$\frac{d}{dt}\left(-ms\dot{\phi}\sin\phi-m_1r\dot{\phi}\sin\phi+m_2\dot{y}\right)=\Sigma F_Y^e=-F_Y \tag{4.4-3}$$

From Eqs. 4.4-2 and 4.4-3, we obtain

$$F_X=(ms+m_1r)(\dot{\phi}^2\sin\phi-\ddot{\phi}\cos\phi) \tag{4.4-4}$$

and

$$F_Y=(ms+m_1r)(\dot{\phi}^2\cos\phi+\ddot{\phi}\sin\phi)-m_2\ddot{y} \tag{4.4-5}$$

Eq. 4.4-5 can be expressed entirely in terms of the angle coordinate ϕ. To this end, we consider a general case where there is a piston pin offset, denoted by e, as shown in Fig. 4.4-2. The piston pin offset is classified into two categories [4.6–4.19]. One is the piston pin offset from the piston center line toward the minor thrust side [see Fig. 4.4-2 (a)], where the crank rotates clockwise while the piston pin is located on the right side of the cylinder's center line. The other is the piston pin offset from the piston center line toward the major thrust side, that is, the crank rotates clockwise while the piston pin is located on the left side of the cylinder's center line [see Fig. 4.4-2 (b)].

An essential step in our analysis is the derivation of an expression for the piston position at any crank angle. To simplify the analysis, the piston position is considered from the crank center instead of from top position. Referring to Fig. 4.4-2 (a), we see that the distance O_1B_1 is

$$O_1B_1=r\cos\phi+L\cos\beta \tag{4.4-6}$$

Also, from Fig. 4.4-2, we have

$$L\sin\beta+e=r\sin\phi, \tag{4.4-7}$$

where, as noted above, e is the measure of the piston pin offset. If the piston pin offset is from the piston center line toward the minor thrust side, as in Fig. 4.3-2 (a), then e is positive. Conversely, if the piston pin offset is from the piston center line toward the major thrust side, as in Fig. 4.4-2 (b), then e is negative. With no offset, e is zero.

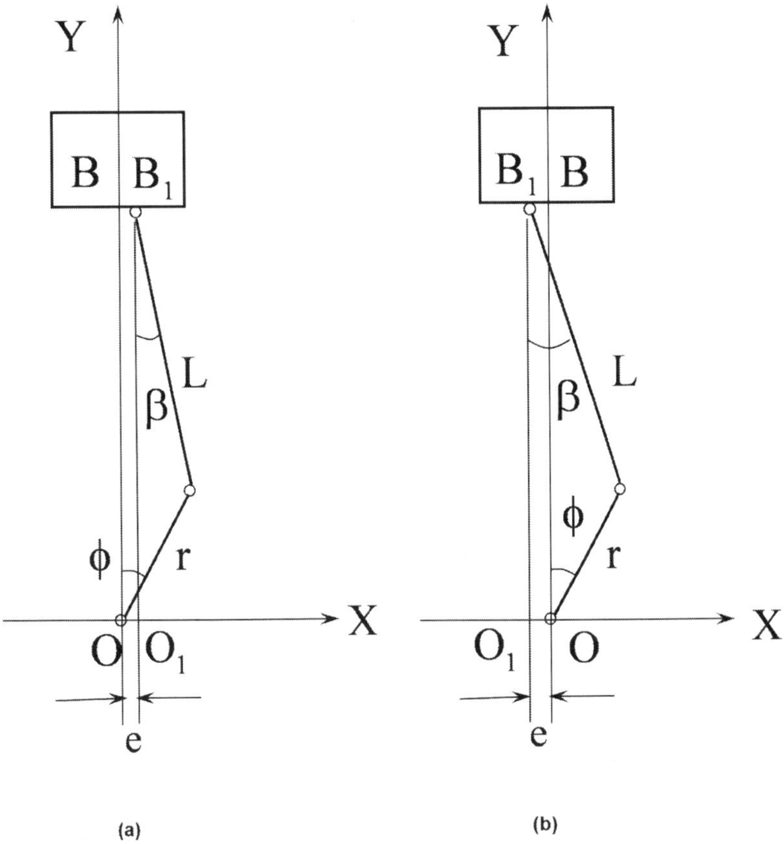

Fig. 4.4-2 A single-cylinder engine with piston pin offset: (a) the pin offset toward the minor thrust side, (b) the pin offset toward the major thrust side.

From Eq. 4.4-7, $\sin\beta$ and $\cos\beta$ are immediately found as

$$\sin\beta = \frac{r\sin\phi - e}{L} \tag{4.4-8}$$

and

$$\cos\beta = \sqrt{1 - \left(\frac{r\sin\phi - e}{L}\right)^2} \tag{4.4-9}$$

Finally, by substituting from Eq. 4.4-9 into Eq. 4.4-6, we obtain

$$O_1B_1 = r\cos\phi + L\sqrt{1 - \left(\frac{r\sin\phi - e}{L}\right)^2} \tag{4.4-10}$$

In practical applications, r and e are usually small compared with L. Thus, by using the binomial theorem, we can expand the radical of Eq. 4.4-10 into a rapidly converging power series. By retaining the first two terms of this series, Eq. 4.4-10 takes the form

$$O_1B_1 = r\cos\phi + L\left[1 - \frac{1}{2}\left(\frac{r\sin\phi - e}{L}\right)^2\right] \tag{4.4-11}$$

or

$$O_1B_1 = r\cos\phi + L - \frac{(r\sin\phi - e)^2}{2L} \tag{4.4-12}$$

Observe again that, in the force components F_X and F_Y of Eqs. 4.4-4 and 4.4-5, we will have the expressions entirely in terms of φ by recognizing that y may be identified with O_1B_1. That is,

$$y = O_1B_1 = r\cos\phi + r\frac{e}{L}\sin\phi - \frac{r^2}{2L}\sin^2\phi + L\left(1 - \frac{e^2}{2L^2}\right), \tag{4.4-13}$$

where the last term, being a constant, will not contribute to the force components. Specifically, by substituting from Eq. 4.4-13 into Eq. 4.4-5 and differentiating, the force components become

$$F_X = (ms + m_1r)(\dot{\phi}^2\sin\phi - \ddot{\phi}\cos\phi) \tag{4.4-14}$$

and

$$F_Y = (ms + m_1r)(\dot{\phi}^2\cos\phi + \ddot{\phi}\sin\phi) + \\ + m_2r\left[\dot{\phi}^2\left(\cos\phi + \frac{e}{L}\sin\phi + \frac{r}{L}\cos 2\phi\right) + \ddot{\phi}\left(\sin\phi - \frac{e}{L}\cos\phi + \frac{r}{2L}\sin 2\phi\right)\right] \tag{4.4-15}$$

In studying engine balance, we are most interested in the case where the engine runs with uniform angular speed ω. For this condition, in Eqs. 4.4-14 and 4.4-15, the terms that contain $\ddot{\phi}$ vanish. Then, by letting $\dot{\phi} = \omega$, we obtain

$$F_X = (ms + m_1r)\omega^2\sin\phi \tag{4.4-16}$$

and

$$F_Y = (ms + m_1r)\omega^2\cos\phi + m_2\omega^2r\left[\cos\phi + \frac{e}{L}\sin\phi + \frac{r}{L}\cos 2\phi\right] \tag{4.4-17}$$

From Eqs. 4.4-16 and 4.4-17, we see that the single-cylinder engine will not transmit forces to its foundation when the following conditions are met:

$$ms + m_1r = 0 \qquad \text{and} \qquad m_2 = 0 \tag{4.4.-18}$$

The first of these conditions can be satisfied by adding a counterweight to the crankshaft so that the mass m of the crank, its crankpin, and the counterweight balances that of the *big-end mass* m_1 of the connecting rod. That is, a counterweight can be added opposite the connecting rod big-end mass so that the mass center of the rotating system consisting of (1) the crank, (2) the crankpin, (3) the counterweight, and (4) the big-end mass of the connecting rod is located at *O*, on the axis of rotation.

Fig. 4.4-3 depicts a counterweight-balanced system so that the first of Eq. 4.4-18 is satisfied. That is, the distance $\hat{s}$ of m (the mass of the crank, the crankpin, and the counterweight) away from the rotation axis is

$$\hat{s} = \frac{m_1}{m} r \qquad (4.4\text{-}19)$$

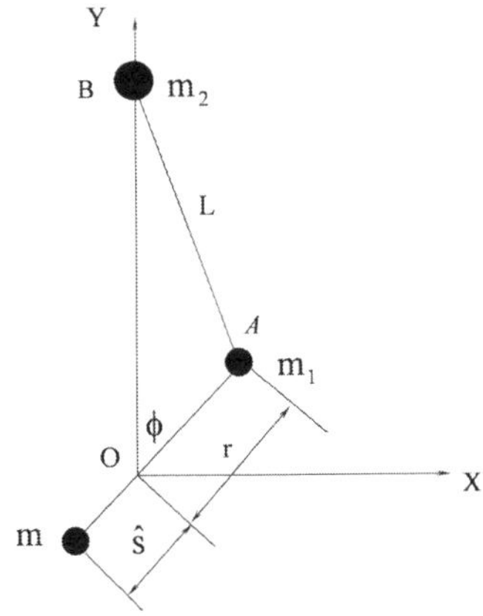

Fig. 4.4-3 A single-cylinder engine with counterweight on the crankshaft [4.4].

When Eq. 4.4-19 is satisfied, the mass center of m and m_1 remains stationary at *O*, and it follows from Eq. 4.4-16 that the transverse force F_X is zero. In practice, due to the errors in manufacture, the counterweighted crankshaft itself of multicylinder engines must be balanced before being put into service. The next sections provide a more detailed discussion of this topic.

To satisfy the second condition of Eq. 4.4-18, the reciprocating mass m_2 must vanish. This condition, however, cannot be physically satisfied. Instead, the most practical approach is to reduce the effect of the small-end mass m_2 of the connecting rod at *B*. This is accomplished by increasing the big-end mass at *A*, as represented in Fig. 4.4-4.

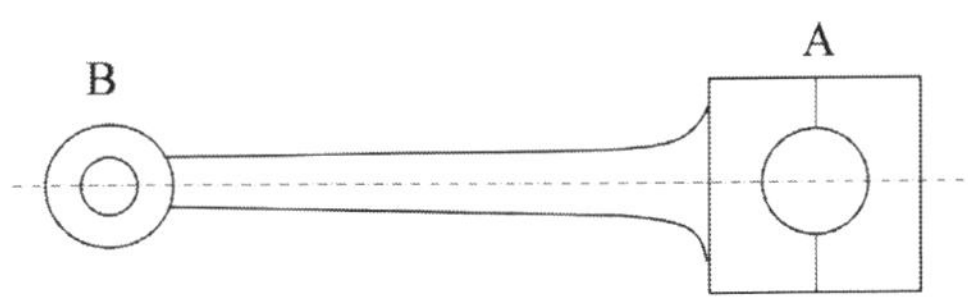

Fig. 4.4-4 A sketch of a modified connecting rod.

We now consider the moment of the inertia forces with respect to the *Z*-axis [4.4]. To this end, we need a general expression for the angular momentum of the system in Fig. 4.4-1. Since the mass m_2 at point *B* moves along the *Y*-axis, its moment of momentum with respect to point *O* vanishes. Therefore, we do not need to consider it further. Concerning the connecting rod, recall that it was replaced by the big-end mass and the small-end mass at its ends *A* and *B*. For translation motion, these two mass particles are dynamically equivalent to the actual rod. To be dynamically equivalent in rotation, the two mass particles at the ends of the line *AB* must have the same moment of inertia as the rod itself about any axis normal to the plane of the figure. For the total moment of inertia with respect to point *B*, this condition requires that

$$m_1 L^2 = I_c + m^* (L - c)^2, \qquad (4.4\text{-}20)$$

where I_c is the moment of inertia of the connecting rod about its centroidal axis normal to the plane of the figure. From this condition, and the expression (a) defining the magnitude of m_1 (in the beginning of this section), we have

$$I_c = m^* c(L-c) \tag{4.4-21}$$

Since $I_c>0$, Eq. 4.4-21 requires that

$$c < L \tag{4.4-22}$$

Next, the total angular moment of momentum of the system with respect to the Z-axis is

$$mk_0^2\dot{\phi}+m_1r^2\dot{\phi},$$

where k_0^2 is the radius of gyration of the crank with respect to the Z-axis. Equating the rate of change of this angular momentum to the corresponding moment of all external forces acting on the system, we obtain

$$\frac{d}{dt}(mk_0^2\dot{\phi}+m_1r^2\dot{\phi})=-M_z,$$

from which we have

$$M_z=-(mk_0^2+m_1r^2)\ddot{\phi} \tag{4.4-23}$$

We see that the moment of Eq. 4.4-23 vanishes for any configuration of the system in Fig. 4.4-1 only if

$$mk_0^2+m_1r^2=0 \tag{4.4-24}$$

This condition can be satisfied only if $m_1<0$ which, as can be seen from the first of expressions (a), requires that

$$c > L \tag{4.4-25}$$

We see that the conditions of Eqs. 4.4-22 and 4.4-25 cannot be satisfied simultaneously. Thus, we can never entirely eliminate M_z by any kind of balancing.

From all of the foregoing, we conclude that a single-cylinder reciprocating engine cannot be perfectly balanced and that it always produces some dynamic loading on its foundation.

4.5 Inertia Forces and Couples of Multicylinder Engines

The equations developed in the preceding section for a single-cylinder engine can be generalized to determine the inertia forces and couples of a multicylinder engine.

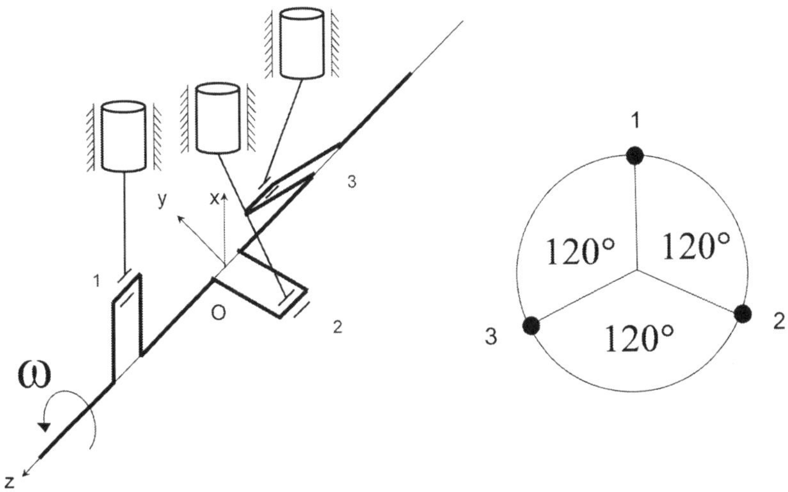

Fig. 4.5-1 An inline three-cylinder engine.

As an example, consider an inline three-cylinder engine with the crank arrangement shown in Fig. 4.5-1. Selecting coordinate axes as shown, by using Eq. 4.4-16, the first-order unbalance force produced by all three cylinders in x-direction is

$$F_X = \sum_{i=1}^{3} m_1 \omega^2 r \sin(\phi + \theta_i), \tag{4.5-1}$$

where the phase angles are

$$\theta_1 = 0,\ \theta_2 = 120^\circ,\ \theta_3 = 240^\circ \tag{4.5-2}$$

Substituting these values into Eq. 4.5-1, we obtain

$$F_X = 0 \tag{4.5-3}$$

We see that with such an arrangement, the engine is balanced relative to forces in the x-direction.

Similarly, from Eq. 4.4-17, we see that the engine is also balanced relative to forces in the y-direction.

The moment produced by the first-order inertia forces about the x-axis can be calculated as follows:

$$M_x^{(1)} = -F_1^y a + F_3^y a, \tag{4.5-4}$$

where F_1^y and F_3^y are the first-order inertia forces produced by the first cylinder and the third cylinder, respectively. From Eq. 4.4-17, noting that the phase angle for the third cylinder is 240°, we have

$$F_1^y = m_1\omega^2 r\cos\phi + m_2\omega^2 r\left[\cos\phi + \frac{e}{L}\sin\phi\right] \tag{4.5-5}$$

$$F_3^y = m_1\omega^2 r\cos(\phi+240^\circ) + m_2\omega^2 r\left[\cos(\phi+240^\circ) + \frac{e}{L}\sin(\phi+240^\circ)\right] \tag{4.5-6}$$

Substituting from Eqs. 4.5-5 and 4.5-6 into Eq. 4.5-4, we obtain

$$M_x^{(1)} = -\sqrt{3}(m_1+m_2)\omega^2 ra\cos(\phi+30^\circ) \; -\sqrt{3}m_2\omega^2 ra(e/L)\sin(\phi+30^\circ) \tag{4.5-7}$$

Next, using the expression of the inertia force in Eq. 4.4-17, we can show that the moment of the second-order inertia force about the x-axis is

$$M_x^{(2)} = -\sqrt{3}m_2\omega^2 ra(r/L)\cos(2\phi+30^\circ) \tag{4.5-8}$$

Eqs. 4.5-7 and 4.5-8 show that, even though the three-cylinder engine is balanced relative to the horizontal and vertical forces, it is nevertheless unbalanced relative to the axial moments.

As a second example, we consider the celebrated V8 engine. Currently, there are three main types of V8 engines in production. They are all similar, differing mainly in their crankshaft configurations. That is, the engines have two banks, each with four cylinders. The banks are oriented at 90° relative to each other, and the crankshafts have four offsets or *throws* with two connecting rods on each throw. One of these rods controls a piston movement on the left bank and the other rod controls a piston movement on the right bank.

There is no commonly accepted procedure or methodology for numbering and labeling the cylinders. For our analyses herein, we follow the methodology defined in [4.20, 4.21]. Specifically, the cylinders in the right bank (looking from behind) are numbered C1, C2, C3, and C4, and those on the left bank are numbered C5, C6, C7, and C8. Observe that with this numbering, the cylinder farthest away from the power output end is identified as C1. Fig. 4.5-2 depicts this numbering scheme.

Note that the numbering method will not affect the conclusion of a dynamic analysis.

Fig. 4.5-3 shows a crank arrangement for the engine. The front and rear cranks are set at 180° to each other and are in the same plane. The two intermediate cranks oppose each other and lie in a plane orthogonal to the plane of the end

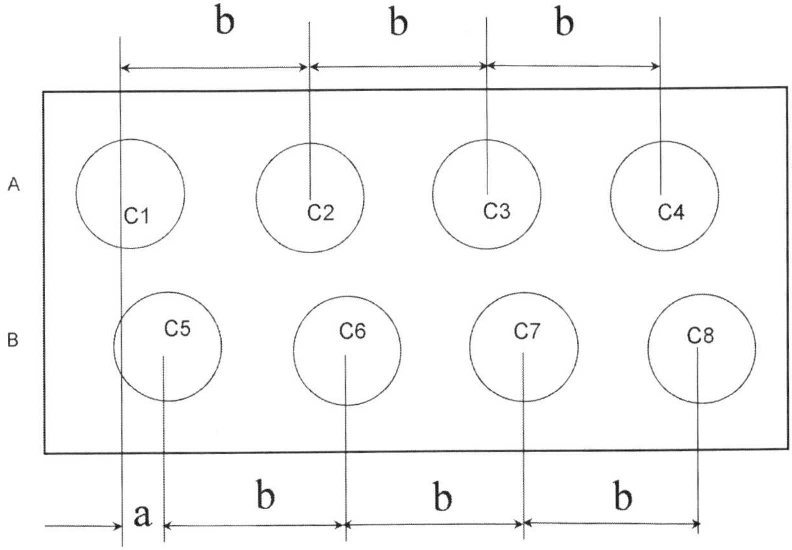

Fig. 4.5-2 A numbered V8 engine.

cranks. Since the angle between the cranks is fixed, if crank 1 is a reference and the angle between cranks 1 and 2 is 90°, then the angle between cranks 3 and 4 is 270°.

The equation developed for a single-cylinder engine can be used to determine the inertia forces and couples of the V8 engine. To this end, it is necessary to find the phase angles between the eight cylinders. We take the cylinder C1 in bank A as a reference; then the phase angles for each cylinder are shown in Figs. 4.5-4 and 4.5-5 [4.20].

Table 4.5-1 summarizes the phase angles between the cylinders.

By using Eqs. 4.4-16 and 4.4-17 as a guide, we can develop the inertia force components for the cylinders as follows:

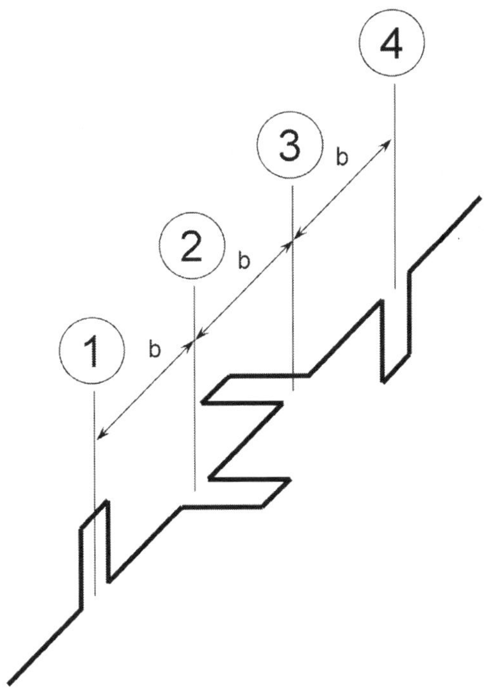

Fig. 4.5-3 Crank arrangement of a V8 engine.

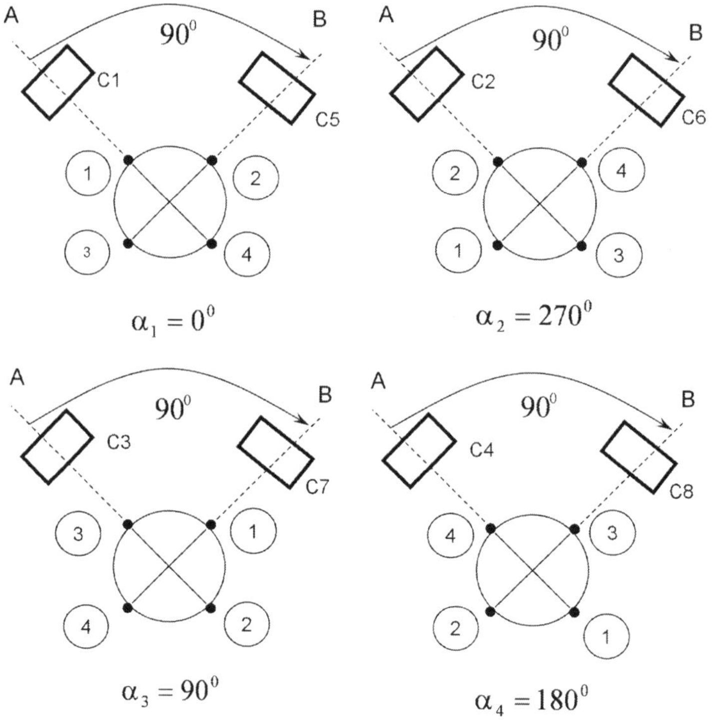

Fig. 4.5-4 Phase angles of cylinders C1 to C4.

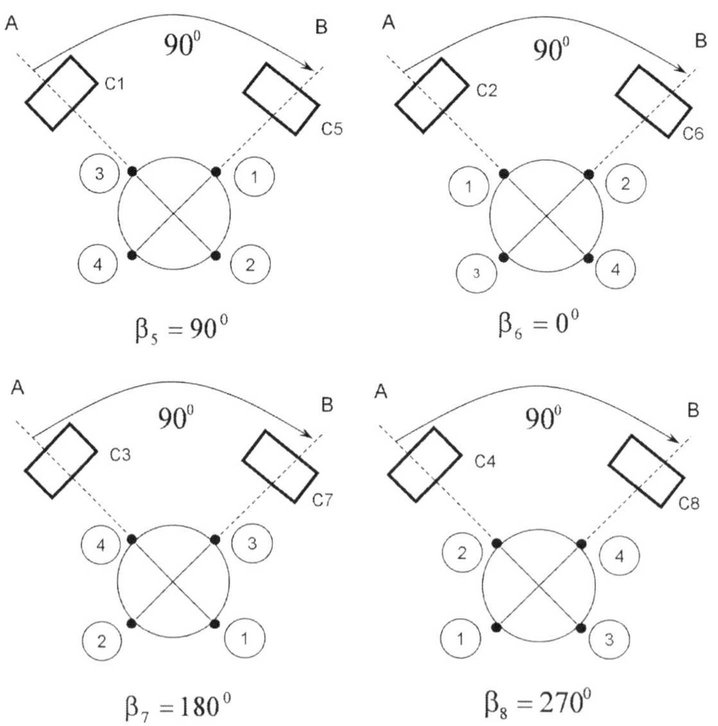

Fig. 4.5-5 Phase angles of cylinders C5 to C8.

Table 4.5-1 The Phase Angles between the Eight Cylinders

Cylinder # in bank A	1	2	3	4
Phase angle symbol	α_1	α_2	α_3	α_4
Phase angle value	0°	270°	90°	180°

Cylinder # in bank B	5	6	7	8
Phase angle symbol	β_5	β_6	β_7	β_8
Phase angle value	90°	0°	180°	270°

In bank A,

$$F_i^x = m_1\omega^2 r\sin(\phi-\alpha_i),\ (i=1,2,3,4) \tag{4.5-9}$$

$$F_i^y = m_1\omega^2 r\cos(\phi-\alpha_i)+m_2\omega^2 r\left[\cos(\phi-\alpha_i)+\frac{e}{L}\sin(\phi-\alpha_i)\right],\ (i=1,2,3,4) \tag{4.5-10}$$

In bank B,

$$F_j^x = m_1\omega^2 r\sin(\phi-\beta_j),\ (j=5,6,7,8) \tag{4.5-11}$$

$$F_j^y = m_1\omega^2 r\cos(\phi-\beta_j)+m_2\omega^2 r\left[\cos(\phi-\beta_j)+\frac{e}{L}\sin(\phi-\beta_j)\right],\ (j=5,6,7,8) \tag{4.5-12}$$

Observe that for a 90° V8 engine, the banks are inclined at 45° to the horizontal and vertical axes. Therefore, the horizontal components of inertia forces can be evaluated as follows:

For cylinders in bank A,

$$F_j^H = F_j^x\cos 45^\circ - F_j^y\cos 45^\circ,\ (j=1,2,3,4) \tag{4.5-13}$$

For cylinders in bank B,

$$F_j^H = F_j^x\cos 45^\circ + F_j^y\cos 45^\circ,\ (j=5,6,7,8) \tag{4.5-14}$$

By carrying out the indicated analysis, we obtain

$$F_1^H = \frac{m_1\omega^2 r}{\sqrt{2}}(\sin\phi-\cos\phi)-\frac{m_2\omega^2 r}{\sqrt{2}}\left(\cos\phi+\frac{e}{L}\sin\phi\right) \tag{4.5-15}$$

$$F_2^H = \frac{m_1\omega^2 r}{\sqrt{2}}(\cos\phi+\sin\phi)+\frac{m_2\omega^2 r}{\sqrt{2}}\left(\sin\phi-\frac{e}{L}\cos\phi\right) \tag{4.5-16}$$

$$F_3^H = -\frac{m_1\omega^2 r}{\sqrt{2}}(\cos\phi+\sin\phi)-\frac{m_2\omega^2 r}{\sqrt{2}}\left(\sin\phi-\frac{e}{L}\cos\phi\right) \tag{4.5-17}$$

$$F_4^H = -\frac{m_1\omega^2 r}{\sqrt{2}}(\sin\phi-\cos\phi)+\frac{m_2\omega^2 r}{\sqrt{2}}\left(\cos\phi+\frac{e}{L}\sin\phi\right) \tag{4.5-18}$$

$$F_5^H = \frac{m_1\omega^2 r}{\sqrt{2}}(\sin\phi-\cos\phi)+\frac{m_2\omega^2 r}{\sqrt{2}}\left(\sin\phi-\frac{e}{L}\cos\phi\right) \tag{4.5-19}$$

$$F_6^H = \frac{m_1\omega^2 r}{\sqrt{2}}(\sin\phi+\cos\phi)+\frac{m_2\omega^2 r}{\sqrt{2}}\left(\cos\phi+\frac{e}{L}\sin\phi\right) \tag{4.5-20}$$

$$F_7^H = -\frac{m_1\omega^2 r}{\sqrt{2}}(\sin\phi+\cos\phi)-\frac{m_2\omega^2 r}{\sqrt{2}}\left(\cos\phi+\frac{e}{L}\sin\phi\right) \tag{4.5-21}$$

$$F_8^H = -\frac{m_1\omega^2 r}{\sqrt{2}}(\sin\phi-\cos\phi)-\frac{m_2\omega^2 r}{\sqrt{2}}\left(\sin\phi-\frac{e}{L}\cos\phi\right) \tag{4.5-22}$$

Similarly, the vertical components of the inertia forces are as follows:

For cylinders in bank A,

$$F_j^V = F_j^x \sin 45° + F_j^y \sin 45°, \ (j=1,2,3,4) \tag{4.5-23}$$

For cylinders in bank B,

$$F_j^V = -F_j^x \sin 45° + F_j^y \sin 45°, \ (j=5,6,7,8) \tag{4.5-24}$$

Carrying out the analysis, we have

$$F_1^V = \frac{m_1\omega^2 r}{\sqrt{2}}(\sin\phi+\cos\phi)+\frac{m_2\omega^2 r}{\sqrt{2}}\left(\cos\phi+\frac{e}{L}\sin\phi\right) \tag{4.5-25}$$

$$F_2^V = \frac{m_1\omega^2 r}{\sqrt{2}}(\cos\phi-\sin\phi)-\frac{m_2\omega^2 r}{\sqrt{2}}\left(\sin\phi-\frac{e}{L}\cos\phi\right) \tag{4.5-26}$$

$$F_3^V = -\frac{m_1\omega^2 r}{\sqrt{2}}(\cos\phi - \sin\phi) + \frac{m_2\omega^2 r}{\sqrt{2}}\left(\sin\phi - \frac{e}{L}\cos\phi\right) \tag{4.5-27}$$

$$F_4^V = -\frac{m_1\omega^2 r}{\sqrt{2}}(\sin\phi + \cos\phi) - \frac{m_2\omega^2 r}{\sqrt{2}}\left(\cos\phi + \frac{e}{L}\sin\phi\right) \tag{4.5-28}$$

$$F_5^V = \frac{m_1\omega^2 r}{\sqrt{2}}(\cos\phi + \sin\phi) + \frac{m_2\omega^2 r}{\sqrt{2}}\left(\sin\phi - \frac{e}{L}\cos\phi\right) \tag{4.5-29}$$

$$F_6^V = -\frac{m_1\omega^2 r}{\sqrt{2}}(\sin\phi - \cos\phi) + \frac{m_2\omega^2 r}{\sqrt{2}}\left(\cos\phi + \frac{e}{L}\sin\phi\right) \tag{4.5-30}$$

$$F_7^V = \frac{m_1\omega^2 r}{\sqrt{2}}(\sin\phi - \cos\phi) - \frac{m_2\omega^2 r}{\sqrt{2}}\left(\cos\phi + \frac{e}{L}\sin\phi\right) \tag{4.5-31}$$

$$F_8^V = -\frac{m_1\omega^2 r}{\sqrt{2}}(\cos\phi + \sin\phi) - \frac{m_2\omega^2 r}{\sqrt{2}}\left(\sin\phi - \frac{e}{L}\cos\phi\right) \tag{4.5-32}$$

By inspection of the various terms of Eqs. 4.5-15 to 4.5-32, we see that

$$\sum_{i=1}^{8} F_i^H = 0 \qquad \text{and} \qquad \sum_{i=1}^{8} F_i^V = 0 \tag{4.5-33}$$

Eq. 4.5-33 shows that with the crank arrangement of Fig.4.5-3, the horizontal and vertical components of the inertia forces are self-balanced. Observe, however, in Fig. 4.5-3, the crankshaft is not symmetric when viewed along axes perpendicular to the rotation axis. Consequently, the inertia forces create rotating moments about these axes.

To see this in greater detail, consider the moment of the inertia forces about the axis of cylinder C1. From Fig. 4.5-2, we see, after routine analysis, that the horizontal and vertical components of the inertia moments are

$$\begin{aligned} M_H = {} & F_2^V \cdot b + F_3^V \cdot 2b + F_4^V \cdot 3b + F_5^V \cdot a \\ & + F_6^V(a+b) + F_7^V(a+2b) + F_8^V(a+3b) \end{aligned} \tag{4.5-34}$$

$$\begin{aligned} M_V = {} & -[F_2^H \cdot b + F_3^H \cdot 2b + F_4^H \cdot 3b + F_5^H a \\ & + F_6^H(a+b) + F_7^H(a+2b) + F_8^H(a+3b)] \end{aligned} \tag{4.5-35}$$

By substituting for F_j^H and F_j^V $(j=2,\cdots,8)$ from Eqs. 4.5-16 to 4.5-22 and Eqs. 4.5-26 to 4.5-32 into Eqs. 4.5-35 and 4.5-34, respectively, we obtain

$$M_H = -\frac{(2m_1+m_2)\omega^2 rb}{\sqrt{2}}(2\sin\phi+4\cos\phi)-\frac{m_2\omega^2 reb}{\sqrt{2}L}(4\sin\phi-2\cos\phi) \quad (4.5\text{-}36)$$

and

$$M_V = \frac{(2m_1+m_2)\omega^2 rb}{\sqrt{2}}(4\sin\phi-2\cos\phi)-\frac{m_2\omega^2 reb}{\sqrt{2}L}(2\sin\phi+4\cos\phi) \quad (4.5\text{-}37)$$

Observe that the inertia moment components M_H and M_V of Eqs. 4.5-36 and 4.5-37 are independent of parameter a (see Fig. 4.5-2). This means that although the configuration of a 90° V8 engine may be different, the inertia moments are the same. For example, if cylinder C1 is in bank B instead of bank A, the expressions for inertia moments are unchanged.

Next, observe that, taken together, Eqs. 4.5-36 and 4.5-37 describe a rotating unbalance inertia torque with constant magnitude, as given by

$$M=\sqrt{(M_H)^2+(M_V)^2}=\sqrt{10[(2m_1+m_2)^2+(m_2e/L)^2]}\,\omega^2 rb, \quad (4.5\text{-}38)$$

where the torque rotation is in the direction of the crank rotation and at the same rate as the crank rotation.

To describe the inertia moments in greater detail, it is helpful to introduce a pair of perpendicular unit vector sets, $\mathbf{N}_H$, $\mathbf{N}_V$, $\mathbf{n}_h$, and $\mathbf{n}_v$, as shown in Fig. 4.5-6. Let $\mathbf{N}_H$ and $\mathbf{N}_V$ be fixed relative to the engine block and parallel to horizontal and vertical directions, respectively. Let the $\mathbf{n}_h$ and $\mathbf{n}_v$ be fixed relative to the crankshaft, with $\mathbf{n}_v$ parallel to crank 1, directed radially outward. Let $\mathbf{n}_h$ and $\mathbf{n}_v$ be, respectively, parallel to $\mathbf{N}_H$ and $\mathbf{N}_V$ when the crank angle ϕ is 45°. Table 4.5-2 then lists the corresponding direction cosines for the two unit vector sets.

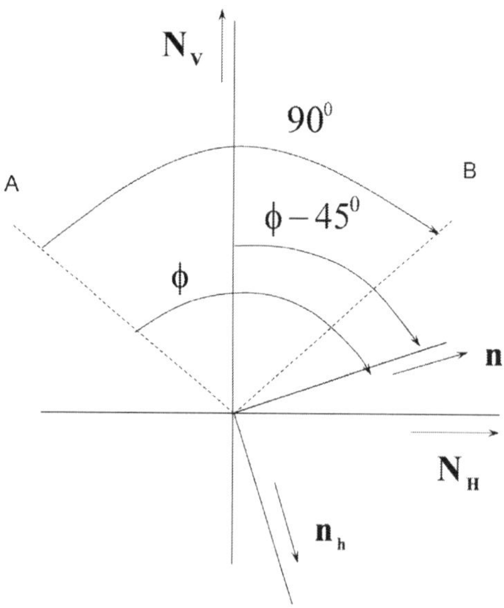

Fig. 4.5-6 Two unit vector sets.

Table 4.5-2 The Direct Cosines for the Two Unit Vector Sets of Fig. 4.5-6

	$\mathbf{N}_V$	$\mathbf{N}_H$
$\mathbf{n}_v$	$\cos(\phi-45°)$	$\sin(\phi-45°)$
$\mathbf{n}_h$	$-\sin(\phi-45°)$	$\cos(\phi-45°)$

Referring to Table 4.5-2, we can see that the components m_h^c and m_v^c of the inertia moments on the crankshaft are

$$m_h^c = -M_V \sin(\phi-45°) + M_H \cos(\phi-45°) \tag{4.5-39}$$

$$m_v^c = M_V \cos(\phi-45°) + M_H \sin(\phi-45°) \tag{4.5-40}$$

By substituting from Eqs. 4.5-36 and 4.5-37 for M_H and M_V into Eqs. 4.5-39 and 4.5-40, we obtain (see Fig. 4.5-7)

$$m_h^c = -\left[3(2m_1+m_2)+\frac{e}{L}m_2\right]\omega^2 rb \tag{4.5-41}$$

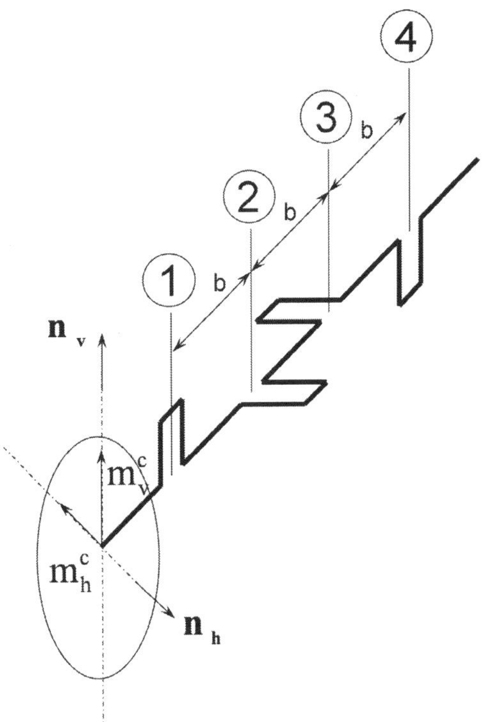

Fig. 4.5-7 Couples of inertia forces expressed in the crankshaft coordinate system.

$$m_v^c = \left[(2m_1 + m_2) - \frac{3e}{L} m_2 \right] \omega^2 rb \tag{4.5-42}$$

Also, the magnitude of the moment is found as

$$M = \sqrt{(m_v^c)^2 + (m_h^c)^2} = \sqrt{10[(2m_1 + m_2)^2 + (m_2 e/L)^2]}\,\omega^2 rb, \tag{4.5-43}$$

which, as expected, is the same as in Eq. 4.5-38.

It should be noted that Eqs. 4.5-41 and 4.5-42 are valid when the pin offset is toward the minor thrust side. When the pin offset is toward the major thrust side, it is necessary to replace e with $-e$ in Eqs. 4.5-41 and 4.5-42.

Table 4.5-3 lists the unbalance forces and moments of engines commonly used in industry.

4.6 Balancing of Crankshafts

The crankshaft is sometimes called *the backbone of the engine* [4.8]. Engine vibration problems usually involve crankshaft vibrations. When the vibration originates in the crankshaft, it is generally due to crankshaft imbalance.

To eliminate imbalance, a crankshaft must be balanced statically and dynamically. But without the piston and connecting rod assemblies, a nonsymmetric crankshaft is not in dynamic balance. Therefore, a practical approach is to balance the crankshaft by adding ring weights on the crankpins.

A question arising there is: How large should the ring weights be? In 1948, Timoshenko [4.4] provided a theoretical solution, where there is no piston pin offset. In practice, however, many engines have offset cylinders to minimize piston thrust against the cylinder walls [4.5, 4.6, 4.17, 4.18, 4.20, 4.22].

In this section, we develop expressions for determining the ring weights for balancing a 90° V8 engine with offset cylinders. *Offset cylinders* have the pins attaching the rods to the pistons (*wrist pins* or *gudgeon pins*) with axes displaced away from the piston axes.

From Eqs. 4.5-41 and 4.5-42, we know that when the engine runs with uniform angular speed ω, the rotating mass m_1, the reciprocating mass m_2, and the pin offset e will produce two inertia couples. Since the inertia couples are independent of the position of the crank, therefore, it is possible to contain the inertia couples by using suitable ring weights fastened on each of the crankpins. Specifically, let two identical ring weights with mass $W_{1,4}$ be fastened on crankpins 1 and 4, and let another two identical ring weights with mass $W_{2,3}$ be fastened on crankpins 2 and 3, as in Fig. 4.6-1.

The objective of the ring weight attachment is as follows: When the crankshaft is spun with a uniform angular speed ω, the ring weights produce the same

Table 4.5-3 Unbalance Forces and Moments of Various Engines*

Engine type	Crankpin arrangement	1st order		2nd order	
		Force	Moment	Force	Moment
Inline 2	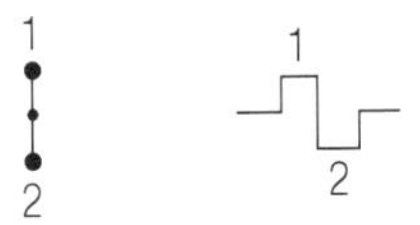	no	yes	yes	no
90° V2		yes	yes	yes	yes
Inline 3	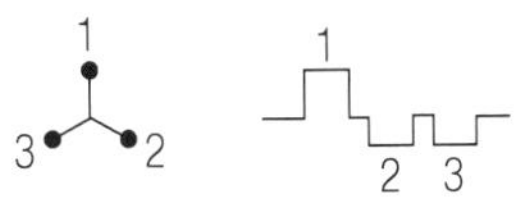	no	yes	no	yes
Inline 4	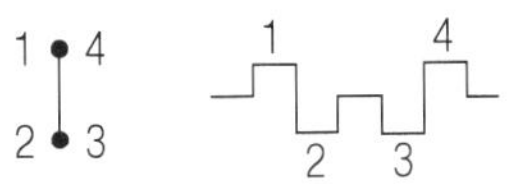	no	no	yes	no
20° V4	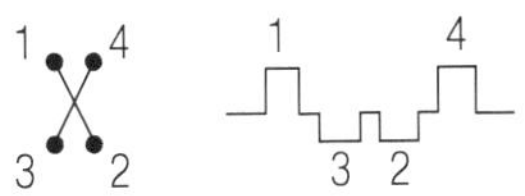	no	yes	yes	yes
30° V4	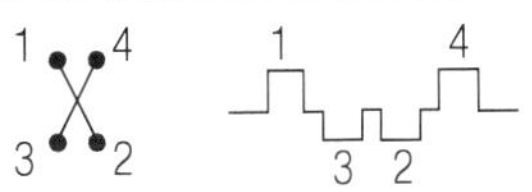	no	yes	yes	yes
60° V4	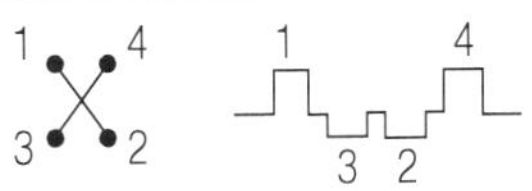	no	yes	yes	yes
90° V4	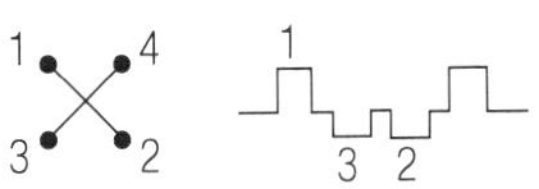	no	yes	yes	yes
Opposed flat 4	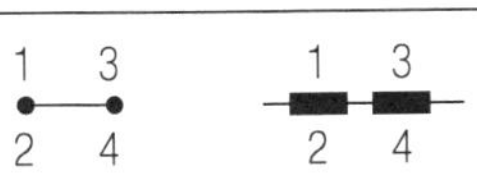	no	yes	no	no
Opposed flat 4	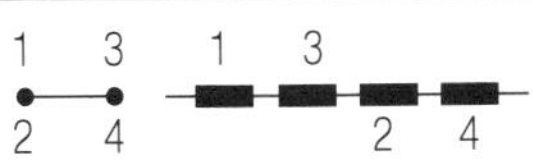	no	no	no	yes
Opposed flat 4	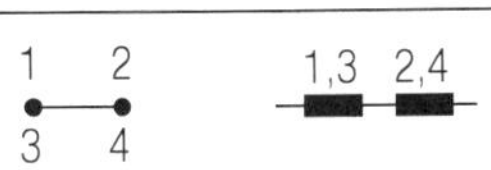	no	yes	no	yes
Opposed flat 4	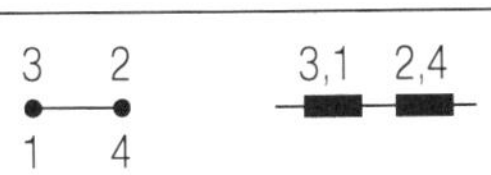	no	yes	no	no

Continued

Table 4.5-3 *Continued*

Engine type	Crankpin arrangement	1st order		2nd order	
		Force	Moment	Force	Moment
Inline 5	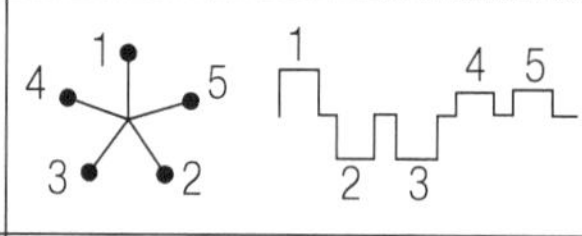	no	yes	no	yes
Inline 5	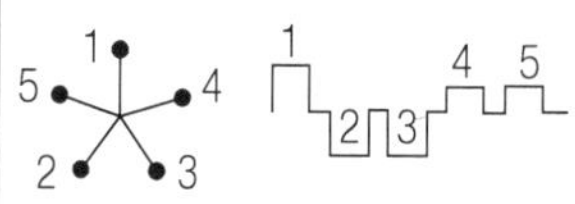	no	yes	no	yes
Inline 6	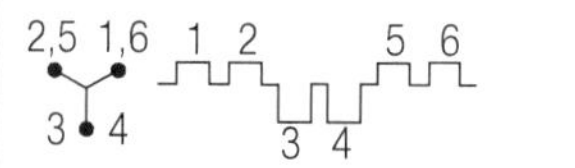	no	no	no	no
30° V6	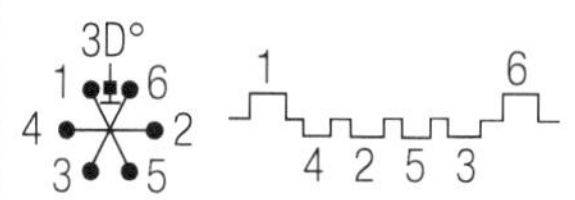	no	yes	no	yes
60° V6	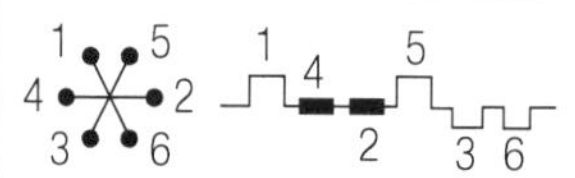	no	yes	no	yes
60° V6	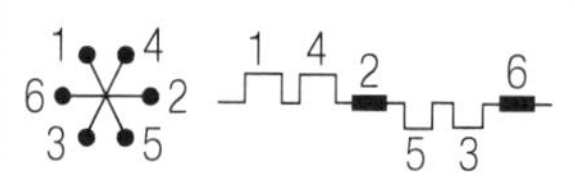	no	yes	no	yes
60° V6	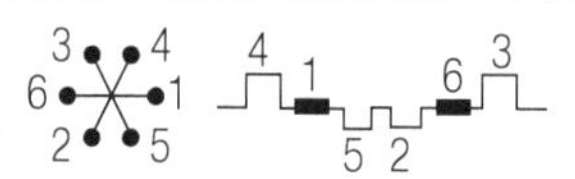	no	yes	no	yes
Opposed flat 6	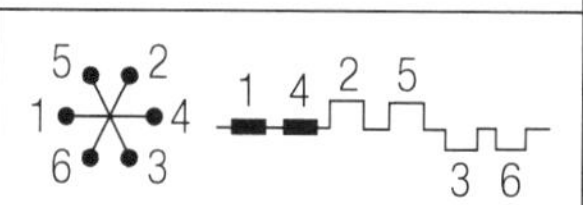	no	no	no	no

**no* means there is no unbalance force or moment, and *yes* means there is an unbalance force or moment.

inertia torques as created by the rotating mass, the reciprocating mass, and the pin offset. This leads to the equations

$$-W_{1,4}\omega^2 r(3b) = m_h^c \tag{4.6-1}$$

$$W_{2,3}\omega^2 rb = m_v^c \tag{4.6-2}$$

Substituting Eqs. 4.5-41 and 4.5-42 for m_h^c and m_v^c into Eqs. 4.6-1 and 4.6-2, we obtain the masses of the ring weights as

$$W_{1,4} = 2\left(m_1 + \frac{m_2}{2}\right) + \frac{e}{3L}m_2 \tag{4.6-3}$$

$$W_{2,3} = 2\left(m_1 + \frac{m_2}{2}\right) - \frac{3e}{L} m_2 \tag{4.6-4}$$

In the particular case where there is no offset (e=0), we have

$$W_{1,4} = W_{2,3} = 2\left(m_1 + \frac{m_2}{2}\right) \tag{4.6-5}$$

Eq. 4.6-5 indicates that if there is no pin offset (e=0), the ring weights for each of the crankpins are the same [4.4]. It should be mentioned that if there is pin offset ($e > 0$), the ring weights must be modified accordingly, as in Eqs. 4.6-3 and 4.6-4. Observe that the modification terms for the $W_{1,4}$ and the $W_{2,3}$ are different.

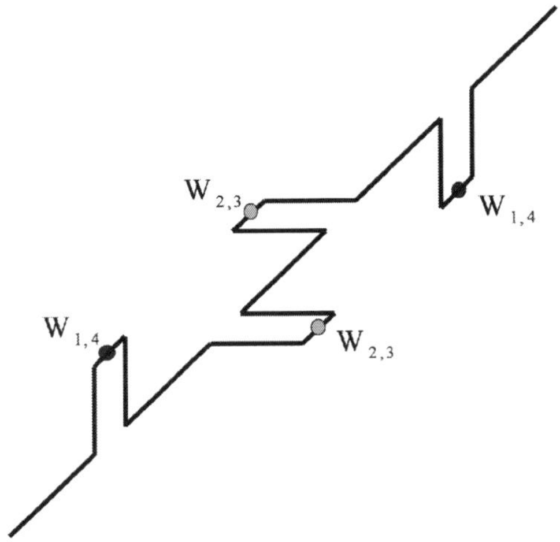

Fig. 4.6-1 Ring weights attached to crankpins.

The ring weights for the case with the piston pin offset toward the major thrust side can be determined by replacing e with $-e$ in Eqs. 4.6-3 and 4.6-4. Specifically, in this case, the $W_{1,4}$ and the $W_{2,3}$ are given by

$$W_{1,4} = 2\left(m_1 + \frac{m_2}{2}\right) - \frac{e}{3L} m_2 \tag{4.6-6}$$

$$W_{2,3} = 2\left(m_1 + \frac{m_2}{2}\right) + \frac{3e}{L} m_2 \tag{4.6-7}$$

At this point, it may be helpful to conclude with a few remarks.

First, to accurately determine the ring weights, the rotating and reciprocating masses must be accurately determined. For the connecting rod, its rotating (big-end) and reciprocating (small-end) masses can be determined experimentally using a scale and a special fixture, as shown in Figs. 4.6-2 and 4.6-3. That is, each end of the rod is weighted in turn, with the other end suspended horizontally. The result may be checked by adding together the rotating and reciprocating masses. The results should be equal to the total mass of the connecting rod. The weights of the piston, piston pin, piston rings, and so forth, can be directly measured on a scale.

Second, the design and manufacture of the ring weights must satisfy the following requirements [4.10, 4.22]. (1) Their weights must be precise.

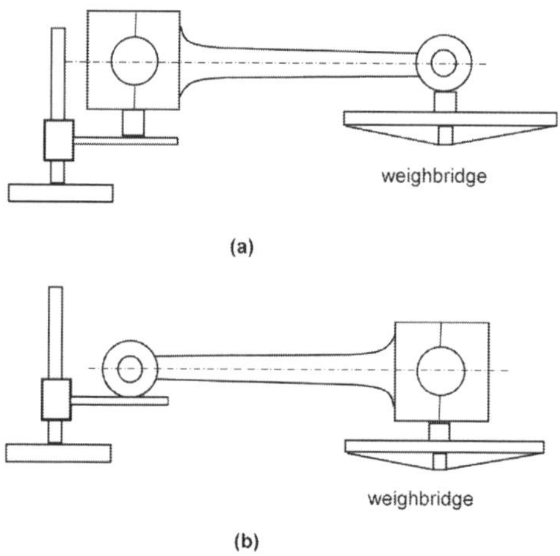

Fig. 4.6-2 Experimental setup for weighing connecting rod mass: (a) for small-end mass; (b) for big-end mass.

Otherwise, all advantage derived from balancing will be of no avail. (2) The rings must be hardened enough to withstand wear and possible deformation if accidentally dropped or impacted by some hard object. (3) Each ring weight itself must be balanced statically and dynamically. The photographs in Figs. 4.6-4 and 4.6-5 show how to install the ring weights on the test crankshaft.

Finally, the two-plane influence coefficient balancing method [4.2] can be used for balancing the crankshafts. A detailed description of the method has been discussed in Section 4.3. In this method, a trial mass is placed on one end plane of the crankshaft, and the vibration responses at two ends of the crankshaft are recorded. From these measurements, two influence coefficients are computed. Next, the trial mass

Fig. 4.6-3 Experimental setup for weighing big-end mass of connecting rod. (Courtesy of Raven Engineering, Inc.)

Fig. 4.6-4 Installation of ring weight on the crankshaft (step one). (Courtesy of Raven Engineering, Inc.)

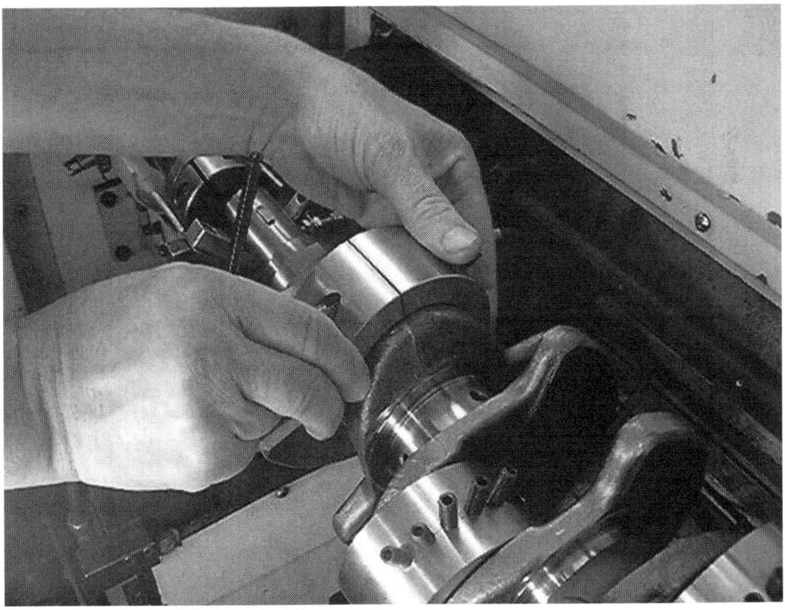

Fig. 4.6-5 Installation of ring weight on the crankshaft (step two). (Courtesy of Raven Engineering, Inc.)

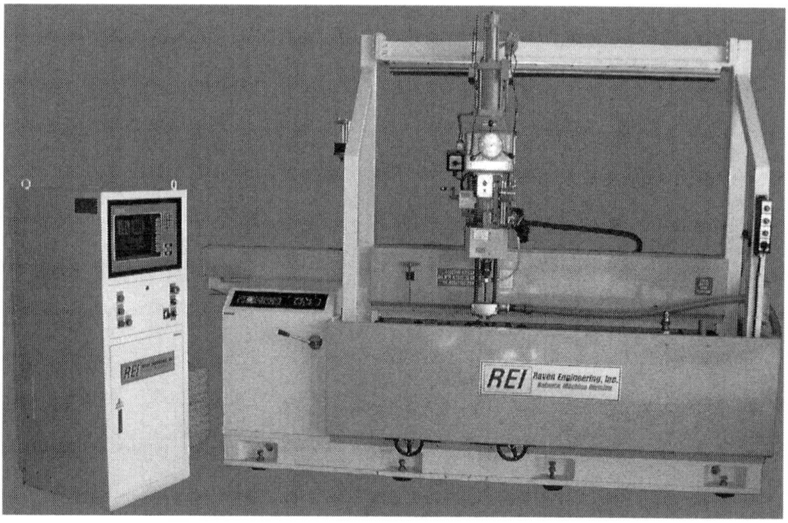

Fig. 4.6-6 Crankshaft balance machine. (Courtesy of Raven Engineering, Inc.)

is removed and placed at the other end plane. The procedure is repeated to generate two more influence coefficients. This then forms a 2×2 matrix of complex influence coefficients, which must be inverted for the imbalance. This method can determine the imbalance magnitudes and relative phase angles on the two planes. To correct for dynamic imbalance, it is necessary to introduce on each plane a radial inertia force that is equal in magnitude but opposite in direction to the one manifested by the balancing machine. This is done, in practice, by removing an equal amount of metal from the two end planes. Fig. 4.6-6 shows a crankshaft balance machine.

Problems

4.1 Consider the idealized system consisting of a disk D with mass m and mass center G mounted on a light shaft supported by two bearings, as represented in Fig. P4.1. Let the mass of D be 10 kg, and let the mass center G of D be displaced

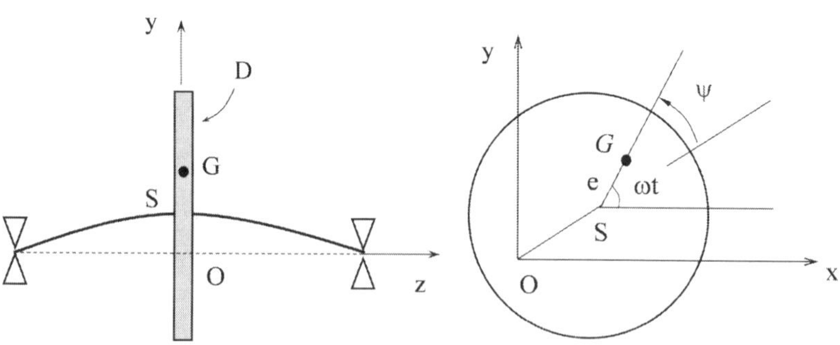

Fig. P4.1

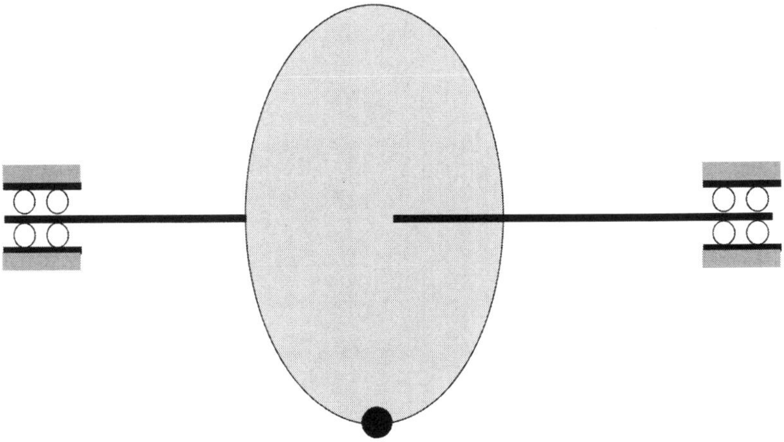

Fig. P4.2

a distance e of 2 mm away from geometric center. Let the shaft stiffness k be 5×10^5 N/m and let the damping ratio ς be 0.1. Determine the amplitudes and phase angles of synchronous whirl of the disk for the following three rotation speeds of the shaft: (1) 1500 rpm, (2) 2135 rpm, (3) 3200 rpm. Explain the results.

Ans. (1) OS=1.9 mm, $\phi=15.5°$; (2) OS=10 mm, $\phi=89.9°$; (3) OS=3.5 mm, $\phi=166.5°$

4.2 For a thin disk shown in Fig. P4.2, the baseline initial vibration is $\mathbf{Z}=-1.251+j1.28$. When a trial weight producing imbalance $\mathbf{U}_t=9e^{j0}$ is placed on the disk, the new response is: $\mathbf{Z}_t=0.456+j1.212$. Find the imbalance of the disk.

Ans. $\mathbf{U}=-6.2879+j6.2567=8.87e^{j135.14°}$

4.3 For the two-plane unbalance system shown in Fig. P4.3, the baseline responses at planes 1 and 2 are given by $\mathbf{Z}_1=5.6e^{j65°}$, $\mathbf{Z}_2=7.5e^{j185°}$. After adding the first trial unbalance $\mathbf{U}_{t1}=10e^{j270°}$ in plane 1, the new responses are

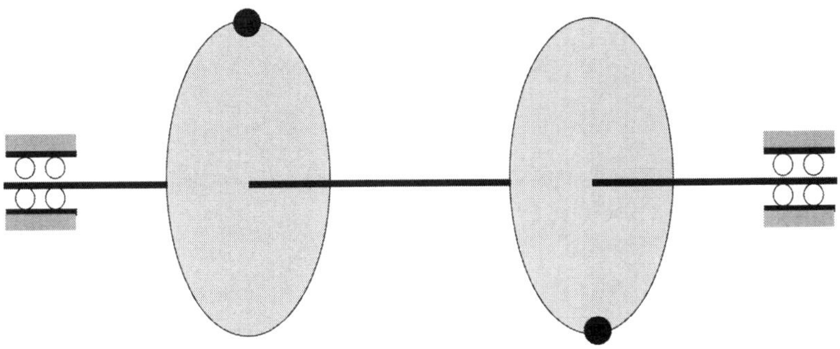

Fig. P4.3

$\mathbf{Z}_{11} = 6.9e^{j123°}$; $\mathbf{Z}_{21} = 5.5e^{j228°}$.

Then, the first trial weight is removed and a second trial weight $\mathbf{U}_{t2} = 12e^{j180°}$ is placed in the second plane. The resulting vibrations are $\mathbf{Z}_{12} = 5.3e^{j36°}$; $\mathbf{Z}_{22} = 10.5e^{j162°}$. Find the imbalance of the system.

Ans. $\mathbf{U}_1 = -11.4471 - j2.6669$; $\mathbf{U}_2 = -16.3862 - j3.4422$

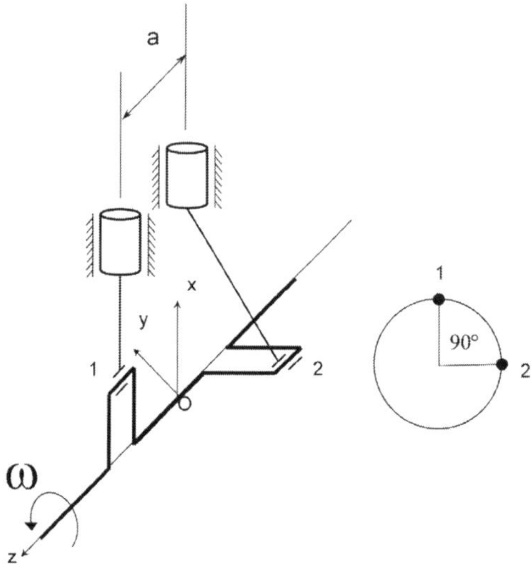

Fig. P4.4

4.4 A two-cylinder engine with cranks 90° apart and spacing a is shown in Fig. P4.4. Assuming a uniform angular velocity ω and taking coordinate axes as shown, the forces F_x and F_y for cylinder 1 can be obtained by using Eqs. 4.4-16 and 4.4-17. The corresponding forces for cylinder 2 can be obtained by substituting in the same equations $\phi - 90$ for ϕ. Calculate the moments of the forces with respect to the x- and y-axes. Assume pin offset e=0.

Ans. $M_x = -\left(1/\sqrt{2}\right)\omega^2 a\left(ms + m_1 r\right)\sin(\phi + 45°)$

$M_y = \left(1/\sqrt{2}\right)\omega^2 a\left(ms + m_1 r + m_2 r\right)\cos(\phi + 45°)$

4.5 Repeat Problem 4.4 for a three-cylinder engine with the crank arrangement shown in Fig. P4.5. If coordinate axes are selected as shown, the forces for crank 1 can be obtained by using Eqs. 4.4-16 and 4.4-17. The corresponding forces for cranks 2 and 3 can be obtained by substituting $\phi + 135°$ and $\phi + 270°$ in the same equation for ϕ, respectively. Assume the pin offset $e = 0$.

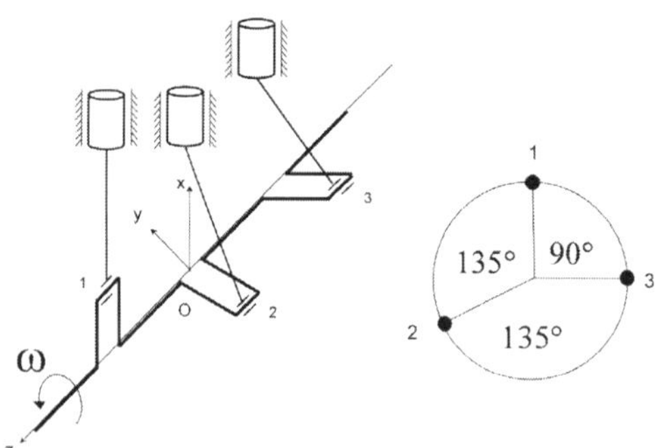

Fig. P4.5

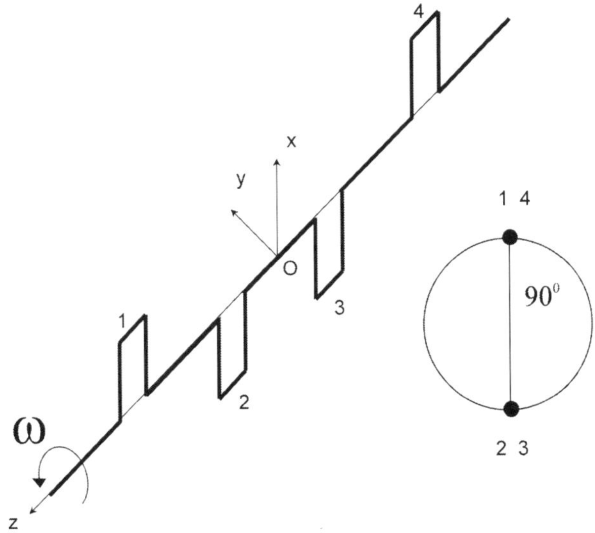

Fig. P4.6

Ans. $M_x = -\omega^2 a(ms + m_1 r)[2\sin\phi + \sin(\phi + 135°)]$

$M_y = \omega^2 a(ms + m_1 r + m_2 r)[2\cos\phi + \cos(\phi + 135°)]$

4.6 Fig. P4.6 shows a possible crank arrangement of an inline four-cylinder engine with uniform cylinder spacing a. Assume the angular velocity ω of the cranks is constant. Using Eqs. 4.4-16 and 4.4-17, and substituting ϕ, $\phi - \pi$, $\phi - \pi$, and φ for cranks 1, 2, 3, and 4, respectively, show that the first-order resultant inertia forces and the moments about the x- and y-axes are completely balanced, that is, F_x=F_y=0, and M_x=M_y=0.

4.7 Repeat Problem 4.6. Show that resultant inertial forces and moments of the second order are

$$F_x^{(2)} = 0,\ F_y^{(2)} = 4m_2\omega^2 r(r/L)\cos 2\phi,\ M_x = M_y = 0$$

4.8 Fig. P4.8 shows a possible crank arrangement of an inline five-cylinder engine with uniform cylinder spacing a. Assume the angular speed ω of the crank is constant and that the pin offset e is 0. Find the resultant forces and moments of the first-order inertia forces.

Ans. $F_x = F_y = 0;$

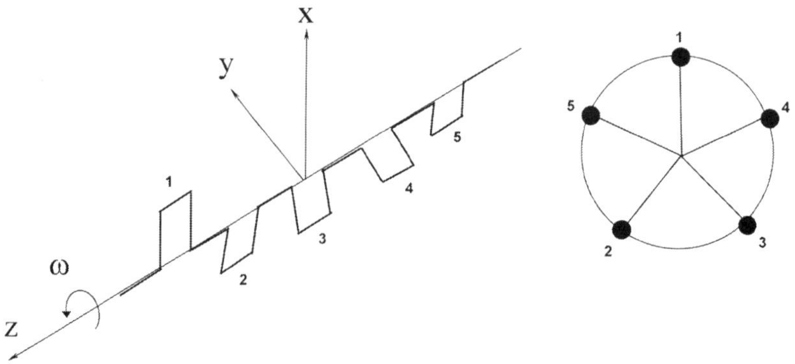

Fig. P4.8

$$M_x = -0.449(ms + m_1 r)\omega^2 a \sin(\phi - 54°)$$

$$M_y = 0.449(ms + m_1 r + m_2 r)\omega^2 a \cos(\phi - 54°)$$

4.9 Repeat Problem 4.8 for the crank arrangement of an inline five-cylinder engine as shown in Fig. P4.9. Show that

$$F_x = F_y = 0$$

$$M_x = -0.499(ms + m_1 r)\omega^2 a \sin(\phi + 54°)$$

$$M_y = 0.499(ms + m_1 r + m_2 r)\omega^2 a \cos(\phi + 54°)$$

4.10 Fig. P4.10 shows the arrangement of a crankshaft for a 90° V8 engine, which has the following parameters: reciprocating mass m_2=685.01 g, rotating mass m_1=478.07 g, piston pin offset e=1.06 mm toward the major thrust side,

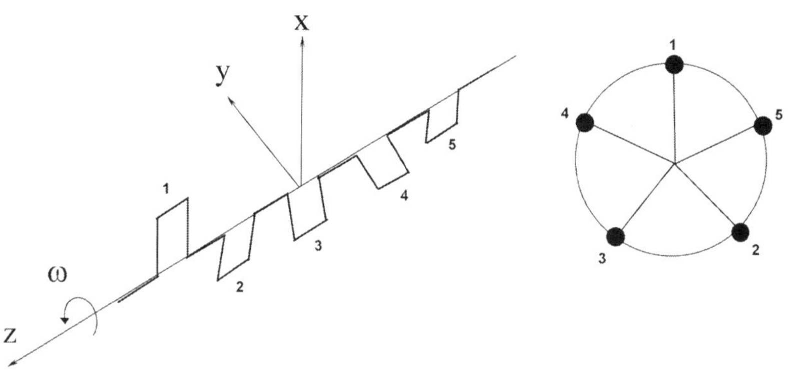

Fig. P4.9

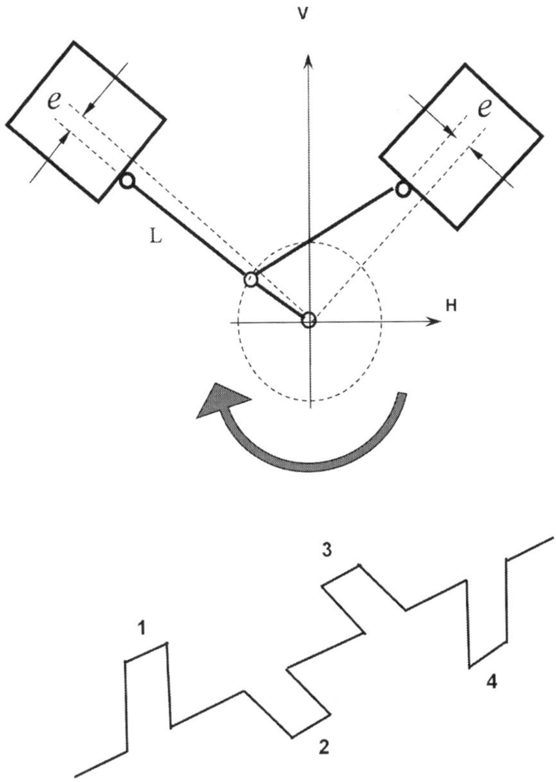

Fig. P4.10

and length of connecting rod L=146.01 mm. Determine the ring weights for balancing the crankshaft.

Ans. $W_{1,4}$=1639.49 g, $W_{2,3}$=1656.07 g

4.11 Repeat Problem 4.10 for the engine pin offset toward the minor thrust side, as shown in Fig. P4.11.

Ans. $W_{1,4}$=1642.81 g, $W_{2,3}$=1626.23 g

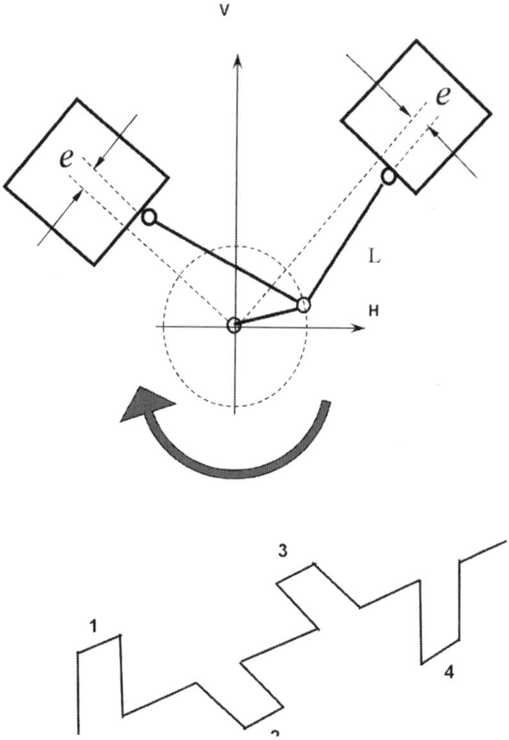

Fig. P4.11

4.12 The joint of an axle flange 1 with a driveshaft end *2* can be modeled as two disks, as shown in Fig. P4.12 (a). The total imbalance of the two disks measured on a balance machine is **A**. After the first test is completed, the two disks are disconnected. The driveshaft is then rotated 180° and connected again as shown in Fig. P4.12 (b). The new total imbalance, measured on the balance machine, is **B**. Find the imbalances of disks 1 and 2.

Ans. $\mathbf{U_1 = \frac{A+B}{2}, U_2 = \frac{A-B}{2}}$

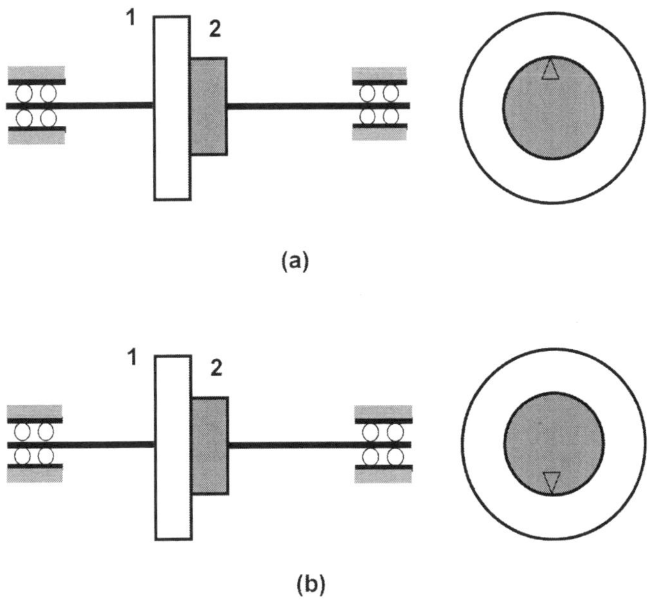

Fig. P4.12

References

4.1. Thomson, W. T., and Dahleh, M. D., *Theory of Vibration with Applications*, 5th ed., Prentice Hall, Upper Saddle River, NJ, 1998.

4.2. Gunter, E. J., and Jackson, C., "Balanging of Rigid and Flexible Rotors," in *Handbook of Rotodynamics*, F. F. Ehrich, ed., Krieger Publishing Company, Malabar, FL, 1999, Ch. 3, pp. 3.1–3.117.

4.3. Tse, F. S., Morse, I. E., and Hinkle, R. T., *Mechanical Vibration, Theory and Applications*, Allyn and Bacon, Inc., Boston, 1979.

4.4. Timoshenko, S. P., *Advanced Dynamics*, McGraw-Hill, New York, 1948.

4.5. Judge, A. W., *Automobile and Aircraft Engines*, Sir Isaac Pitman & Sons, Ltd., London, 1936.

4.6. Angle, G. D., *Engine Dynamics and Crankshaft Design*, Airplane Engine Encyclopedia Co., General Motors Building, Detroit, MI, 1925.

4.7. Holowenko, A. R., *Dynamics of Machinery*, John Wiley & Sons, Inc., New York, 1955.

4.8. Kuns, R. F., and Plumridge, T. C., *Automobile Engines*, American Technical Society, Chicago, 1946.

4.9. Seaholm, E. W., "Balance of the Cadillac V-63 Engine," *SAE Journal*, January, 1924, pp. 70–73.

4.10. Anderson, D. E., "Practical Balancing of a V-type Engine Crankshaft," *SAE Transactions*, Part I, 1924, pp. 277–294.

4.11. Robert, L. L., "Practical Methods of Engine Balancing," *SAE Transactions*, Part I, 1924, pp. 263–276.

4.12. Platt, M., "Engine Balancing Calculations," reprint, *The Automobile Engineer*, August and September, 1924.

4.13. Smith, J., *Crankshaft Design and Manufacture*, Whitehall Technical Press, Ltd., London, 1946.

4.14. Wilson, W. K., *The Balancing of Oil Engines*, J. B. Lippincott Company, Philadelphia, 1929.

4.15. Thomson, W., *Fundamental of Automotive Engine Balance*, Mechanical Engineering Publications, Ltd., London, 1978.

4.16. Heisler, H., *Advanced Engine Technology*, SAE International, Warrendale, PA, 1995.

4.17. Burrell, G., and Butler, G., "A Study in Applied Physics: Locating the Piston Pin to Minimize Piston Slap," *General Motors Engineering Journal*, Index to Vol. 1, General Motors Corporation, June–July 1953 to November–December 1954, pp. 38–43.

4.18. Swan, J. B., "Offset Cylinder and Pistons," *The Automobile Engineer*, March, 1922, pp. 72–76.

4.19. Weertman, W. L., and Lecher, R. J., "Chrysler Corporation's New Hemi Head High Performance Engines," in *American Iron: A History of the OHV V8, 1961–1970*, STS Press, Service Technical Society, Warrendale, PA, pp. 87–112.

4.20. Schenek Trebel Corp., *Fundamentals of Balancing*, Deer Park, Long Island, NY, 1980.

4.21. Adier, U., ed., Automotive Handbook, Robert Bosh Corporation, Gerlingen, Germany, 1986.

4.22. Liu, C. Q., and Orzechowski, J., "Theoretical and Practical Aspects of Balancing a V8 Engine Crankshaft," SAE paper number 2005-01-2454, SAE International, Warrendale, PA, 2005.

Chapter Five

Multi-Degree-of-Freedom Systems

5.1 Introduction

When a system requires two or more variables to define its motion, the system is said to be a *multi-degree-of-freedom* system. Since two-degree-of-freedom (DOF) systems are the simplest of these, it is useful to begin our study with two-DOF systems. Also, since two-DOF systems model the behavior of larger multi-degree-of-freedom systems, there is no great loss of generality by focusing upon two-DOF systems.

5.2 Two-Degree-of-Freedom Systems

(1) Equations of Motion

Consider the system as shown in Fig. 5.2-1. If x_1 and x_2 are taken as independent coordinates, and the static equilibrium positions are taken as origins, the equations of motion of the system are readily seen to be

$$\begin{aligned} 2m\ddot{x}_1 &= -4kx_1 + 2k(x_2 - x_1) \\ m\ddot{x}_2 &= -2k(x_2 - x_1) - 2kx_2 \end{aligned} \tag{5.2-1}$$

or

$$\begin{aligned} 2m\ddot{x}_1 + 6kx_1 - 2kx_2 &= 0 \\ m\ddot{x}_2 - 2kx_1 + 4kx_2 &= 0, \end{aligned} \tag{5.2-2}$$

where notationally k and m are spring stiffness and mass values, respectively, as represented by Fig. 5.2-1.

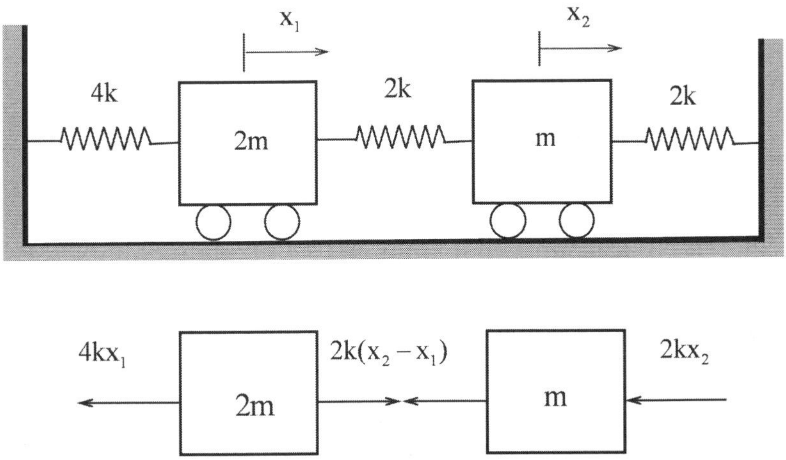

Fig. 5.2-1 A two-degree-of-freedom system.

(2) Characteristic Equation and Natural Frequency

To solve Eq. 5.2-2, we seek harmonic solutions of the forms

$$x_1 = X_1 e^{j\omega t}, \quad x_2 = X_2 e^{j\omega t} \tag{5.2-3}$$

Recall from Euler's formula that $e^{j\omega t}$ may be written as $\cos\omega t + j\sin\omega t$, with again j being the imaginary $\sqrt{-1}$. Therefore, the solution forms of Eqs. 5.2-3 describe vibratory motion. For a multi-degree-of-freedom system, a vibratory motion, as described by Eqs. 5.2-3, satisfying the governing dynamical equations, is called *normal mode vibration* [5.1]. With normal mode vibration, each point has the same frequency harmonic motion, with each simultaneously passing through the static equilibrium position.

By substituting from Eqs. 5.2-3 into Eqs. 5.2-2, we have

$$\begin{aligned} (6k - 2m\omega^2)X_1 - 2kX_2 &= 0 \\ -2kX_1 + (4k - m\omega^2)X_2 &= 0 \end{aligned} \tag{5.2-4}$$

Eqs. 5.2-4 form a pair of simultaneous, homogeneous, linear, algebraic equations for the amplitudes X_1 and X_2. Nonzero values of X_1 and X_2 then exist only if the determinant of the coefficients is zero. That is,

$$\Delta = \begin{vmatrix} 6k - 2m\omega^2 & -2k \\ -2k & 4k - m\omega^2 \end{vmatrix} = 0 \tag{5.2-5}$$

By letting λ be ω^2, Eq. 5.2-5 becomes

$$\lambda^2 - 7\left(\frac{k}{m}\right)\lambda + 10\left(\frac{k}{m}\right)^2 = 0 \tag{5.2-6}$$

Eq. 5.2-6 is called the *characteristic equation* of the vibrating system. Its roots (λ_1, λ_2) correspond to the *natural frequencies* of the system. They are

$$\lambda_1 = \omega_1^2 = 2\frac{k}{m}, \quad \lambda_2 = \omega_2^2 = 5\frac{k}{m} \tag{5.2-7}$$

The natural frequencies ω_1 and ω_2 are then

$$\omega_1 = \sqrt{2\frac{k}{m}}, \quad \omega_2 = \sqrt{5\frac{k}{m}} \tag{5.2-8}$$

Observe that ω_1 and ω_2 are circular natural frequencies, related to natural frequencies f_1 and f_2 as $\omega_1 = 2\pi f_1$ and $\omega_2 = 2\pi f_2$.

(3) Mode Shapes

Since the determinant of the coefficients of Eqs. 5.2-4 is zero, we cannot determine specific values of the amplitudes X_1 and X_2. Nevertheless, by substituting frequency values into either one[1] of the equations, we can find the ratios of the amplitudes. Ratios found, using the two frequencies, determine the *normal mode shapes*, or *real mode shapes*, or *normal modal vectors*. (For an undamped system, the mode shapes are always normal modes.)

The physical interpretation of a normal mode is that, if the system is excited in the shape of a normal mode, each point of the system undergoes harmonic motion and each point simultaneously passes through its static equilibrium position.

To illustrate these concepts, if we substitute $\omega_1^2 = 2\frac{k}{m}$ into the first of Eqs. 5.2-4, we have

$$\frac{X_1}{X_2} = 1 \tag{5.2-9a}$$

With the individual magnitudes of X_1 and X_2 being unspecified (in the absence of initial values), we can express X_1 and X_2 as elements of the column vector

$$\mathbf{u}_1 = \begin{bmatrix} X_1 \\ X_2 \end{bmatrix} = \begin{bmatrix} 1 \\ 1 \end{bmatrix}, \tag{5.2-9b}$$

where $\mathbf{u}_1$ is called the first *modal vector* or *mode shape* (corresponding to the first frequency).

Similarly, using $\omega_2^2 = 5\frac{k}{m}$, we obtain the second modal vector

$$\mathbf{u}_2 = \begin{bmatrix} X_1 \\ X_2 \end{bmatrix} = \begin{bmatrix} 1 \\ -2 \end{bmatrix} \tag{5.2-10}$$

Fig. 5.2-2 provides a graphical representation of the two modes. The physical significance of these results is as follows: In the first mode, the two masses vibrate in the same direction with frequency ω_1, through the same distance. The coupling spring is neither stretched nor compressed in this motion. In the

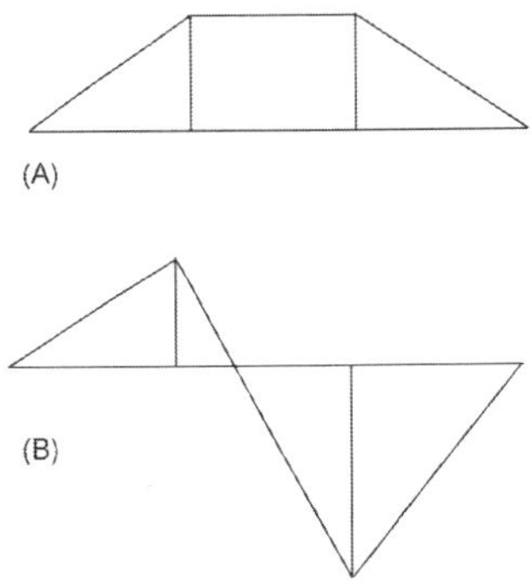

Fig. 5.2-2 Mode shapes of system in Fig. 5.2-1. (A) $\omega_1 = \sqrt{2\frac{k}{m}}$; (B) $\omega_2 = \sqrt{5\frac{k}{m}}$.

[1] Note that in view of Eq. 5.2-5, with the determinant of the coefficients being zero, Eqs. 5.2-4 are "dependent." They are, in a sense, the same equation.

second mode, the two masses vibrate with frequency ω_2, but in opposition to each other, and the ratio of the amplitudes of vibration is 1:2.

5.3 Eigenvalues and Eigenvectors

In this section, we use a matrix method to illustrate another way of determining the frequencies and mode shapes.

As an example, consider again the system as shown in Fig. 5.2-1. The equation of motion of the system can be written as

$$M\ddot{\mathbf{x}} + K\mathbf{x} = 0, \tag{5.3-1}$$

where M, K, and $\mathbf{x}$ are

$$M = m\begin{bmatrix} 2 & 0 \\ 0 & 1 \end{bmatrix} = \text{mass matrix} \tag{5.3-2}$$

$$K = k\begin{bmatrix} 6 & -2 \\ -2 & 4 \end{bmatrix} = \text{stiffness matrix} \tag{5.3-3}$$

$$\mathbf{x} = \begin{bmatrix} x_1 \\ x_2 \end{bmatrix} = \text{displacement vector} \tag{5.3-4}$$

For an oscillatory harmonic motion, $\ddot{\mathbf{x}}$ will be negatively proportional to $\mathbf{x}$ as $-\omega^2\mathbf{x}$, or $-\lambda\mathbf{x}$. Then, by substituting into Eq. 5.3-1, we have

$$K\mathbf{x} = \lambda M\mathbf{x} \tag{5.3-5}$$

Finding the solution of Eq. 5.3-5 is an eigenvalue problem. To illustrate the procedure for obtaining the solution, we rewrite Eq. 5.3-5 as

$$[K - \lambda M]\mathbf{x} = 0 \tag{5.3-6}$$

A nonzero (nontrivial) solution for $\mathbf{x}$ will occur only if the determinant of the multiplying matrix is zero. That is,

$$\Delta = |K - \lambda M| = 0 \tag{5.3-7}$$

By substituting the mass matrix of Eq. 5.3-2 and the stiffness matrix of Eq. 5.3-3 into Eq. 5.3-7, we obtain the *characteristic equation*

$$\lambda^2 - 7\left(\frac{k}{m}\right)\lambda + 10\left(\frac{k}{m}\right)^2 = 0 \tag{5.3-8}$$

Solving Eq. 5.3-8, we obtain the *eigenvalues* (leading to the natural frequencies of the system) as

$$\lambda_1 = \omega_1^2 = 2\frac{k}{m}, \quad \lambda_2 = \omega_2^2 = 5\frac{k}{m} \tag{5.3-9}$$

Substituting the eigenvalues into Eq. 5.3-6, we obtain the *eigenvectors* (normal mode shapes).

The eigenvectors can also be found by using the adjoint matrix of the system matrix as follows: first, the coefficient matrix of Eq. 5.3-6, called the *system impedance matrix*, is expressed as

$$Z=[K-\lambda M] \tag{5.3-10}$$

For the system shown in Fig. 5.2-1, we have Z as

$$Z=\begin{bmatrix} 6k-2m\omega^2 & -2k \\ -2k & 4k-m\omega^2 \end{bmatrix} \tag{5.3-11}$$

Next, the cofactor of Z is found as follows: the element C_{ij} of the cofactor is simply the determinant of the matrix obtained by deleting the i-th row and the j-th column of Z, and then multiplying the result by $(-1)^{i+j}$. Thus, the matrix of cofactors of the matrix of Eq. 5.3-11 is found as

$$C=\begin{bmatrix} 4k-m\omega^2 & 2k \\ 2k & 6k-2m\omega^2 \end{bmatrix} \tag{5.3-12}$$

The transpose C^T of the matrix of cofactors is called the *adjoint* matrix. (In this example, $C^T=C$). Thus, the adjoint of Z is

$$adj(Z)=C^T=\begin{bmatrix} 4k-m\omega^2 & 2k \\ 2k & 6k-2m\omega^2 \end{bmatrix} \tag{5.3-13}$$

An important property of the adjoint matrix is stated in the following theorem [5.2, 5.3]:

Theorem: Any column of the adjoint matrix is an eigenvector of the system matrix.

This theorem can be proved as follows: the inverse Z^{-1} of Z can be calculated from its adjoint matrix as

$$Z^{-1}=\frac{adj(Z)}{|Z|} \tag{5.3-14}$$

Multiplying Eq. 5.3-14 by $|Z|Z$, we obtain

$$|Z|I=Z\,adj(Z) \tag{5.3-15}$$

That is,

$$|K-\lambda M|I=[K-\lambda M]adj(Z) \tag{5.3-16}$$

When λ is equal to the eigenvalue, the left side of Eq. 5.3-16 is zero. Hence, its right side is also zero. That is, $[K-\lambda M]adj(Z)=0$. Comparing this with $[K-\lambda M]\mathbf{x}=0$, we see that the adjoint matrix consists of columns, each of which is the eigenvector (multiplied by an arbitrary constant).

Substituting $\lambda_1 = \omega_1^2 = 2\dfrac{k}{m}$ into the first column of Eq. 5.3-13, we obtain the first mode shape

$$\begin{bmatrix} 4k - m\omega^2 \\ 2k \end{bmatrix} = \begin{bmatrix} 2k \\ 2k \end{bmatrix} \Rightarrow \mathbf{u}_1 = \begin{bmatrix} 1 \\ 1 \end{bmatrix} \qquad (5.3\text{-}17)$$

Similarly, using $\lambda_2 = \omega_2^2 = 5\dfrac{k}{m}$, we obtain the second mode shape

$$\begin{bmatrix} 4k - m\omega^2 \\ 2k \end{bmatrix} = \begin{bmatrix} -k \\ 2k \end{bmatrix} \Rightarrow \mathbf{u}_2 = \begin{bmatrix} 1 \\ -2 \end{bmatrix} \qquad (5.3\text{-}18)$$

The *modal matrix* is then given by

$$S = \begin{bmatrix} \mathbf{u}_1 & \mathbf{u}_2 \end{bmatrix} = \begin{bmatrix} 1 & 1 \\ 1 & -2 \end{bmatrix} \qquad (5.3\text{-}19)$$

Thus, we obtain the same results as in Section 5.2.

For large systems, the computation of eigenvalue problems can be performed easily with most programmable calculators as well as in the mathematics software packages. For example, in Matlab, using the following command

```
[V,D]=eig(K,M);,
```

one can obtain the solution, where the diagonal elements of D are the eigenvalues, that is, $\lambda = \omega^2$; and the columns of V are the associated eigenvectors. For vibration, the eigenvalues are the squares of the natural frequencies (ω^2), and the eigenvectors determine the mode shapes.

Example 5.3-1 Determine natural frequencies and mode shapes of the system shown in Fig. 5.3-1.

Solution: We can readily obtain the governing equations of motion by examining free-body diagrams of the three masses. The results are

$$m\ddot{x}_1 + 2kx_1 - kx_2 = 0 \qquad (a)$$

$$m\ddot{x}_2 - kx_1 + 2kx_2 - kx_3 = 0 \qquad (b)$$

$$m\ddot{x}_3 - kx_2 + 2kx_3 = 0 \qquad (c)$$

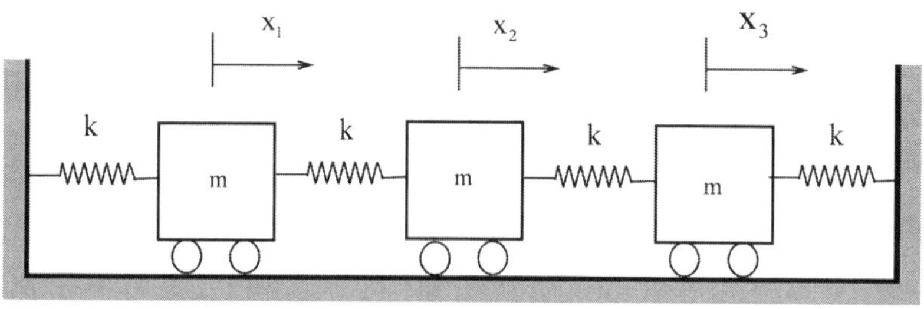

Fig. 5.3-1 A three-degree-of-freedom system.

The mass and stiffness matrices are

$$M=\begin{bmatrix} m & 0 & 0 \\ 0 & m & 0 \\ 0 & 0 & m \end{bmatrix},\quad K=\begin{bmatrix} 2k & -k & 0 \\ -k & 2k & -k \\ 0 & -k & 2k \end{bmatrix} \tag{d}$$

The system impedance matrix is

$$Z=\left[K-\omega^2 M\right]=\begin{bmatrix} 2k-\omega^2 m & -k & 0 \\ -k & 2k-\omega^2 m & -k \\ 0 & -k & 2k-\omega^2 m \end{bmatrix} \tag{e}$$

By setting the determinant of Z equal to zero, we obtain the characteristic equation

$$\left(\omega^2-\frac{2k}{m}\right)\left[\omega^4-4\frac{k}{m}\omega^2+2\left(\frac{k}{m}\right)^2\right]=0 \tag{f}$$

By solving for ω^2, the frequencies (eigenvalues) are seen to be

$$\omega_1^2=\left(2-\sqrt{2}\right)\frac{k}{m},\quad \omega_3^2=\frac{2k}{m},\quad \omega_2^2=\left(2+\sqrt{2}\right)\frac{k}{m} \tag{g}$$

As before, we can obtain the modal vectors by examining the columns of the adjoint matrix of the system impedance matrix Z. To illustrate the procedure, observe that the first column of the adjoint matrix of Z is

$$\text{1st column}=\begin{bmatrix} (2k-m\omega^2)^2-k^2 \\ -k(2k-m\omega^2) \\ k^2 \end{bmatrix} \tag{h}$$

Then, by substituting the values of the ω_i^2 $(i=1,2,3)$ into Eq. h, we obtain the modal vectors

$$\mathbf{u}_1=\begin{bmatrix} x_1 \\ x_2 \\ x_3 \end{bmatrix}=\begin{bmatrix} 1 \\ \sqrt{2} \\ 1 \end{bmatrix},\quad \mathbf{u}_2=\begin{bmatrix} x_1 \\ x_2 \\ x_3 \end{bmatrix}=\begin{bmatrix} 1 \\ 0 \\ -1 \end{bmatrix},\quad \mathbf{u}_3=\begin{bmatrix} x_1 \\ x_2 \\ x_3 \end{bmatrix}=\begin{bmatrix} 1 \\ -\sqrt{2} \\ 1 \end{bmatrix} \tag{i}$$

5.4 Orthogonal Properties of the Modal Vectors

Interestingly, the modal vectors (or eigenvectors) are orthogonal to one another with respect to the mass and stiffness matrices. To see this, let the equation defining the i-th mode be

$$K\mathbf{u}_i=\lambda_i M\mathbf{u}_i \tag{5.4-1}$$

By pre-multiplying by the transpose of mode j, we obtain

$$\mathbf{u}_j^T K\mathbf{u}_i = \lambda_i(\mathbf{u}_j^T M\mathbf{u}_i) \tag{5.4-2}$$

Similarly, we have

$$\mathbf{u}_i^T K\mathbf{u}_j = \lambda_j(\mathbf{u}_i^T M\mathbf{u}_j) \tag{5.4-3}$$

Since M and K are symmetric matrices, the following relationships hold:

$$\mathbf{u}_j^T M\mathbf{u}_i = \mathbf{u}_i^T M\mathbf{u}_j \tag{5.4-4}$$

and

$$\mathbf{u}_j^T K\mathbf{u}_i = \mathbf{u}_i^T K\mathbf{u}_j \tag{5.4-5}$$

Therefore, by subtracting Eq. 5.4-3 from Eq. 5.4-2, we obtain

$$0 = (\lambda_i - \lambda_j)\mathbf{u}_i^T M\mathbf{u}_j \tag{5.4-6}$$

If $\lambda_i \neq \lambda_j$, the above equation requires that

$$\mathbf{u}_i^T M\mathbf{u}_j = 0 \tag{5.4-7}$$

As a consequence of Eq. 5.4-7, from Eq. 5.4-3 we also have

$$\mathbf{u}_i^T K\mathbf{u}_j = 0 \tag{5.4-8}$$

Eqs. 5.4-7 and 5.4-8 define the orthogonal character of the modal vectors. It is important to note that the modal vectors are orthogonal with respect to the mass and stiffness matrices, respectively, but they are not generally orthogonal to each other.

Finally, if $i=j$, Eq. 5.4-6 is satisfied for any finite value of the products of $\mathbf{u}_i^T M\mathbf{u}_j$ and $\mathbf{u}_i^T K\mathbf{u}_j$. We therefore define $\overline{m}_i$ and $\overline{k}_i$ as

$$\mathbf{u}_i^T M\mathbf{u}_i = \overline{m}_i \tag{5.4-9}$$

$$\mathbf{u}_i^T K\mathbf{u}_i = \overline{k}_i, \tag{5.4-10}$$

where $\overline{m}_i$ is called the *modal mass* (or the *generalized mass*), and where $\overline{k}_i$ is called the *modal stiffness* (or the *generalized stiffness*). From Eq. 5.4-2, we have

$$\lambda_i = \omega_i^2 = \frac{\mathbf{u}_i^T K\mathbf{u}_i}{\mathbf{u}_i^T M\mathbf{u}_i} = \frac{\overline{k}_i}{\overline{m}_i} \tag{5.4-11}$$

If a modal vector is normalized such that the modal mass $\overline{m}_i = 1$, the resulting vector is called a *normalized modal vector*. It is evident that if each of the modal vectors is divided by the square root of the corresponding modal mass and designated as $\hat{\mathbf{u}}_i$ such that

$$\hat{\mathbf{u}}_i = \frac{1}{\sqrt{\overline{m}_i}}\mathbf{u}_i, \tag{5.4-12}$$

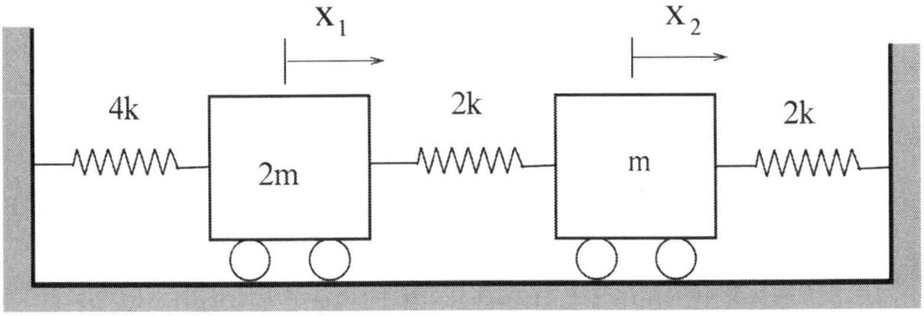

Fig. 5.4-1 Two-degree-of-freedom system for Example 5.4-1.

then the resulting modal vector is normalized, since

$$\hat{\mathbf{u}}_i^T M \hat{\mathbf{u}}_i = 1 \tag{5.4-13}$$

For the normalized modal vectors, we then also have the relations

$$\hat{\mathbf{u}}_i^T K \hat{\mathbf{u}}_i = \omega_i^2 \tag{5.4-14}$$

It is important to note that if the mass and stiffness matrices are not symmetric, we no longer have the orthogonality relations of Eqs. 5.4-7 and 5.4-8. Although the nonsymmetric eigenvalue problem is an interesting subject, it is beyond the scope of this book. The interested reader is referred to the texts by Meirovitch [5.4, 5.5].

Example 5.4-1 Determine the normalized modal vectors for the system shown in Fig. 5.4-1.

Solution: The frequencies and modes of the system have been obtained in Section 5.2. These are

$$\omega_1^2 = 2\frac{k}{m}, \quad \omega_2^2 = 5\frac{k}{m} \tag{a}$$

$$\mathbf{u}_1 = \begin{bmatrix} 1 \\ 1 \end{bmatrix}, \quad \mathbf{u}_2 = \begin{bmatrix} 1 \\ -2 \end{bmatrix} \tag{b}$$

The mass and stiffness matrices are

$$M = m\begin{bmatrix} 2 & 0 \\ 0 & 1 \end{bmatrix}, \quad K = k\begin{bmatrix} 6 & -2 \\ -2 & 4 \end{bmatrix} \tag{c}$$

Thus, the modal masses are

$$\bar{m}_1 = \mathbf{u}_1^T M \mathbf{u}_1 = 3m, \quad \bar{m}_2 = \mathbf{u}_2^T M \mathbf{u}_2 = 6m \tag{d}$$

Therefore, the normalized modal vectors are

$$\hat{\mathbf{u}}_1 = \frac{1}{\sqrt{3m}}\begin{bmatrix} 1 \\ 1 \end{bmatrix}, \quad \hat{\mathbf{u}}_2 = \frac{1}{\sqrt{6m}}\begin{bmatrix} 1 \\ -2 \end{bmatrix} \tag{e}$$

It is easy to verify the following results:

$$\hat{\mathbf{u}}_1^T M \hat{\mathbf{u}}_1 = 1, \quad \hat{\mathbf{u}}_2^T M \hat{\mathbf{u}}_2 = 1 \tag{f}$$

$$\hat{\mathbf{u}}_1^T K \hat{\mathbf{u}}_1 = \omega_1^2, \quad \hat{\mathbf{u}}_2^T K \hat{\mathbf{u}}_2 = \omega_2^2 \tag{g}$$

Comment

Once the modal vectors are known, the mode shapes, which define the geometric patterns of the vibrations, are immediately seen by inspection of the components of the modal vectors.

5.5 Zero Eigenvalues and Repeated Roots

(1) Zero Eigenvalues

On occasion, the characteristic equation yields zero valued roots. This implies that the system under consideration has zero valued frequencies. When this occurs, the masses undergo identical motion. That is, the distance between the masses remains constant, and the system moves (if it moves at all) as a rigid body. Consequently, this is called the *rigid body mode.*

To illustrate this, consider the system represented in Fig. 5.5-1. By using the static equilibrium positions as the origins of the coordinates, the matrix equation is readily seen to be

$$\begin{bmatrix} m & 0 \\ 0 & m \end{bmatrix}\begin{bmatrix} \ddot{x}_1 \\ \ddot{x}_2 \end{bmatrix} + \begin{bmatrix} k & -k \\ -k & k \end{bmatrix}\begin{bmatrix} x_1 \\ x_2 \end{bmatrix} = 0 \quad \text{or} \quad M\ddot{\mathbf{x}} + K\mathbf{x} = 0 \tag{5.5-1}$$

The stiffness matrix in Eq. 5.5-1 is singular (its determinant is zero). This kind of system is called a *semidefinite system.* (If the mass matrix is singular, it is also called a semidefinite system.) It can be shown that a semidefinite system always has zero eigenvalues.

The system impedance matrix Z for the system of Fig. 5.5-1 is

$$Z = \left[K - \omega^2 M\right] = \begin{bmatrix} k - \omega^2 m & -k \\ -k & k - \omega^2 m \end{bmatrix} \tag{5.5-2}$$

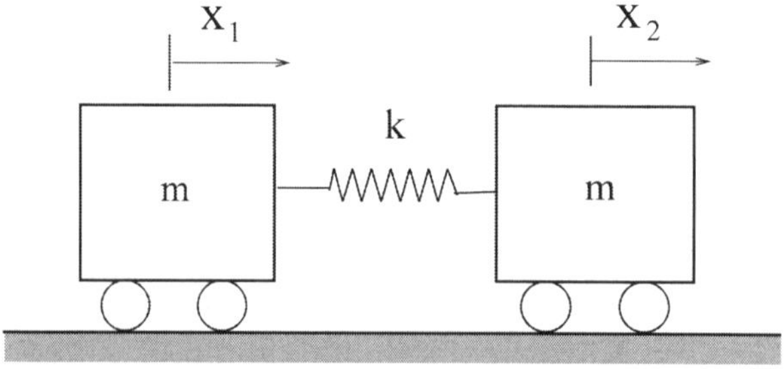

Fig. 5.5-1 A semidefinite system.

Setting the determinant of Z equal to zero, we obtain the characteristic equation

$$\omega^2\left(\omega^2-2\frac{k}{m}\right)=0 \tag{5.5-3}$$

The eigenvalues are then

$$\lambda_1=\omega_1^2=0,\quad \lambda_2=\omega_2^2=2\frac{k}{m} \tag{5.5-4}$$

The adjoint matrix of the system impedance matrix is then seen to be

$$adj(Z)=\begin{bmatrix} k-\omega^2 m & k \\ k & k-\omega^2 m \end{bmatrix} \tag{5.5-5}$$

Substituting the eigenvalues into the first column of the adjoint matrix, we obtain the modal vectors

$$\mathbf{u}_1=\begin{bmatrix} 1 \\ 1 \end{bmatrix},\quad \mathbf{u}_2=\begin{bmatrix} -1 \\ 1 \end{bmatrix} \tag{5.5-6}$$

With the first eigenvalue being zero, the corresponding mode is called the rigid body mode. This results from the stiffness matrix being singular. The second frequency is not zero. The system then undergoes harmonic vibration with this frequency. In this case, the two masses move in opposite directions.

(2) Repeated Roots

In our previous discussions, we assumed that the roots of the characteristic equation were distinct. Under this assumption, we discovered that the modal vectors were orthogonal with respect to both the mass and stiffness matrices.

Let us now remove this assumption of distinct valued roots. Specifically, let us consider a *degenerate system* where two or more of the eigenvalues are equal to each other—so-called *repeated roots*. In this case, the corresponding modal vectors are not unique, and a linear combination of such modal vectors will also satisfy the equations of motion. To illustrate this, let X_1 and X_2 be modal vectors belonging to a common eigenvalue $\lambda_1=\lambda_2\equiv\lambda_0$. We can then write [5.1]

$$AX_1=\lambda_0 X_1 \tag{5.5-7}$$

and

$$AX_2=\lambda_0 X_2 \tag{5.5-8}$$

By multiplying Eq. 5.5-7 by a constant b and adding it to Eq. 5.5-8, we obtain

$$A(X_1+bX_2)=\lambda_0(X_1+bX_2) \tag{5.5-9}$$

Thus, a new eigenvector $X_{12} = X_1 + bX_2$, which is a linear combination of the two, satisfies the basic equation

$$AX_{12} = \lambda_0 X_{12} \tag{5.5-10}$$

Hence, no unique modal vector, or mode, exist for λ_0. We have, however, the freedom of replacing the modal vectors corresponding to the repeated roots by linear combinations of themselves. Using this freedom, we can always perform the Gram-Schmidt type of orthogonalization and find a set of modal vectors that are both complete and orthogonal, as in the nondegenerate case.

Example 5.5-1 Consider the system shown in Fig. 5.5-2, where a bar connecting the masses is rigid and negligible in weight. Find the natural frequencies (eigenvalues) and modal vectors.

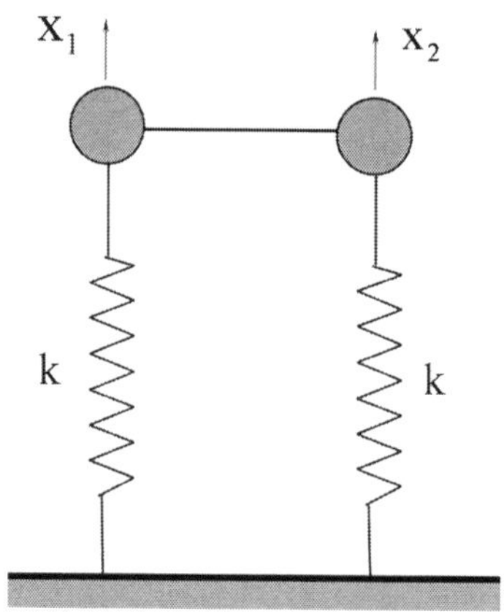

Fig. 5.5-2 A two-degree-of-freedom system for Example 5.5-1.

Solution: If we taking the static equilibrium position as the origin of the coordinate system, the equations of motion for the system may be written as

$$m\ddot{x}_1 + kx_1 = 0 \tag{a}$$

and

$$m\ddot{x}_2 + kx_2 = 0 \tag{b}$$

or in matrix form as

$$\begin{bmatrix} m & 0 \\ 0 & m \end{bmatrix} \begin{bmatrix} \ddot{x}_1 \\ \ddot{x}_2 \end{bmatrix} + \begin{bmatrix} k & 0 \\ 0 & k \end{bmatrix} \begin{bmatrix} x_1 \\ x_2 \end{bmatrix} = 0 \tag{c}$$

As before, let us look for a solution of these equations having the forms $x_1 = X_1 e^{j\omega t}$, $x_2 = X_2 e^{j\omega t}$. By substituting into Eq. c, we have

$$\begin{bmatrix} k - m\omega^2 & 0 \\ 0 & k - m\omega^2 \end{bmatrix} \begin{bmatrix} X_1 \\ X_2 \end{bmatrix} = 0 \tag{d}$$

The characteristic equation is then

$$\Delta = \begin{vmatrix} k - m\omega^2 & 0 \\ 0 & k - m\omega^2 \end{vmatrix} = 0 \tag{e}$$

Hence, the eigenvalues are found to be

$$\omega_1^2 = \omega_2^2 = \frac{k}{m} \tag{f}$$

Finally, the adjoint matrix of the system impedance matrix is

$$adj(Z) = \begin{bmatrix} k-\omega^2 m & 0 \\ 0 & k-\omega^2 m \end{bmatrix} \tag{g}$$

In this case, we cannot use the adjoint matrix to find the modal vectors, since, when substituting the eigenvalue $\omega^2 = k/m$ into Eq. g, each element of the matrix becomes zero. But the modal vector cannot be zero!

Substituting the eigenvalue $\omega^2 = k/m$ into Eq. d, we obtain

$$\begin{bmatrix} 0 & 0 \\ 0 & 0 \end{bmatrix} \begin{bmatrix} X_1 \\ X_2 \end{bmatrix} = 0 \tag{h}$$

Obviously, we can choose X_1 and X_2 arbitrarily. This means that the modal vectors are not unique. As an example, we might take the first modal vector $\mathbf{u}_1$ as

$$\mathbf{u}_1 = \begin{bmatrix} 1 \\ 0 \end{bmatrix} \tag{i}$$

Then we can readily find another vector $\mathbf{u}_2$ orthogonal to $\mathbf{u}_1$. For example, we might take $\mathbf{u}_2$ as

$$\mathbf{u}_2 = \begin{bmatrix} 0 \\ 1 \end{bmatrix} \tag{j}$$

As represented in Fig. 5.5-3 (a), two natural motions are possible [5.6]: (1) the left end is pulled up and down with the frequency $\omega = \sqrt{k/m}$, while the right

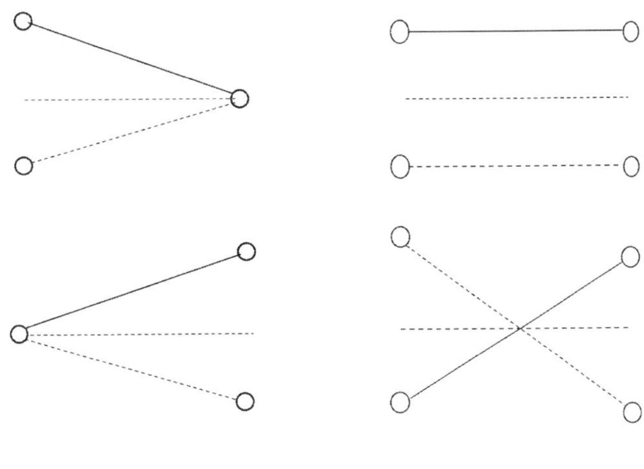

(a) (b)

Fig. 5.5-3 Mode shapes of the system shown in Fig. 5.5-2.

end is kept in place; (2) the right end is pulled up and down with the same frequency while the left end is kept in place.

Next, we might take

$$\mathbf{u}_1 = \begin{bmatrix} 1 \\ 1 \end{bmatrix}, \quad \mathbf{u}_2 = \begin{bmatrix} 1 \\ -1 \end{bmatrix} \tag{k}$$

as the orthogonal modal vectors, where each is a linear combination of the modal vectors given in Eqs. i and j. Specifically, we have

$$\begin{bmatrix} 1 \\ 1 \end{bmatrix} = \begin{bmatrix} 1 \\ 0 \end{bmatrix} + \begin{bmatrix} 0 \\ 1 \end{bmatrix}, \quad \text{and} \quad \begin{bmatrix} 1 \\ -1 \end{bmatrix} = \begin{bmatrix} 1 \\ 0 \end{bmatrix} - \begin{bmatrix} 0 \\ 1 \end{bmatrix} \tag{l}$$

As represented in Fig. 5.5-3 (b), two possible motions are (1) a bobbing up and down parallel to itself with the frequency $\omega = \sqrt{k/m}$, and (2) a rocking with the same frequency about the center of the connecting bar. Observe that each mode shape shown in Fig. 5.5-3 (a) can also be expressed as a linear combination of the mode shapes shown in Fig. 5.5-3 (b). That is,

$$\begin{bmatrix} 1 \\ 0 \end{bmatrix} = \begin{bmatrix} 1 \\ 1 \end{bmatrix} + \begin{bmatrix} 1 \\ -1 \end{bmatrix}, \quad \text{and} \quad \begin{bmatrix} 0 \\ 1 \end{bmatrix} = \frac{1}{2}\begin{bmatrix} 1 \\ 1 \end{bmatrix} - \frac{1}{2}\begin{bmatrix} 1 \\ -1 \end{bmatrix} \tag{m}$$

5.6 Response of a System to Initial Conditions

When the initial displacements and initial velocities of the motion of a system are known, it is possible (at least, in theory) to determine a detailed response of the system. There are two principal methods for finding the response: The first, called *mode summation* or the *expansion method* [5.7], is based upon linear algebraic procedures for vector representation in n-dimensional space. The second is the use of numerical methods to obtain a representation of the response.

In this section, we discuss the linear algebra approach. We will consider the second method in the next chapter (see Section 6.3).

Let ω_i and $\mathbf{u}_i$ $(i=1,2,\cdots,n)$ be the natural frequencies and associated modal vectors. Then the response of the system to initial conditions can be written as

$$\mathbf{x} = \sum_{i=1}^{n} C_i \mathbf{u}_i \sin(\omega_i t + \psi_i), \tag{5.6-1}$$

where $\mathbf{x} = [x_1, x_2, \cdots, x_n]^T$, C_i, and $\psi_i (i=1,2,\cdots,n)$ are $2n$ constants, which can be determined by initial conditions (n initial displacements and n initial velocities). The constants C_i are called *modal participation factors*, which represent the contribution of each mode to the total response. The larger the C_i, the more the i-th mode effects the response. The validity of Eq. 5.6-1 is sometimes called the *expansion theorem* [5.8].

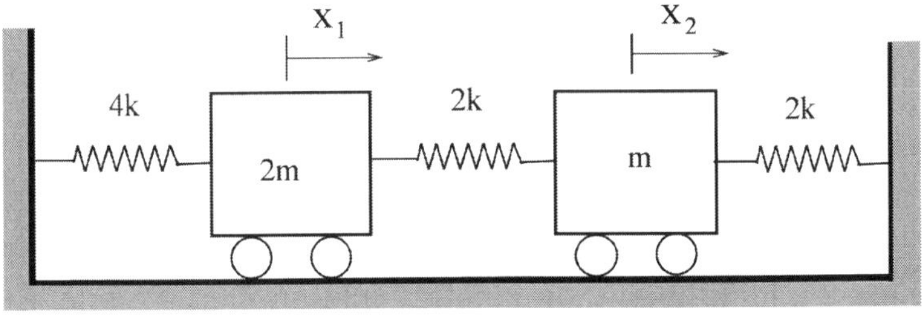

Fig. 5.6-1 A two-degree-of-freedom system.

Example 5.6-1 Consider the system shown in Fig. 5.6-1. Determine the response of the system under the following initial conditions:

$$x_1(0)=5.0 \text{ mm}, \; x_2(0)=0$$
$$\dot{x}_1(0)=0, \; \dot{x}_2(0)=0$$

Solution: In Example 5.4-1, we obtained the frequencies and modal vectors as

$$\omega_1=\sqrt{\frac{2k}{m}}, \; \omega_2=\sqrt{\frac{5k}{m}} \tag{a}$$

$$\mathbf{u}_1=\begin{bmatrix} 1 \\ 1 \end{bmatrix}, \; \mathbf{u}_2=\begin{bmatrix} 1 \\ -2 \end{bmatrix} \tag{b}$$

According to the mode summation method, the general solution of the system can be written as a combination of the modal vectors with associated frequencies as follows:

$$\mathbf{x}=A\mathbf{u}_1 \sin(\omega_1 t+\psi_1)+B\mathbf{u}_2 \sin(\omega_2 t+\psi) \tag{c}$$

or

$$\begin{bmatrix} x_1 \\ x_2 \end{bmatrix}=A\begin{bmatrix} 1 \\ 1 \end{bmatrix}\sin(\omega_1 t+\psi_1)+B\begin{bmatrix} 1 \\ -2 \end{bmatrix}\sin(\omega_2 t+\psi_2), \tag{d}$$

where A, B, ψ_1, and ψ_2 are constants, which are determined by the initial conditions. At $t=0$, let x_1 be 5 and x_2 be 0. Then we have

$$\begin{bmatrix} 5 \\ 0 \end{bmatrix}=\begin{bmatrix} A\sin\psi_1+B\sin\psi_2 \\ A\sin\psi_1-2B\sin\psi_2 \end{bmatrix}, \tag{e}$$

from which we obtain

$$A\sin\psi_1=10/3, \; B\sin\psi_2=5/3 \tag{f}$$

Next, differentiating in Eq. d and letting $t=0$ in the result, we have

$$\begin{bmatrix} 0 \\ 0 \end{bmatrix}=\begin{bmatrix} \omega_1 A\cos\psi_1+\omega_2 B\cos\psi_2 \\ \omega_1 A\cos\psi_1-2\omega_2 B\cos\psi_2 \end{bmatrix}, \tag{g}$$

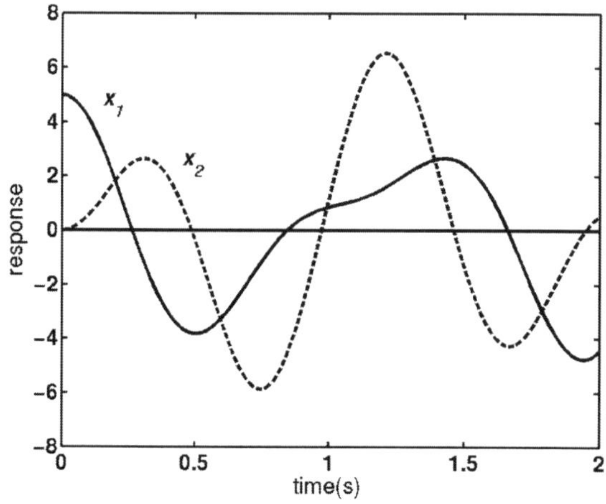

Fig. 5.6-2 Plots of the responses of Example 5.6-1.

from which we obtain

$$\begin{aligned} \cos\psi_1 &= 0, \ \psi_1 = 90^\circ \\ \cos\psi_2 &= 0, \ \psi_2 = 90^\circ \end{aligned} \tag{h}$$

By using Eq. f, we then obtain

$$A = 10/3, \ B = 5/3 \tag{i}$$

Substituting into Eq. d, we obtain the solution

$$\begin{bmatrix} x_1 \\ x_2 \end{bmatrix} = \frac{10}{3}\begin{bmatrix} 1 \\ 1 \end{bmatrix}\cos\sqrt{\frac{2k}{m}}t + \frac{5}{3}\begin{bmatrix} 1 \\ -2 \end{bmatrix}\cos\sqrt{\frac{5k}{m}}t \tag{j}$$

It is easy to show that this solution satisfies the equation of motion and the initial conditions. Fig. 5.6-2 shows the solution for $k = 25$ N/m and $m = 2$ kg.

Note that, in this case, the response of each mass contains both frequencies of the system. That is, the responses for both x_1 and x_2 are combinations of signals containing the two frequencies ω_1 and ω_2. Note also that the modal vectors determine the relative magnitudes of these two harmonic signals.

5.7 Coordinate Coupling

In the most general case, the equations of motion of a multi-degree-of-freedom system are coupled. That is, different coordinate variables appear in the same equation. For example, the equations of motion of a two-degree-of-freedom system can be expressed in matrix form as

$$\begin{bmatrix} m_{11} & m_{12} \\ m_{21} & m_{22} \end{bmatrix}\begin{bmatrix} \ddot{x}_1 \\ \ddot{x}_2 \end{bmatrix} + \begin{bmatrix} k_{11} & k_{12} \\ k_{21} & k_{22} \end{bmatrix}\begin{bmatrix} x_1 \\ x_1 \end{bmatrix} = 0, \tag{5.7-1}$$

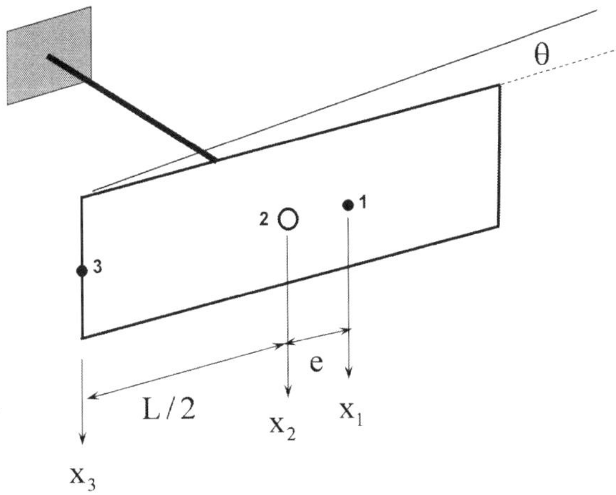

Fig. 5.7-1 Vibration model of a vehicle rearview mirror housing.

which reveals the type of coupling present. If the mass matrix is not diagonal, it is said that the system has *dynamic coupling*, or that the system is *dynamically coupled*. If the stiffness matrix is not diagonal, it is said that the system has *static coupling*, or that the system is *statically coupled*. It is also possible to have both forms of coupling.

In general, we can choose any coordinate system to describe the motion of a system. However, different choices of coordinates will result in different equations of motion. That is, the choice of coordinates will define the type of coupling. To see this more clearly, consider a vibration model of a vehicle rearview mirror housing, as represented in Fig. 5.7-1. The mirror housing is modeled as a bar attached to the ground (the vehicle body) through a shaft. Let k and k_t represent, respectively, the vertical bending stiffness and torsional stiffness of the shaft. Let the length of the bar be L. The mass center of the bar is located at point 1.

Static Coupling. Choosing x_1 and θ as the generalized coordinates, where x_1 is the linear vertical displacement of the mass center, and θ is the angular displacement of the bar about the shaft. Both x_1 and θ are measured from the bar's static equilibrium position. The equations of motion of the mirror housing are then

$$m\ddot{x}_1 = -k(x_1 - e\theta) \tag{5.7-2}$$

$$J_1\ddot{\theta} = -k_t\theta + k(x_1 - e\theta)e, \tag{5.7-3}$$

where J_1 is the moment of inertia of the bar about point 1 (its mass center). The system has static coupling as shown by the matrix equation

$$\begin{bmatrix} m & 0 \\ 0 & J_1 \end{bmatrix}\begin{bmatrix} \ddot{x}_1 \\ \ddot{\theta} \end{bmatrix} + \begin{bmatrix} k & -ke \\ -ke & k_t + ke^2 \end{bmatrix}\begin{bmatrix} x_1 \\ \theta \end{bmatrix} = \begin{bmatrix} 0 \\ 0 \end{bmatrix} \tag{5.7-4}$$

Dynamic Coupling. If, instead, we choose x_2 and θ as the generalized coordinates, where x_2 is the vertical displacement of point 2 of the bar. Then the equations of motion are

$$m(\ddot{x}_2+e\ddot{\theta})=-kx_2 \tag{5.7-5}$$

$$J_2\ddot{\theta}+(m\ddot{x}_2)e=-k_t\theta, \tag{5.7-6}$$

where J_2 is the moment of inertia of the bar about point 2. The system then has dynamic coupling as shown by the matrix equation

$$\begin{bmatrix} m & me \\ me & J_2 \end{bmatrix}\begin{bmatrix} \ddot{x}_2 \\ \ddot{\theta} \end{bmatrix}+\begin{bmatrix} k & 0 \\ 0 & k_t \end{bmatrix}\begin{bmatrix} x_2 \\ \theta \end{bmatrix}=\begin{bmatrix} 0 \\ 0 \end{bmatrix} \tag{5.7-7}$$

Static and Dynamic Coupling. Finally, if we choose x_3 and θ as the generalized coordinates, where x_3 is the vertical displacement of point 3 of the bar. The equations of motion become

$$m[\ddot{x}_3+(L/2+e)\ddot{\theta}]=-k\left[x_3+(L/2)\theta\right] \tag{5.7-8}$$

$$J_3\ddot{\theta}+(m\ddot{x}_3)(L/2+e)=-k_t\theta-k\left[x_3+(L/2)\theta\right](L/2) \tag{5.7-9}$$

or in matrix form

$$\begin{bmatrix} m & m(L/2+e) \\ m(L/2+e) & J_3 \end{bmatrix}\begin{bmatrix} \ddot{x}_3 \\ \ddot{\theta} \end{bmatrix}$$

$$+\begin{bmatrix} k & k(L/2) \\ k(L/2) & k_t+k(L/2)^2 \end{bmatrix}\begin{bmatrix} x_3 \\ \theta \end{bmatrix}=\begin{bmatrix} 0 \\ 0 \end{bmatrix} \tag{5.7-10}$$

It is seen from Eq. 5.7-10 that both static and dynamic coupling are now present.

The above example illustrates that the choice of coordinates may result in static or dynamic coupling or both. It is desirable, of course, to find a coordinate system that makes the equations of motion uncoupled. Such uncoupled coordinates are desirable since each equation can be solved independently of the others.

It is important to note that the choice of coordinates may change the type of coupling, but it does not change the natural frequencies of the system. In the following section, we will see that the modal matrix of the system can be used to decouple the equations of motion. Each of the uncoupled equations is called a *modal equation*. Once the solutions of the modal equations have been determined, the modal matrix can be used to transform the solutions back to the original coordinate system.

5.8 Undamped Systems and Coordinate Decoupling

(1) Forced Response and Coordinate Decoupling

When the n modal vectors (eigenvectors) are assembled into a square matrix with each normal mode represented by a column, we called it the *modal matrix* S. The columns may be arranged in any order, but, by convention, they are usually placed in order of increasing frequency of the corresponding modes. Thus, for a two-degree-of-freedom system, the modal matrix can be written as

$$S=\begin{bmatrix} \mathbf{u}_1 & \mathbf{u}_2 \end{bmatrix} \tag{5.8-1}$$

The modal matrix makes it possible to decouple the equations of motion [5.1, 5.4–5.12]. Specifically, the equations of motion of an undamped n-degree-of-freedom system may be written as

$$M\ddot{\mathbf{x}}+K\mathbf{x}=\mathbf{f} \tag{5.8-2}$$

This equation can be decoupled, or it can be reduced into a set of n decoupled modal equations. To see this, let us introduce a new coordinate vector $\mathbf{y}$, called a *modal coordinate vector*, as

$$\mathbf{x}=S\mathbf{y} \tag{5.8-3}$$

Substituting $\mathbf{x}=S\mathbf{y}$ into Eq. 5.8-2 and pre-multiplying the result by S^T, we have

$$S^T M S\ddot{\mathbf{y}}+S^T KS\mathbf{y}=S^T\mathbf{f} \tag{5.8-4}$$

Due to the orthogonality of the modal vectors, Eq. 5.8-4 becomes n independent (decoupled) equations

$$\bar{m}_i\ddot{y}_i+\bar{k}_i y_i=f_i \quad (i=1,2,\cdots,n), \tag{5.8-5}$$

where $\bar{m}_i$ is the generalized mass, $\bar{k}_i$ is the generalized stiffness, and f_i is the generalized force associated with the i-th generalized coordinate, defined as

$$\bar{m}_i=\mathbf{u}_i^T M\mathbf{u}_i,\ \ \bar{k}_i=\mathbf{u}_i^T K\mathbf{u}_i,\ \ f_i=\mathbf{u}_i^T\mathbf{f} \tag{5.8-6}$$

Suppose that we express $\mathbf{x}$ in Eq. 5.8-3 as

$$\mathbf{x}=\bar{S}\mathbf{y}, \tag{5.8-7}$$

where $\bar{S}$ is the normalized modal matrix. Then, by substituting for $\mathbf{x}$ into Eq. 5.8-2 and then pre-multiplying by $\bar{S}^T$, Eq. 5.8-4 takes the form

$$\ddot{y}_i+\omega_i^2 y_i=\bar{f}_i \quad (i=1,\cdots,n), \tag{5.8-8}$$

where the ω_i $(i=1,\cdots,n)$ are the natural frequencies of the system.

Observe that if the exciting force vector $\mathbf{f}$ is orthogonal to the modal vector $\mathbf{u}_i$, such that $\mathbf{u}_i^T\mathbf{f}=0$, then the i-th mode has no force associated with it. Consequently, the i-th mode cannot experience resonance.

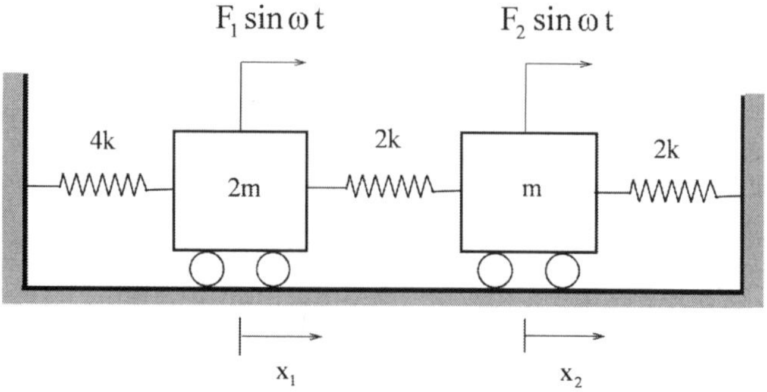

Fig. 5.8-1 Two-degree-of-freedom system for Example 5.8-1.

For harmonic excitation with frequency ω, the solution of Eq. 5.8-8 can be evaluated by Eq. 3.2-14. Then we have

$$y_i = \frac{\bar{f}_i}{\omega_i^2 - \omega^2} \sin \omega t \quad (i = 1, 2, \cdots, n) \tag{5.8-9}$$

Once the modal solutions are known, the normalized modal matrix $\bar{S}$ can be used with the vector $\mathbf{y} = [y_1, \cdots, y_n]^T$ to find the solutions $\mathbf{x}$ in the original coordinate system. That is,

$$\mathbf{x} = \bar{S}\mathbf{y} \tag{5.8-10}$$

Example 5.8-1 Find the decoupled equations of motion of the system as shown in Fig. 5.8-1.

Solution: If we use free-body diagrams, we find the equations of motion for the system to be

$$m\begin{bmatrix} 2 & 0 \\ 0 & 1 \end{bmatrix}\begin{bmatrix} \ddot{x}_1 \\ \ddot{x}_2 \end{bmatrix} + k\begin{bmatrix} 6 & -2 \\ -2 & 4 \end{bmatrix}\begin{bmatrix} x_1 \\ x_2 \end{bmatrix} = \begin{bmatrix} F_1 \\ F_2 \end{bmatrix} \sin \omega t \tag{a}$$

The mass and stiffness matrices are

$$M = m\begin{bmatrix} 2 & 0 \\ 0 & 1 \end{bmatrix}, \quad K = k\begin{bmatrix} 6 & -2 \\ -2 & 4 \end{bmatrix} \tag{b}$$

If we follow the procedure in Example 5.4-1, we find the normalized mode vectors to be

$$\hat{\mathbf{u}}_1 = \frac{1}{\sqrt{3m}}\begin{bmatrix} 1 \\ 1 \end{bmatrix}, \quad \hat{\mathbf{u}}_2 = \frac{1}{\sqrt{6m}}\begin{bmatrix} 1 \\ -2 \end{bmatrix} \tag{c}$$

Thus, the normalized modal matrix may be written as

$$\bar{S} = \frac{1}{\sqrt{6m}}\begin{bmatrix} \sqrt{2} & 1 \\ \sqrt{2} & -2 \end{bmatrix} \tag{d}$$

The reader should verify that the $\bar{S}$ is not orthogonal, but it is orthogonal with respect to the mass matrix, that is, $\bar{S}^T M\bar{S} = \text{diag}(1,1)$. To decouple the original equations of motion, we introduce the following transformation:

$$\mathbf{x} = \bar{S}\mathbf{y} = \frac{1}{\sqrt{6m}}\begin{bmatrix} \sqrt{2} & 1 \\ \sqrt{2} & -2 \end{bmatrix}\begin{bmatrix} y_1 \\ y_2 \end{bmatrix}, \tag{e}$$

where $\mathbf{y}$ is called the modal coordinate vector. Then, with the mass and stiffness matrices (b), Eq. 5.8-8 becomes

$$\begin{bmatrix} 1 & 0 \\ 0 & 1 \end{bmatrix}\begin{bmatrix} \ddot{y}_1 \\ \ddot{y}_2 \end{bmatrix} + \frac{k}{m}\begin{bmatrix} 2 & 0 \\ 0 & 5 \end{bmatrix}\begin{bmatrix} y_1 \\ y_2 \end{bmatrix} = \begin{bmatrix} \dfrac{F_1+F_2}{\sqrt{3m}} \\ \dfrac{F_1-2F_2}{\sqrt{6m}} \end{bmatrix}\sin\omega t \tag{f}$$

The original equation of motion has thus been transformed into the uncoupled Eq. f by the coordinate transformation of Eq. e. It is easy to find the solutions y_1 and y_2 from the modal Eq. f. Once the modal solutions are obtained, and substituting them back into Eq. e, the solutions in the original coordinate system can be found.

It should be noted that if $F_1 = 2F_2$, the excitation force vector is orthogonal to the second modal vector. The second mode then has no force associated with it. In this case, even when the exciting frequency ω is equal to the second natural frequency of the system, it cannot cause resonance. This result will be verified in Example 6.3-3 by solving for the time response directly, using numerical integration.

(2) Initial Value Problem and Coordinate Decoupling

Consider the equation of motion

$$M\ddot{\mathbf{x}} + K\mathbf{x} = 0, \tag{5.8-11}$$

subject to the initial conditions

$$\mathbf{x}(0) = \mathbf{x}_0, \;\; \dot{\mathbf{x}}(0) = \dot{\mathbf{x}}_0, \tag{5.8-12}$$

where $\mathbf{x}_0 = [x_1(0), \cdots, x_n(0)]^T$ is the vector of initial displacements, and $\dot{\mathbf{x}}(0) = [\dot{x}_1(0), \dot{x}_2(0), \ldots, \dot{x}_n(0)]^T$ is the vector of initial velocities. As outlined in the previous section, the normalized modal matrix $\bar{S}$ can be used to decouple Eq. 5.8-11. Letting $\mathbf{x} = \bar{S}\mathbf{y}$ in Eq. 5.8-11 and pre-multiplying the result by $\bar{S}^T$ yields the following n decoupled equations:

$$\ddot{y}_i + \omega_i^2 y_i = 0 \quad (i = 1, \cdots, n), \tag{5.8-13}$$

where ω_i $(i = 1, \cdots, n)$ are the natural frequencies of the system.

Eqs. 5.8-13 are called the modal equations. Being decoupled, they are easy to integrate. The initial conditions, however, are expressed in terms of $\mathbf{x}$, as apposed to the $\mathbf{y}$ of these modal equations. Therefore, it is necessary to express the initial conditions in terms of $\mathbf{y}$. To this end, using the foregoing transformation matrix $\bar{S}$, we obtain

$$\mathbf{y}_0 = \bar{S}^{-1}\mathbf{x}_0, \ \dot{\mathbf{y}}_0 = \bar{S}^{-1}\dot{\mathbf{x}}_0 \tag{5.8-14}$$

From a computational perspective, it happens that Eqs. 5.8-14 do not provide the most efficient means for expressing the initial conditions in terms of $\mathbf{y}$. Indeed, the computation of the inverse of $\bar{S}$ can be avoided. Recall that $\bar{S}$ is orthogonal with respect to the mass matrix. That is,

$$I = \bar{S}^T M \bar{S}, \tag{5.8-15}$$

where, as before, I is the identity matrix. Post-multiplying Eq. 5.8-15 by $\bar{S}^{-1}$, we obtain

$$\bar{S}^{-1} = \bar{S}^T M \tag{5.8-16}$$

Substituting into Eq. 5.8-14, the initial conditions become

$$\mathbf{y}_0 = \bar{S}^T M \mathbf{x}_0, \ \dot{\mathbf{y}}_0 = \bar{S}^T M \dot{\mathbf{x}}_0 \tag{5.8-17}$$

If we use these expressions, we avoid the computation of the inverse of $\bar{S}$.

From Eq. 1.4-9, the solution of the modal equation (Eq. 5.8-13) is then simply

$$y_i = \frac{\sqrt{\omega_i^2 y_{i0}^2 + \dot{y}_{i0}^2}}{\omega_i} \sin\left(\omega_i t + \tan^{-1}\frac{\omega_i y_{i0}}{\dot{y}_{i0}}\right), \quad (i = 1, \cdots, n) \tag{5.8-18}$$

Once the y_i are obtained, the solutions in the original coordinate system are obtained by the transformation

$$\mathbf{x} = \bar{S}\mathbf{y} \tag{5.8-19}$$

Example 5.8-2 Calculate the solution of the two-degree-of-freedom system of Example 5.6-1, mass matrix M, stiffness matrix K, initial displacement array $\mathbf{x}_0$, and initial velocity array $\dot{\mathbf{x}}_0$ being

$$M = m\begin{bmatrix} 2 & 0 \\ 0 & 1 \end{bmatrix}, \ K = k\begin{bmatrix} 6 & -2 \\ -2 & 2 \end{bmatrix}, \ \mathbf{x}_0 = \begin{bmatrix} 5 \\ 0 \end{bmatrix}, \ \dot{\mathbf{x}}_0 = \begin{bmatrix} 0 \\ 0 \end{bmatrix}$$

Solution: From Examples 5.4-1 and 5.6-1, the natural frequencies ω_1, ω_2, and the normalized modal matrix $\bar{S}$ are

$$\omega_1 = \sqrt{\frac{2k}{m}}, \ \omega_2 = \sqrt{\frac{5k}{m}}, \quad \text{and} \quad \bar{S} = \frac{1}{\sqrt{6m}}\begin{bmatrix} \sqrt{2} & 1 \\ \sqrt{2} & -2 \end{bmatrix} \tag{a}$$

The inverse of $\bar{S}$ is given by Eq. 5.8-16 as

$$\bar{S}^{-1} = \bar{S}^T M = \sqrt{\frac{m}{6}}\begin{bmatrix} 2\sqrt{2} & \sqrt{2} \\ 2 & -2 \end{bmatrix} \tag{b}$$

Hence, the initial conditions in the modal coordinate system are

$$\mathbf{y}_0 = \bar{S}^{-1}\mathbf{x}_0 = \sqrt{\frac{m}{6}}\begin{bmatrix} 2\sqrt{2} & \sqrt{2} \\ 2 & -\sqrt{2} \end{bmatrix}\begin{bmatrix} 5 \\ 0 \end{bmatrix} = \begin{bmatrix} 10\sqrt{m/3} \\ 10\sqrt{m/6} \end{bmatrix} \tag{c}$$

and

$$\dot{\mathbf{y}}_0 = \bar{S}^{-1}\dot{\mathbf{x}}_0 = \begin{bmatrix} 0 \\ 0 \end{bmatrix} \tag{d}$$

The response of the system in the modal coordinate system is given by Eq. 5.8-18

$$y_1 = 10\sqrt{\frac{m}{3}}\sin\omega_1 t = 10\sqrt{\frac{m}{3}}\sin\sqrt{\frac{2k}{m}}t \tag{e}$$

and

$$y_2 = 10\sqrt{\frac{m}{6}}\sin\omega_2 t = 10\sqrt{\frac{m}{6}}\sin\sqrt{\frac{5k}{m}}t \tag{f}$$

The solution in the original coordinate system is obtained by the transformation $\mathbf{x} = \bar{S}\mathbf{y}$. From this we obtain

$$\begin{bmatrix} x_1 \\ x_2 \end{bmatrix} = \frac{10}{3}\begin{bmatrix} 1 \\ 1 \end{bmatrix}\cos\sqrt{\frac{2k}{m}}t + \frac{5}{3}\begin{bmatrix} 1 \\ -2 \end{bmatrix}\cos\sqrt{\frac{5k}{m}}t \tag{g}$$

This solution is identical to the solution obtained in Example 5.6-1.

5.9 Damped Systems and Coordinate Decoupling

Although we can decouple the equations of motion for undamped systems, we cannot always decouple the equations when damping is present. The difficulty with damped systems is that the damping matrix provides additional coupling between the variables, which cannot always be decoupled by modal transformation.

If, however, the damping matrix is proportional to either the mass or the stiffness matrix, or to a linear combination of these matrices, it is possible to decouple the equations using the modal matrix. Otherwise, the equations can be decoupled only in a state space. We explore this later in Section 6.2.

(1) Coordinate Decoupling with Forced Response

The equations of motion of an n-degree-of-freedom system with viscous damping and an arbitrary excitation force may be written as

$$M\ddot{\mathbf{x}} + C\dot{\mathbf{x}} + K\mathbf{x} = \mathbf{f} \tag{5.9-1}$$

If the damping matrix can be written as a linear combination of the mass and stiffness matrix, that is, if

$$C = aM + bK, \tag{5.9-2}$$

where a and b are constants, then the damping is called *proportional damping.*

With proportional damping, Eq. 5.9-1 can be decoupled by either the modal matrix S or the normalized modal matrix $\overline{S}$ of the corresponding undamped system. To illustrate this, using $\overline{S}$, let a new coordinate vector $\mathbf{y}$ be introduced, as before, as

$$\mathbf{x} = \overline{S}\mathbf{y} \tag{5.9-3}$$

Substituting this expression into Eq. 5.9-1 and then pre-multiplying by $\overline{S}^T$, we obtain

$$\overline{S}^T M\overline{S}\ddot{\mathbf{y}} + \overline{S}^T C\overline{S}\dot{\mathbf{y}} + \overline{S}^T K\overline{S}\mathbf{y} = \overline{S}^T \mathbf{f}, \tag{5.9-4}$$

with C given by Eq. 5.9-2. Recall that

$$\overline{S}^T M\overline{S} = I \quad \text{and} \quad \overline{S}^T K\overline{S} = \text{diag}(\omega_1^2, \cdots, \omega_n^2) \equiv \Lambda \tag{5.9-5}$$

Thus, Eq. 5.9-4 becomes

$$I\ddot{\mathbf{y}} + (aI + b\Lambda)\dot{\mathbf{y}} + \Lambda\mathbf{y} = \mathbf{h}, \tag{5.9-6}$$

where $\mathbf{h}$ is $\overline{S}^T\mathbf{f}$.

Since all the coefficients on the left side of Eq. 5.9-6 are diagonal matrices, we have a set of n decoupled second-order equations of the form

$$\ddot{y}_i + 2\varsigma_i\omega_i\dot{y} + \omega_i^2 y = h_i \quad (i = 1, \cdots, n), \tag{5.9-7}$$

where $2\varsigma_i\omega_i$ and ς_i are

$$2\varsigma_i\omega_i = a + b\omega_i^2 \quad \text{and} \quad \varsigma_i = \frac{a}{2\omega_i} + \frac{b\omega_i^2}{2}(i = 1, \cdots, n) \tag{5.9-8}$$

The steady state (or particular) solution of Eqs. 5.9-7 is given by Eq. 3.2-14 as

$$y_i = \frac{h_i}{\sqrt{(\omega_i^2 - \omega^2)^2 + (2\varsigma_i\omega_i\omega)^2}} \sin\left(\omega t - \tan^{-1}\frac{2\varsigma_i\omega_i\omega}{\omega_i^2 - \omega^2}\right) \quad (i = 1, \cdots, n) \tag{5.9-9}$$

The solutions in the original coordinate system may now be calculated using the transformation $\mathbf{x} = \overline{S}\mathbf{y}$ (see Eq. 5.9-3).

Example 5.9-1 Consider the two-degree-of-freedom system governed by the matrix equation

$$\begin{bmatrix} 4 & 0 \\ 0 & 1 \end{bmatrix}\ddot{\mathbf{x}} + \begin{bmatrix} 0.8 & -0.2 \\ -0.2 & 0.2 \end{bmatrix}\dot{\mathbf{x}} + \begin{bmatrix} 8 & -2 \\ -2 & 2 \end{bmatrix}\mathbf{x} = \begin{bmatrix} 0 \\ F_2(t) \end{bmatrix}, \tag{a}$$

where $F_2(t) = 2\sin\sqrt{3}t$. Here the damping matrix is propositional to the stiffness matrix; that is, $C = 0.1K$. Calculate the steady-state response.

Solution: The natural frequencies and modal vectors of the corresponding undamped system can be found from the equation $\left[K-\lambda M\right]\mathbf{x}=0$. That is,

$$\begin{bmatrix} -4\lambda+8 & -2 \\ -2 & -\lambda+2 \end{bmatrix}\mathbf{x}=0 \tag{b}$$

This yields the eigenvalues λ_1 and λ_2 and the normalized modal matrix $\bar{S}$ as

$$\lambda_1=1,\ \ \lambda_2=3,\ \ \bar{S}=\frac{1}{\sqrt{8}}\begin{bmatrix} 1 & -1 \\ 2 & 2 \end{bmatrix} \tag{c}$$

Thus, the natural frequencies of the system are $\omega_1=1$ and $\omega_2=\sqrt{3}$.

The modal matrix can now be used to decouple Eq. a. To this end, let $\mathbf{x}=\bar{S}\mathbf{y}$. Substituting into Eq. a, and multiplying by $\bar{S}^T$, we obtain the decoupled modal equations

$$\ddot{y}_1+0.1\dot{y}_1+y_1=\sqrt{2}\sin\sqrt{3}t \tag{d}$$

and

$$\ddot{y}_2+0.3\dot{y}_2+3y_2=\sqrt{2}\sin\sqrt{3}t \tag{e}$$

Comparing the coefficient of $\dot{y}_i$ in each case to $2\varsigma_i\omega_i$ yields the damping ratios

$$\varsigma_1=\frac{0.1}{2(1)}=0.05 \tag{f}$$

and

$$\varsigma_2=\frac{0.3}{2\sqrt{3}}=0.087 \tag{g}$$

Note that, although the exciting force F_2 is applied to only one mass, it becomes engaged with both coordinates when transformed to modal coordinates.

The steady-state response is the particular solution of Eqs. d and e. These solutions are given by Eq. 3.2-14 as

$$\begin{aligned} y_{1p}&=\frac{\sqrt{2}}{\sqrt{(1-3)^2+[2(0.05)\sqrt{3}(1)]^2}}\sin\left(\sqrt{3}t-\tan^{-1}\frac{2(0.05)\sqrt{3}(1)}{1-3}\right) \\ &=0.7045\sin(\sqrt{3}t-175^\circ) \end{aligned} \tag{h}$$

and

$$\begin{aligned} y_{2p}&=\frac{\sqrt{2}}{\sqrt{(3-3)^2+[2(0.087)(\sqrt{3})(\sqrt{3})]^2}}\sin\left(\sqrt{3}t-\tan^{-1}\frac{2(0.087)(\sqrt{3})(\sqrt{3})}{3-3}\right) \\ &=2.7092\sin(\sqrt{3}t-\pi/2)=-2.7092\cos\sqrt{3}t \end{aligned} \tag{i}$$

Recall now that the transformation relation used is $\mathbf{x}=\bar{S}\mathbf{y}$. Thus, the solution in the original coordinate system is given by

$$\mathbf{x}=\bar{S}\mathbf{y}=\frac{1}{\sqrt{8}}\begin{bmatrix} 1 & -1 \\ 2 & 2 \end{bmatrix}\begin{bmatrix} 0.7045\sin(\sqrt{3}t-175^\circ) \\ -2.7092\cos\sqrt{3}t \end{bmatrix}$$

Then, in the steady state, we have

$$x_1=0.2491\sin\left(\sqrt{3}t-175^\circ\right)+0.9578\cos\sqrt{3}t \tag{j}$$

and

$$x_2=0.4982\sin\left(\sqrt{3}t-175^\circ\right)-1.9156\cos\sqrt{3}t \tag{k}$$

Note that the second term of each expression represents the contribution of the second mode to the response of the system. Because the exciting frequency ($\omega=\sqrt{3}$) is equal to the second natural frequency of the system, and the vector of the force distribution is not orthogonal to the second modal vector, this causes a large portion of the vibration near the resonance frequency.

(2) Coordinate Decoupling with Initial Value Problems

Consider the equations of motion of an n-degree-of-freedom system

$$M\ddot{\mathbf{x}}+C\dot{\mathbf{x}}+K\mathbf{x}=0, \tag{5.9-10}$$

subject to the initial conditions

$$\mathbf{x}(0)=\mathbf{x}_0,\quad \dot{\mathbf{x}}(0)=\dot{\mathbf{x}}_0 \tag{5.9-11}$$

Also, assume the damping is proportional such that

$$C=aM+bK, \tag{5.9-12}$$

where a and b are constants.

Let $\bar{S}$ be the normalized modal matrix of the corresponding undamped system. Following the procedure of the previous section, let $\mathbf{x}$ be $\bar{S}\mathbf{y}$ in Eq. 5.9-10 and then pre-multiply by $\bar{S}^T$. This yields the n decoupled modal equations

$$\ddot{y}_i+2\varsigma_i\omega_i\dot{y}+\omega_i^2 y=0 \quad (i=1,\cdots,n), \tag{5.9-13}$$

where

$$2\varsigma_i\omega_i=a+b\omega_i^2,\quad \text{or}\quad \varsigma_i=\frac{a}{2\omega_i}+\frac{b\omega_i^2}{2} \quad (i=1,\cdots,n) \tag{5.9-14}$$

The initial conditions in the $\mathbf{y}$ coordinate system can be obtained by transformation (see Eq. 5.8-17) as

$$\mathbf{y}_0=\bar{S}^T M\mathbf{x}_0 \quad \text{and} \quad \dot{\mathbf{y}}_0=\bar{S}^T M\dot{\mathbf{x}}_0 \tag{5.9-15}$$

With reference to Eq. 2.5-15, the solutions of Eq. 5.9-13 are given by

$$y_i = A_i e^{-\varsigma_i \omega_i t} \sin(\omega_i \sqrt{1-\varsigma_i^2}\, t + \varphi_i), \tag{5.9-16}$$

where the A_i and φ_i are

$$A_i = \sqrt{y_{i0}^2 + \left(\frac{\dot{y}_{i0} + \varsigma_i \omega_i y_{i0}}{\omega_i \sqrt{1-\varsigma_i^2}} \right)^2}, \quad \varphi_i = \tan^{-1} \frac{y_{i0}\omega_i \sqrt{1-\varsigma_i^2}}{\dot{y}_{i0} + \varsigma \omega_i y_{i0}} \tag{5.9-17}$$

In the above, we assume that the matrices M and K are positive definite. For positive semidefinite systems, some eigenvalues are zero. To study such systems, let the first r eigenvalues be zero. Then the first r equations in Eq. 5.9-13 become

$$\ddot{y}_i = 0 \quad (i = 1, 2, \cdots, r) \tag{5.9-18}$$

The solutions for these equations are

$$y_i = y_{i0} + \dot{y}_{i0} t \quad (i = 1, 2, \cdots, r) \tag{5.9-19}$$

Once the solutions in the modal coordinate system are obtained, the solutions in the original coordinate system can be found from the simple transformation $\mathbf{x} = \overline{S}\mathbf{y}$.

Example 5.9-2 Consider the two-degree-of-freedom system governed by the matrix equation

$$\begin{bmatrix} 4 & 0 \\ 0 & 1 \end{bmatrix} \ddot{\mathbf{x}} + \begin{bmatrix} 0.8 & -0.2 \\ -0.2 & 0.2 \end{bmatrix} \dot{\mathbf{x}} + \begin{bmatrix} 8 & -2 \\ -2 & 2 \end{bmatrix} \mathbf{x} = \begin{bmatrix} 0 \\ 0 \end{bmatrix} \tag{a}$$

Calculate the response of the system with the initial conditions

$$\mathbf{x}_0 = \begin{bmatrix} 0 \\ 0 \end{bmatrix}, \quad \dot{\mathbf{x}}_0 = \begin{bmatrix} 1 \\ -1 \end{bmatrix} \tag{b}$$

Solution: From Example 5.9-1, the eigenvalues λ_1, λ_2 and the normalized modal matrix $\overline{S}$ are

$$\lambda_1 = 1, \quad \lambda_2 = 3, \quad \overline{S} = \frac{1}{\sqrt{8}} \begin{bmatrix} 1 & -1 \\ 2 & 2 \end{bmatrix} \tag{c}$$

Thus, the natural frequencies of the system are $\omega_1 = 1$ and $\omega_2 = \sqrt{3}$.

Let $\mathbf{x} = \overline{S}\mathbf{y}$. Substituting into Eq. a, and multiplying by $\overline{S}^T$, we obtain the decoupled modal equations

$$\ddot{y}_1 + 0.1\dot{y}_1 + y_1 = 0 \tag{d}$$

and

$$\ddot{y}_2 + 0.3\dot{y}_2 + 3y_2 = 0 \tag{e}$$

Comparing the coefficient of y_i in each case to $2\varsigma_i\omega_i$ yields the damping ratios

$$\varsigma_1 = \frac{0.1}{2(1)} = 0.05 \tag{f}$$

and

$$\varsigma_2 = \frac{0.3}{2\sqrt{3}} = 0.087 \tag{g}$$

The transformed initial conditions are then given by Eq. 5.8-17 as

$$\mathbf{y}_0 = \bar{S}^T M\mathbf{x}_0 = 0 \tag{h}$$

$$\dot{\mathbf{y}}_0 = \bar{S}^T M\dot{\mathbf{x}}_0 = \frac{1}{\sqrt{8}}\begin{bmatrix} 1 & 2 \\ -1 & 2 \end{bmatrix}\begin{bmatrix} 4 & 0 \\ 0 & 1 \end{bmatrix}\begin{bmatrix} 1 \\ -1 \end{bmatrix} = \frac{1}{\sqrt{8}}\begin{bmatrix} 2 \\ -6 \end{bmatrix} \tag{i}$$

The responses in the modal coordinate system are then obtained from Eq. 5.9-16 as

$$y_1 = 0.7080e^{-0.05t}\sin(0.9975t) \tag{j}$$

and

$$y_2 = -1.2294e^{-0.1507t}\sin(1.7255t) \tag{k}$$

The responses in the original coordinate system are then given by $\mathbf{x} = \bar{S}\mathbf{y}$. That is,

$$x_1 = 0.2503e^{-0.05t}\sin(0.9975t) + 0.4346e^{-0.15t}\sin(1.7255t) \tag{l}$$

and

$$x_2 = 0.5006e^{-0.05t}\sin(0.9975t) - 0.8692e^{-0.15t}\sin(1.7255t) \tag{m}$$

Examples 5.9-1 and 5.9-2 illustrate the basic concepts and procedures. It happens, however, that for large systems, the computation of the frequencies and the modal matrix can be quite laborious. It is generally easier to use established software to make the computations. We explore this approach in Chapter Six.

5.10 Lagrange's Equations

Up to this point, in our illustrations, we have considered relatively simple systems, with only a few degrees of freedom (usually only one or two). As such, we have been able to obtain the governing equations of motion by inspection of free-body diagrams (or envisioned free-body diagrams) and then by enforcing Newton's laws. For large systems, however, the number of free-body diagrams increases, and the number of variables and interactive forces increases. The use of Newton's laws then becomes increasingly cumbersome.

An alternative approach is to use Lagrange's equations to obtain the equations of motion. With Lagrange's equations, interactive forces and constraint forces across frictionless surfaces are automatically eliminated from the analysis. These forces then need not be considered further. Lagrange's equations produce the equations of motion simply via differentiation of scalar energy functions (kinetic and potential energy) and by using the work done by the applied forces.

In this section, we provide a brief description of Lagrange's method. References [5.13–5.15] provide more details for interested readers.

(1) Generalized Coordinates and Virtual Displacements

To define the motion of a system, it is convenient to introduce variables describing the degrees of freedom. Any variable (such as a rectangular coordinate, an angle, or a distance) that uniquely defines configuration of a system may be used. Taken together, such variables are called *generalized coordinates*.

If a system S has n degrees of freedom, n generalized coordinates q_r $(r=1,2,\cdots,n)$ are needed to uniquely define the configuration of S, and consequently the motion of S.

Consider next a system S of particles. The displacement $\mathbf{r}_j$ of any particle j of S can be expressed by the n independent generalized coordinates q_r and the time t as

$$\mathbf{r}_j=\mathbf{r}_j(q_1,q_2,\cdots,q_n,t) \tag{5.10-1}$$

A *virtual displacement* $\delta\mathbf{r}_j$ is an infinitesimal change in the displacement $\mathbf{r}_j$ that does not violate the constraints, expressed as

$$\delta\mathbf{r}_j=\sum_r\frac{\partial\mathbf{r}_j}{\partial q_r}\delta q_r, \tag{5.10-2}$$

where the time t is not involved, and where δq_r $(r=1,2,\cdots,n)$ is called the virtual change of the generalized coordinate q_r.

(2) Virtual Work and Generalized Forces

Using generalized coordinates and the concept of virtual displacements, we can define *generalized forces*. To this end, consider an n-degree-of-freedom system S acted upon by an arbitrary set of forces $\mathbf{F}_j$ $(j=1,2,\cdots,N)$ acting on particles P_j of S. As before, let $\delta\mathbf{r}_j$ be the virtual displacement of P_j. Then the *virtual work* δW done by the forces $\mathbf{F}_j$ on the system S is defined as

$$\delta W=\sum_j\mathbf{F}_j\cdot\delta\mathbf{r}_j=\sum_j\sum_r\mathbf{F}_j\cdot\frac{\partial\mathbf{r}_j}{\partial q_r}\delta q_r \tag{5.10-3}$$

By interchanging the order of summation, the *generalized force* Q_r associated with the generalized coordinate q_r is defined as

$$Q_r=\sum_j\mathbf{F}_j\cdot\frac{\partial\mathbf{r}_j}{\partial q_r}\quad(r=1,2,\cdots,n) \tag{5.10-4}$$

The virtual work on the system then takes the form

$$\delta W = \sum_r Q_r \delta q_r = Q_1 \delta q_1 + Q_2 \delta q_2 + \cdots + Q_n \delta q_n \qquad (5.10\text{-}5)$$

Consequently, the generalized forces Q_r may be expressed as

$$Q_r = \frac{\delta W}{\delta q_r} \quad (r = 1, 2, \cdots, n) \qquad (5.10\text{-}6)$$

That is, once the expression for the virtual work is obtained, the coefficient of δq_r in the expression is the corresponding generalized force.

(3) Lagrange's Equations

Lagrange's equations are fundamentally equivalent to Newton's laws, but they are usually derived using a variational principle minimizing an energy function (see, for example, [5.14], pp. 156–159). Specifically, for a vibrating system *S*, Lagrange's equations are found to be

$$\frac{d}{dt}\left(\frac{\partial T}{\partial \dot{q}_r}\right) - \frac{\partial T}{\partial q_r} = Q_r \quad (r = 1, 2, \cdots, n), \qquad (5.10\text{-}7)$$

where T is the kinetic energy of the system, and Q_r is the generalized force corresponding to the generalized coordinate q_r.

There are several variations in the form of Lagrange's equations. If there is no dissipation of the total energy of the system, that is, if the system is *conservative*, possessing a potential energy function $U(q_1, q_2, \cdots, q_n)$, then the work W done by external forces on the system may be expressed as

$$W = -U(q_1, q_2, \cdots, q_n) + C, \qquad (5.10\text{-}8)$$

where C is a constant determined by an arbitrarily chosen reference level for U. For example, the reference level for U may be selected such that C is zero.

From Eqs. 5.10-5 and 5.10-8, we see that the virtual work δW may be expressed in terms of U as

$$\delta W = -\sum_r \frac{\partial U}{\partial q_r} \delta q_r \qquad (5.10\text{-}9)$$

Thus, the generalized force Q_r can be identified and also expressed in terms of U as

$$Q_r = -\frac{\partial U}{\partial q_r} \quad (r = 1, 2, \cdots, n) \qquad (5.10\text{-}10)$$

In view of Eqs. 5.10-7 and 5.10-10, Lagrange's equation may be written in the form

$$\frac{d}{dt}\left(\frac{\partial T}{\partial \dot{q}_r}\right) - \frac{\partial T}{\partial q_r} + \frac{\partial U}{\partial q_r} = 0 \quad (r = 1, 2, \cdots, n) \qquad (5.10\text{-}11)$$

At this point, it is convenient to introduce a *Lagrangian function L* as

$$L = T - U \tag{5.10-12}$$

By observing that U is not a function of the $\dot{q}_r$, we can write Eq. 5.10-11 as

$$\frac{d}{dt}\left(\frac{\partial L}{\partial \dot{q}_r}\right) - \frac{\partial L}{\partial q_r} = 0 \quad (r = 1, 2, \cdots, n) \tag{5.10-13}$$

Eq. 5.10-13 represents the standard form of Lagrange's equation for conservative systems.

Example 5.10-1 [5.16] Two identical pin-connected and pin-supported uniform rods are moving in a vertical plane, as shown in Fig. 5.10-1. Let the orientations of the rods be defined by the angles θ_1 and θ_2, as shown. Let each rod have length b and mass m. Assume the pin connections are frictionless. Use the Lagrangian method to derive the differential equations of motion and obtain the natural frequencies for small vibration by assuming small oscillation.

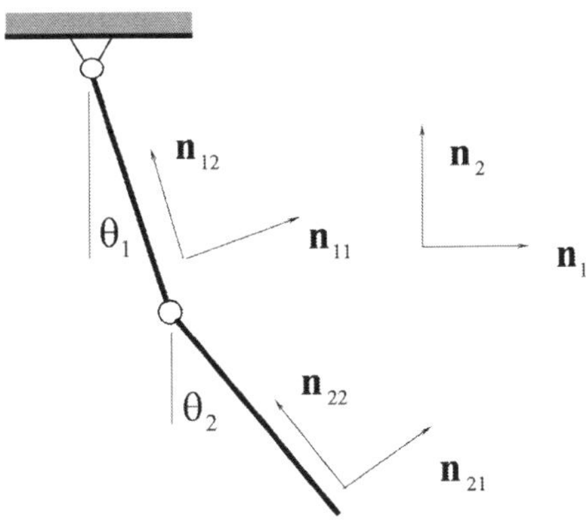

Fig. 5.10-1 A double-rod pendulum [5.16].

Solution: Let the rods be called B_1 and B_2, and let their mass centers be G_1 and G_2. Introducing unit vectors $\mathbf{n}_{11}$, $\mathbf{n}_{12}$, $\mathbf{n}_{21}$, and $\mathbf{n}_{22}$, as shown, the mass center velocities of the rods can be expressed as

$$\mathbf{V}^{G_1} = (b/2)\dot{\theta}_1 \mathbf{n}_{11} \quad \text{and} \quad \mathbf{V}^{G_2} = b\dot{\theta}_1 \mathbf{n}_{11} + (b/2)\dot{\theta}_2 \mathbf{n}_{21} \tag{a}$$

The angular velocities of the rods are

$$\boldsymbol{\omega}^{B_1} = \dot{\theta}_1 \mathbf{n}_3 \quad \text{and} \quad \boldsymbol{\omega}^{B_2} = \dot{\theta}_2 \mathbf{n}_3, \tag{b}$$

where the unit vector $\mathbf{n}_3$ is normal to the plane of motion. The kinetic energy of the system is seen to be

$$T = \frac{1}{2}m\mathbf{V}^{G_1} \cdot \mathbf{V}^{G_1} + \frac{1}{2}I\,\boldsymbol{\omega}^{B_1} \cdot \boldsymbol{\omega}^{B_1} + \frac{1}{2}m\mathbf{V}^{G_2} \cdot \mathbf{V}^{G_2} + \frac{1}{2}I\,\boldsymbol{\omega}^{B_2} \cdot \boldsymbol{\omega}^{B_2}, \tag{c}$$

where I, equal to $(1/12)mb^2$, is the moment of inertia of the rod about its mass center.

Substituting Eqs. a and b into Eq. c, and noting that $\mathbf{n}_{11} \cdot \mathbf{n}_{21} = \cos(\theta_2 - \theta_1)$, we obtain

$$T = \frac{1}{2} m b^2 \left[\frac{4}{3} \dot{\theta}_1^2 + \frac{1}{3} \dot{\theta}_2^2 + \dot{\theta}_1 \dot{\theta}_2 \cos(\theta_2 - \theta_1) \right] \tag{d}$$

Choosing the suspension point as the reference level for potential energy, we obtain the potential energy U as

$$U = -mg[(b/2)\cos\theta_1 + b\cos\theta_1 + (b/2)\cos\theta_2] \tag{e}$$

Therefore, the Lagrangian function is

$$L = T - U = \frac{1}{2} m b^2 \left[\frac{4}{3} \dot{\theta}_1^2 + \frac{1}{3} \dot{\theta}_2^2 + \dot{\theta}_1 \dot{\theta}_2 \cos(\theta_2 - \theta_1) \right] + mgb \left[\frac{3}{2} \cos\theta_1 + \frac{1}{2} \cos\theta_2 \right] \tag{f}$$

The θ_1 equation is obtained from the expressions

$$\frac{\partial L}{\partial \dot{\theta}_1} = m b^2 \left[\frac{4}{3} \dot{\theta}_1 + \frac{1}{2} \dot{\theta}_2 \cos(\theta_2 - \theta_1) \right]$$

$$\frac{d}{dt} \frac{\partial L}{\partial \dot{\theta}_1} = m b^2 \left[\frac{4}{3} \ddot{\theta}_1 + \frac{1}{2} \ddot{\theta}_2 \cos(\theta_2 - \theta_1) - \frac{1}{2} \dot{\theta}_2 (\dot{\theta}_2 - \dot{\theta}_1) \sin(\theta_2 - \theta_1) \right]$$

and

$$\frac{\partial L}{\partial \theta_1} = \frac{1}{2} m b^2 \dot{\theta}_1 \dot{\theta}_2 \sin(\theta_2 - \theta_1) - \frac{3}{2} mgb \sin\theta_1$$

After substitution into Lagrange's equation, we obtain

$$\frac{4}{3} \ddot{\theta}_1 + \frac{1}{2} \ddot{\theta}_2 \cos(\theta_2 - \theta_1) - \frac{1}{2} \dot{\theta}_2^2 \sin(\theta_2 - \theta_1) + \frac{3}{2} \frac{g}{b} \sin\theta_1 = 0 \tag{g}$$

In a similar fashion, the θ_2 equation is obtained from the expressions

$$\frac{\partial L}{\partial \dot{\theta}_2} = m b^2 \left[\frac{1}{3} \dot{\theta}_2 + \frac{1}{2} \dot{\theta}_1 \cos(\theta_2 - \theta_1) \right]$$

$$\frac{d}{dt} \frac{\partial L}{\partial \dot{\theta}_2} = m b^2 \left[\frac{1}{3} \ddot{\theta}_2 + \frac{1}{2} \ddot{\theta}_1 \cos(\theta_2 - \theta_1) - \frac{1}{2} \dot{\theta}_1 (\dot{\theta}_2 - \dot{\theta}_1) \sin(\theta_2 - \theta_1) \right]$$

and

$$\frac{\partial L}{\partial \theta_2} = -\frac{1}{2} m b^2 \dot{\theta}_1 \dot{\theta}_2 \sin(\theta_2 - \theta_1) - \frac{1}{2} mgb \sin\theta_2,$$

which, with Lagrange's equation, yields

$$\frac{1}{3} \ddot{\theta}_2 + \frac{1}{2} \ddot{\theta}_1 \cos(\theta_2 - \theta_1) + \frac{1}{2} \dot{\theta}_1^2 \sin(\theta_2 - \theta_1) + \frac{1}{2} \frac{g}{b} \sin\theta_2 = 0 \tag{h}$$

We now linearize the equations for the case of small vibration. That is, we assume that θ_1, θ_2, and their time derivatives are much smaller than one. With these assumptions, we can use the approximations

$$\cos(\theta_2-\theta_1)\approx 1,\ \sin(\theta_2-\theta_1)\approx\theta_2-\theta_1,\ \sin\theta_1\approx\theta_1,\ \text{and}\ \sin\theta_2\approx\theta_2$$

The equations of motion *g* and *h* then become

$$8\ddot{\theta}_1+3\ddot{\theta}_2+9(g/b)\theta_1=0 \tag{i}$$

and

$$3\ddot{\theta}_1+2\ddot{\theta}_2+3(g/b)\theta_2=0, \tag{j}$$

where we neglected higher-order terms in the small quantities.

The mass and stiffness matrices for the small vibration are immediately seen to be

$$M=\begin{bmatrix} 8 & 3 \\ 3 & 2 \end{bmatrix} \quad \text{and} \quad K=\frac{g}{b}\begin{bmatrix} 9 & 0 \\ 0 & 3 \end{bmatrix} \tag{k}$$

The characteristic equation is then

$$|K-\lambda M|=0 \tag{l}$$

By expanding the determinant, we have

$$7\lambda^2-42(g/b)\lambda+27(g/b)^2=0 \tag{m}$$

The roots of the characteristic equation are

$$\lambda_1=\frac{g}{b}\left(3-\frac{6}{\sqrt{7}}\right) \quad \text{and} \quad \lambda_2=\frac{g}{b}\left(3+\frac{6}{\sqrt{7}}\right) \tag{n}$$

Therefore, the natural frequencies for small vibration of the system are

$$\omega_1=\sqrt{\frac{g}{b}\left(3-\frac{6}{\sqrt{7}}\right)} \quad \text{and} \quad \omega_2=\sqrt{\frac{g}{b}\left(3+\frac{6}{\sqrt{7}}\right)} \tag{o}$$

5.11 Kane's Equations

There are numerous laws, principals, and equations that may be used to obtain equations of motion for mechanical systems. Although they are all essentially equivalent to Newton's laws, their methodologies and ease of use vary considerably—particularly for large systems. With large systems (systems with many degrees of freedom), some of the methodologies become virtually intractable. The best of the methodologies for ease of use and computational efficiency appears to be with Kane's equations. Indeed, Kane's equations and their associated procedures are increasingly becoming the method of choice for studying the dynamics of large mechanical systems.

In this section, we briefly summarize the procedures of Kane's equations, with an emphasis upon vibrating systems. References [5.16–5.18] provide more details.

The advantages with Kane's equations include (1) the avoidance of the use of virtual quantities such as virtual displacements and virtual work, (2) the avoidance of the use of energy functions and consequently avoiding the need to differentiate them, and (3) the use of an efficient procedure to calculate generalized forces and thereby eliminate interactive constraint forces between bodies of the system. In place of virtual displacements and rotations, Kane's equations use *partial velocity* and *partial angular velocity* vectors.

We begin our discussion with a review of these vectors.

(1) Partial Velocities and Partial Angular Velocities

Consider a mechanical system S with n degrees of freedom measured by generalized coordinates q_r $(r=1,2,\cdots,n)$. Consider a typical body B_i of S, and let G_i be the mass center of B_i. Let $\mathbf{V}^{G_i}$ be the velocity of G_i and $\boldsymbol{\omega}^{B_i}$ be the angular velocity of B_i in an inertial reference frame R.

From the principles of kinematics, it is readily seen that $\mathbf{V}^{G_i}$ and $\boldsymbol{\omega}^{B_i}$ are linear functions of the generalized coordinate derivatives $\dot{q}_r$ $(r=1,2,\cdots,n)$. Specifically, $\mathbf{V}^{G_i}$ and $\boldsymbol{\omega}^{B_i}$ may be written in the forms [5.16–5.18]

$$\mathbf{V}^{G_i} = \mathbf{V}_t^{G_i} + \sum_{r=1}^{n} \mathbf{V}_{\dot{q}_r}^{G_i} \dot{q}_r \tag{5.11-1}$$

and

$$\boldsymbol{\omega}^{B_i} = \boldsymbol{\omega}_t^{B_i} + \sum_{r=1}^{n} \boldsymbol{\omega}_{\dot{q}_r}^{B_i} \dot{q}_r, \tag{5.11-2}$$

where the coefficients $\mathbf{V}_t^{G_i}$, $\mathbf{V}_{\dot{q}_r}^{G_i}$, $\boldsymbol{\omega}_t^{B_i}$, and $\boldsymbol{\omega}_{\dot{q}_r}^{B_i}$, are, in general, functions of the q_r and time t. The coefficients of the $\dot{q}_r$, $\mathbf{V}_{\dot{q}_r}^{G_i}$, and $\boldsymbol{\omega}_{\dot{q}_r}^{B_i}$ are called *partial velocities* and *partial angular velocities* of G_i and B_i, respectively, for the generalized coordinate q_r. Alternatively, the partial velocity of G_i and the partial angular velocity of B_i, for q_r, may be defined as

$$\mathbf{V}_{\dot{q}_r}^{G_i} = \frac{\partial \mathbf{V}^{G_i}}{\partial \dot{q}_r} \tag{5.11-3}$$

and

$$\boldsymbol{\omega}_{\dot{q}_r}^{B_i} = \frac{\partial \boldsymbol{\omega}^{B_i}}{\partial \dot{q}_r} \tag{5.11-4}$$

(2) Inertia Forces

If we consider the inertia forces on all of the particles of a typical body B_i, we have an extremely large set of forces. To simplify the analysis, we can replace this set of forces by an equivalent force system consisting of only a single force

$\mathbf{R}_i^*$, passing through the mass center G_i, together with a couple with torque $\mathbf{T}_i^*$. From [5.16–5.18], $\mathbf{R}_i^*$ and $\mathbf{T}_i^*$ are seen to be

$$\mathbf{R}_i^* = -m\mathbf{a}^{G_i} \tag{5.11-5}$$

and

$$\mathbf{T}_i^* = -\boldsymbol{\alpha}^{B_i} \cdot \mathbf{I}^{B_i} - \boldsymbol{\omega}^{B_i} \times (\mathbf{I}^{B_i} \cdot \boldsymbol{\omega}^{B_i}), \tag{5.11-6}$$

where $\mathbf{a}^{G_i}$ is the acceleration of the mass center G_i in an inertia reference frame R; $\boldsymbol{\alpha}^{B_i}$ and $\boldsymbol{\omega}^{B_i}$ are the angular acceleration and angular velocity, respectively, of B_i in R; and $\mathbf{I}^{B_i}$ is the central dyadic of B_i.

Let $\mathbf{n}_1$, $\mathbf{n}_2$, and $\mathbf{n}_3$ be mutually perpendicular unit vectors parallel to principal inertia directions of B_i for G_i. Let $\mathbf{T}_i^*$, $\mathbf{I}^{B_i}$, $\boldsymbol{\omega}^{B_i}$, and $\boldsymbol{\alpha}^{B_i}$ be expressed as

$$\mathbf{T}_i^* = T_1^*\mathbf{n}_1 + T_2^*\mathbf{n}_2 + T_3^*\mathbf{n}_3 \tag{5.11-7}$$

$$\mathbf{I}^{B_i} = I_{11}\mathbf{n}_1\mathbf{n}_1 + I_{22}\mathbf{n}_2\mathbf{n}_2 + I_{33}\mathbf{n}_3\mathbf{n}_3 \tag{5.11-8}$$

$$\boldsymbol{\omega}^{B_i} = \omega_1\mathbf{n}_1 + \omega_2\mathbf{n}_2 + \omega_3\mathbf{n}_3 \tag{5.11-9}$$

$$\boldsymbol{\omega}^{B_i} = \alpha_1\mathbf{n}_1 + \alpha_2\mathbf{n}_2 + \alpha_3\mathbf{n}_3 \tag{5.11-10}$$

Then Eq. 5.11-6 may be expressed in component form as

$$T_1^* = -\alpha_1 I_{11} + \omega_2\omega_3(I_{22} - I_{33}) \tag{5.11-11}$$

$$T_2^* = -\alpha_2 I_{22} + \omega_3\omega_1(I_{33} - I_{11}) \tag{5.11-12}$$

$$T_3^* = -\alpha_3 I_{33} + \omega_1\omega_1(I_{11} - I_{22}) \tag{5.11-13}$$

Next, if body B_i has planar motion such that $\boldsymbol{\alpha}^{B_i}$ and $\boldsymbol{\omega}^{B_i}$ have the simplified forms

$$\boldsymbol{\omega}^{B_i} = \omega\mathbf{n}_3 \quad \text{and} \quad \boldsymbol{\alpha}^{B_i} = \alpha\mathbf{n}_3 = \dot{\omega}\mathbf{n}_3, \tag{5.11-14}$$

then the term $\boldsymbol{\omega}^{B_i} \times (\mathbf{I}^{B_i} \cdot \boldsymbol{\omega}^{B_i})$ of Eq. 5.11-6 becomes zero, and the components of $\mathbf{T}_i^*$ are

$$T_1^* = T_2^* = 0 \quad \text{and} \quad T_3^* = -\alpha I_{33} \tag{5.11-15}$$

Finally, the generalized inertia force associated with the generalized coordinate q_r for the system consisting of N bodies is

$$F_{q_r}^* = \sum_{i=1}^{N} \mathbf{R}_i^* \cdot \mathbf{V}_{\dot{q}_r}^{G_i} + \sum_{i=1}^{N} \mathbf{T}_i^* \cdot \boldsymbol{\omega}_{\dot{q}_r}^{B_i} \quad (r = 1, 2, \cdots, n), \tag{5.11-16}$$

where $\mathbf{V}_{\dot{q}_r}^{G_i}$ and $\boldsymbol{\omega}_{\dot{q}_r}^{B_i}$ are the partial velocity for G_i and the partial angular velocity for B_i (see Eqs. 5.11-3 and 5.11-4).

(3) Generalized Applied Forces

For computational purposes, it is convenient to replace the applied (or *active*) force system on body B_i by an equivalent force system, consisting of a single force $\mathbf{F}_i$ passing through the mass center G_i together with a couple with torque $\mathbf{M}_i$. Then the generalized force associated with the generalized coordinate q_r is

$$F_{q_r} = \sum_{i=1}^{N} \mathbf{F}_i \cdot \mathbf{V}_{\dot{q}_r}^{G_i} + \sum_{i=1}^{N} \mathbf{M}_i \cdot \boldsymbol{\omega}_{\dot{q}_r}^{B_i} \quad (r = 1, 2, \cdots, n), \tag{5.11-17}$$

where, again, $\mathbf{V}_{\dot{q}_r}^{G_i}$ and $\boldsymbol{\omega}_{\dot{q}_r}^{B_i}$ are the partial velocity for G_i and the partial angular velocity for B_i (see Eqs. 5.11-3 and 5.11-4).

(4) Kane's Equations

Kane's equations state that the sum of the generalized inertia force and the generalized applied force of the system is zero for each generalized coordinate q_r, namely,

$$F_{q_r}^{*} + F_{q_r} = 0 \quad (r = 1, 2, \cdots, n) \tag{5.11-18}$$

Example 5.11-1 Repeat Example 5.10-1, using Kane's method to derive the equations of motion of the system.

Solution: As before, let the rods be called B_1 and B_2, and let their mass centers be G_1 and G_2. Introducing unit vectors $\mathbf{n}_{11}$, $\mathbf{n}_{12}$, $\mathbf{n}_{21}$, and $\mathbf{n}_{22}$, as shown in Fig. 5.10-1, the mass center velocities of the rods may be expressed as

$$\mathbf{V}^{G_1} = (b/2)\dot{\theta}_1 \mathbf{n}_{11} \qquad \text{and} \qquad \mathbf{V}^{G_2} = b\dot{\theta}_1 \mathbf{n}_{11} + (b/2)\dot{\theta}_2 \mathbf{n}_{21} \tag{a}$$

The angular velocities of the rods are

$$\boldsymbol{\omega}^{B_1} = \dot{\theta}_1 \mathbf{n}_3 \qquad \text{and} \qquad \boldsymbol{\omega}^{B_2} = \dot{\theta}_2 \mathbf{n}_3, \tag{b}$$

where $\mathbf{n}_3$ is a unit vector perpendicular to the plane of motion.

The mass center accelerations and the angular accelerations of the rods are

$$\mathbf{a}^{G_1} = (b/2)\ddot{\theta}_1 \mathbf{n}_{11} + (b/2)\dot{\theta}_1^2 \mathbf{n}_{12} \tag{c}$$

$$\mathbf{a}^{G_2} = b\ddot{\theta}_1 \mathbf{n}_{11} + b\dot{\theta}_1^2 \mathbf{n}_{12} + (b/2)\ddot{\theta}_2 \mathbf{n}_{21} + (b/2)\dot{\theta}_2^2 \mathbf{n}_{22} \tag{d}$$

$$\boldsymbol{\alpha}^{B_1} = \ddot{\theta}_1 \mathbf{n}_3 \tag{e}$$

$$\boldsymbol{\alpha}^{B_2} = \ddot{\theta}_2 \mathbf{n}_3 \tag{f}$$

From Eqs. a and b, the partial velocities of G_1 and G_2 for $\dot{\theta}_1$ and $\dot{\theta}_2$ are seen to be

$$\mathbf{V}_{\dot{\theta}_1}^{G_1} = (b/2)\mathbf{n}_{11} \qquad \text{and} \qquad \mathbf{V}_{\dot{\theta}_2}^{G_1} = 0 \tag{g}$$

$$\mathbf{V}_{\dot{\theta}_1}^{G_2} = b\mathbf{n}_{11} \qquad \text{and} \qquad \mathbf{V}_{\dot{\theta}_2}^{G_2} = (b/2)\mathbf{n}_{21} \tag{h}$$

From Eq. b, the partial angular velocities of B_1 and B_2 are seen to be

$$\boldsymbol{\omega}^{B_1}_{\dot{\theta}_1} = \mathbf{n}_3 \qquad \text{and} \qquad \boldsymbol{\omega}^{B_1}_{\dot{\theta}_2} = 0 \tag{i}$$

$$\boldsymbol{\omega}^{B_2}_{\dot{\theta}_1} = 0 \qquad \text{and} \qquad \boldsymbol{\omega}^{B_2}_{\dot{\theta}_2} = \mathbf{n}_3 \tag{j}$$

Since the rods have planar motion, the term $\boldsymbol{\omega}\times(\mathbf{I}^G\cdot\boldsymbol{\omega})$ of the inertia torque is zero. Therefore, the generalized inertia forces can be expressed as

$$F^*_{\theta_1} = (-m\mathbf{a}^{G_1})\cdot\mathbf{V}^{G_1}_{\dot{\theta}_1} + (-m\mathbf{a}^{G_2})\cdot\mathbf{V}^{G_2}_{\dot{\theta}_1} - \boldsymbol{\alpha}^{B_1}\cdot\mathbf{I}^{B_1}\cdot\boldsymbol{\omega}^{B_1}_{\dot{\theta}_1} - \boldsymbol{\alpha}^{B_2}\cdot\mathbf{I}^{B_2}\cdot\boldsymbol{\omega}^{B}_{\dot{\theta}_1} \tag{k}$$

$$F^*_{\theta_2} = (-m\mathbf{a}^{G_1})\cdot\mathbf{V}^{G_1}_{\dot{\theta}_2} + (-m\mathbf{a}^{G_2})\cdot\mathbf{V}^{G_2}_{\dot{\theta}_2} - \boldsymbol{\alpha}^{B_1}\cdot\mathbf{I}^{B_1}\cdot\boldsymbol{\omega}^{B_1}_{\dot{\theta}_2} - \boldsymbol{\alpha}^{B_2}\cdot\mathbf{I}^{B_2}\cdot\boldsymbol{\omega}^{B_2}_{\dot{\theta}_2} \tag{l}$$

Note that

$$\mathbf{I}^{B_1} = \mathbf{I}^{B_2} = \frac{1}{12}mb^2\mathbf{n}_3\mathbf{n}_3,\ \mathbf{n}_{21}\cdot\mathbf{n}_{11} = \cos(\theta_2-\theta_1), \quad \text{and} \quad \mathbf{n}_{22}\cdot\mathbf{n}_{11} = -\sin(\theta_2-\theta_1)$$

From Eqs. k and l, we obtain

$$F^*_{\theta_1} = -mb^2\left[\frac{4}{3}\ddot{\theta}_1 + \frac{1}{2}\ddot{\theta}_2\cos(\theta_2-\theta_1) - \frac{1}{2}\dot{\theta}_2^2\sin(\theta_2-\theta_1)\right] \tag{m}$$

and

$$F^*_{\theta_2} = -mb^2\left[\frac{1}{3}\ddot{\theta}_2 + \frac{1}{2}\ddot{\theta}_1\cos(\theta_2-\theta_1) + \frac{1}{2}\dot{\theta}_1^2\sin(\theta_2-\theta_1)\right] \tag{n}$$

The only nonzero (*working*) active forces are due to gravity. These are represented by forces $-mg\mathbf{n}_2$ through G_1 and G_2. The generalized applied forces are then

$$F_{\theta_1} = (-mg\mathbf{n}_2)\cdot\mathbf{V}^{G_1}_{\dot{\theta}_1} + (-mg\mathbf{n}_2)\cdot\mathbf{V}^{G_2}_{\dot{\theta}_1} = -\frac{3}{2}mgb\sin\theta_1 \tag{o}$$

and

$$F_{\theta_2} = (-mg\mathbf{n}_2)\cdot\mathbf{V}^{G_1}_{\dot{\theta}_2} + (-mg\mathbf{n}_2)\cdot\mathbf{V}^{G_2}_{\dot{\theta}_2} = -\frac{1}{2}mgb\sin\theta_1 \tag{p}$$

Kane's equations state that

$$F^*_{\theta_1} + F_{\theta_1} = 0 \tag{q}$$

and

$$F^*_{\theta_2} + F_{\theta_2} = 0 \tag{r}$$

Hence, we obtain the equations of motion of the system

$$\frac{4}{3}\ddot{\theta}_1 + \frac{1}{2}\ddot{\theta}_2\cos(\theta_2-\theta_1) - \frac{1}{2}\dot{\theta}_2^2\sin(\theta_2-\theta_1) + \frac{3}{2}\frac{g}{b}\sin\theta_1 = 0 \tag{s}$$

and

$$\frac{1}{3}\ddot{\theta}_2+\frac{1}{2}\ddot{\theta}_1\cos(\theta_2-\theta_1)+\frac{1}{2}\dot{\theta}_1^2\sin(\theta_2-\theta_1)+\frac{1}{2}\frac{g}{b}\sin\theta_2=0 \tag{t}$$

We see that these equations are the same as in Example 5.10-1, but the computational effort is considerably less.

Problems

5.1 Find the natural frequency of the system shown in Fig. P5.1.

Ans. $\omega_1=0,\ \omega_2=\sqrt{\dfrac{3k}{2m}}$

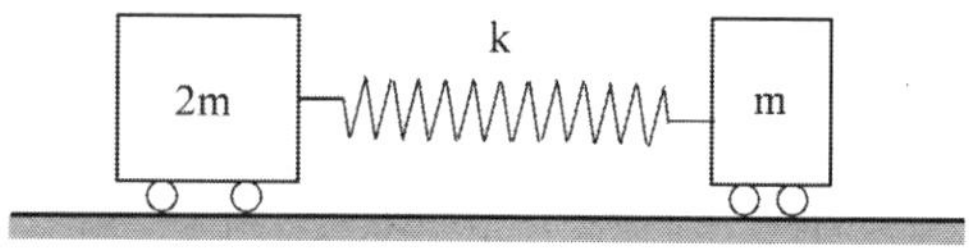

Fig. P5.1

5.2 Find the natural frequencies of torsional vibration of the system shown in Fig. P5.2. Assume that the torsional stiffness of the shaft is K and that the inertia effect of the shaft is negligible compared with J_1 and J_2.

Ans. $\omega=\sqrt{K\left(\dfrac{1}{J_1}+\dfrac{1}{J_2}\right)}$

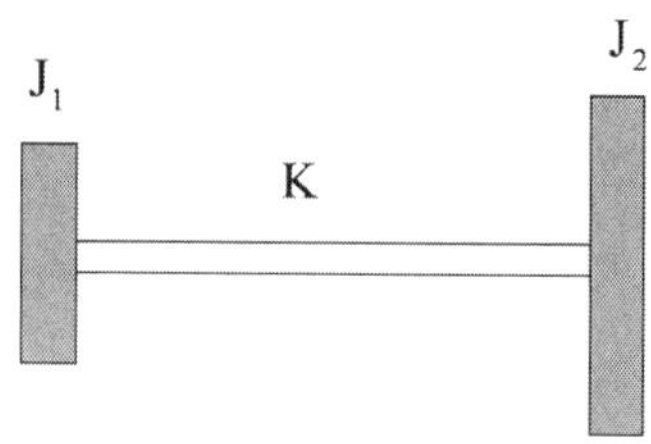

Fig. P5.2

5.3 Use the Lagrangian method to derive Eqs. 5.7-4, 5.7-7, and 5.7-10, which described the coordinate coupling of the system in Fig. 5.7-1.

5.4 [5.16] A triple-rod pendulum consists of three, identical pin-connected rods moving in a vertical plane, as shown in Fig. P5.4. Let each rod have length b. Derive equations of motion for small vibrations of the system.

Ans.

$$(7/3)\ddot{\theta}_1+(3/2)\ddot{\theta}_2+(1/2)\ddot{\theta}_3+(5/2)(g/b)\theta_1=0$$
$$(3/2)\ddot{\theta}_1+(4/3)\ddot{\theta}_2+(1/2)\ddot{\theta}_3+(3/2)(g/b)\theta_2=0$$
$$(1/2)\ddot{\theta}_1+(1/2)\ddot{\theta}_2+(1/3)\ddot{\theta}_3+(1/2)(g/b)\theta_3=0$$

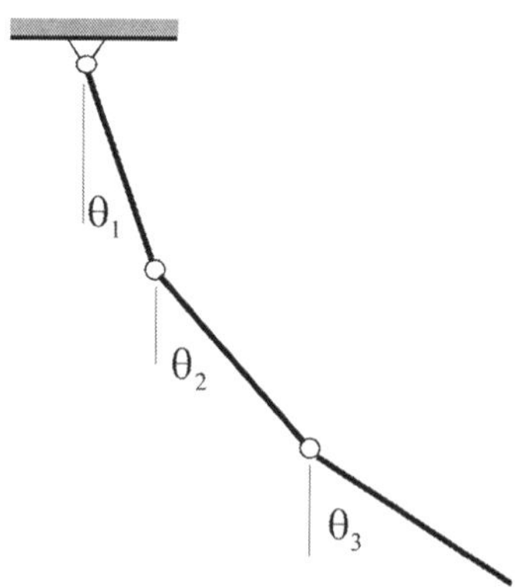

Fig. P5.4

5.5 Derive the equations of motion for small oscillations in the systems shown in Fig. P5.5.

Ans. $(m_1+m_2)\ddot{x}+m_2L\ddot{\theta}+2kx=0$

$$\ddot{x}/L+\ddot{\theta}+(g/L)\theta=0$$

5.6 A uniform bar of length L and mass m is hinged at O and rotated with constant angular velocity Ω about the vertical axis, as in Fig. P5.6. Derive the equation of motion. Find the angle θ_0 for the steady state and the frequency of small oscillation about θ_0.

Ans. $\ddot{\theta}-(1/2)\Omega^2\sin 2\theta+(3g/2L)\sin\theta=0,$

$$\omega=\sqrt{(3g/2L)\cos\theta_0-\Omega^2\cos 2\theta_0}$$

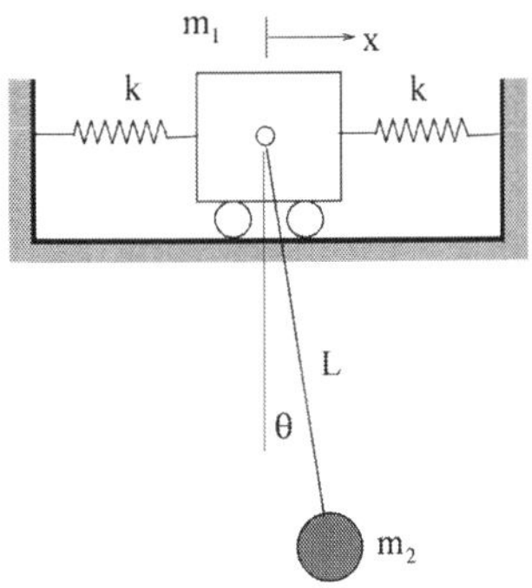

Fig. P5.5

5.7 [5.19] Derive the equations of motion for the system shown in Fig. P5.7. Find the frequency equation for small vibration.

Ans.

$$(M+2m)L_1^2\ddot{\theta}+mL_1L_2(\ddot{\phi}_1+\ddot{\phi}_2)+(M+2m)gL_1\theta=0$$

$$mL_1L_2\ddot{\theta}+mL_2^2\ddot{\phi}_1+mgL_2\phi_1=0,\; mL_1L_2\ddot{\theta}+mL_2^2\ddot{\phi}_2+mgL_2\phi_2=0$$

$$(L_2\omega^2-g)[ML_1L_2\omega^4-(M+2m)(L_1+L_2)g\omega^2+(M+2m)g^2]=0$$

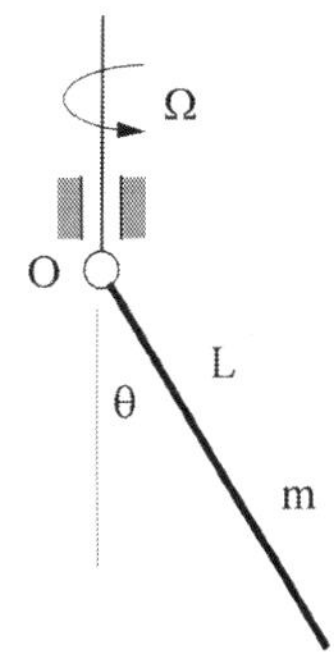

Fig. P5.6

5.8 Consider the system shown in Fig. P5.8. (a) Find the natural frequencies. (b) Find the modal matrix. (c) Find the uncoupled equations of motion.

Ans. $\omega_1=\sqrt{4k/m},\quad \omega_2=\sqrt{8k/m},\quad S=\begin{bmatrix}1 & 1\\ 2 & -2\end{bmatrix}$

$$\ddot{y}_1+\omega_1^2y_1=(1/m)\sin\omega t,$$

$$\ddot{y}_2+\omega_2^2y_2=(1/m)\sin\omega t$$

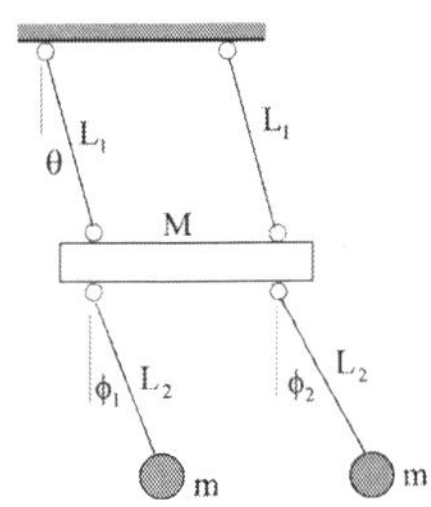

Fig. P5.7

5.9 Referring to the system in Problem 5.8, find the transient response for each of the following initial conditions:

(a) $\mathbf{x}(0)=0;\ \dot{\mathbf{x}}(0)=0;$

(b) $\mathbf{x}(0)=0,\quad \dot{\mathbf{x}}(0)=\begin{bmatrix}2 & 1\end{bmatrix}^T$

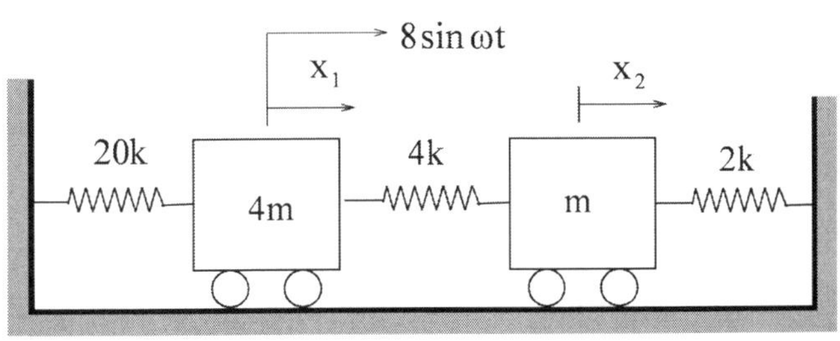

Fig. P5.8

Ans. (a) $y_1 = -\frac{\omega/\omega_1}{m(\omega_1^2-\omega^2)}\sin\omega_1 t + \frac{1}{m(\omega_1^2-\omega^2)}\sin\omega t$

$$y_2 = -\frac{\omega/\omega_2}{m(\omega_2^2-\omega^2)}\sin\omega_2 t + \frac{1}{m(\omega_2^3-\omega^2)}\sin\omega t$$

$$x_1 = y_1 + y_2,\ x_2 = 2(y_1 - y_2)$$

(b) $y_1 = \left[\frac{5}{4\omega_1} - \frac{\omega/\omega_1}{m(\omega_1^2-\omega^2)}\right]\sin\omega_1 t + \frac{1}{m(\omega_1^2-\omega^2)}\sin\omega t$

$$y_2 = \left[\frac{3}{4\omega_2} - \frac{\omega/\omega_2}{m(\omega_2^2-\omega^2)}\right]\sin\omega_2 t + \frac{1}{m(\omega_2^3-\omega^2)}\sin\omega t$$

$$x_1 = y_1 + y_2,\ x_2 = 2(y_1 - y_2)$$

5.10 Find the natural frequencies and modes of the system shown in Fig. P5.10.

Ans. $\omega_1 = 0.445\sqrt{k/m},\quad \omega_2 = 1.247\sqrt{k/m},\quad \omega_3 = 1.802\sqrt{k/m}$

$$S = \begin{bmatrix} 1.000 & 1.000 & 1.000 \\ 1.802 & 0.445 & -1.247 \\ 2.247 & -0.802 & 0.555 \end{bmatrix}$$

5.11 [5.16] Repeat Problem 5.4 for small vibration of the *N*-rod pendulum, which consists of *N* identical pin-connected rods moving in a vertical plane.

Ans. $\sum_{j=1}^{N}\left[m_{ij}\ddot{\theta}_j + (g/L)k_{ij}\theta_j\right] = 0,\ i = 1, 2, \cdots, N$

$m_{ij} = (1/2)[1+2(N-k)$, if $i \neq j$ and k is the larger of i and j.

$$m_{ii} = N - i + 1/3$$

$k_{ij} = 0$, if $i \neq j$; $k_{ii} = N - i + 1/2$

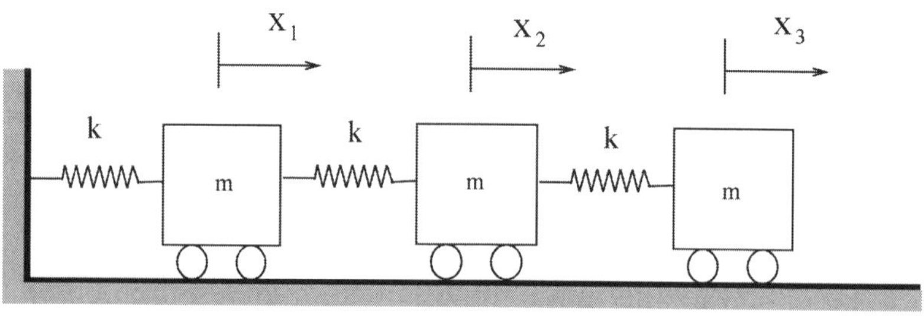

Fig. P5.10

References

5.1. Thomson, W. T., and Dahleh, M. D., *Theory of Vibration with Applications*, 5th ed., Prentice Hall, Inc., Upper Saddle River, NJ, 1998.

5.2. Allemang, R. J., and Brown, D. L., "Experimental Modal Analysis," *Shock and Vibration Handbook*, C. M. Harris, ed., 3rd ed., McGraw-Hill, New York, 1987, pp. 21-1 to 21-34.

5.3. Allemang, R. J., *Vibrations: Analytical and Experimental Modal Analysis*, UC-SDRLL-CN-20-263-691, Department of Mechanical, Industrial, and Nuclear Engineering, University of Cincinnati, Cincinnati, OH, 1992.

5.4. Meirovitch, L., *Analytical Methods in Vibrations*, Macmillan Company, New York, 1967.

5.5. Meirovitch, L., *Principles and Techniques of Vibration*, Prentice Hall, Upper Saddle River, NJ, 1995.

5.6. Den Hartog, J. P., *Mechanical Vibrations*, 4th ed., McGraw-Hill, New York, 1956. (Reprinted by Dover, New York, 1985.)

5.7. Inman, D. J., *Engineering Vibration*, Prentice Hall, Inc., Upper Saddle River, NJ, 1994.

5.8. Tse, F. S., Morse, I. E., and Hinkle, R. T., *Mechanical Vibration, Theory and Applications*, Allyn and Bacon, Inc., Boston, 1979.

5.9. Timoshenko, S. P., Young, D. H., and Weaver, W., Jr., *Vibration Problems in Engineering*, 4th ed., John Wiley & Sons, New York, 1967.

5.10. Harris, C. M., ed., *Shock and Vibration Handbook*, 3rd ed., McGraw-Hill, New York, 1987.

5.11. Craig, R. R., Jr., *Structural Dynamics: An Introduction to Computer Methods*, Wiley, New York, 1981.

5.12. Harker, R. J., *Generalized Method of Vibration Analysis*, John Wiley & Sons, New York, 1983.

5.13. Goldstein, H., *Classical Mechanics*, Addison-Wesley, Reading, MA, 1980.

5.14. Greenwood, D. L., *Classical Dynamics*, Prentice Hall, Englewood Cliffs, NJ, 1977.

5.15. Rosenberg, R. M., *Analytical Dynamics of Discrete Systems*, Plenum Press, New York, 1977.

5.16. Huston, R. L., and Liu, C. Q., *Formula for Dynamic Analysis*, Marcel Dekker, Inc., New York, 2001.

5.17. Huston, R. L., *Multibody Dynamics*, Butterworth-Heinemann, Boston, 1990.

5.18. Kane, T. R., and Levinson, D. A., *Dynamics: Theory and Applications*, McGraw-Hill, New York, 1985.

5.19. Zheng, Z. C., ed., *Mechanical Vibrations*, Mechanical Industry Press, Beijing, China, 1986 (in Chinese).

Chapter Six

Numerical Methods

6.1 Introduction

Most vibration problems of practical importance arise with large mechanical systems—that is, systems with many degrees of freedom. Unfortunately, with large systems, the procedures of the foregoing chapters become tedious and prone to routine errors. In this chapter we consider more efficient methods for finding natural frequencies and mode shapes for these large systems.

If we can develop the problem of finding frequencies and mode shapes into the algebraic eigenvalue-eigenvector problem, we can use one or more of the many available computer programs to determine the frequencies and mode shapes. We begin with a review of various eigenvalue analyses, looking for ways to relate them to vibration problems. References [6.1–6.4] provide more details for interested readers.

6.2 Various Eigenvalue Analyses

(1) The Generalized Eigenvalue Problem

Consider the small vibration of an undamped system with n degrees of freedom. The equation of motion may be written in the matrix form as

$$M\ddot{\mathbf{x}} + K\mathbf{x} = 0, \tag{6.2-1}$$

where, as before, $\mathbf{x}$ is an $n \times 1$ displacement vector, M is the $n \times n$ mass matrix, and K is the $n \times n$ stiffness matrix. Both M and K are symmetric.

If we assume a solution $\mathbf{x}$ of the form $e^{j\omega t}\mathbf{u}$, Eq. 6.2-1 becomes the *generalized eigenvalue problem*

$$K\mathbf{u} = \lambda M\mathbf{u}, \tag{6.2-2}$$

where λ is ω^2. There are, in general, n values λ_i $(i = 1, 2, \cdots, n)$ of λ (the *eigenvalues*) and, correspondingly, n vectors $\mathbf{u}_i$ $(i = 1, 2, \cdots, n)$ (the *eigenvectors*) providing nonzero solutions of Eq. 6.2-2. Consequently, the solutions of Eq. 6.2-1 may be written in the form

$$\mathbf{x}(t) = \sum_i C_i \mathbf{u}_i \sin(\omega_i t + \phi_i), \tag{6.2-3}$$

where the C_i and ϕ_i are constants to be determined by initial conditions.

This approach is direct or *straightforward*, but unfortunately not numerically efficient.

(2) The Algebraic Eigenvalue Problem

In addition to being symmetric, the mass matrix M is positive definite. That is, its determinant is positive. Therefore, M is non-singular and its inverse M^{-1} exists. By multiplying Eq. 6.2-1 by M^{-1} and looking for a solution in the form $\mathbf{x} = e^{j\omega t}\mathbf{u}$, we obtain, after simplification,

$$\left(M^{-1}K\right)\mathbf{u} = \lambda\,\mathbf{u} \tag{6.2-4}$$

Eq. 6.2-4 is the fundamental equation of the standard *algebraic eigenvalue problem*. As before, the solution yields n eigenvalues λ_i $(i = 1, 2, \cdots, n)$ and n eigenvectors $\mathbf{u}_i$ $(i = 1, 2, \cdots, n)$. The solution $\mathbf{x}$ again has the form

$$\mathbf{x}(t) = \sum C_i \mathbf{u}_i \sin(\omega_i t + \phi_i), \tag{6.2-5}$$

where, as before, C_i and ϕ_i $(i = 1, 2, \cdots, n)$ are constants to be determined by initial conditions.

Observe that the matrix $M^{-1}K$ is not symmetric. In general, the eigenvalues of a non-symmetric real matrix will include pairs of complex conjugate values. But since the eigenvalues of the generalized eigenvalue problem of Eq. 6.2-2 are real, and since the eigenvalues of a physical system are dependent upon the physical properties of the system, the eigenvalues of Eqs. 6.2-2 and 6.2-4 will be the same. Similarly, the associated eigenvectors will be the same.

Observe further that if we use Eq. 6.2-4 to obtain the eigenvalues and eigenvectors, we need to evaluate the elements of the inverse matrix M^{-1}. If the matrix M happens to be diagonal [that is, $M = \mathrm{diag}(m_i)$], the inverse is also diagonal, [that is, $M^{-1} = \mathrm{diag}(1/m_i)$]. In this case, we can obtain $M^{-1}K$ by a simple matrix multiplication. If, however, M is not diagonal (that is, if the physical system is dynamically coupled), we need to find convenient means for finding M^{-1}.

Fortunately, there are numerous computer routines available to efficiently invert real symmetric matrices [6.5]. For example, in Matlab, the inverse is computed by the simple command inv(M). Nevertheless, even if M^{-1} is efficiently evaluated, the product $M^{-1}K$ will generally not be symmetric. This asymmetry then leads to inefficiencies in the numerical completion of the eigenvalue problem. This latter difficulty can be overcome by formulating a symmetric eigenvalue problem, as in the following paragraphs.

(3) The Symmetric Eigenvalue Problem

The most efficient way to find eigenvalues and eigenvectors (mode shapes) of a vibration problem (6.2-1) is to cast it in the form of a symmetric eigenvalue problem, due to the advantageous properties associated with symmetry. Indeed, the symmetric eigenvalue problem has several advantages [6.1]. First, the eigenvalues are all real numbers. If the symmetric matrix is positive definite, the eigenvalues are positive real numbers. Second, the eigenvectors are real valued and are orthogonal to each other, even for repeated eigenvalues. Third, the eigenvectors are linearly independent and complete. Thus any n-dimensional vector can be represented as a linear combination of the eigenvectors. In addition, computing the eigenvalues of a symmetric matrix is more efficient than with an asymmetric matrix [6.6].

One of the methods to transform the vibration problem (6.2-1) into a symmetric eigenvalue problem is to use the Cholesky decomposition of the mass matrix. Observe that the mass matrix M is positive definite. Hence, the product $\mathbf{x}^T M\mathbf{x}$ is positive for every nonzero vector $\mathbf{x}$. Since M is symmetric and positive definite, it may be factored into two terms as

$$M = LL^T, \tag{6.2-6}$$

where L is a lower triangular matrix with zeros in every position above the diagonal. This factorization L is called the *Cholesky decomposition.*

In the special case where M happens to be diagonal, the Cholesky decomposition becomes simply the *matrix square root*, denoted by $M^{1/2}$, such that $M^{1/2}M^{1/2} = M$. For example, for the diagonal mass matrix,

$$M = \begin{bmatrix} m_1 & 0 \\ 0 & m_2 \end{bmatrix}; \tag{6.2-7}$$

the Cholesky decomposition is the matrix square root. That is,

$$L = M^{1/2} = \begin{bmatrix} \sqrt{m_1} & 0 \\ 0 & \sqrt{m_2} \end{bmatrix} \tag{6.2-8}$$

This factors M into $M = M^{1/2}M^{1/2} = LL^T$ in the common case of a diagonal mass matrix.

The inverse of $M^{1/2}$, denoted by $M^{-1/2}$, becomes simply

$$L^{-1} = M^{-1/2} = \begin{bmatrix} \dfrac{1}{\sqrt{m_1}} & 0 \\ 0 & \dfrac{1}{\sqrt{m_2}} \end{bmatrix} \tag{6.2-9}$$

If the matrix M is not diagonal, the Cholesky factors can be computed via various available software. In Matlab, for example, this can be done via the simple command `chol(M)`.

With the Cholesky decomposition, the vibration problem (6.2-1) can readily be transformed into a symmetric eigenvalue problem. Specifically, let the vector **x** in Eq. 6.2-1 be replaced with

$$\mathbf{x} = \left(L^T\right)^{-1}\mathbf{q} \tag{6.2-10}$$

Then multiplying the resulting equation by L^{-1} yields

$$L^{-1}M\left(L^T\right)^{-1}\ddot{\mathbf{q}} + L^{-1}K\left(L^T\right)^{-1}\mathbf{q} = 0 \tag{6.2-11}$$

Note that $M = LL^T$ and thus $L^{-1}M\left(L^T\right)^{-1} = I$, the identity matrix. Hence, Eq. 6.2-11 becomes

$$I\ddot{\mathbf{q}} + \tilde{K}\mathbf{q} = 0 \tag{6.2-12}$$

The matrix $\tilde{K} = L^{-1}K\left(L^T\right)^{-1}$ is a symmetric matrix, called the *mass normalized stiffness matrix*.

Eq. 6.2-12 is solved by assuming a solution of the form $\mathbf{q} = \mathbf{v}e^{j\omega t}$, where **v** is a vector of constants. Substituting this form into Eq. 6.2-12, we obtain

$$\tilde{K}\mathbf{v} = \lambda\mathbf{v}, \tag{6.2-13}$$

where $\lambda = \omega^2$ is called the eigenvalue and **v** is called the eigenvector. Since the matrix $\tilde{K}$ is symmetric, this is called the *symmetric eigenvalue problem*. Again, there are n eigenvalues and n eigenvectors. It is known that the eigenvectors of a symmetric matrix are orthogonal and can always be normalized [6.6].

Let V be the normalized modal matrix composed of the orthogonal eigenvectors of $\tilde{K}$, as defined in Eq. 6.2-13 such that

$$V = \begin{bmatrix}\mathbf{v}_1 & \mathbf{v}_2 & \cdots & \mathbf{v}_n\end{bmatrix} \tag{6.2-14}$$

The modal matrix V has the unique property that $V^TV = I$ and $V^{-1} = V^T$. (Note that the eigenvector matrix obtained either from the generalized eigenvalue problem or from the algebraic eigenvalue problem is not orthogonal.)

With the transformation of Eq. 6.2-10, the modal matrix in the original coordinate system is given by

$$S = \left(L^T\right)^{-1}V \tag{6.2-15}$$

The time response in the original coordinate system is then obtained through the expression

$$\mathbf{x}(t)=\left(L^T\right)^{-1}\mathbf{q}=\sum_i C_i\left(L^T\right)^{-1}\mathbf{v}_i\sin(\omega_i t+\phi_i), \qquad (6.2\text{-}16)$$

where, as before, C_i and ϕ_i $(i=1,2,\cdots,n)$ are constants to be determined by initial conditions.

The numerical advantage here is that the vibration problem (6.2-1) is transformed into a symmetric eigenvalue problem (6.2-12). Thus, more efficient numerical algorithms are available to solve it. Computationally, this is the most efficient method of the five possible formulations presented in this Section.

(4) The Standard Eigenvalue Problem in State Space

Consider now the damped vibration problem of the form

$$M\ddot{\mathbf{x}}+C\dot{\mathbf{x}}+K\mathbf{x}=0, \qquad (6.2\text{-}17)$$

where M, C, and K, are the mass, damping, and stiffness matrices, respectively. Their dimensions are $n\times n$. The mass matrix M is symmetric and positive definite. (Thus, M^{-1} exists.) The matrices C and K are also symmetric but assumed only to be positive semidefinite.

Multiplying Eq. 6.2-17 by M^{-1} yields

$$\ddot{\mathbf{x}}+M^{-1}C\dot{\mathbf{x}}+M^{-1}K\mathbf{x}=0 \qquad (6.2\text{-}18)$$

We can transform Eq. 6.2-18 into a first-order matrix differential equation by introducing a pair of $n\times 1$ arrays $\mathbf{y}_1$ and $\mathbf{y}_2$ as $\mathbf{x}$ and $\dot{\mathbf{x}}$, respectively. Then, by differentiation, we immediately have

$$\dot{\mathbf{y}}_1=\dot{\mathbf{x}}=\mathbf{y}_2 \qquad (6.2\text{-}19a)$$

$$\dot{\mathbf{y}}_2=\ddot{\mathbf{x}}=-M^{-1}C\dot{\mathbf{x}}-M^{-1}K\mathbf{x}=-M^{-1}K\mathbf{y}_1-M^{-1}C\mathbf{y}_2 \qquad (6.2\text{-}19b)$$

Eq. 6.2-19 can be written as a simple first-order matrix equation as

$$\dot{\mathbf{y}}=A\mathbf{y}, \qquad (6.2\text{-}20)$$

where the multiplying matrix A, called the *state matrix*, may be written in the partitioned form as

$$A = \begin{bmatrix} 0 & I \\ -M^{-1}K & -M^{-1}C \end{bmatrix}, \tag{6.2-21}$$

where 0 denotes an $n \times n$ matrix of zeros, I is the $n \times n$ identity matrix, and $\mathbf{y}$ is the $2n \times 1$ state vector defined as

$$\mathbf{y} = \begin{bmatrix} \mathbf{y}_1 \\ \mathbf{y}_2 \end{bmatrix} = \begin{bmatrix} \mathbf{x} \\ \dot{\mathbf{x}} \end{bmatrix} \tag{6.2-22}$$

We can solve Eq. 6.2-20 by looking for a solution in the form $\mathbf{y} = \mathbf{z}e^{\lambda t}$, where $\mathbf{z}$ is a $2n \times 1$ nonzero vector of constants, and λ is a scalar. Then, by substituting into Eq. 6.2-20, we have

$$A\mathbf{z} = \lambda \mathbf{z} \tag{6.2-23}$$

Eq. 6.2-23 is the fundamental equation of the standard algebraic eigenvalue problem in state space. This eigenvalue problem, however, has dimension $2n$. Thus, there are $2n$ eigenvalues and, correspondingly, $2n$ eigenvectors. It is readily seen that the $2n$ eigenvalues occur as complex conjugate pairs in the form

$$\lambda_i = -\varsigma_i \omega_i + j\omega_i \sqrt{1 - \varsigma_i^2} \tag{6.2-24a}$$

$$\lambda_i^* = -\varsigma_i \omega_i - j\omega_i \sqrt{1 - \varsigma_i^2}, \tag{6.2-24b}$$

where, as before, $j = \sqrt{-1}$, ω_i is the i-th natural frequency, and ς_i is the i-th modal damping ratio associated with the i-th mode. Thus, the natural frequency and damping ratio can be calculated from the eigenvalues as

$$\omega_i = \sqrt{\left(\mathrm{Re}(\lambda_i)\right)^2 + \left(\mathrm{Im}(\lambda_i)\right)^2} \tag{6.2-25}$$

$$\varsigma_i = \frac{-\mathrm{Re}(\lambda_i)}{\omega_i}, \tag{6.2-26}$$

where $\mathrm{Re}(\lambda_i)$ and $\mathrm{Im}(\lambda_i)$ are the real and imaginary parts of the λ_i, respectively. It is important to note that in this case the eigenvalues are not equal to the squares of the natural frequencies.

The corresponding $2n$ eigenvectors will also occur in complex conjugate pairs as

$$\mathbf{z}_i = \begin{bmatrix} \mathbf{u}_i \\ \lambda_i \mathbf{u}_i \end{bmatrix} \tag{6.2-27a}$$

$$\mathbf{z}_i^* = \begin{bmatrix} \mathbf{u}_i^* \\ \lambda_i \mathbf{u}_i^* \end{bmatrix} \tag{6.2-27b}$$

In general, the physical mode shapes $\mathbf{u}_i$ are complex valued, and appear in complex conjugate pairs. They are referred to as *complex modes*. The physical interpretation of a complex mode is that each point of the system will not pass through the equilibrium position at the same time when the system is excited at that mode only. Recall that, in normal mode vibration, every point of the system undergoes harmonic motion and passes through the equilibrium position simultaneously.

The time response of the system in the original coordinate system can be expressed as

$$x(t) = \sum_i C_i \mathbf{u}_i e^{\lambda_i t} + \sum_i C_i^* \mathbf{u}_i^* e^{\lambda_i^* t}, \tag{6.2-28}$$

where C_i and C_i^* are $2n$ constants to be determined by the initial conditions.

The principal advantage of formulating the problem as in Eq. 6.2-20 is that the governing equation is in a form suitable for numerical solution. Thus, we can numerically study the damped multi-degree-of-freedom vibration problem. A disadvantage of Eq. 6.2-20, however, is that the multiplying matrix A is not symmetric, and it has dimension $2n \times 2n$. Thus, the corresponding eigenvalue expression of Eq. 6.2-23 is numerically more costly than with an n-dimension system without damping.

(5) The Generalized Eigenvalue Problem in State Space

Eq. 6.2-17 can also be transformed into the generalized eigenvalue problem in state space. To this end, introduce the identity

$$M\dot{\mathbf{x}} - M\dot{\mathbf{x}} = 0 \tag{6.2-29}$$

Combining Eqs. 6.2-17 and 6.2-29 into matrix form yields

$$\begin{bmatrix} C & M \\ M & 0 \end{bmatrix} \begin{bmatrix} \dot{\mathbf{x}} \\ \ddot{\mathbf{x}} \end{bmatrix} + \begin{bmatrix} K & 0 \\ 0 & -M \end{bmatrix} \begin{bmatrix} \mathbf{x} \\ \dot{\mathbf{x}} \end{bmatrix} = \begin{bmatrix} 0 \\ 0 \end{bmatrix} \tag{6.2-30}$$

Again, introduce the state vector $\mathbf{y}$ defined by the $2n\times 1$ vector:

$$\mathbf{y}=\begin{bmatrix}\mathbf{y}_1\\ \mathbf{y}_2\end{bmatrix}=\begin{bmatrix}\mathbf{x}\\ \dot{\mathbf{x}}\end{bmatrix} \tag{6.2-31}$$

By combining Eqs. 6.2-30 and 6.2-31, we obtain the first-order matrix differential equation

$$A\dot{\mathbf{y}}+B\mathbf{y}=0, \tag{6.2-32}$$

where

$$A=\begin{bmatrix}C & M\\ M & 0\end{bmatrix},\quad B=\begin{bmatrix}K & 0\\ 0 & -M\end{bmatrix} \tag{6.2-33}$$

We can look for a solution to Eq. 6.2-32 by assuming the form $\mathbf{y}=\mathbf{z}e^{\lambda t}$, where, as before, $\mathbf{z}$ is a nonzero vector array of constants and λ is a scalar. By substituting into Eq. 6.2-32, we obtain, after simplification,

$$B\mathbf{z}=-\lambda A\mathbf{z} \tag{6.2-34}$$

Eq. 6.2-34 is the fundamental equation for the generalized eigenvalue problem in state space.

There are $2n$ eigenvalues and $2n$ eigenvectors. Again the $2n$ eigenvalues λ_i will appear as complex conjugate pairs, as in Eq. 6.2-24. The corresponding $2n$ eigenvectors will also appear as complex conjugate pairs, as in Eq. 6.2-27, and repeated here:

$$\mathbf{z}_i=\begin{bmatrix}\mathbf{u}_i\\ \lambda_i\mathbf{u}_i\end{bmatrix} \tag{6.2-35a}$$

$$\mathbf{z}_i^*=\begin{bmatrix}\mathbf{u}_i^*\\ \lambda_i\mathbf{u}_i^*\end{bmatrix} \tag{6.2-35b}$$

These eigenvectors can be seen to be orthogonal with respect to the matrices A and B, as follows:

$$\mathbf{z}_i^T A\mathbf{z}_j=0 \quad (i\neq j) \tag{6.2-36}$$

$$\mathbf{z}_i^T B\mathbf{z}_j=0 \quad (i\neq j) \tag{6.2-37}$$

$$\mathbf{z}_i^T A\mathbf{z}_i=a_i \tag{6.2-38}$$

and

$$\mathbf{z}_i^T B \mathbf{z}_i = b_i \tag{6.2-39}$$

Next, from Eq. 6.2-34 we have

$$\frac{\mathbf{z}_i^T B \mathbf{z}_i}{\mathbf{z}_i^T A \mathbf{z}_i} = \frac{b_i}{a_i} = -\lambda_i \quad (i = 1, \cdots, n) \tag{6.2-40}$$

When the $2n$ pairs of complex conjugate eigenvectors are assembled into a $2n \times 2n$ square matrix with each eigenvector represented by a column, we called it the *complex modal matrix P*. That is,

$$P = \begin{bmatrix} \mathbf{z}_1 & \cdots & \mathbf{z}_n & \mathbf{z}_1^* & \cdots & \mathbf{z}_n^* \end{bmatrix} = \begin{bmatrix} S & S^* \\ S\Lambda & S^*\Lambda^* \end{bmatrix}, \tag{6.2-41}$$

where

$$\Lambda = \text{diag}(\lambda_1, \cdots, \lambda_n) \tag{6.2-42}$$

$$\Lambda^* = \text{diag}(\lambda_1^*, \cdots, \lambda_n^*) \tag{6.4-43}$$

$$S = \begin{bmatrix} \mathbf{u}_1 & \cdots & \mathbf{u}_n \end{bmatrix} \tag{6.2-44}$$

That is, the Λ array is a diagonal matrix composed of the eigenvalues, and the S array is the modal matrix in the original coordinate system.

If we now form the product $P^T AP$ or $P^T BP$, the result will be a diagonal matrix, since the off-diagonal terms simply express the orthogonality relations, which are zero. Thus, the orthogonality relations can be written as

$$P^T AP = \text{diag}(a_1, \cdots, a_n, a_1^*, \cdots, a_n^*) \tag{6.2-45}$$

$$P^T BP = \text{diag}(b_1, \cdots, b_n, b_1^*, \cdots, b_n^*) \tag{6.2-46}$$

The main advantage of Eq. 6.2-32 is that it can be decoupled, provided we know beforehand the mode shapes (eigenvectors) of the system. To see this more clearly, let us introduce new coordinate vectors $\mathbf{q}$ and $\mathbf{q}^*$ (the complex conjugate of $\mathbf{q}$) such that

$$\begin{bmatrix} \mathbf{q} \\ \mathbf{q}^* \end{bmatrix} = P^{-1}\mathbf{y} \quad \text{or} \quad \mathbf{y} = P\begin{bmatrix} \mathbf{q} \\ \mathbf{q}^* \end{bmatrix} \tag{6.2-47}$$

Substituting **y** from Eq. 6.2-47 into Eq. 6.2-32 and pre-multiplying the result by P^T, we obtain the following decoupled first order differential equations:

$$\dot{q}_i = \lambda_i q_i \quad (i=1,\cdots,n) \tag{6.2-48}$$

$$\dot{q}_i^* = \lambda_i^* q_i^* \quad (i=1,\cdots,n) \tag{6.2-49}$$

Eqs. 6.2-48 and 6.2-49 are called the *modal equations*. Their solutions have the forms

$$q_i = q_i(0)e^{\lambda_i t} \quad (i=1,\cdots,n) \tag{6.2-50}$$

$$q_i^* = q_i^*(0)e^{\lambda_i^* t} \quad (i=1,\cdots,n), \tag{6.2-51}$$

where $q_i(0)$ and $q_i^*(0)$ are the initial values of q_i and q_i^* (sometimes called the *modal initial conditions*), which can be expressed by the initial conditions in the original coordinate system as

$$\begin{bmatrix} \mathbf{q}(0) \\ \mathbf{q}^*(0) \end{bmatrix} = P^{-1}\mathbf{y}(0) = P^{-1}\begin{bmatrix} \mathbf{x}(0) \\ \dot{\mathbf{x}}(0) \end{bmatrix} \tag{6.2-52}$$

The time response in the original coordinate system is given by

$$x_r(t) = \sum_{s=1}^{n} S_{rs} q_s + \sum_{s=1}^{n} S_{rs}^* q_s^* \tag{6.2-53}$$

By using Eqs. 6.2-50 and 6.2-51, we obtain

$$x_r(t) = \sum_{s=1}^{n} A_{rs} e^{\lambda_s t} + \sum_{s=1}^{n} A_{rs}^* e^{\lambda_s^* t} \quad (r=1,\cdots,n), \tag{6.2-54}$$

where the A_{rs} taken as elements form an $n \times n$ matrix, called the *residual matrix*. We will discuss this in greater detail in Chapter Eleven. It is easy to show that Eqs. 6.2-54 and 6.2-28 are exactly the same.

Several approaches to solving for the natural frequencies and mode shapes have been presented in this section. The question arising is: Which one is the best? In theory, the symmetric eigenvalue problem is the best for large systems. For small systems, the difference may be ignored. However, for large systems, the difference is significant.

Example 6.2-1 Use Matlab to solve the following eigenvalue problem:

$$\begin{bmatrix} 7 & 0 \\ 0 & 2 \end{bmatrix}\begin{bmatrix} \ddot{x}_1 \\ \ddot{x}_2 \end{bmatrix} + \begin{bmatrix} 6 & -2 \\ -2 & 4 \end{bmatrix}\begin{bmatrix} x_1 \\ x_2 \end{bmatrix} = 0$$

Solution: The calculated eigenvalues are $\lambda_1 = 2$, $\lambda_2 = 5$. The natural frequencies are $\omega_1 = 1.4142$, $\omega_2 = 2.2361$. The corresponding mode shapes are

$$\mathbf{u}_1 = \begin{bmatrix} 1 \\ 1 \end{bmatrix}, \quad \mathbf{u}_2 = \begin{bmatrix} -1 \\ 2 \end{bmatrix}$$

The Matlab code is as follows:

```
clear all
M=[2,0;0,1]; %...............mass matrix
K=[6,-2;-2,4];%..............stiffness matrix
[V,D]=eig(K,M); % ...........generalized eigenvalue
                             problem
[V,D]=eig(inv(M)*K); % .......algebraic eigenvalue
                             problem
Khat=inv(chol(M))*K*inv(chol(M));
[V,D]=eig(Khat); %...........symmetric eigenvalue
                             problem
A=[zeros(2), eye(2); -inv(M)*K,zeros(2)];
[V,D]=eig(A); % ....algebraic eigenvalue problem in state
                    space
A1=[zeros(2),M; M,zeros(2)];
B1=[K,zeros(2);zeros(2),-M];
[V,D]=eig(B1,-A1); %
generalized eigenvalue problem in state space
```

Example 6.2-2 Consider the following system and compute the natural frequencies and damping ratios. The system is given by

$$M\ddot{\mathbf{x}} + C\dot{\mathbf{x}} + K\mathbf{x} = 0,$$

where

$$M = \begin{bmatrix} 4 & 0 \\ 0 & 2 \end{bmatrix}, \quad C = \begin{bmatrix} 1 & -0.3 \\ -0.3 & 0.3 \end{bmatrix}, \quad K = \begin{bmatrix} 5 & -1 \\ -1 & 1 \end{bmatrix}$$

Solution 1: First, compute the state matrix:

$$A = \begin{bmatrix} 0 & I \\ -M^{-1}K & -M^{-1}C \end{bmatrix}$$

The Matlab command for this is

```
A=[zeros(2),eye(2);-M\K,-M\C];
```

Next, compute the eigenvalues and eigenvectors of A with the Matlab command

```
[V,D]=eig(A);
```

In the result, D is a diagonal matrix, where the diagonal elements are the eigenvalues. The natural frequencies ω_i and damping ratios ς_i are then calculated as

$$\omega_i = \sqrt{\left(\mathrm{Re}\left(\lambda_i\right)\right)^2 + \left(\mathrm{Im}\left(\lambda_i\right)\right)^2}$$

$$\varsigma_i = \frac{-\mathrm{Re}\left(\lambda_i\right)}{\omega_i}$$

The Matlab code for this is

```
w=sqrt(real(D).^2+imag(D).^2);
zeta=-Re(D)./w;
```

The calculated results are as follows:

$$w(\omega) = \begin{bmatrix} 1.1781 & 0 & 0 & 0 \\ 0 & 11781 & 0 & 0 \\ 0 & 0 & 0.6002 & 0 \\ 0 & 0 & 0 & 0.6002 \end{bmatrix}$$

$$zeta(\varsigma) = \begin{bmatrix} 1.312 & 0 & 0 & 0 \\ 0 & 1.312 & 0 & 0 \\ 0 & 0 & 0.0757 & 0 \\ 0 & 0 & 0 & 0.0757 \end{bmatrix}$$

Solution 2: These results can be obtained using the generalized eigenvalue problem procedures of the matrices A and B defined as

$$A = \begin{bmatrix} C & M \\ M & 0 \end{bmatrix}, \quad B = \begin{bmatrix} K & 0 \\ 0 & -M \end{bmatrix}$$

These matrices can be computed by the Matlab commands

```
A=[C,M; M,zeros(2)]; B=[K,zeros(2);zeros(2),-M];
```

Next, compute the eigenvalues and eigenvectors by the command

```
[V,D]=eig(B,-A);
```

Again, the diagonal elements of the matrix D are the eigenvalues. Once the eigenvalues are obtained, the natural frequencies and damping ratios can be calculated as before.

The Matlab code for this calculation is

```
clear all
M=[4,0;0,2]; %............... mass matrix
```

```
K=[5,-1;-1,1];%............. stiffness matrix
C=[1,-0.3;-0.3,0.3];
A=[zeros(2),eye(2);-M\K,-M\C];
[V,D]=eig(A) % algebraic eigenvalue problem
w=sqrt(real(D).^2+imag(D).^2)
zeta=-real(D)./w
A1=[C,M;M,zeros(2)];
B1=[K, zeros(2);zeros(2), -M];
[V1,D1]=eig(B1,-A1) % generalized eigenvalue problem
w1=sqrt(real(D1).^2+imag(D1).^2)
zeta1=-real(D1)./w1
```

Example 6.2-3 Compute the time response of the undamped system defined by the mass and stiffness matrices M and K with the initial conditions $\mathbf{x}(0)$ and $\dot{\mathbf{x}}(0)$, where M, K, $\mathbf{x}(0)$, and $\dot{\mathbf{x}}(0)$ are

$$M = m\begin{bmatrix} 2 & 0 \\ 0 & 1 \end{bmatrix}, \quad K = k\begin{bmatrix} 6 & -2 \\ -2 & 4 \end{bmatrix}, \quad \mathbf{x}(0) = \begin{bmatrix} 5 \\ 0 \end{bmatrix}, \quad \dot{\mathbf{x}}(0) = \begin{bmatrix} 0 \\ 0 \end{bmatrix}$$

Compare the result to that obtained in Example 5.8-2 for the same system and initial conditions.

Solution: Since the mass matrix is diagonal, the Cholesky decomposition L is simply $M^{1/2}$:

$$L = M^{1/2} = \begin{bmatrix} \sqrt{2m} & 0 \\ 0 & \sqrt{m} \end{bmatrix}, \quad L^{-1} = M^{-1/2} = \begin{bmatrix} \dfrac{1}{\sqrt{2m}} & 0 \\ 0 & \dfrac{1}{\sqrt{m}} \end{bmatrix} \tag{a}$$

The matrix $\tilde{K}$ in Eq. 6.2-13 then becomes

$$\tilde{K} = L^{-1}K\left(L^{T}\right)^{-1} = \frac{k}{m}\begin{bmatrix} 3 & -\sqrt{2} \\ -\sqrt{2} & 4 \end{bmatrix} \tag{b}$$

Rewriting Eq. 6.2-13 yields the eigenvalue problem

$$\left(\tilde{K} - \lambda I\right)\mathbf{v} = 0$$

or

$$\begin{bmatrix} \dfrac{3k}{m} - \lambda & -\dfrac{\sqrt{2}k}{m} \\ -\dfrac{\sqrt{2}k}{m} & \dfrac{4k}{m} - \lambda \end{bmatrix}\mathbf{v} = 0 \tag{c}$$

The characteristic equation is then

$$\lambda^2 - 7\frac{k}{m}\lambda + 10\frac{k}{m} = 0 \tag{d}$$

Eq. d has two roots:

$$\lambda_1 = \omega_1^2 = 2\frac{k}{m}, \quad \lambda_2 = \omega_2^2 = 5\frac{k}{m} \tag{e}$$

Following the procedure in Example 5.4-1, the normalized eigenvectors are found to be

$$\mathbf{v}_1 = \frac{1}{\sqrt{3}}\begin{bmatrix} \sqrt{2} \\ 1 \end{bmatrix}, \quad \mathbf{v}_2 = \frac{1}{\sqrt{3}}\begin{bmatrix} -1 \\ \sqrt{2} \end{bmatrix} \tag{f}$$

Thus, the normalized modal matrix can be written as

$$V = \frac{1}{\sqrt{3}}\begin{bmatrix} \sqrt{2} & -1 \\ 1 & \sqrt{2} \end{bmatrix} \tag{g}$$

The modal matrix in the original coordinate system is then given by

$$S = \left(L^T\right)^{-1} V = \frac{1}{\sqrt{6m}}\begin{bmatrix} \sqrt{2} & -1 \\ \sqrt{2} & 2 \end{bmatrix} \tag{h}$$

The solution in modal coordinate system can now be expressed as

$$\mathbf{q} = c_1\mathbf{v}_1 \sin(\omega_1 t + \phi_1) + c_2\mathbf{v}_2 \sin(\omega_2 t + \phi_2) \tag{i}$$

or in matrix form as

$$\begin{bmatrix} q_1 \\ q_2 \end{bmatrix} = c_1\begin{bmatrix} \sqrt{2/3} \\ \sqrt{1/3} \end{bmatrix}\sin(\omega_1 t + \phi_1) + c_2\begin{bmatrix} -\sqrt{1/3} \\ \sqrt{2/3} \end{bmatrix}\sin(\omega_2 t + \phi_2), \tag{j}$$

where the c_1, c_2, ϕ_1, and ϕ_2 are constants to be determined by the initial conditions. The modal initial conditions then become

$$\mathbf{q}(0) = L^T\mathbf{x}(0) = \begin{bmatrix} \sqrt{2m} & 0 \\ 0 & \sqrt{m} \end{bmatrix}\begin{bmatrix} 5 \\ 0 \end{bmatrix} = \begin{bmatrix} 5\sqrt{2m} \\ 0 \end{bmatrix} \tag{k}$$

and

$$\dot{\mathbf{q}}(0) = L^T\dot{\mathbf{x}}(0) = \begin{bmatrix} \sqrt{2m} & 0 \\ 0 & \sqrt{m} \end{bmatrix}\begin{bmatrix} 0 \\ 0 \end{bmatrix} = \begin{bmatrix} 0 \\ 0 \end{bmatrix} \tag{l}$$

Substituting Eq. k into Eq. j yields

$$\begin{bmatrix} 5\sqrt{2m} \\ 0 \end{bmatrix} = \begin{bmatrix} c_1\sqrt{2/3}\sin\phi_1 - c_2\sqrt{1/3}\sin\phi_2 \\ c_1\sqrt{1/3}\sin\phi_1 + c_2\sqrt{2/3}\sin\phi_2 \end{bmatrix}, \tag{m}$$

from which we have

$$c_1\sin\phi_1 = 10\sqrt{m/3},\ \ c_2\sin\phi_2 = -5\sqrt{2m/3} \tag{n}$$

Next, differentiating Eq. j and letting $t=0$ in the result, we have

$$\begin{bmatrix} 0 \\ 0 \end{bmatrix} = \begin{bmatrix} \sqrt{2/3}c_1\omega_1\cos\phi_1 - \sqrt{1/3}c_2\omega_2\cos\phi_2 \\ \sqrt{1/3}c_1\omega_1\cos\phi_1 + \sqrt{2/3}c_2\omega_2\cos\phi_2 \end{bmatrix} \tag{o}$$

From Eq. o, we have

$$\cos\phi_1 = 0,\ \phi_1 = 90^\circ \tag{p}$$

and

$$\cos\phi_2 = 0,\ \phi_2 = 90^\circ \tag{q}$$

By using Eq. n, we then obtain

$$c_1 = 10\sqrt{m/3},\ c_2 = -5\sqrt{2m/3} \tag{r}$$

Using the values of c_1, c_2, ϕ_1, and ϕ_2 in Eq. j yields the modal solution

$$\begin{bmatrix} q_1 \\ q_2 \end{bmatrix} = \begin{bmatrix} \dfrac{10\sqrt{2m}}{3} \\ \dfrac{10\sqrt{m}}{3} \end{bmatrix}\cos\omega_1 t + \begin{bmatrix} \dfrac{5\sqrt{2m}}{3} \\ -\dfrac{10\sqrt{m}}{3} \end{bmatrix}\cos\omega_2 t \tag{s}$$

Using the transformation $\mathbf{x} = \left(L^T\right)^{-1}\mathbf{q}$ yields the solution in the original coordinate system as

$$\begin{bmatrix} x_1 \\ x_2 \end{bmatrix} = \frac{10}{3}\begin{bmatrix} 1 \\ 1 \end{bmatrix}\cos\sqrt{\frac{2k}{m}}t + \frac{5}{3}\begin{bmatrix} 1 \\ -2 \end{bmatrix}\cos\sqrt{\frac{5k}{m}}t \tag{t}$$

This solution, as expected, is identical to the solution obtained in Example 5.8-2.

Alternatively, the solution may be obtained as follows: Using the modal matrix (h), the solution can be expressed as

$$\begin{bmatrix} x_1 \\ x_2 \end{bmatrix} = C_1 \begin{bmatrix} \sqrt{\dfrac{1}{3m}} \\ \dfrac{1}{\sqrt{3m}} \end{bmatrix} \sin(\omega_1 t + \phi_1) + C_2 \begin{bmatrix} \dfrac{-1}{\sqrt{6m}} \\ \dfrac{2}{\sqrt{6m}} \end{bmatrix} \sin(\omega_2 t + \phi_2), \tag{u}$$

where C_1, C_2, ϕ_1, and ϕ_2 are constants to be determined by the initial conditions in the original coordinate system. This procedure will also yield the same result.

Example 6.2-4 Use the symmetric eigenvalue problem technique to calculate the steady-state response of the two-degree-of-freedom system of Example 5.9-1. That is, solve the system

$$\begin{bmatrix} 4 & 0 \\ 0 & 1 \end{bmatrix} \ddot{x} + \begin{bmatrix} 0.8 & -0.2 \\ -0.2 & 0.2 \end{bmatrix} \dot{x} + \begin{bmatrix} 8 & -2 \\ -2 & 2 \end{bmatrix} x = \begin{bmatrix} 0 \\ F_2(t) \end{bmatrix},$$

where $F_2(t) = 2\sin\sqrt{3}t$. Observe that the damping matrix C is propositional to the stiffness matrix K, that is, $C = 0.1K$.

Solution: The matrices $L = M^{1/2}$ and $L^{-1} = M^{-1/2}$ become

$$L = \begin{bmatrix} 2 & 0 \\ 0 & 1 \end{bmatrix} \quad L^{-1} = \begin{bmatrix} 1/2 & 0 \\ 0 & 1 \end{bmatrix} \tag{a}$$

Substituting the transformation $\mathbf{x} = \left(L^T\right)^{-1} \mathbf{q}$ into the equations of motion and multiplying the result by L^{-1}, we obtain

$$I\ddot{\mathbf{q}} + \tilde{C}\dot{\mathbf{q}} + \tilde{K}\mathbf{q} = \mathbf{f}(t), \tag{b}$$

where

$$\tilde{C} = \begin{bmatrix} 0.2 & -0.1 \\ -0.1 & 0.2 \end{bmatrix}, \quad \tilde{K} = \begin{bmatrix} 2 & -1 \\ -1 & 2 \end{bmatrix}, \quad \text{and} \quad \mathbf{f}(t) = \begin{bmatrix} 0 \\ F_2(t) \end{bmatrix} \tag{c}$$

The eigenvalue problem for the symmetric matrix $\tilde{K}$ yields the eigenvalues λ_1 and λ_2 and the normalized eigenvector matrix V as

$$\lambda_1 = 1, \quad \lambda_2 = 3, \quad V = \frac{1}{\sqrt{2}} \begin{bmatrix} 1 & -1 \\ 1 & 1 \end{bmatrix} \tag{d}$$

Thus, the natural frequencies of the system are $\omega_1 = 1$ and $\omega_2 = \sqrt{3}$. Eq. b can be decoupled by the transformation $\mathbf{q} = V\mathbf{r}$. Substituting this into Eq. b, and multiplying the result by V^T, we obtain the decoupled modal equations

$$\ddot{r}_1 + 0.1\dot{r}_1 + r_1 = \sqrt{2}\sin\sqrt{3}t \tag{e}$$

$$\ddot{r}_2 + 0.3\dot{r}_2 + 3r_2 = \sqrt{2}\sin\sqrt{3}t \tag{f}$$

Comparing the coefficient of $\dot{r}_i$ in each case to $2\varsigma_i\omega_i$ yields the damping ratios

$$\varsigma_1 = \frac{0.1}{2(1)} = 0.05 \tag{g}$$

$$\varsigma_2 = \frac{0.3}{2\sqrt{3}} = 0.087 \tag{h}$$

Note that although the eigenvalues and eigenvectors of the system can be calculated by the generalized eigenvalue problem as shown in Eq. 6.2-2, and although the eigenvectors can also be used for decoupling the equations of motion as in Example 5.9-1, we nevertheless still transform the equations into those of the symmetric eigenvalue problem, since computing the eigenvalues of a symmetric matrix is faster and more efficient than computing the eigenvalues of an asymmetric matrix. This difference is significant for large systems.

The steady-state response is the particular solution of Eqs. e and f. This solution is given by Eq. 3.2-14 as

$$\begin{aligned} r_{1p} &= \frac{\sqrt{2}}{\sqrt{(1-3)^2 + [2(0.05)\sqrt{3}(1)]^2}} \sin\left(\sqrt{3}t - \tan^{-1}\frac{2(0.05)\sqrt{3}(1)}{1-3}\right) \\ &= 0.7045\sin(\sqrt{3}t - 175^0) \end{aligned} \tag{i}$$

and

$$\begin{aligned} r_{2p} &= \frac{\sqrt{2}}{\sqrt{(3-3)^2 + [2(0.087)(\sqrt{3})(\sqrt{3})]^2}} \sin\left(\sqrt{3}t - \tan^{-1}\frac{2(0.087)(\sqrt{3})(\sqrt{3})}{3-3}\right) \\ &= 2.7092\sin(\sqrt{3}t - \pi/2) = -2.7092\cos\sqrt{3}t \end{aligned} \tag{j}$$

Recall that the transformation relations used were $\mathbf{x} = \left(L^T\right)^{-1}\mathbf{q}$ and $\mathbf{q} = V\mathbf{r}$. Thus, the solution expressed in the original coordinate system is

$$\mathbf{x} = \left(L^T\right)^{-1} V\mathbf{r} = \begin{bmatrix} 1/2 & 0 \\ 0 & 1 \end{bmatrix} \begin{bmatrix} 1/\sqrt{2} & -1/\sqrt{2} \\ 1/\sqrt{2} & 1/\sqrt{2} \end{bmatrix} \begin{bmatrix} 0.7045\sin(\sqrt{3}t - 175^\circ) \\ -2.7092\cos\sqrt{3}t \end{bmatrix} \qquad \text{(k)}$$

Therefore, in the steady-state, we have

$$x_1 = 0.2491\sin\left(\sqrt{3}t - 175^\circ\right) + 0.9578\cos\sqrt{3}t \qquad \text{(l)}$$

$$x_2 = 0.4982\sin\left(\sqrt{3}t - 175^\circ\right) - 1.9156\cos\sqrt{3}t \qquad \text{(m)}$$

These solutions are identical to those obtained in Example 5.9-1.

Example 6.2-5 Numerically compute the coefficients of the modal equations for the problem given in Example 6.2-4.

Solution: The matrices $\tilde{C}$ and $\tilde{K}$ are

$$\tilde{C} = \begin{bmatrix} 0.2 & -0.1 \\ -0.1 & 0.2 \end{bmatrix}, \quad \tilde{K} = \begin{bmatrix} 2 & -1 \\ -1 & 2 \end{bmatrix}$$

$\tilde{C}$ and $\tilde{K}$ can be obtained numerically by using the Matlab code:

```
clear all
% input matrice
M=[4,0;0,1];
C=[0.8,-0.2;-0.2,0.2];
K=[8,-2;-2,2];
L=chol(M); % Cholesky decomposition
Kt=inv(L)*K*inv(L')
Kt=
 2 -1
 -1 2
Ct=inv(L)*C*inv(L')
Ct=
    0.2000 -0.1000
    -0.1000 0.2000
% compute eigensolutions
[V,D]=eig(Kt)
```

```
V=
 -0.7071 -0.7071
 -0.7071 0.7071
D=
 1 0
 0 3
V'*Ct*V % c matrix in modal equation
ans =
   0.1000 0.0000
   0.0 0.3000
V'*Kt*V % k matrix in modal equation
ans =
 1.0000 0.0000
 0.0000 3.0000
```

Observe that the first eigenvector is the negative of that produced in Example 6.2-4. This is of no consequence since both are correct. Recall that an eigenvector may always be multiplied by a scalar (in this case -1) without changing its orientation. Moreover, the value -1 does not change the magnitude. When the modal equations are transformed back and expressed in terms of physical coordinates, the signs recombine yielding the same solution.

6.3 Numerical Evaluation of the Time Response

Consider the forced response of a damped linear system. The most general initial value problem can be written as

$$M\ddot{\mathbf{x}} + C\dot{\mathbf{x}} + K\mathbf{x} = \mathbf{f}, \ \ \mathbf{x}(0) = \mathbf{x}_0, \ \ \dot{\mathbf{x}}(0) = \dot{\mathbf{x}}_0 \tag{6.3-1}$$

To solve for the forced response numerically, Eq. 6.3-1 needs to be transformed into the first-order, or state space, form. To this end, multiplying Eq. 6.3-1 by M^{-1} (since M is positive definite, its inversion exists) leads to

$$\ddot{\mathbf{x}} + M^{-1}C\dot{\mathbf{x}} + M^{-1}K\mathbf{x} = M^{-1}\mathbf{f} \tag{6.3-2}$$

Introducing two new vectors $\mathbf{y}_1 = \mathbf{x}$ and $\mathbf{y}_2 = \dot{\mathbf{x}}$, we have

$$\dot{\mathbf{y}}_1 = \dot{\mathbf{x}} = \mathbf{y}_2 \tag{6.3-3}$$

$$\dot{\mathbf{y}}_2 = \ddot{\mathbf{x}} = -M^{-1}K\mathbf{y}_1 - M^{-1}C\mathbf{y}_2 + M^{-1}\mathbf{f}, \tag{6.3-4}$$

where the expression for $\ddot{\mathbf{x}}$ in Eq. 6.3-4 is replaced from Eq. 6.3-2 by moving the terms $M^{-1}C\dot{\mathbf{x}}$ and $M^{-1}K\mathbf{x}$ to the right side.

Eq. 6.3-3 and 6.3-4 can be written as

$$\dot{\mathbf{y}} = A\mathbf{y} + \mathbf{b}, \tag{6.3-5}$$

where

$$A = \begin{bmatrix} 0_{n\times n} & I_{n\times n} \\ -M^{-1}K & -M^{-1}C \end{bmatrix} \tag{6.3-6}$$

$$\mathbf{b} = \begin{bmatrix} 0_{n\times 1} \\ M^{-1}\mathbf{f} \end{bmatrix} \tag{6.3-7}$$

and

$$\mathbf{y} = \begin{bmatrix} \mathbf{y}_1 \\ \mathbf{y}_2 \end{bmatrix} = \begin{bmatrix} \mathbf{x} \\ \dot{\mathbf{x}} \end{bmatrix} \tag{6.3-8}$$

Eq. 6.3-5 has a form suitable for numerical integration. The following examples illustrate how to use Matlab to perform such integration.

Example 6.3-1 Numerically integrate and graphically represent the response of a damped single-degree-of-freedom system with a harmonic force applied to the mass as indicated in Fig. 6.3-1.

Solution: The equation of motion is

$$m\ddot{x} + c\dot{x} + kx = F_0 \sin \omega t \tag{a}$$

To transform this into a state space form, we introduce two new variables $y_1 = x$ and $y_2 = \dot{x}$. Differentiating these two variables yields

$$\dot{y}_1 = \dot{x} = y_2 \tag{b}$$

and

$$\dot{y}_2 = \ddot{x} = -m^{-1}c\dot{x} - m^{-1}kx = -m^{-1}ky_1 - m^{-1}cy_2 \tag{c}$$

These expressions can be combined into the first-order vector differential equation as

$$\dot{\mathbf{y}} = A\mathbf{y} + \mathbf{b}, \tag{d}$$

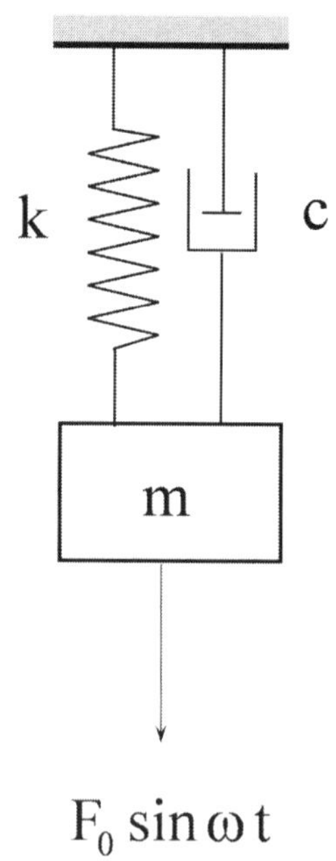

Fig. 6.3-1 A damped system excited by a harmonic force.

where

$$A=\begin{bmatrix} 0 & 1 \\ -k/m & -c/m \end{bmatrix}, \quad \mathbf{b}=\begin{bmatrix} 0 \\ \dfrac{F_0}{m}\sin\omega t \end{bmatrix} \tag{e}$$

Eq. d can be solved numerically. To compare the result obtained by this numerical method with the analytical solution in Section 3.2, we choose the same parameters as follows: (1) the mass $m=2$ kg, (2) the spring stiffness $k=3500$ N/m, (3) the damping coefficient $c=13$ Ns/m, (4) the force amplitude $F_0=65$ N, (5) the excitation frequency $\omega=5$ rad/s, (6) the initial displacement $x_0=0.03$ m, and (7) the initial velocity $\dot{x}_0=1.5$ m/s.

Fig. 6.3-2 shows the solution, which is the same as the analytical solution given in Fig. 3.2-3. The Matlab code for solving for the response is

```
clear all
x0=[0.03;1.500];
ts=[0 2.5];
[t,x]=ode23('f',ts,x0);
h=plot(t,x(:,1),'k',t1,x1,'k');
set(h,'LineWidth',2);
hx=get(gca,'Xlabel');
hy=get(gca,'Ylabel');
set(hx,'String','time(s)');
```

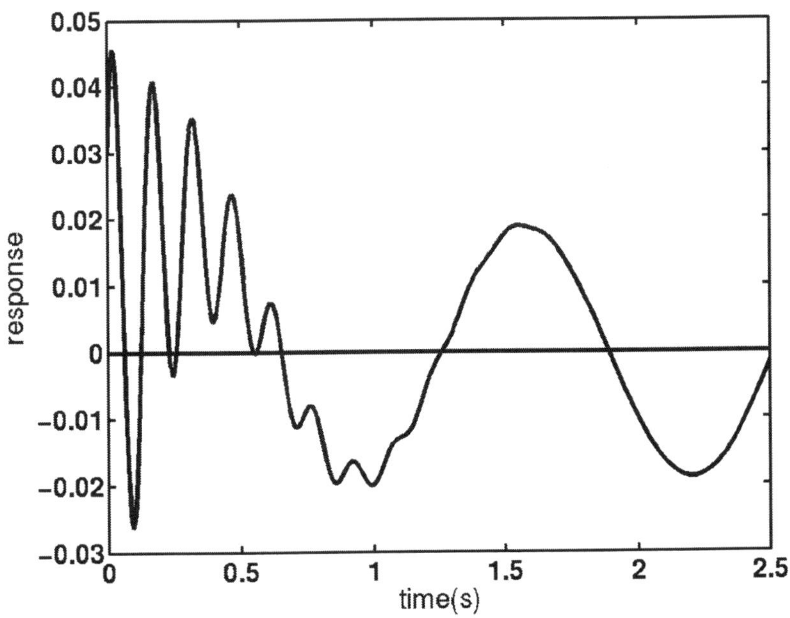

Fig. 6.3-2 The numerical solution for the system of Example 6.3-1.

```
set(hy,'String','response (m)');
set(hx,'FontSize',14);
set(hy,'FontSize',14);
%
function v=f(t,x)
k=3500;% ..................stiffness, N/m
c=13; %....................damping, Ns/m
m=2; %.....................mass, Kg
F0=65;%....................force amplitude, N
A=[0,1;-k/m, -c/m];
w=5;
b=[0;F0/m*sin(w*t)];
v=A*x+b;
```

Fig. 6.3-3 shows the graphical representation of the velocity $y\ (=\dot{x})$ as a function of the displacement x.

In Fig. 6.3-3, the $x-y$ plane is called the *phase plane*. The coordinates (x, y) then define the *state* of the system. As the state of the system changes, the point with coordinates (x, y) in the phase plane moves and traces out a curve, called the *trajectory*.

From Fig. 6.3-3, we see that initially $(t=0)$ neither the displacement x, nor its slope $\dot{x}$ (=y), is zero. Therefore, in the phase plane the trajectory does not

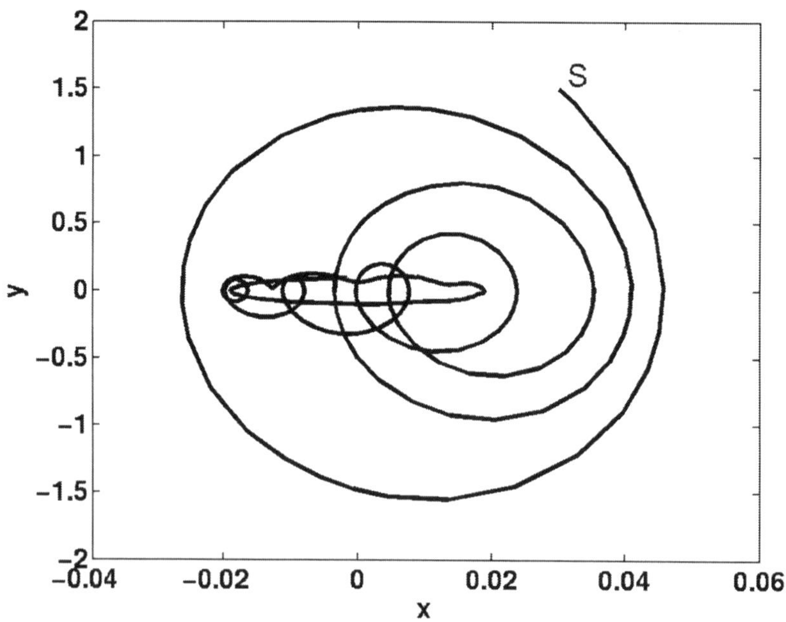

Fig. 6.3-3 Phase plane plot for the system of Example 6.3-1.

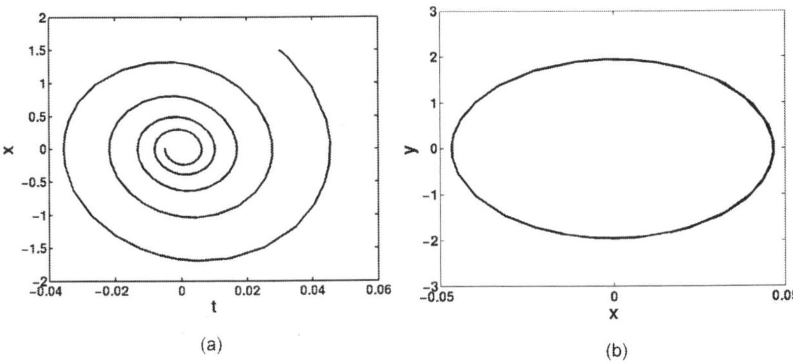

Fig. 6.3-4 Phase plane representation for the system of Example 6.3-1: (a) decay oscillation ($F_0 = 0$, $c \neq 0$); (b) harmonic oscillation ($F_0 = 0$, $c = 0$)

begin at the origin. Instead it starts at the point S in Fig. 6.3-3. The trajectory then continues along a decaying spiral path approaching an elongated ellipse. The decay is due to energy being dissipated during oscillation. Once the transient oscillation is dissipated, the steady-state vibration remains.

Next, Fig. 6.3-4 shows two special case phase planes (a) and (b). In (a), the excitation force is zero, but the damping is not zero ($F_0 = 0$, $c \neq 0$). In this case, the phase point moves along a decaying spiral path. In (b), both the excitation force and the damping are zero ($F_0 = 0$, $c = 0$). But since the initial conditions are not zero, the system undergoes a harmonic vibration, and the phase plane trajectory is a repeating ellipse.

Example 6.3-2 Numerically integrate and graph the response of the system in Example 5.6-1, where the mass is $m = 2$ kg, the stiffness constant is $k = 25$ N/m, and the initial conditions are $x_1(0) = 5$ mm, $x_2(0) = 0$, and $\dot{x}_1(0) = \dot{x}_2(0) = 0$.

Solution: The equation of motion is

$$\begin{bmatrix} 4 & 0 \\ 0 & 2 \end{bmatrix} \begin{bmatrix} \ddot{x}_1 \\ \ddot{x}_2 \end{bmatrix} + 25 \begin{bmatrix} 6 & -2 \\ -2 & 4 \end{bmatrix} \begin{bmatrix} x_1 \\ x_2 \end{bmatrix} = \begin{bmatrix} 0 \\ 0 \end{bmatrix} \tag{a}$$

Thus, by inspection, M, K, and $\mathbf{f}$ are

$$M = \begin{bmatrix} 4 & 0 \\ 0 & 2 \end{bmatrix}, \quad K = 25 \begin{bmatrix} 6 & -2 \\ -2 & 4 \end{bmatrix}, \quad \mathbf{f} = \begin{bmatrix} 0 \\ 0 \end{bmatrix} \tag{b}$$

Fig. 6.3-5 shows the graphs of the solution using Matlab. The results obtained here are identical to those in Example 5.6-1, obtained using the mode summation method.

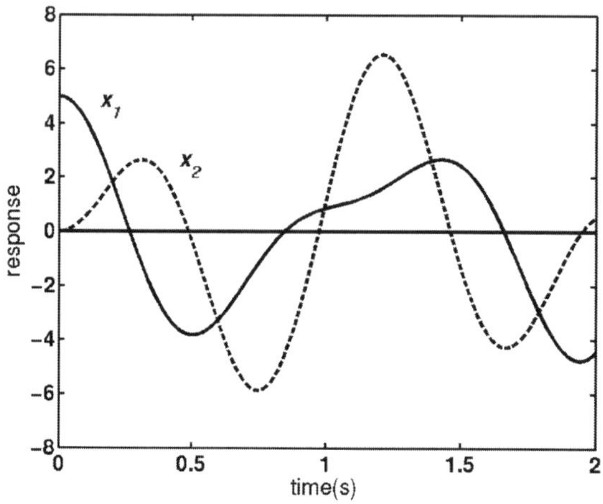

Fig. 6.3-5 Plots of the solutions of Example 6.3-2.

The Matlab code to produce the solution is

```
clear all
ymax=8;
y0=[0.005;0;0;0];
ts=[0 2];
[t,y]=ode45('myf',ts,y0);
h=plot(t,1000*y(:,1),'k',t,1000*y(:,2),'--k'); % change
 unit to mm
set(gca,'Ylim',[-ymax,ymax]);
set(h,'LineWidth',2);
hx=get(gca,'Xlabel');
hy=get(gca,'Ylabel');
set(hx,'String','time(s)');
set(hy,'String','response (mm)');
text(0.1,4.5,'itx_1');
text(0.4,2.3,'itx_2');
set(hx,'FontSize',14);
set(hy,'FontSize',14);
%
function v=myf(t,y);
M=[4,0;0,2];
K=25*[6,-2;-2,4];
```

```
A=[zeros(2),eye(2);-inv(M)*K, zeros(2)];
v=A*y;
```

Example 6.3-3 Numerically integrate and plot the response of the system of Example 5.8-1 shown again in Fig. 6.3-6, where $m = 2$ kg, $k = 25$ N/m, and $\omega = \sqrt{5k/m}$, with zero initial displacements and initial velocities. Consider the following two cases and determine if a resonance response occurs.

Case 1: $F_1 = 2,\ F_2 = 1$

Case 2: $F_1 = 1,\ F_2 = 1$

Solution: From equilibrium of the forces on each mass, we readily obtain the equations of motion in matrix form as

$$\begin{bmatrix} 4 & 0 \\ 0 & 2 \end{bmatrix} \begin{bmatrix} \ddot{x}_1 \\ \ddot{x}_2 \end{bmatrix} + 25 \begin{bmatrix} 6 & -2 \\ -2 & 4 \end{bmatrix} \begin{bmatrix} x_1 \\ x_2 \end{bmatrix} = \begin{bmatrix} F_1 \\ F_2 \end{bmatrix} \sin \omega t \tag{a}$$

In Section 5.2, we obtained the natural frequencies and associated mode shapes of the system. They are

$$\omega_1 = \sqrt{\frac{2k}{m}}, \quad \omega_2 = \sqrt{\frac{5k}{m}} \tag{b}$$

$$\mathbf{u}_1 = \begin{Bmatrix} 1 \\ 1 \end{Bmatrix}, \quad \mathbf{u}_2 = \begin{Bmatrix} 1 \\ -2 \end{Bmatrix} \tag{c}$$

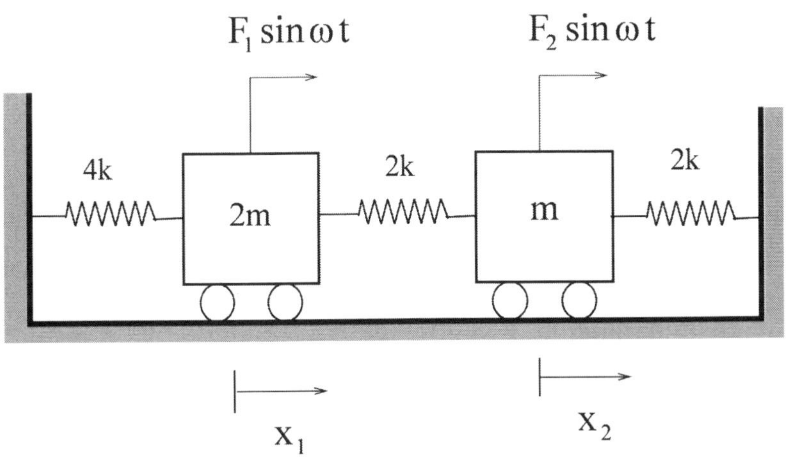

Fig. 6.3-6 Two-degree-of-freedom system for Example 6.3-3.

In this example, the excitation frequency is $\omega = \sqrt{5k/m}$. This is equal to the second natural frequency of the system. In general, this will cause an unbounded oscillation in the undamped case and a response with a maximum amplitude in the damped case. In this example, however, in case 1, the force vector is orthogonal to the modal vector of the exciting frequency, that is,

$$\begin{bmatrix} F_1 & F_2 \end{bmatrix}^T \mathbf{u}_2 = \begin{bmatrix} 2 & 1 \end{bmatrix} \begin{bmatrix} 1 \\ -2 \end{bmatrix} = 0 \tag{d}$$

Thus, the system does not resonate. In case 2, with the force vector not being orthogonal to the modal vector $\mathbf{u}_2$, the system resonates.

The responses of the system for the two cases are shown in Figs. 6.3-7 and 6.3-8, respectively. The Matlab code to produce the solutions is

```
clear all
x0=[0;0;0;0];
ts=[0 12];
[t,x]=ode45('f',ts,x0);
h=plot(t,x(:,1),'k',t,x(:,2),'--k');
set(h,'LineWidth',2);
hx=get(gca,'Xlabel');
hy=get(gca,'Ylabel');
set(hx,'String','time(s)');
set(hy,'String','response (m)');
set(hx,'FontSize',14);
set(hy,'FontSize',14);
function v=f(t,x)
M=[4,0;0,2];
K=25*[6,-2;-2,4];
%f=[2/3;1/3]; %................case 1
f=[1;1]; %.....................case 2
w=sqrt(5*25/2); %............. w=sqrt(5*k/m)
A=[zeros(2),eye(2);-inv(M)*K,zeros(2)];
b=inv(M)*f;
v=A*x+[0;0;b]*sin(w*t);
```

Finally, let us explore why the two responses are identical in case 1. Consider the equation of motion in modal coordinates given in Eq. f of Example 5.8-1 and repeated here as

$$\begin{bmatrix} 1 & 0 \\ 0 & 1 \end{bmatrix} \begin{bmatrix} \ddot{y}_1 \\ \ddot{y}_2 \end{bmatrix} + \frac{k}{m} \begin{bmatrix} 2 & 0 \\ 0 & 5 \end{bmatrix} \begin{bmatrix} y_1 \\ y_2 \end{bmatrix} = \begin{bmatrix} \dfrac{F_1 + F_2}{\sqrt{3m}} \\ \dfrac{F_1 - 2F_2}{\sqrt{6m}} \end{bmatrix} \sin \omega t \tag{e}$$

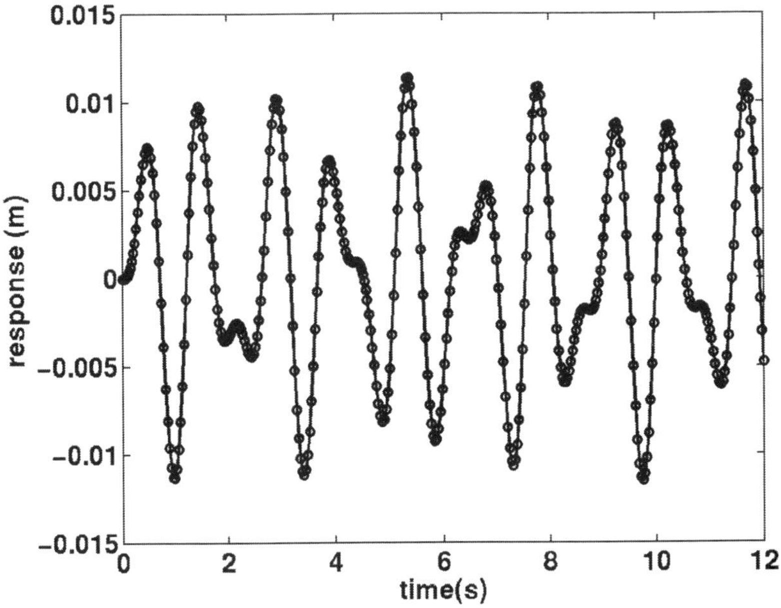

Fig. 6.3-7 The numerical solutions for case 1 in Example 6.3-3. x_1 and x_2 are identical.

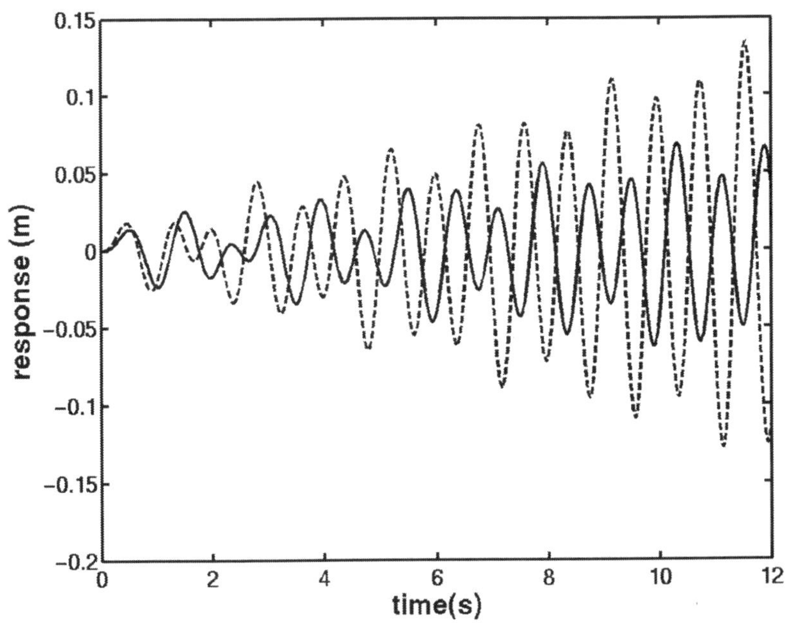

Fig. 6.3-8 The numerical solutions for case 2 in Example 6.3-3. Solid line: x_1; dotted line: x_2. Resonance occurs.

When $F_1 = 2F_2$, then the second equation of Eq. e becomes

$$\ddot{y}_2 + \frac{5k}{m} y_2 = 0 \tag{f}$$

When both the initial displacement and the initial velocity are zero, the solution of this equation is $y_2 = 0$. The solution in the original coordinate system **x** is given by

$$\begin{bmatrix} x_1 \\ x_2 \end{bmatrix} = \overline{S} \begin{bmatrix} y_1 \\ y_2 \end{bmatrix}, \tag{g}$$

where $\overline{S}$ is the normalized modal matrix given by Eq. d of Example 5.8-1, and repeated here as

$$\overline{S} = \frac{1}{\sqrt{6m}} \begin{bmatrix} \sqrt{2} & 1 \\ \sqrt{2} & -2 \end{bmatrix} \tag{h}$$

Substitution of Eq. h into Eq. g yields $x_1 = x_2$.

Example 6.3-4 Numerically integrate and graph the solution of the Van der Pol equation

$$\ddot{x} - \mu(1 - x^2)\dot{x} + x = 0, \tag{a}$$

where μ is a constant. Consider solution subject to the following initial conditions:

Case 1: $x_0 = 0$; $\dot{x}_0 = 0.5$

Case 2: $x_0 = -2$; $\dot{x}_0 = 2$

Solution: This equation has been studied extensively (see [6.7–6.11]). The interested reader is referred to the original papers of Van der Pol et al. for quite a number of applications [6.12]. Here we simply show how we can use software to solve the equation numerically.

Observe that the parameter μ controls the velocity (damping) term's significance. Observe also that the damping term is nonlinear in that it depends on both the velocity and the displacement. For $x < 1$, the damping is negative, and the amplitude will increase with time. For $x > 1$, the damping is positive, and the amplitude will diminish with time. For any initial condition, the amplitude will increase or decrease, depending on whether x is small or large, and it will finally reach a stable state known as the *limit cycle*.

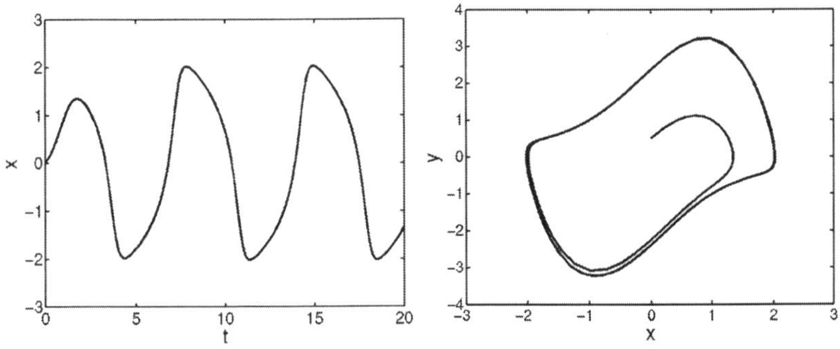

(a) Time plane (b) Phase plane

Fig. 6.3-9 The solution of the Van der Pol equation with $\mu = 1.5$, $x_0 = 0, \dot{x}_0 = 0.5$. (The phase point spirals out to the limit cycle.)

To integrate Eq. a numerically, we need to transform it into the first-order state space form. To this end, let $y_1 = x$, $y_2 = \dot{x}$. This leads to the first order equations:

$$\dot{y}_1 = y_2 \tag{b}$$

$$\dot{y}_2 = \mu(1 - y_1^2)y_2 - y_1 \tag{c}$$

We can conveniently solve the equation using Matlab. Fig. 6.3-9 shows the results for the initial condition of Case 1 with μ=1.5. In the phase plane, the trajectory begins inside a limit cycle curve, spiraling out to it.

Fig. 6.3-10 shows the results for the initial condition of Case 2, with μ again being 1.5. Here, in the phase plane, the trajectory begins outside the limit cycle, spiraling into it.

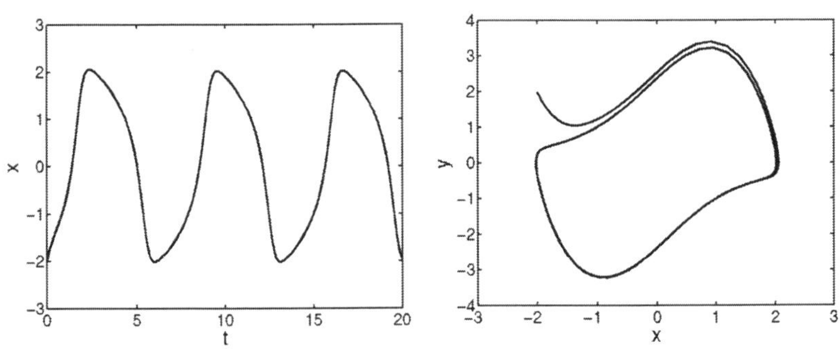

(a) Time plane (b) Phase plane

Fig. 6.3-10 The solution of the Van der Pol equation with $\mu = 1.5$, $x_0 = -2$, $\dot{x}_0 = 2$. The phase point approaches to the limit cycle from outside.

The Matlab code to produce the solutions is

```
clear all
%y0=[0.0;0.5]; % case 1, phase point spirals out to limit
  cycle
y0=[-2;2]; % case 2,phase point decays to limit cycle
ts=[0 20];
[t,y]=ode45('vdpol',ts,y0);
h=plot(t,y(:,1),'k'); % time response plot
%h=plot(y(:,1),y(:,2),'k'); % phase plane plot
set(h,'LineWidth',2);
hx=get(gca,'Xlabel');
hy=get(gca,'Ylabel');
set(hx,'String','time (s)');
set(hy,'String','displacement');
%set(hx,'String','displacement');
%set(hy,'String','velocity');
set(hx,'FontSize',14);
set(hy,'FontSize',14);
%
function ydot=vdpol(t,y)
mu=1.5;
ydot=[y(2); mu*(1-y(1)^2)*y(2)-y(1)];
```

Example 6.3-5 Compute the time response of the system given by

$$\begin{bmatrix} 2 & 0 \\ 0 & 1 \end{bmatrix}\ddot{\mathbf{x}} + \begin{bmatrix} 3 & -0.3 \\ -0.3 & 0.3 \end{bmatrix}\dot{\mathbf{x}} + \begin{bmatrix} 5 & -2 \\ -2 & 2 \end{bmatrix}\mathbf{x} = \begin{bmatrix} 1 \\ 1 \end{bmatrix}\sin 3t$$

Let the initial conditions be

$$\mathbf{x}_0 = \begin{bmatrix} 0.1 \\ 0.1 \end{bmatrix}; \quad \dot{\mathbf{x}}_0 = \begin{bmatrix} 0.5 \\ 1 \end{bmatrix}$$

Solution: Here we have

$$M = \begin{bmatrix} 2 & 0 \\ 0 & 1 \end{bmatrix}, \quad C = \begin{bmatrix} 3 & -0.3 \\ -0.3 & 0.3 \end{bmatrix}, \quad K = \begin{bmatrix} 5 & -2 \\ -2 & 2 \end{bmatrix}, \quad \mathbf{f} = \begin{bmatrix} 1 \\ 1 \end{bmatrix}\sin 3t \qquad \text{(a)}$$

Next, we write the equations of motion in the form of Eq. 6.3-5, that is,

$$\dot{\mathbf{y}} = A\mathbf{y} + \mathbf{b}, \qquad \text{(b)}$$

where A, $\mathbf{b}$, and $\mathbf{y}$ are

$$A = \begin{bmatrix} 0_{2\times 2} & I_{2\times 2} \\ -M^{-1}K & -M^{-1}C \end{bmatrix} \tag{c}$$

$$\mathbf{b} = \begin{bmatrix} 0_{2\times 1} \\ M^{-1}\mathbf{f} \end{bmatrix} \tag{d}$$

and

$$\mathbf{y} = \begin{bmatrix} \mathbf{y}_1 \\ \mathbf{y}_2 \end{bmatrix} = \begin{bmatrix} \mathbf{x} \\ \dot{\mathbf{x}} \end{bmatrix} \tag{e}$$

We again use Matlab to obtain the solution. Fig. 6.3-11 shows the results. We see that after about 10 sec, the solution approaches that of a steady-state forced vibration.

The Matlab code for this solution is

```
clear all
y0=[0.1;0.1;0.5;1];
ts=[0 20];
[t,y]=ode45('f',ts,y0);
```

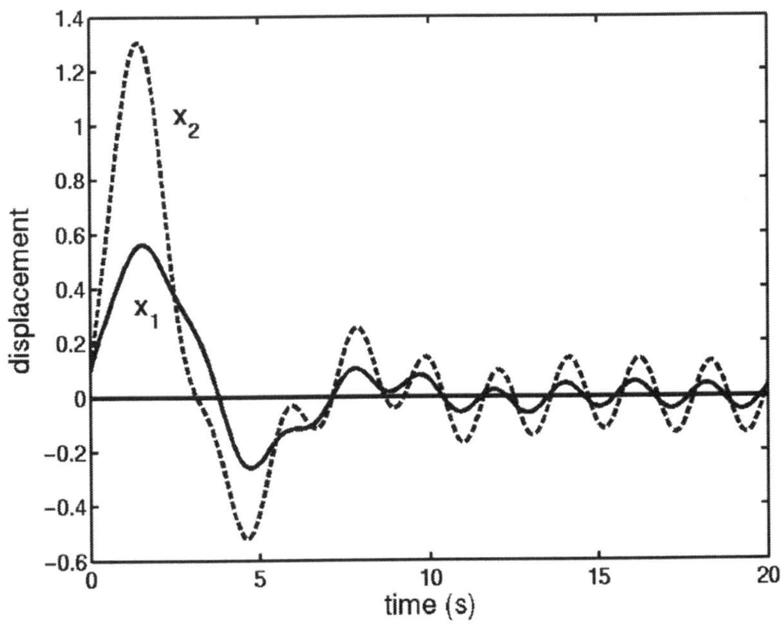

Fig. 6.3-11 Time response of the system of Example 6.3-5.

```
for ii=1:200
tt(ii)=0.1*(ii-1);
xx(ii)=0;
end
h=plot(t,y(:,1),'k',t,y(:,2),'--k',tt,xx,'k');
set(h,'LineWidth',2);
hx=get(gca,'Xlabel');
hy=get(gca,'Ylabel');
set(hx,'String','time (s)');
set(hy,'String','displacement');
set(hx,'FontSize',14);
set(hy,'FontSize',14);
text(1.3,0.4, 'x_1');
text(2.2,1,'x_2');
%
function ydot=f(t,y)
M=[2 0;0 1];
K=[5 -2;-2 2];
C=[3 -0.3;-0.3 0.3];
F=[1;1];
b=[0;0;inv(M)*F];
w=3;
A=[zeros(2), eye(2);-inv(M)*K, -inv(M)*C];
ydot=A*y+b*sin(w*t);
```

The foregoing examples illustrate the method of using computer software to solve vibration problems numerically. Although the examples are all simple, the method and routines work for general multi-degree-of-freedom systems, limited only by the array size for a particular code. Moreover, this method can be used to solve nonlinear differential equations, as indicated in Example 6.3-4.

Problems

6.1 Numerically calculate the natural frequencies and modal matrix of the system shown in Fig. P6.1.

Ans. $\omega_1 = 0.445\sqrt{k/m}$, $\omega_2 = 1.247\sqrt{k/m}$, $\omega_3 = 1.802\sqrt{k/m}$

$$S = \begin{bmatrix} 1.000 & 1.000 & 1.000 \\ 1.802 & 0.445 & -1.247 \\ 2.247 & -0.802 & 0.555 \end{bmatrix}$$

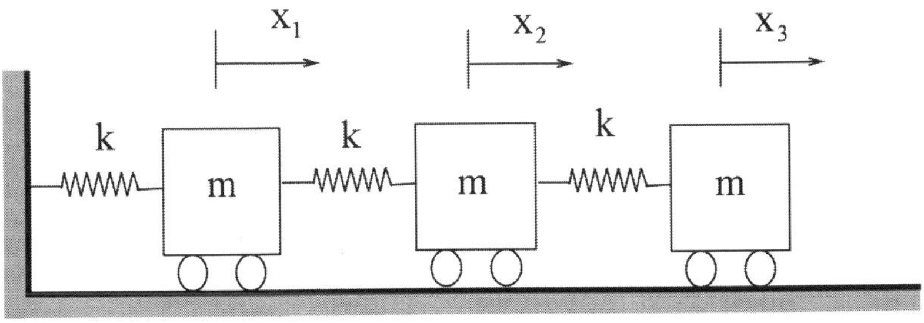

Fig. P6.1

6.2 Determine the general form of the solution to the system in Problem 6.1.

Ans.

$$\begin{bmatrix} x_1 \\ x_2 \\ x_3 \end{bmatrix} = C_1 \begin{bmatrix} 1.000 \\ 1.802 \\ 2.247 \end{bmatrix} \sin\left(0.445\sqrt{\frac{k}{m}}t + \phi_1\right) + C_2 \begin{bmatrix} 1.000 \\ 0.445 \\ -0.802 \end{bmatrix} \sin\left(1.247\sqrt{\frac{k}{m}}t + \phi_2\right)$$

$$+ C_3 \begin{bmatrix} 1.000 \\ -1.247 \\ 0.555 \end{bmatrix} \sin\left(1.802\sqrt{\frac{k}{m}}t + \phi_3\right)$$

6.3 Convert the following into that of a symmetric algebraic eigenvalue problem:

$$M = m\begin{bmatrix} 4 & 0 \\ 0 & 1 \end{bmatrix}, \quad K = k\begin{bmatrix} 5 & -2 \\ -2 & 4 \end{bmatrix}$$

Ans. $\tilde{K} = \dfrac{k}{m}\begin{bmatrix} 5/4 & -1 \\ -1 & 4 \end{bmatrix}, \quad (\tilde{K} - \lambda I)\mathbf{v} = 0$

6.4 In Fig. P6.1, assume a force F is slowly applied to the mass m_3 in the x direction and then released with zero initial velocity. Find the response of the system.

Ans.

$$\begin{bmatrix} x_1 \\ x_2 \\ x_3 \end{bmatrix} = \frac{F}{k}\begin{bmatrix} 1.220\cos\omega_1 t - 0.280\cos\omega_2 t + 0.0597\cos\omega_3 t \\ 2.199\cos\omega_1 t - 0.125\cos\omega_2 t - 0.0744\cos\omega_3 t \\ 2.742\cos\omega_1 t + 0.225\cos\omega_2 t + 0.0331\cos\omega_3 t \end{bmatrix}$$

$$\omega_1 = 0.445\sqrt{k/m}, \quad \omega_2 = 1.247\sqrt{k/m}, \quad \omega_3 = 1.802\sqrt{k/m}$$

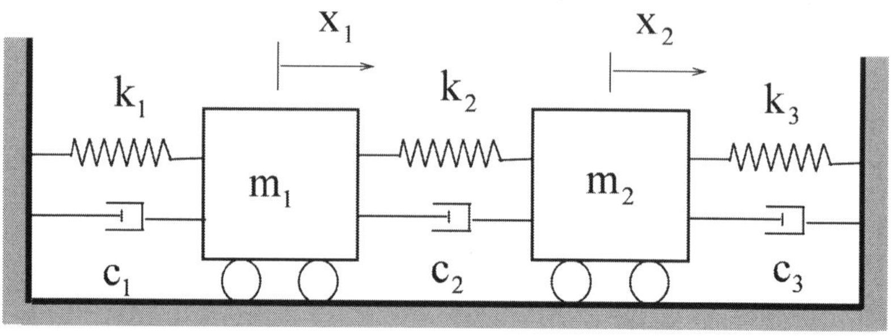

Fig. P6.5

6.5 For the system as shown in Fig. P6.5, $m_1=4$, $m_2=2$, $c_1=2$, $c_2=c_3=1$, $k_1=4$, $k_2=k_3=2$. Find (1) the coefficients of the modal equations, (2) the undamped natural frequencies, and (3) the damping ratios.

Ans. $\ddot{y}_1+0.5\dot{y}_1+y_1=0,\ \ \ddot{y}_2+1.25\dot{y}_2+2.5y_2=0$

$$\omega_1=1.0,\ \ \omega_2=1.5811;\ \ \varsigma_1=0.2500,\ \ \varsigma_2=0.3953$$

6.6 Numerically compute the coefficients of the modal equations for the system in Problem 6.5. (See Example 6.2-4.)

6.7 Calculate the steady-state response of the following two-degree-of-freedom system. (See Example 6.3-5.)

$$\begin{bmatrix} 5 & 0 \\ 0 & 2 \end{bmatrix}\begin{bmatrix} \ddot{x}_1 \\ \ddot{x}_2 \end{bmatrix}+\begin{bmatrix} 0.7 & -0.3 \\ -0.3 & 0.3 \end{bmatrix}\begin{bmatrix} \dot{x}_1 \\ \dot{x}_2 \end{bmatrix}+\begin{bmatrix} 7 & -3 \\ -3 & 3 \end{bmatrix}\begin{bmatrix} x_1 \\ x_2 \end{bmatrix}=\begin{bmatrix} 0 \\ 2\sin\sqrt{3}t \end{bmatrix},$$

with the initial conditions

$$\mathbf{x}_0=\begin{bmatrix} 0.1 \\ 0.1 \end{bmatrix}\text{m};\ \ \dot{\mathbf{x}}_0=\begin{bmatrix} 0.5 \\ 1 \end{bmatrix}\text{m/s}$$

6.8 Fig. P6.8 shows a torsional system with three degrees of freedom. (a) Determine the equations of motion and frequency equation. (b) Find the response of the system to the initial conditions:

$$\{\phi\}=\begin{bmatrix} 0 & 0 & 0 \end{bmatrix}^T,\ \ \{\dot{\phi}\}=\begin{bmatrix} \dot{\theta} & \dot{\theta} & \dot{\theta} \end{bmatrix}^T$$

Ans. $\omega^6-5(k_r/I)\omega^4+6(k_r/I)^2\omega^2-(k_r/I)^3=0$

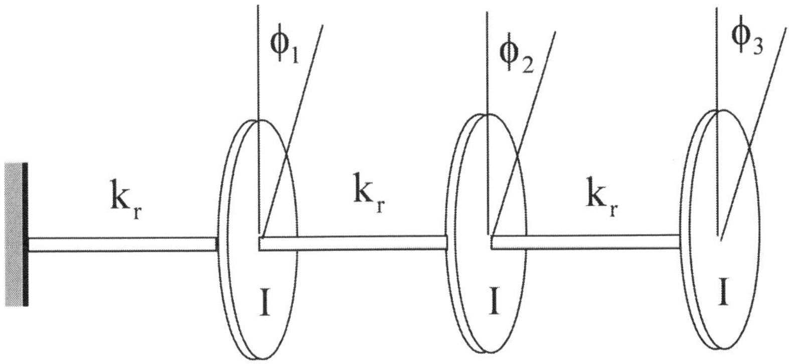

Fig. P6.8

$$\omega_1 = 0.445\sqrt{k_r / I},\ \ \omega_2 = 1.247\sqrt{k_r / I},\ \ \omega_3 = 1.802\sqrt{k_r / I}$$

$$\begin{bmatrix} \phi_1 \\ \phi_2 \\ \phi_3 \end{bmatrix} = \begin{bmatrix} \left(0.543\frac{\sin\omega_1 t}{\omega_1} + 0.349\frac{\sin\omega_2 t}{\omega_2} + 0.108\frac{\sin\omega_3 t}{\omega_3}\right)\dot{\theta} \\ \left(0.979\frac{\sin\omega_1 t}{\omega_1} + 0.155\frac{\sin\omega_2 t}{\omega_2} - 0.134\frac{\sin\omega_3 t}{\omega_3}\right)\dot{\theta} \\ \left(1.220\frac{\sin\omega_1 t}{\omega_1} - 0.280\frac{\sin\omega_2 t}{\omega_2} + 0.060\frac{\sin\omega_3 t}{\omega_3}\right)\dot{\theta} \end{bmatrix}$$

6.9 Numerically integrate and graph the solution of the Van der Pol equation

$$\ddot{x} - 1.5(1 - x^2)\dot{x} + x = 0,$$

with the following initial conditions: (1) $x_0 = 0; \dot{x}_0 = 0.2$; (2) $x_0 = -1.5; \dot{x}_0 = 2$.

6.10 For the system shown in Fig. P6.10, find the natural frequencies and the response of the system to the initial conditions:

$$\mathbf{x}_0 = \begin{bmatrix} 1 & 0 & 0 \end{bmatrix}^T,\ \ \dot{\mathbf{x}}_0 = \begin{bmatrix} 0 & 0 & 1 \end{bmatrix}^T$$

Ans. $\omega_1 = 0.5469\sqrt{k / m},\ \ \omega_2 = 1.1420\sqrt{k / m},\ \ \omega_3 = 1.6010\sqrt{k / m}$

$$x_1 = 0.1501\cos\omega_1 t + 0.1465\frac{\sin\omega_1 t}{\omega_1} + 0.4421\cos\omega_2 t - 0.2634\frac{\sin\omega_2 t}{\omega_2}$$
$$+ 0.4078\cos\omega_3 t + 0.1170\frac{\sin\omega_3 t}{\omega_3}$$

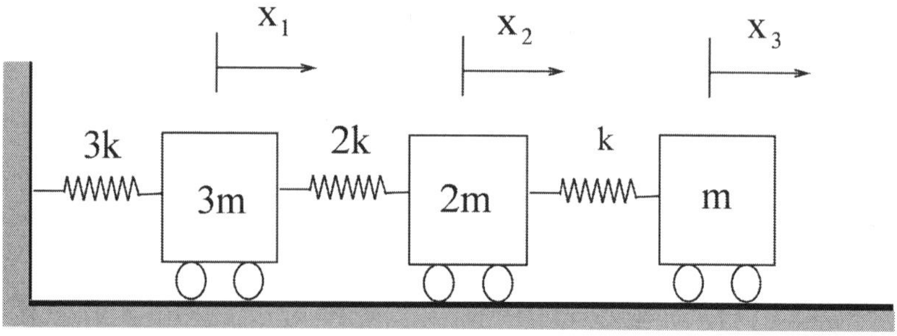

Fig. P6.10

$$x_2 = 0.3080\cos\omega_1 t + 0.3004\frac{\sin\omega_1 t}{\omega_1} + 0.2404\cos\omega_2 t - 0.1432\frac{\sin\omega_2 t}{\omega_2} - 0.5484\cos\omega_3 t - 0.1573\frac{\sin\omega_3 t}{\omega_3}$$

$$x_3 = 0.4394\cos\omega_1 t + 0.4286\frac{\sin\omega_1 t}{\omega_1} - 0.7902\cos\omega_2 t + 0.4707\frac{\sin\omega_2 t}{\omega_2} + 0.3508\cos\omega_3 t + 0.1006\frac{\sin\omega_3 t}{\omega_3}$$

References

6.1. Inman, D. J., *Engineering Vibration*, Prentice Hall, Inc., Upper Saddle River, New Jersey, 1994.

6.2. Craig, R. R., Jr., *Structural Dynamics: A Introduction to Computer Methods*, Wiley, New York, 1981.

6.3. Harker, R. J., *Generalized Method of Vibration Analysis*, John Wiley & Sons, New York, 1983.

6.4. Meirovitch, L., *Analytical Methods in Vibrations*, Macmillan Company, New York, 1967.

6.5. Press, W. H., et al., *Numerical Recipes: The Art of Scientific Computing*, Cambridge University Press, New York, 1986.

6.6. Golub, G. H., and Van Loan, C. F., *Matrix Computation*, The John Hopkins University Press, Baltimore and London, 1989.

6.7. Thomson, W. T., and Dahleh, M. D., *Theory of Vibration with Applications*, 5th ed., Prentice Hall, Inc., Upper Saddle River, NJ, 1998.

6.8. Abramson, H. N., "Nonlinear Vibration," in *Shock and Vibration Handbook*, C. M. Harris, ed., 3rd ed., McGraw-Hill, New York, 1987.

6.9. McConnell, K. G., *Vibration Testing, Theory and Practice*, John Wiley & Sons, Inc., New York, 1995.

6.10. Tse, F. S., Morse, I. E., and Hinkle, R. T., *Mechanical Vibration, Theory and Applications*, Allyn and Bacon, Inc., Boston, 1979.

6.11. Den Hartog, J. P., *Mechanical Vibrations*, 4th ed., McGraw-Hill, New York, 1956. (Reprinted by Dover, New York, 1985.)

6.12. Van der Pol, B., "On Relaxation-Oscillations," *The London, Edinburgh and Dublin Philosophical Magazine and Journal of Science*, Vol. 2, 1927, pp. 978–992.

Chapter Seven

Transfer Functions and Frequency Response Functions

7.1 Introduction

A comprehensive vibration analysis of a given mechanical system ultimately rests upon the solution of the system's governing differential equation. The Laplace transform and the Fourier transform provide useful methods of solving differential equations by *transferring* differential equations into algebraic equations. That is, by use of tabulated transforms of elementary functions, the solution of a differential equation (and hence, a vibration problem) becomes simply a routine algebraic analysis. It also happens that these transform methods provide a basis for experimental techniques.

7.2 The Laplace Transformation and the Fourier Transformation

(1) The Laplace Transform

Let $x(t)$ be an integrable function such that $x(t)=0$ for $t<0$. The Laplace transform $L[x(t)]$ of $x(t)$ is defined as [7.1]

$$L\left[x(t)\right]=X(s)=\int_0^\infty x(t)e^{-st}\,dt, \tag{7.2-1}$$

where the variable s is, in general, a complex quantity. The complex plane $s=\sigma+j\omega$ is referred as the *s plane*, or the *s domain*, or the *Laplace domain.*

If the Laplace transform $X(s)$ is known, the determination of the originating function $x(t)$ can be accomplished by using the *inverse Laplace transform*, denoted by $L^{-1}\left[X(s)\right]$. The inverse Laplace transform is

$$L^{-1}\left[X(s)\right]=x(t)=\frac{1}{2\pi j}\int_{-\infty}^{+\infty} X(s)e^{st}\,ds \quad \left(j=\sqrt{-1}\right) \tag{7.2-2}$$

Taken together, Eqs. 7.2-1 and 7.2-2 form the *Laplace transform pair.*

The function $x(t)$ must satisfy certain conditions, for otherwise the Laplace transform will not exist. Fortunately, however, most functions describing physical phenomena satisfy these conditions. (See, for example, a mathematics text such as [7.1]).

The Laplace transform has several properties useful in vibration analysis. We list a few of them here:

(a) The shift theorem. Let a be a real or complex number. Let $X(s)$ be the Laplace transform of $x(t)$. Then the Laplace transform of $e^{at}x(t)$ is $X(s-a)$. Hence, the effect of multiplying $x(t)$ by e^{at} in the time domain is to shift the transform of $x(t)$ by a in the s domain.

(b) The transform of derivatives. Let $X(s)$ be the Laplace transform of $x(t)$. Then the Laplace transform of the derivative of $x(t)$ is

$$L\left[dx(t)/dt\right]=sX(s)-x(0), \tag{7.2-3}$$

where $x(0)$ is the value of $x(t)$ at $t=0$.

(c) The transform of convolution integrals. The following integrals

$$\int_0^t f(t-\tau)h(\tau)d\tau=\int_0^t f(\tau)h(t-\tau)d\tau \tag{7.2-4}$$

are known as the convolution integrals and are denoted as $f(t)*h(t)$. The Laplace transform of a convolution integral is equal to the product of the Laplace transforms of each function. That is,

$$L\left[f(t)*h(t)\right]=L\left[f(t)\right]\cdot L\left[h(t)\right] \tag{7.2-5}$$

Table 7.2-1 lists some common Laplace transforms. More complete tables can be found in various mathematics handbooks.

Table 7.2-1 Common Laplace Transforms with Zero Initial Conditions

Time domain	Laplace domain
$f(t)$	$F(s)$
$\delta(0)$	1
1	$1/s$
e^{-at}	$\dfrac{1}{s+a}$
$\sin\omega_n t$	$\dfrac{\omega_n}{s^2+\omega_n^2}$
$\cos\omega_n t$	$\dfrac{s}{s^2+\omega_n^2}$
$\delta(t-a)$	e^{-as}
$e^{at}f(t)$	$F(s-a)$

Example 7.2-1 Find the inverse Laplace transform of the transfer function defined as

$$H(s)=\frac{1}{ms^2+cs+k}, \quad (c^2<4mk)$$

Solution: The transfer function can be written using the partial fraction expansion as

$$H(s)=\frac{1}{ms^2+cs+k}=\frac{1}{j2m\omega_d}\left(\frac{1}{s-\lambda_1}-\frac{1}{s-\lambda_2}\right),$$

where λ_1 and λ_2 are the roots of the characteristic equation $ms^2+cs+k=0$. That is,

$$\lambda_1=-\frac{c}{2m}+j\sqrt{\frac{c^2}{4m^2}-\frac{k}{m}}\equiv-\varsigma\omega_n+j\omega_d \quad \left(j=\sqrt{-1}\right)$$

$$\lambda_2=-\frac{c}{2m}-j\sqrt{\frac{c^2}{4m^2}-\frac{k}{m}}\equiv-\varsigma\omega_n-j\omega_d \quad \left(j=\sqrt{-1}\right)$$

Using Table 7.2-1, we obtain

$$h(t)=L^{-}(H(s))=\frac{1}{j2m\omega_d}\left(e^{\lambda_1 t}-e^{\lambda_2 t}\right)$$

or

$$h(t)=\frac{e^{-\varsigma\omega_n t}}{m\omega_d}\sin\left(\omega_d t\right)$$

$h(t)$ is commonly called the *impulse response function* [7.2, 7.3].

(2) The Fourier Transform

As in the previous case with the Laplace domain, equivalent information can be presented in the *frequency domain* using the Fourier transform, as discussed in Section 1.5. The Fourier transform of a function $x(t)$ is denoted by $X(\omega)$ and the *Fourier transform pair* is given by [7.4, 7.5]

$$X(\omega)=\int_{-\infty}^{+\infty}x(t)e^{-j\omega t}dt \tag{7.2-6}$$

$$x(t)=\frac{1}{2\pi}\int_{-\infty}^{+\infty}X(\omega)e^{j\omega t}d\omega \tag{7.2-7}$$

Observe that the definitions of the Fourier transform and the Laplace transform are similar. The principal difference is that the Fourier transform is defined from negative infinity to positive infinity, whereas the Laplace

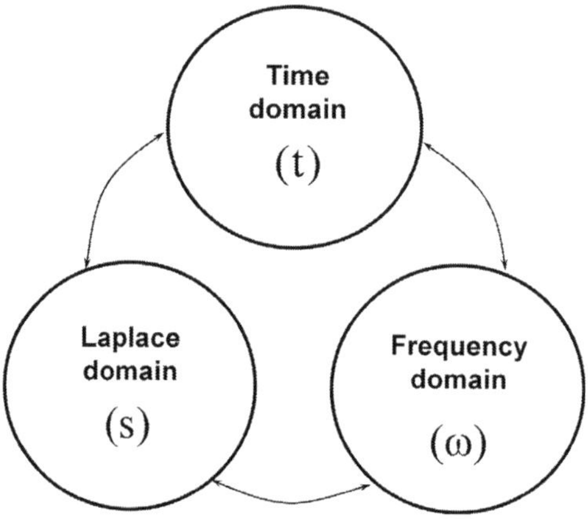

Fig. 7.2-1 Three function domains.

transform is defined from zero to positive infinity. For a time function $x(t)$ such that $x(t)=0$ for $t<0$, the form of the Fourier transform pair given in Eqs. 7.2-6 and 7.2-7 can be obtained by substituting $s=j\omega$ in the Laplace transform pair given by Eqs. 7.2-1 and 7.2-2. Although this is not mathematically rigorous, it does provide a connection between the two types of transformations, and the results are the same.

A time function can be described in different domains. The Laplace transform changes the domain of the function from the time domain to the Laplace domain, whereas the Fourier transform changes the domain of the function from the time domain to the frequency domain. The three domains can be transformed from one to another, as represented in Fig. 7.2-1 [7.2, 7.3].

7.3 Laplace Domain: Transfer Function

If a differential equation of motion is operated on by a Laplace transform, the relation between the transform of the forcing function and that of the displacement function is a simple algebraic expression. To illustrate this, consider the equation of motion of a forced single-degree-of-freedom system:

$$m\ddot{x}+c\dot{x}+kx=f(t) \tag{7.3.1}$$

By taking the Laplace transform and, for simplicity, by assuming zero initial conditions, we obtain

$$\left(ms^2+cs+k\right)X(s)=F(s), \tag{7.3-2}$$

where $X(s)$ and $F(s)$ are the Laplace transforms of $x(t)$ and $f(t)$, respectively. The ratio of $X(s)$ and $F(s)$ is then simply

$$H_d(s)=\frac{X(s)}{F(s)}=\frac{1}{ms^2+cs+k} \tag{7.3-3}$$

The ratio $X(s)/F(s)$, denoted by $H_d(s)$, is variously called the *displacement transfer function*, the *receptance*, the *dynamic compliance*, or the *admittance*.

By setting the denominator equal to zero in Eq. 7.3-3, we obtain the *characteristic equation*

$$ms^2 + cs + k = 0 \tag{7.3-4}$$

Experimentalists use several different transfer functions, depending upon the type of measurement being taken: displacement, velocity, or acceleration (see [7.2, 7.6]). For velocity, the ratio of the transform of the velocity to that of the forcing function, denoted by $H_v(s)$, is

$$H_v(s) = sH_d(s) = \frac{s}{ms^2 + cs + k} \tag{7.3-5}$$

[Observe that with zero initial conditions, in view of Eq. 7.2-3, $H_v(s)$ is $sH_d(s)$.]

The velocity transfer function $H_v(s)$ is called the *mobility*.

Similarly, with zero initial conditions, the ratio of the transform of the acceleration to that of the forcing function, denoted by $H_a(s)$, is

$$H_a(s) = s^2 H_d(s) = \frac{s^2}{ms^2 + cs + k} \tag{7.3-6}$$

$H_a(s)$ is alternately called the *acceleration transfer function*, or the *accelerance*, or the *inertance*.

Since transfer functions are complex-valued functions of the complex independent variable s, they have real and imaginary parts. Thus, a transfer function can be viewed as a function of two variables and represented as a surface.

Fig. 7.3-1 shows a typical graphical representation of the receptance transfer function of a single-degree-of-freedom system [7.2], where the mass, damping, and stiffness values are $m = 10$, $c = 1.5$, and $k = 3$; and where σ and ω are, respectively, the real part and imaginary part of the Laplace variable s. That is, $s = \sigma + j\omega$. The Matlab code for exhibiting the surface is

```
clear all
m=10; c=1.5; k=3;
a=-c/(2*m);
b=sqrt(k/m-c^2/(4*m^2));
rot=60;
lambda(1)=a+j*b;
lambda(2)=a-j*b;
[sigma,omega]=meshgrid(-0.2:0.01:-0.05,-1:0.02:1);
s=sigma+j*omega;
H=(1.0/10).*(1.0./((s-lambda(1)).*(s-lambda(2))));
view=[rot,30];
```

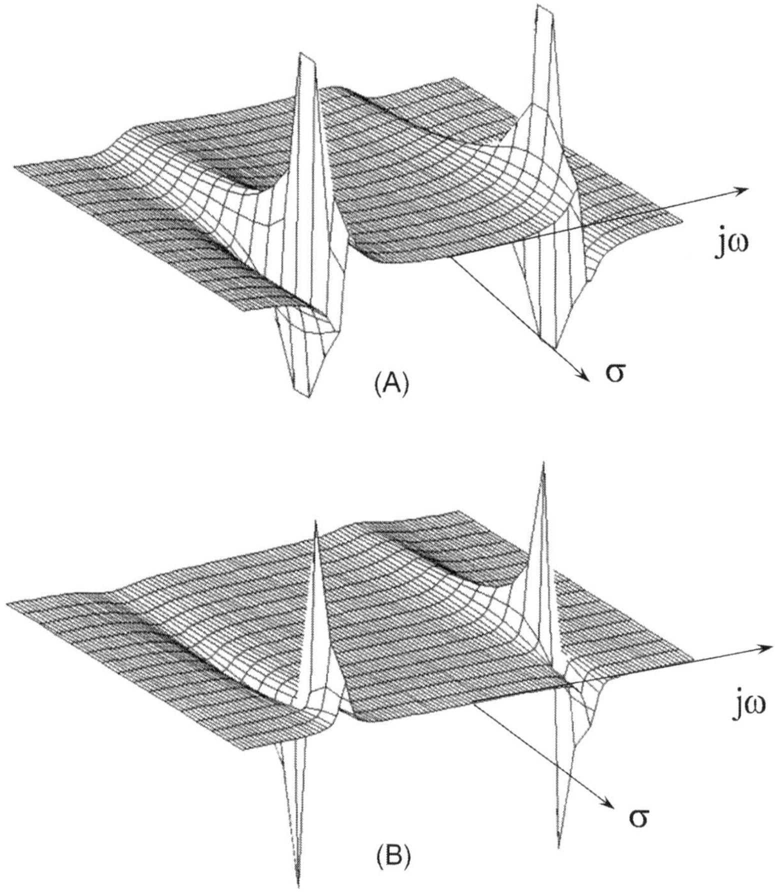

Fig. 7.3-1 Laplace domain representation of a single-degree-of-freedom system, where $m = 10$, $c = 1.5$, $k = 3$: (A) real part, (B) imaginary part [7.2]. (This graph is reproduced with permission of McGraw-Hill Companies.)

```
mesh(real(H),view);
pause
mesh(imag(H),view)
```

The definitions of undamped natural frequency, damped natural frequency, and the damping ratio are all relative to information represented by Fig. 7.3-1 [7.2]. The projection of the pole locations onto the plane of zero amplitude yields the damping coefficient, damped natural frequency, undamped natural frequency, and the damping ratio. Recall that, once the poles are known, the undamped natural frequency and damping ratio can be calculated from Eqs. 6.2-25 and 6.2-26. That is,

$$\omega_n = \sqrt{\left(\mathrm{Re}(\lambda_r)\right)^2 + \left(\mathrm{Im}(\lambda_r)\right)^2} \qquad (7.3\text{-}7)$$

and

$$\varsigma = -\frac{\mathrm{Re}(\lambda_r)}{\omega_n} \tag{7.3.8}$$

Finally, it should be noted that the transfer function is directly related to the impulse response function. To see this clearly, recall that the impulse response function, denoted $h(t)$, is the response of a system with zero initial conditions to a unit impulse applied at $t=0$. The motion of the system in this case can be described by Eq. 7.3-1, with $f(t)=\delta(t)$, where $\delta(t)$ is the Dirac delta function. The Laplace transform of $\delta(t)$ is

$$L\left[\delta(t)\right]=F(s)=1, \tag{7.3-9}$$

and it follows from Eq. 7.3-3 that

$$L\left[h(t)\right]=\frac{1}{ms^2+cs+k} \tag{7.3-10}$$

Eq. 7.3-10 indicates that the receptance transfer function is merely the Laplace transform of the impulse response function. This result will be used repeatedly in the sequel.

7.4 Frequency Domain: Frequency Response Function

We can obtain the frequency domain representations of the system by substituting $s=j\omega$ into expressions of the transfer functions as in Eqs. 7.3-3, 7.3-5, and 7.3-6. Thus, we obtain the corresponding *frequency response functions* (denoted FRF). They are

(1) Receptance (or Admittance, or Dynamic Compliance)

$$H_d(\omega)=\frac{X(\omega)}{F(\omega)}=\frac{1}{k-m\omega^2+jc\omega} \tag{7.4-1}$$

(2) Mobility

$$H_v(\omega)=j\omega H_d(\omega)=\frac{j\omega}{k-m\omega^2+jc\omega} \tag{7.4-2}$$

(3) Accelerance (or Inertance)

$$H_a(\omega)=-\omega^2 H_d(\omega)=-\frac{\omega^2}{k-m\omega^2+jc\omega} \tag{7.4-3}$$

The receptance frequency response function is denoted by $H_d(\omega)$, meaning displacement FRF. It is sometimes denoted as x/F. Similarly, $H_v(\omega)$

represents the velocity FRF, and it is sometimes denoted as v/F. Finally, $H_a(\omega)$ represents the acceleration FRF, and it is sometimes denoted as a/F.

From these relationships, we can construct three other ratios: (1) the *apparent stiffness* (force per unit of displacement), (2) the *mechanical impedance* (force per unit of velocity), and (3) the *apparent mass* (force per unit of acceleration). The name of *mechanical impedance* is also given to the ratios F/x and F/a.

Unfortunately, with all this terminology and notation, there may occur some confusion during the reading of research papers and other books. Therefore, we need to be careful to keep the various quantities clear and distinct.

Observe that when $\omega=0$, $H_d(0)$ is $1/k$. This provides an estimate of static stiffness.

Since the receptance frequency response function $H_d(\omega)$ in Eq. 7.4-1 is a complex-valued function of a real-valued independent variable ω, it has real and imaginary parts given by

$$\begin{aligned} H_d(\omega) &= \frac{k-m\omega^2}{(k-m\omega^2)^2+(c\omega)^2} + j\frac{-\omega c}{(k-m\omega^2)^2+(c\omega)^2} \\ &\equiv H_d^R(\omega) + jH_d^I(\omega) \end{aligned} \tag{7.4-4}$$

Next, let us define the frequency ratio r and the damping ratio ς as

$$r = \frac{\omega}{\sqrt{k/m}} = \frac{\omega}{\omega_n} \quad \text{and} \quad \varsigma = \frac{c}{2\sqrt{km}} \tag{7.4-5}$$

We can now write the receptance FRF in the dimensionless form

$$H_d(r) = \frac{1}{k}\frac{1-r^2}{(1-r^2)^2+(2\varsigma r)^2} + j\frac{1}{k}\frac{-2\varsigma r}{(1-r^2)^2+(2\varsigma r)^2} \tag{7.4-6}$$

$$\equiv H_d^R(r) + jH_d^I(r)$$

Fig. 7.4-1 presents graphs of the real and imaginary parts of the receptance of a single-degree-of-freedom system [7.6]. Several points of interest are noted: At point D_1, the real part is zero, corresponding to the natural frequency of the system. Point D_2, the corresponding point of the imaginary part is at a minimum value. Points B_1 and B_2 correspond to the largest difference in the real part. They are called *upper half power points*. The corresponding points C_1 and C_2 on the imaginary graph are called *lower half power points*.

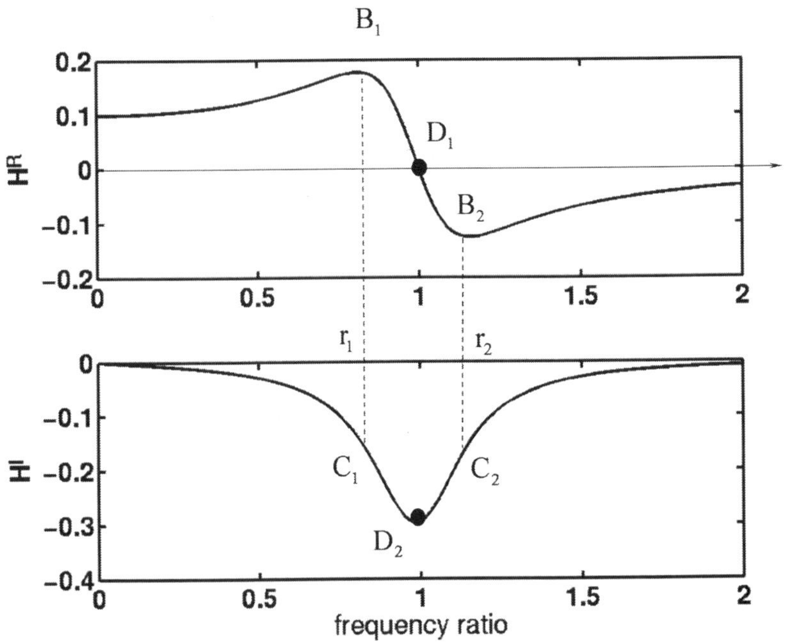

Fig. 7.4-1 Frequency domain representation of a single-degree-of-freedom system. H^R: real part; H^I: imaginary part [7.6].

As in the Laplace domain, the definitions of undamped natural frequency, damped natural frequency, and the damping ratio are all relative to the information represented by Fig. 7.4-1.

Consider the imaginary part of the receptance. Differentiating H_d^I with respect to r and setting the result equal to zero, we have

$$\frac{dH_d^I}{dr}=0 \qquad (7.4.7)$$

We thus obtain

$$r=\sqrt{1-\varsigma^2} \qquad \text{or} \qquad \omega=\sqrt{1-\varsigma^2}\,\omega_n \qquad (7.4\text{-}8)$$

Eq. 7.4-8 indicates that the receptance imaginary part has an extremum at the damped natural frequency. For small values of the damping ratio ς, $\omega \approx \omega_n$. Thus, for a small damping ratio, the undamped natural frequency of the system is approximately equal to the frequency that corresponds to point D_2 on the imaginary part (see Fig. 7.4-1).

Next, let r_1 and r_2 be the corresponding frequency ratios at points B_1 and B_2 on the receptance real part. They can be determined by differentiating H_d^R with respect to r and setting the result to zero. That is,

$$\frac{dH_d^R}{dr} = 0 \tag{7.4-9}$$

Then, from Eq. 7.4-9, we obtain

$$r_1^2 = 1 - 2\varsigma, \quad r_2^2 = 1 + 2\varsigma \tag{7.4-10}$$

Subtracting the two expressions yields

$$4\varsigma = r_2^2 - r_1^2 = \frac{\omega_2^2 - \omega_1^2}{\omega_n^2} \tag{7.4.11}$$

Therefore, the damping ratio can be evaluated simply as

$$\varsigma = \frac{\omega_2^2 - \omega_1^2}{4\omega_n^2} \approx \frac{(\omega_2 - \omega_1)}{2\omega_n} \tag{7.4-12}$$

(4) The Bode Diagrams

A Bode diagram is a graph of an FRF magnitude and phase as functions of the frequency. Fig. 7.4-2 shows a typical receptance Bode diagram of a

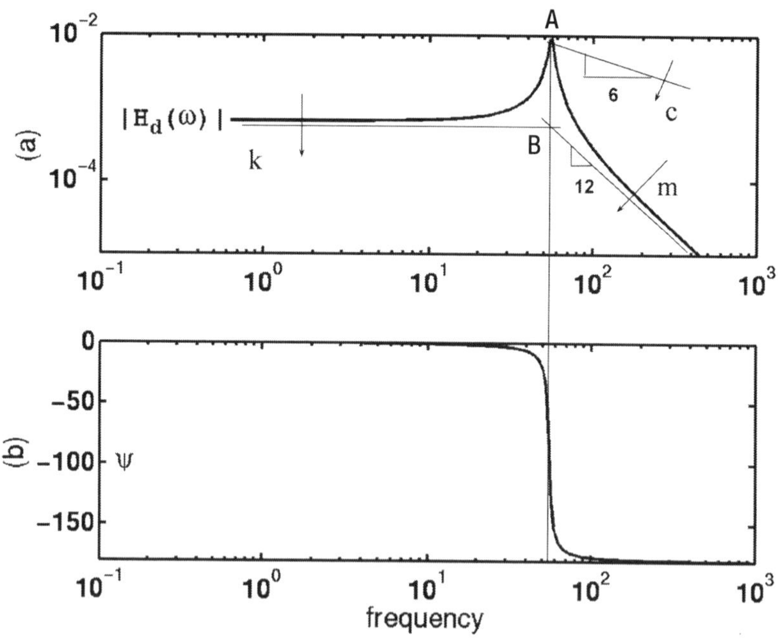

Fig. 7.4-2 Receptance Bode diagram of a single-degree-of-freedom system, where $m = 0.5$, $c = 2$, and $k = 1500$: (a) amplitude; (b) phase [7.6].

single-degree-of-freedom system [7.6], where the mass, damping coefficient, and spring constant values are $m=0.5$, $c=2$, and $k=1500$.

In Fig. 7.4-2, the magnitude and phase angle of $H_d(\omega)$ are

$$|H_d(\omega)|=\sqrt{H(\omega)H(-\omega)}=\frac{1}{k}\frac{1}{\sqrt{\left(1-\omega^2/\omega_n^2\right)^2+\left(2\varsigma\omega/\omega_n\right)^2}} \tag{7.4-13}$$

and

$$\psi=\tan^{-1}\left(\frac{H_d^I}{H_d^R}\right)=\tan^{-1}\left(\frac{2\varsigma\omega/\omega_n}{\omega^2/\omega_n^2-1}\right) \tag{7.4-14}$$

Observe in Fig. 7.4-2 that log scales are used. This is a convenience for several reasons. First, the log scales enable us to maintain detail for large ranges of the variables. Next, the use of log scales converts the curves into asymptotic straight lines. Finally, the log scales enable the development of the *dB* variable defined as [7.6]

$$dB=20\log(x/x_r), \tag{7.4-15}$$

where x_r is a reference value.

In Fig. 7.4-2, the amplitude curve can be used to estimate system parameters. For example, when ω is small compared with the natural frequency ω_n (that is, $\omega<<\omega_n$), Eq. 7.4-13 shows that $|H_d(\omega)|$ approaches $1/k$, where k is the spring constant. Thus, in Fig. 7.4-2 (a), we have the nearly horizontal *spring line* which we can use to evaluate $1/k$ and, hence, the stiffness k itself. The direction of increasing stiffness is shown by the downward pointing arrow on the left side of the graph.

Alternatively, when ω is large compared with ω_n (that is, $\omega>>\omega_n$), Eq. 7.4-13 shows that $|H_d(\omega)|$ approaches $1/m\omega^2$ (see also Eq. 7.4-1). Thus, on a log scale graph, $|H_d(\omega)|$ becomes a straight line. Recall that $\log(1/m\omega^2)$ may be expressed as [7.6]

$$\log\left(\frac{1}{m\omega^2}\right)=-\log(m)-2\log(\omega) \tag{7.4-16}$$

In Fig. 7.4-2, this line is called the *mass line*. Its slope is 12 dB/octave, since $20\left[2\log(2)\right]$ is approximately 12. The direction of increasing mass is shown by the downward pointing arrow on the right side of the graph.

The spring line and the mass line intersect at point *B*, corresponding to the natural frequency of the system.

Recall that the spring force and inertia force balance when the system oscillates at the natural frequency ω_n. Thus, for frequency near ω_n, Eq. 7.4-1 shows that $\left[H_d(\omega)\right]$ is nearly $1/c\omega$. Thus, on the log scale, we have a *damping line* since $\log(1/c\omega)$ is [7.6]

$$\log\left(\frac{1}{c\omega}\right) = -\log(c) - \log(\omega) \tag{7.4-17}$$

Since $20\left[\log(2)\right]$ is approximately 6, Eq. 7.4-17 indicates that the damping line has a slope of approximately 6dB/octave. The peak value is determined by the intersection of the natural frequency line and the damping line [point *A* in Fig. 7.4-2 (a)].

The phase angle ψ is seen to vary from 0 to $-180°$ as the frequency passes through resonance. The phase angle is $-90°$ at resonance, as shown in Fig. 7.4-2 (b).

(5) The Nyquist Diagrams

A Nyquist diagram is a graphical representation of the real part of an FRF as a function of the imaginary part for increasing values of frequency [7.6, 7.7]. Fig. 7.4-3 presents a typical Nyquist diagram for the receptance (admittance or dynamic compliance) frequency response function of a single-degree-of-freedom system, where the mass, damping, and stiffness values are $m = 0.1$, $c = 2$, and $k = 1000$.

Nyquist diagrams are sometimes called *Nyquist circles*. It happens, however, that only the Nyquist diagram of mobility FRF is a perfect circle. The Nyquist diagrams of the receptance and accelerance (inertance) FRFs are not exact circles.

It is readily seen that the real and imaginary parts of the FRFs of the receptance, the mobility, and the accelerance satisfy the following relations, respectively [7.7]:

$$\left[H_d^R(\omega)\right]^2 + \left[H_d^I(\omega) + \frac{1}{2\omega c}\right]^2 = \frac{1}{(2\omega c)^2} \tag{7.4-18}$$

$$\left[H_v^R(\omega) - \frac{1}{2c}\right]^2 + \left[H_v^I(\omega)\right]^2 = \left(\frac{1}{2c}\right)^2 \tag{7.4-19}$$

$$\left(H_a^R(\omega)\right)^2 + \left[H_a^I(\omega) - \frac{\omega}{2c}\right]^2 = \left(\frac{\omega}{2c}\right)^2, \tag{7.4-20}$$

where $H_x^R(\omega)$ and $H_x^I(\omega)$ $(x = d, v, a)$ are, respectively, the real and imaginary parts of the corresponding FRFs.

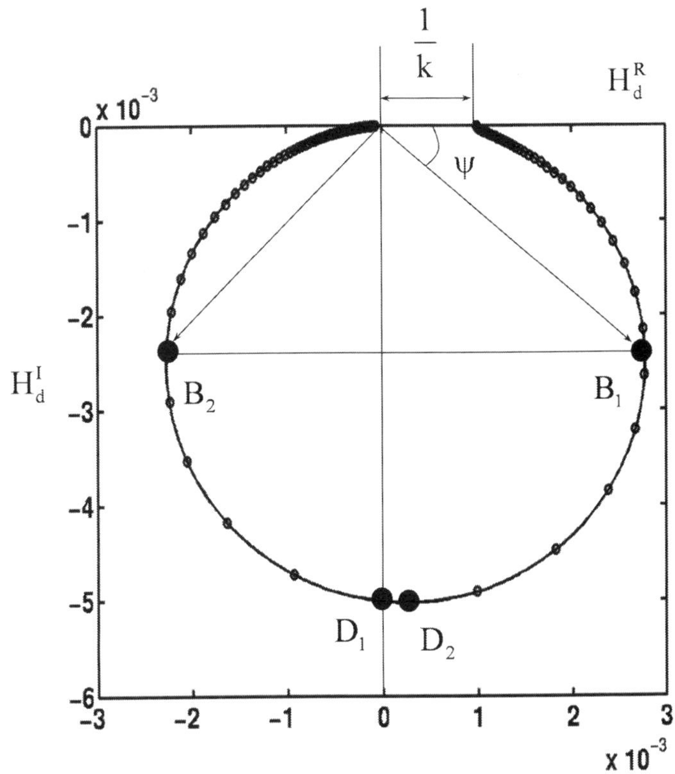

Fig. 7.4-3 Nyquist graph of the receptance [7.6, 7.7].

Observe that with ω being a variable, only Eq. 7.4-19 is an equation of a circle. The circle has radius $1/(2c)$ and center at $(H_v^I(\omega)=0, H_v^R=1/2c)$. Eqs. 7.4-18 and 7.4-20 are not circle equations since the right sides, defining the circle radii, are functions of the frequency ω, a variable.

The Nyquist diagram of Fig. 7.4-3, when viewed as a polar coordinate representation, provides the magnitude of $H_d(\omega)$ as the radial coordinate. Fig. 7.4-3 also shows the phase angle ψ. Once the diagram is constructed, the magnitude $|H_d(\omega)|$ and the phase angle ψ are obtained without reference to the frequency ω.

The points (dots) on the curve (near circle) of Fig. 7.4-3 correspond to equal increments of frequencies. Observe that equal increments of frequency do not correspond to equally spaced dots around the near circle.

Next, observe that when ω is zero, the Nyquist graph begins at a distance $1/k$ away from the vertical axis.

Finally, refer to Fig. 7.4-1: the points B_1, B_2, D_1, and D_2 in the real and imaginary parts of Fig. 7.4-1 are also shown in Fig. 7.4-3. Observe, in the two figures, that the real part, $H_d^R(\omega)$, is zero at point D_1, corresponding to the natural frequency (resonance) of the system. Point D_2 represents the corresponding imaginary part, $H_d^I(\omega)$, at resonance. Then, in view of Eq. 7.4-18, we can determine the damping from the radial coordinate of D_2. Note that for small values of damping, D_1 and D_2 are close to each other. Note also that D_1 and D_2 are located between the two frequencies forming the largest arc length on the near circle.

7.5 Time Domain: Impulse Response Function

The transfer function $H_d(s)$ in Eq. 7.4-1 can be expressed in terms of partial fractions as

$$H_d(s)=\frac{1}{ms^2+cs+k}=\frac{1}{m(s-\lambda_1)(s-\lambda_2)}=\frac{A_1}{s-\lambda_1}+\frac{A_2}{s-\lambda_2}, \tag{7.5-1}$$

where λ_1 and λ_2 are the roots of the quadratic equation $ms^2+cs+k=0$. For a vibrating system, these roots are complex conjugate poles.

For an underdamped system, λ_1 and λ_2 may be expressed in the form

$$\lambda_r=\alpha_r+j\beta_r \quad (r=1,2), \tag{7.5.2}$$

where α_r is the damping factor and β_r is the damped natural frequency.

Next, in Eq. 7.5-1, A_1 and A_2 are called the *residues* of the transfer function. In general, A_1 and A_2 form a complex conjugate pair. For a single-degree-of-freedom system, they are purely imaginary (see Example 7.5-1).

The impulse response function, $h(t)$, of the single-degree-of-freedom system can be determined from Eq. 7.5-1, assuming that the initial conditions are zero and that the system excitation is a unit impulse. The inverse Laplace transform of Eq. 7.5-1 is the impulse response of the system. Hence, the impulse response function can be written as [7.2, 7.3]

$$h(t)=A\exp(\lambda_1 t)+A^*\exp(\lambda_1^* t) \tag{7.5-3}$$

or

$$h(t)=\exp(\alpha_1 t)\left[A\exp(+j\beta_1 t)+A^*\exp(-j\beta_1 t)\right], \tag{7.5-4}$$

where

$\lambda_1=\alpha_1+j\beta_1=$ complex pole of the system

$\alpha_1=-\varsigma\omega_n=$ damping factor

$\beta_1=\sqrt{1-\varsigma^2}\,\omega_n\equiv\omega_d=$ damped natural frequency

$\lambda_1^* =$ complex conjugate of λ_1

$A =$ complex residue

$A^* =$ complex conjugate of residue A

Thus, the residue A controls the amplitude of the impulse response, the real part of the pole is the decay rate, and the imaginary part of the pole is the frequency of oscillation.

Example 7.5-1 Show that for an underdamped single-degree-of-freedom system, the residues A_1 and A_2 are purely imaginary.

Solution: Multiplying the left and right sides of Eq. 7.5-1 by $(s-\lambda_1)$ and then letting $s=\lambda_1$, we obtain

$$A_1 = \frac{1}{m(\lambda_1 - \lambda_2)} \tag{a}$$

Since, for an underdamped single-degree-of-freedom system, λ_1 and λ_2 are complex conjugates of one another, we have $\lambda_1 = \alpha + j\beta$ and $\lambda_2 = \alpha - j\beta$. Substituting into Eq. a, we obtain

$$A_1 = \frac{1}{m(\lambda_1 - \lambda_2)} = \frac{1}{j2m\beta} \tag{b}$$

Similarly , we see that A_2 is also purely imaginary. That is,

$$A_2 = \frac{1}{m(\lambda_2 - \lambda_1)} = -\frac{1}{j2m\beta} \tag{c}$$

Example 7.5-2 Starting from Eq. 7.5-4, show that for an underdamped single-degree-of-freedom system, the impulse response function is given by

$$h(t) = \frac{1}{m\omega_d} e^{-\varsigma\omega_n t} \sin(\omega_d t), \tag{a}$$

where ω_n is the natural frequency of the system, and $\omega_d = \sqrt{1-\varsigma^2}\,\omega_n$ is the damped natural frequency of the system.

Solution: Let the complex poles be written as

$$\lambda_1 = -\varsigma\omega_n + j\omega_d \quad \text{and} \quad \lambda_2 = -\varsigma\omega_n - j\omega_d \tag{b}$$

From Eqs. b and c of Example 7.5-1, the residues are given by

$$A = \frac{1}{j2m\omega_d} \quad \text{and} \quad A^* = -\frac{1}{j2m\omega_d} \tag{c}$$

Substituting these expressions into Eq. 7.5-4 yields the desired result.

It is important to find the poles and residues of a system for several reasons. First, the definition of undamped natural frequency, damped natural frequency, and damping ratio are all related to the poles. Second, the residues are directly related to the mode shapes of the system. Finally, once the poles and residues are known, the time domain representation of the system can be obtained using Eq. 7.5-4.

There are several ways to find the poles and residues of a system. One is to convert the Laplace domain polynomial quotient into a pole-residue representation. Specifically, suppose $a(s)$ and $b(s)$ are polynomials. Then the ratio $b(s)/a(s)$ may be written in the form

$$\frac{b(s)}{a(s)}=\frac{b_1 s^m+b_2 s^{m-1}+\cdots+b_{m+1}}{a_1 s^n+a_2 s^{n-1}+\cdots+a_{n+1}}=\frac{A_1}{s-\lambda_1}+\frac{A_2}{s-\lambda_2}+\cdots+\frac{A_n}{s-\lambda_n}+K(s), \tag{7.5-5}$$

where $K(s)$ is called the *direct term*. (The direct term does not appear if $m<n$.)

Calculating the poles and residues is a simple command in most computational software programs. In Matlab, for example, the command `[R,P,K]=residue(b,a)` can immediately return the poles, residues, and direct term. The residues are returned in a column vector `R`, the pole locations in a column vector `P`, and the direct terms in row vector `K`. [The direct term vector is empty if `length(b) < length(a)`.]

This concept can be directly used for finding the poles and residues of the transfer function. Indeed, the receptance transfer function of a single-degree-of-freedom system can be written as a partial fraction expansion of the ratio $b(s)/a(s)$ as

$$H_d(s)=\frac{1}{ms^2+cs+k}=\frac{b(s)}{a(s)}=\frac{A_1}{s-\lambda_1}+\frac{A_2}{s-\lambda_2}, \tag{7.5-6}$$

where

$$b(s)=0.s^2+0.s+1 \tag{7.5-7}$$

$$a(s)=ms^2+cs+k \tag{7.5-8}$$

The following example illustrates the calculation of the poles, residues, and impulse response function using Matlab software.

Example 7.5-3 Given $m=10$, $k=16000$, and $c=20$ for a single-degree-of-freedom system, find the poles and residues of the system, and graph the impulse response function.

Solution: The receptance transfer function can be expressed as the ratio of two polynomials $b(s)/a(s)$, where $b(s)=0.s^2+0.s+1$ and $a(s)=ms^2+cs+k$. With the command `[R,P,K]=residue(b,a)`, the poles are found to be

$$\lambda_1=-1.0+j39.9875, \quad \lambda_2=\lambda_1^*=-1.0-j39.9875$$

and the residues are

$$A_1=-j0.00125, \quad A_2=A_1^*=j0.00125$$

Thus, the impulse response function is given by Eq. 7.5-4. That is,

$$h(t)=\exp(-1.0t)\left[(-j0.00125\cdot\exp(j39.9875t)+j0.00125\cdot\exp(-j39.9875t)\right]$$

or

$$h(t)=0.0025e^{-t}\sin(39.9875t)$$

Fig. 7.5-1 shows the results, where the solid line is calculated based on Eq. 7.5-4, and the small circle points are calculated by using Eq. a of

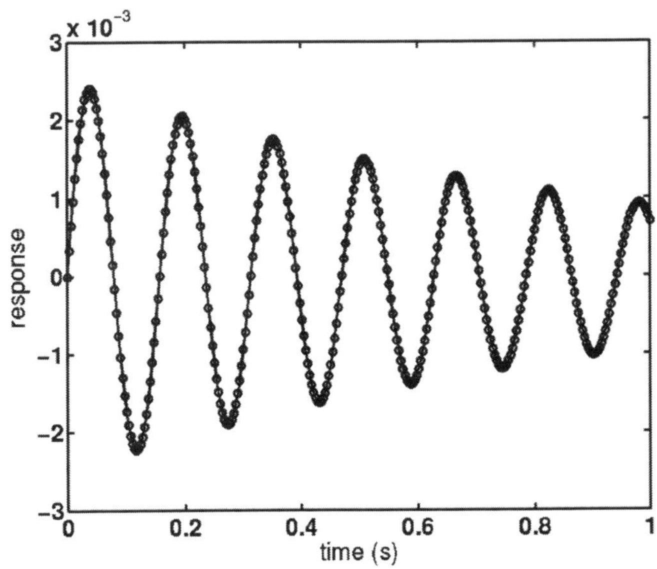

Fig. 7.5-1 Graphs of the impulse response function. Solid line: based on Eq. 7.5-4; small circles: based on Eq. a of Example 7.5-2. (The two results are identical.)

Example 7.5-2. They are, as expected, identical. The Matlab code for this calculation is

```
clear all
mass=10;
stiff=16000;
damp=20;
a=[mass, damp, stiff];
b=[0,0,1];
[r,p,k]=residue(b,a); % .......find poles and residues
residue=r(1);
lambda=p(1);
t=linspace(0,1,300);
xt=residue*exp(lambda.*t)+residue'.*exp(lambda'.*t);
h=plot(t,xt);
```

7.6 FRFs of Multi-Degree-of-Freedom Systems

The concepts of transfer functions and frequency response functions for single-degree-of-freedom systems can readily be extended to multi-degree-of-freedom systems. To this end, consider an equation of motion in matrix form:

$$M\ddot{\mathbf{x}}+C\dot{\mathbf{x}}+K\mathbf{x}=\mathbf{f}, \tag{7.6-1}$$

where, as before, M, C, and K are the $n\times n$ mass, damping, and stiffness matrices; $\mathbf{x}$ is the $n\times 1$ displacement vector; and $\mathbf{f}$ is the $n\times 1$ excitation force vector.

For multi-degree-of-freedom systems, the central issue is the effect of damping on the frequencies and modal vectors. There are principally three types of damping: (1) structural damping, (2) viscous damping, and (3) coulomb damping. It is often difficult to ascertain which type of damping is present in a particular structure. Most structures exhibit damping characteristics resulting from a combination of damping mechanisms.

To avoid the difficulty of numerous mathematical representations for the various damping phenomena, modal analysis methods usually assume that the damping matrix can be approximated as an equivalent viscous proportional damping matrix. Under this assumption, the equivalent damping matrix is equal to a linear combination of the system mass and stiffness matrices. That is, the damping matrix can be expressed as

$$C=aM+bK, \tag{7.6-2}$$

where a and b are constants. For proportional damping, the coordinate transformation that diagonalizes the system mass and stiffness matrices thus also diagonalizes the system damping matrix.

(1) Laplace Domain: Transfer Function Matrix

Taking the Laplace transform in Eq. 7.6-1, and assuming that all initial conditions are zero, we have

$$(Ms^2+Cs+K)X(s)=F(s) \tag{7.6-3}$$

To simplify Eq. 7.6-3, the concept of a *system impedance matrix*, $Z(s)$, is defined as

$$Z(s)=Ms^2+Cs+K \tag{7.6-4}$$

Eq. 7.6-3 then becomes

$$Z(s)X(s)=F(s) \tag{7.6-5}$$

The receptance transfer function matrix $H(s)$ can now be defined as

$$H(s)=Z^{-1}(s)=(Ms^2+Cs+K)^{-1} \tag{7.6-6}$$

(For the sake of brevity, the subscript d for the receptance transfer function matrix is dropped.) Therefore, the transfer function matrix can be expressed as

$$H(s)=[Z(s)]^{-1}=\frac{adj(Z(s))}{|Z(s)|}=\frac{adj(Z(s))}{D(s-\lambda_1)(s-\lambda_2)\cdots(s-\lambda_{2n})}, \tag{7.6-7}$$

where, as indicated in Section 5.3, the matrix $adj(Z(s))$ is the adjoint of matrix $Z(s)$, λ_r $(r=1,2,\cdots,n)$ are the poles of the system, and D is a constant.

The importance of the transfer function matrix is that it is directly related to the modal parameters: natural frequencies, damping factors, and modal vectors. These relationships provide a theoretical basis for experimental modal analysis. In this method, frequency response functions are measured using excitation at single or multiple points. The frequency response functions are then used as input to a modal parameter estimation process to determine the modal parameters.

There are several ways to find the relationship between the transfer function matrix and the modal parameters. A simple way is as follows: Let S be the modal matrix of the corresponding undamped system. That is, S is a square matrix with each normal mode represented by a column. Since SS^{-1} and $S^{-T}S^T$ are $n\times n$ identity matrices, we may multiply the right side of Eq. 7.6-6 by such identity matrices. That is,

$$H(s)=SS^{-1}\left(Ms^2+Cs+K\right)^{-1}S^{-T}S^T \tag{7.6-8a}$$

$$H(s)=S\left[S^{-1}\left(Ms^2+Cs+K\right)^{-1}S^{-T}\right]S^T \tag{7.6-8b}$$

$$H(s)=S\left[S^T\left(Ms^2+Cs+K\right)S\right]^{-1}S^T \tag{7.6-8c}$$

Since the S is the modal matrix of the system, by virtue of orthogonality, we have

$$S^T MS = diag(m_r) \tag{7.6-9}$$

$$S^T CS = \mathrm{diag}(c_r) \tag{7.6-10}$$

and

$$S^T KS = \mathrm{diag}(k_r) \tag{7.6-11}$$

In these equations, the off-diagonal terms are zero (see Section 5.4). The m_r, c_r, and k_r (r=1,2,···,n) are then the modal mass, the modal damping, and the modal stiffness, respectively.

In view of Eqs. 7.6-9 to 7.6-11, we have

$$S^T(Ms^2+Cs+K)S = \mathrm{diag}(m_r s^2 + c_r s + k_r) \tag{7.6-12}$$

Since this is a diagonal matrix, the inverse becomes simply

$$\left[S^T\left(Ms^2+Cs+K\right)S\right]^{-1} = diag\left(\frac{1}{m_r s^2 + c_r s + k_r}\right) \tag{7.6-13}$$

Therefore, it follows from Eqs. 7.6-8c and 7.6-13 that the transfer function matrix can be written as

$$H(s) = S\,\mathrm{diag}\left(\frac{1}{m_r s^2 + c_r s + k_r}\right)S^T = \sum_{r=1}^{n}\frac{\mathbf{u}_r\mathbf{u}_r^T}{m_r s^2 + c_r s + k_r} \tag{7.6-14}$$

or

$$H(s) = \sum_{r=1}^{n}\frac{\mathbf{u}_r\mathbf{u}_r^T}{m_r(s^2 + 2\varsigma_r\omega_r s + \omega_r^2)}, \tag{7.6-15}$$

where

$$\omega_r = \frac{k_r}{m_r} = \text{undamped natural frequency for mode } r$$

$$\varsigma_r = \frac{c_r}{2\sqrt{k_r m_r}} \quad = \text{damping ratio for mode } r$$

$\mathbf{u}_r$ = modal vector for mode r

For an underdamped system, ς_r is small. Therefore, $s^2 + 2\varsigma_r \omega_r s + \omega_r^2$ has two roots, which are complex conjugates of one another. Letting λ_r and λ_r^* be the two complex roots (the poles of the system), Eq. 7.6-15 can be written as

$$H(s) = \sum_{r=1}^{n} \frac{\mathbf{u}_r \mathbf{u}_r^T}{m_r (s-\lambda_r)(s-\lambda_r^*)} \tag{7.6-16}$$

Next, Eq. 7.6-16 can be written as a partial fraction expansion as [7.2, 7.3]

$$H(s) = \sum_{r=1}^{n} \frac{\mathbf{u}_r \mathbf{u}_r^T}{m_r (s-\lambda_r)(s-\lambda_r^*)} = \sum_{r=1}^{n} \left[\frac{A_{1r}}{s-\lambda_r} + \frac{A_{2r}}{s-\lambda_r^*} \right] \tag{7.6-17}$$

The matrices A_{1r} and A_{2r} can be found as follows: Multiplying both sides of Eq. 7.6-17 by $(s-\lambda_r)$, and letting $s=\lambda_r$, we obtain

$$A_{1r} = \frac{\mathbf{u}_r \mathbf{u}_r^T}{m_r(\lambda_r - \lambda_r^*)} \tag{7.6-18}$$

Since λ_r and λ_r^* are complex conjugates of one another, without any loss of generality, let λ_r and λ_r^* be expressed as

$$\lambda_r = \alpha_r + j\beta_r, \quad \lambda_r^* = \alpha_r - j\beta_r \tag{7.6-19}$$

Eq. 7.6-18 then becomes

$$A_{1r} = \frac{\mathbf{u}_r \mathbf{u}_r^T}{m_r(\lambda_r - \lambda_r^*)} = \frac{1}{j2m_r\beta_r} \mathbf{u}_r \mathbf{u}_r^T \equiv Q_r \mathbf{u}_r \mathbf{u}_r^T, \tag{7.6-20}$$

where

$$Q_r = \frac{1}{j2m_r\beta_r}, \tag{7.6-21}$$

which is called the modal scaling factor. It is evident that the Q_r is purely imaginary.

Similarly, multiplying both sides of Eq. 7.6-17 by $(s-\lambda_r^*)$ and letting $s=\lambda_r^*$, we obtain

$$A_{2r} = \frac{\mathbf{u}_i \mathbf{u}_i^T}{m_r(\lambda_r^* - \lambda_r)} = Q_r^* \mathbf{u}_r \mathbf{u}_r^T, \tag{7.6-22}$$

where Q_r^* is the complex conjugate of Q_r.

Eqs. 7.6-20 and 7.6-22 indicate that the matrices A_{1r} and A_{2r} are complex conjugates of one another. Thus, the transfer function matrix 7.6-17 can be written as

$$H(s)=\sum_{r=1}^{n}\left[\frac{A_r}{s-\lambda_r}+\frac{A_r^*}{s-\lambda_r^*}\right], \tag{7.6-23}$$

where A_r $(r=1,2,\cdots,n)$ is called the *residue matrix* for mode r, and A_r^* is the complex conjugate of A_r.

The importance of the residue matrix is that it is directly related to the modal vectors of the system. Indeed, by virtue of Eqs. 7.6-20 and 7.6-22, the transfer function matrix may be written as

$$H(s)=\sum_{r=1}^{n}\left[\frac{Q_r\mathbf{u}_r\mathbf{u}_r^T}{s-\lambda_r}+\frac{Q_r^*\mathbf{u}_r\mathbf{u}_r^T}{s-\lambda_r^*}\right], \tag{7.6-24}$$

where

$\lambda_r = \alpha_r + j\beta_r$ = system pole

α_r = damping factor

β_r = damped natural frequency

s = Laplace variable

$\mathbf{u}_r$ = modal vector for mode r

Q_r = modal scaling factor

* = complex conjugate

n = number of modal frequencies

Finally, let $h_{pq}(s)$ be the element in the position of the p-th row and the q-th column of the transfer function matrix. It then follows from Eq. 7.6-23 that

$$h_{pq}(s)=\sum_{r=1}^{n}\left[\frac{A_{pqr}}{s-\lambda_r}+\frac{A_{pqr}^*}{s-\lambda_r^*}\right], \tag{7.6-25}$$

where A_{pqr} is called the residue, which is the element in the (p, q)th position of the matrix A_r.

(2) Frequency Domain: Frequency Response Function Matrix

The transfer function matrix in the Laplace domain includes, as a subset, the frequency response function matrix. Indeed, letting $s = j\omega$ in Eqs. 7.6-23 and 7.6-24 leads to the frequency response function matrix [7.2, 7.3]

$$H(\omega)=\sum_{r=1}^{n}\left[\frac{A_r}{j\omega-\lambda_r}+\frac{A_r^*}{j\omega-\lambda_r^*}\right] \tag{7.6-26}$$

or

$$H(\omega)=\sum_{r=1}^{n}\left[\frac{Q_r\mathbf{u}_r\mathbf{u}_r^T}{j\omega-\lambda_r}+\frac{Q_r^*\mathbf{u}_r\mathbf{u}_r^T}{j\omega-\lambda_r^*}\right], \quad (7.6\text{-}27)$$

where

$\lambda_r = \alpha_r + j\beta_r$ = system pole

$j = \sqrt{-1}$

α_r = damping factor

β_r = damped natural frequency

ω = angular frequency variable

A_r = residue matrix for mode r

$\mathbf{u}_r$ = modal vector for mode r

Q_r = modal scaling factor

* = complex conjugate

n = number of modal frequencies

Eq. 7.6-27 indicates that each column of the residue matrix is an estimate to within a constant of the modal vector for that pole. Since the residue matrix is symmetric, each row is then also an estimate of the modal vector of that pole to within a constant. These statements hold for each column and each row of the transfer function matrix.

Next, it follows from Eq. 7.6-26 that the element in the p-th row and q-th column of the frequency response function matrix can be expressed as

$$h_{pq}(\omega)=\sum_{r=1}^{n}\left[\frac{A_{pqr}}{j\omega-\lambda_r}+\frac{A_{pqr}^*}{j\omega-\lambda_r^*}\right], \quad (7.6\text{-}28)$$

where, again, A_{pqr} is called the residue, which is the product of the modal deformations at the input point p and response point q with a modal scaling factor for mode r. When p is equal to q, $h_{pp}(\omega)$ is called the driving-point FRF. That is, the input and output are being measured at the same point on the system. When p is not equal to q, $h_{pq}(\omega)$ is called the cross-point FRF. That is, the input is measured at point p, while the output is measured at point q. By virtue of Maxwell-Betti relations, $h_{pq}(\omega)$ is equal to $h_{qp}(\omega)$.

To illustrate these concepts, consider the simple three-degree-of-freedom system shown in Fig. 7.6-1. For specificity, let the masses m_r, spring constants k_r, and damping coefficients c_r be

m_1=4 kg, m_2=8 kg, m_3=5 kg,

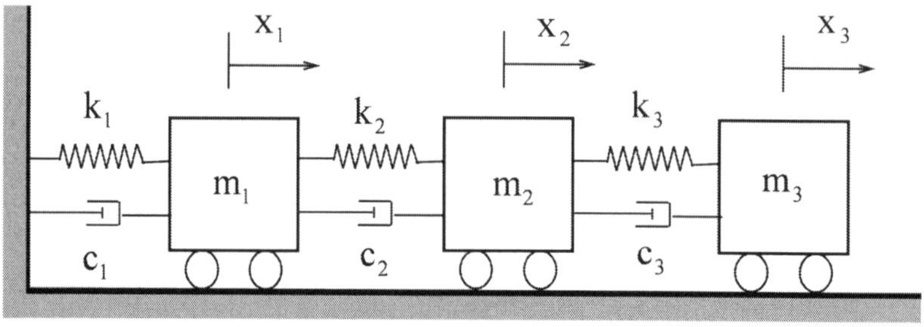

Fig. 7.6-1 A three-degree-of-freedom system.

k_1= 0000 N/m, k_2=30000 N/m, k_3=40000 N/m,

$c_r = 0.0002 k_r \ (r = 1, 2, 3)$.

Fig. 7.6-2 shows the driving-point FRF, $h_{11}(\omega)$, and the phase angle. Fig. 7.6-3 shows the cross-point FRF, $h_{12}(\omega)$, and the corresponding phase angle.

Finally, note that the poles are the roots of $s^2 + 2\varsigma_r \omega_r s + \omega_r^2$ (see Eq. 7.6-15). Thus, the poles are directly related to the natural frequency and damping ratio as

$$\lambda_r \equiv \alpha_r + j\beta_r = -\varsigma_r \omega_r + j\sqrt{1-\varsigma_r^2}\,\omega_r \tag{7.6-29}$$

$$\lambda_r^* \equiv \alpha_r - j\beta_r = -\varsigma_r \omega_r - j\sqrt{1-\varsigma_r^2}\,\omega_r \tag{7.6-30}$$

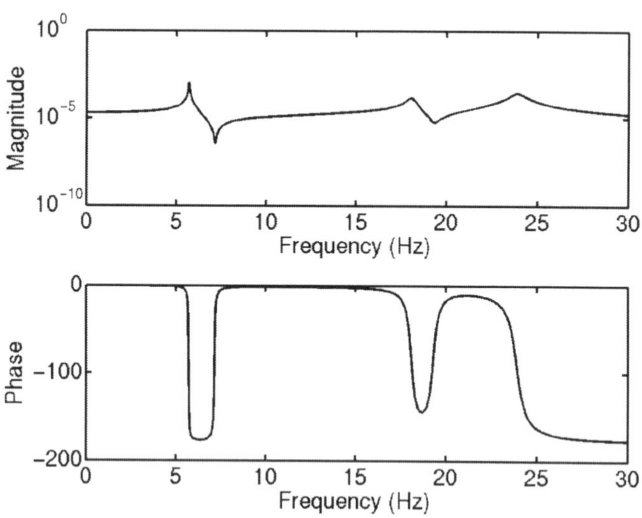

Fig. 7.6-2 Driving-point FRF h_{11} and phase angle of the three-degree-of-freedom system of Fig. 7.6-1.

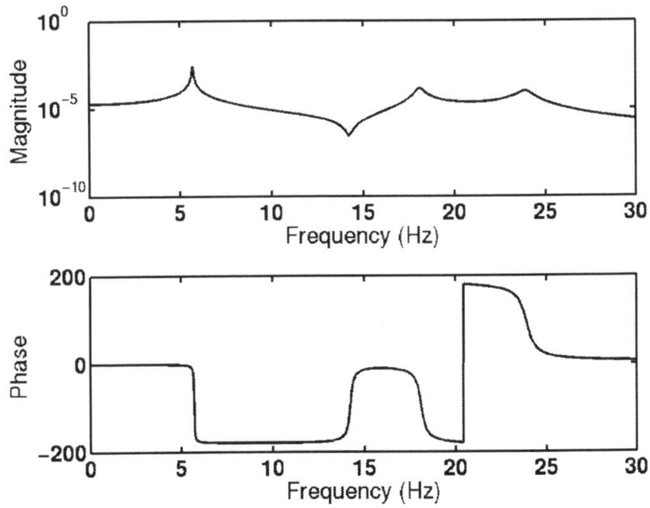

Fig. 7.6-3 Cross-point FRF h_{12} and phase angle of the three-degree-of-freedom system of Fig. 7.6-1.

Once the poles are determined, the undamped natural frequency, ω_r, is given by

$$\omega_r = \sqrt{\alpha_r^2 + \beta_r^2} \tag{7.6-31}$$

Similarly, the damping ratio, ς_r, is given by

$$\varsigma_r = -\frac{\alpha_r}{\omega_r} \tag{7.6-32}$$

(3) Time Domain: Impulse Response Functions

The impulse response function $h_{pq}(t)$ is the inverse Laplace transform of $h_{pq}(s)$, or, alternatively, the inverse Fourier transform of $h_{pq}(\omega)$. Thus, the equation for the impulse response function may be expressed as [7.2]

$$h_{pq}(t) = \sum_{r=1}^{n}\left[A_{pqr}\exp(\lambda_r t) + A_{pqr}^{*}\exp(\lambda_r^{*} t)\right], \tag{7.6-33}$$

where

t = time variable

λ_r = system pole

p = measured degree-of-freedom (input)

q = measured degree-of-freedom (output)

r = modal vector number

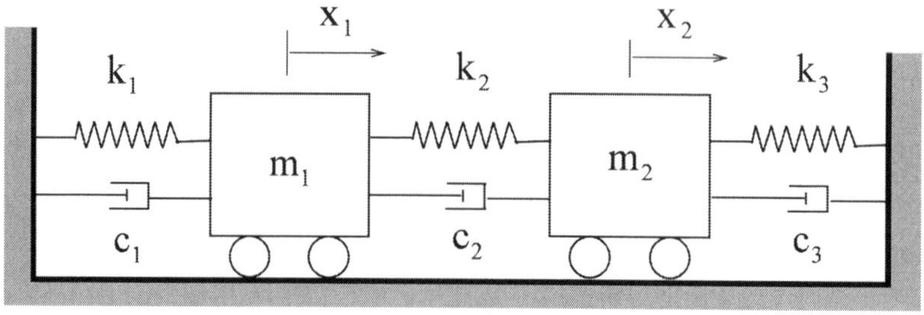

Fig. 7.6-4 Two-degree-of-freedom system for Example 7.6-1.

A_{pqr} = residue

* = complex conjugate

n = number of modal frequencies

Example 7.6-1[1] Consider the two-degree-of-freedom system shown in Fig. 7.6-4. Let the mass, damping, and stiffness values be $m_1 = 4$, $m_2 = 2$, $c_1 = 2$, $c_2 = c_3 = 1$, $k_1 = 4$, and $k_2 = k_3 = 2$. Determine the system poles and the residue matrices, and write the transfer function matrix in the form of Eq. 7.6-23.

Solution: The mass, damping, and stiffness matrices are

$$M = \begin{bmatrix} m_1 & 0 \\ 0 & m_2 \end{bmatrix} = \begin{bmatrix} 4 & 0 \\ 0 & 2 \end{bmatrix}, \; C = \begin{bmatrix} c_1 + c_2 & -c_2 \\ -c_2 & c_2 + c_3 \end{bmatrix} = \begin{bmatrix} 3 & -1 \\ -1 & 2 \end{bmatrix}$$

$$K = \begin{bmatrix} k_1 + k_2 & -k_2 \\ -k_2 & k_2 + k_3 \end{bmatrix} = \begin{bmatrix} 6 & -2 \\ -2 & 4 \end{bmatrix}$$

The system impedance matrix is then given by

$$Z(s) = Ms^2 + Cs + K = \begin{bmatrix} 4s^2 + 3s + 6 & -s - 2 \\ -s - 2 & 2s^2 + 2s + 4 \end{bmatrix} \tag{a}$$

Setting $|Z(s)| = 0$ yields the characteristic equation

$$8s^4 + 14s^2 + 33s^2 + 20s + 20 = 0 \tag{b}$$

or

$$8\left(s^2 + \frac{5}{4}s + \frac{5}{2}\right)\left(s^2 + \frac{1}{2}s + 1\right) = 0 \tag{c}$$

[1]R. J. Allemang provided a similar example in [7.3].

The poles of the system are then found to be

$$\lambda_1 = -\frac{1}{4} + j\frac{\sqrt{15}}{4},\ \lambda_1^* = -\frac{1}{4} - j\frac{\sqrt{15}}{4} \quad \left(j=\sqrt{-1}\right) \tag{d}$$

$$\lambda_2 = -\frac{5}{8} + j\frac{3\sqrt{15}}{8},\ \lambda_2^* = -\frac{5}{8} - j\frac{3\sqrt{15}}{8} \quad \left(j=\sqrt{-1}\right) \tag{e}$$

The transfer function matrix is then given by:

$$H(s) = \frac{adj(Z(s))}{|Z(s)|} \tag{f}$$

That is,

$$H(s) = \frac{\begin{bmatrix} 2s^2+2s+4 & s+2 \\ s+2 & 4s^2+3s+6 \end{bmatrix}}{8(s-\lambda_1)(s-\lambda_1^*)(s-\lambda_2)(s-\lambda_2^*)} \tag{g}$$

Eq. g can be written in the form of a partial fraction expansion as

$$H(s) = \frac{A_1}{s-\lambda_1} + \frac{A_2}{s-\lambda_2} + \frac{A_1^*}{s-\lambda_1^*} + \frac{A_2^*}{s-\lambda_2^*} \tag{h}$$

The residue matrices can be found by multiplying Eq. g by $(s-\lambda_r)$ and letting the Laplace variable s be equal to λ_r in the result. When this is done, the transfer function matrix becomes

$$H(s) = \frac{-j\frac{\sqrt{15}}{45}\begin{bmatrix} 1 & 1 \\ 1 & 1 \end{bmatrix}}{s-\lambda_1} + \frac{-j\frac{\sqrt{15}}{135}\begin{bmatrix} 1 & -2 \\ -2 & 4 \end{bmatrix}}{s-\lambda_2} \tag{i}$$

$$+\frac{j\frac{\sqrt{15}}{45}\begin{bmatrix} 1 & 1 \\ 1 & 1 \end{bmatrix}}{s-\lambda_1^*} + \frac{j\frac{\sqrt{15}}{135}\begin{bmatrix} 1 & -2 \\ -2 & 4 \end{bmatrix}}{s-\lambda_2^*}$$

Each column or row of the residue matrix can be regarded as the modal vector for that pole to within a constant factor. Taking the first column of the residue matrices as modal vectors, we have

$$\mathbf{u}_1 = \begin{bmatrix} 1 \\ 1 \end{bmatrix},\ \mathbf{u}_2 = \begin{bmatrix} 1 \\ -2 \end{bmatrix} \tag{j}$$

The modal scaling factors in this case are

$$Q_1 = -\frac{j\sqrt{15}}{45},\ Q_1^* = \frac{j\sqrt{15}}{45} \tag{k}$$

and

$$Q_2 = -\frac{j\sqrt{15}}{135},\ Q_2^* = \frac{j\sqrt{15}}{135} \tag{l}$$

The transfer function matrix can then be written as

$$H(s) = \sum_{r=1}^{2}\left[\frac{Q_r \mathbf{u}_r \mathbf{u}_r^T}{s-\lambda_r} + \frac{Q_r^* \mathbf{u}_r \mathbf{u}_r^T}{s-\lambda_r^*}\right] \tag{m}$$

Next, from Eq. i, the driving-point FRF $h_{11}(\omega)$ and the cross-point FRF $h_{12}(\omega)$ are given by

$$h_{11}(\omega) = \frac{-j\dfrac{\sqrt{15}}{45}}{j\omega-\lambda_1} + \frac{-j\dfrac{\sqrt{15}}{135}}{j\omega-\lambda_2} + \frac{j\dfrac{\sqrt{15}}{45}}{j\omega-\lambda_1^*} + \frac{j\dfrac{\sqrt{15}}{135}}{j\omega-\lambda_2^*} \tag{n}$$

$$h_{12}(\omega) = \frac{-j\dfrac{\sqrt{15}}{45}}{j\omega-\lambda_1} + \frac{-j\dfrac{\sqrt{15}}{135}(-2)}{j\omega-\lambda_2} + \frac{j\dfrac{\sqrt{15}}{45}}{j\omega-\lambda_1^*} + \frac{j\dfrac{\sqrt{15}}{135}(-2)}{j\omega-\lambda_2^*} \tag{o}$$

The impulse response functions $h_{11}(t)$ and $h_{12}(t)$ are given by

$$\begin{aligned} h_{11}(t) = &-j\frac{\sqrt{15}}{45}\left[\exp(\lambda_1 t) - \exp(\lambda_1^* t)\right] \\ &-j\frac{\sqrt{15}}{135}\left[\exp(\lambda_2 t) - \exp(\lambda_2^* t)\right] \end{aligned} \tag{p}$$

and

$$\begin{aligned} h_{12}(t) = &-j\frac{\sqrt{15}}{45}\left[\exp(\lambda_1 t) - \exp(\lambda_1^* t)\right] \\ &+j\frac{2\sqrt{15}}{135}\left[\exp(\lambda_2 t) - \exp(\lambda_2^* t)\right] \end{aligned} \tag{q}$$

Substituting the values of the poles from Eqs. d and e into Eqs. p and q leads to

$$h_{11}(t) = \frac{2\sqrt{15}}{45} e^{-\frac{1}{4}t} \sin\left(\frac{\sqrt{15}}{4}t\right) + \frac{2\sqrt{15}}{135} e^{-\frac{5}{8}t} \sin\left(\frac{3\sqrt{15}}{8}t\right) \tag{r}$$

and

$$h_{12}(t)=\frac{2\sqrt{15}}{45}e^{-\frac{1}{4}t}\sin\left(\frac{\sqrt{15}}{4}t\right)-\frac{4\sqrt{15}}{135}e^{-\frac{5}{8}t}\sin\left(\frac{3\sqrt{15}}{8}t\right) \tag{s}$$

Finally, for the system with nonproportional damping, the modal vectors are complex and Eq. 7.6.24 has the form

$$H(\omega)=\sum_{r=1}^{N}\left[\frac{Q_r\mathbf{u}_r\mathbf{u}_r^T}{j\omega-\lambda_r}+\frac{Q_r^*\mathbf{u}_r^*\mathbf{u}_r^{*T}}{j\omega-\lambda_r^*}\right], \tag{7.6-34}$$

where $\mathbf{u}_r^*$ is the complex conjugate of $\mathbf{u}_r$. This result can be validated by following the procedure in Problem 7.8. The following is an illustrative example.

Example 7.6-2 Consider the system as shown in Fig. 7.6-4, where the mass, damping and stiffness are as follows: $m_1=4$, $m_2=2$, $c_1=c_2=c_3=1$, $k_1=4$, $k_2=k_3=2$. Determine the system poles and residue matrices, and write the transfer function matrix in the form of Eq. 7.6-34.

Solution: The mass, damping, and stiffness matrices are

$$M=\begin{bmatrix} m_1 & 0 \\ 0 & m_2 \end{bmatrix}=\begin{bmatrix} 4 & 0 \\ 0 & 2 \end{bmatrix} \tag{a}$$

$$C=\begin{bmatrix} c_1+c_2 & -c_2 \\ -c_2 & c_2+c_3 \end{bmatrix}=\begin{bmatrix} 2 & -1 \\ -1 & 2 \end{bmatrix} \tag{b}$$

$$K=\begin{bmatrix} k_1+k_2 & -k_2 \\ -k_2 & k_2+k_3 \end{bmatrix}=\begin{bmatrix} 6 & -2 \\ -2 & 4 \end{bmatrix} \tag{c}$$

Observe that the damping matrix is not proportional to the mass and stiffness matrices. The system impedance matrix is given by

$$Z(s)=Ms^2+Cs+K=\begin{bmatrix} 4s^2+2s+6 & -s-2 \\ -s-2 & 2s^2+2s+4 \end{bmatrix} \tag{d}$$

The characteristic equation is

$$|Z(s)|=8\left(s^4+\frac{3}{2}s^3+\frac{31}{8}s^2+2s+\frac{5}{2}\right)=0 \tag{e}$$

With Matlab, the roots are found to be

$$\lambda_1=-0.1657+j0.9904,\quad \lambda_1^*=-0.1657-j0.9904$$

$$\lambda_2=-0.5843+j1.4622,\quad \lambda_2^*=-0.5843-j1.4622,$$

where, as before, j is $\sqrt{-1}$. Thus, the transfer matrix can be written as

$$H(s)=\frac{adjZ(s)}{|Z(s)|}=\frac{\begin{bmatrix} 2s^2+2s+4 & s+2 \\ s+2 & 4s^2+2s+6 \end{bmatrix}}{8(s-\lambda_1)(s-\lambda_2)(s-\lambda_1^*)(s-\lambda_2^*)} \tag{f}$$

Also, this can be written as a partial fraction expansion

$$H(s)=\frac{A_1}{s-\lambda_1}+\frac{A_2}{s-\lambda_2}+\frac{A_1^*}{s-\lambda_1^*}+\frac{A_2^*}{s-\lambda_2^*} \tag{g}$$

The residue matrices can be found by multiplying Eq. g by $(s-\lambda_r)$ and letting the Laplace variable s be equal to λ_r in the result. To express this more clearly, let

$$A_1=\begin{bmatrix} A_{111} & A_{121} \\ A_{211} & A_{221} \end{bmatrix}, \quad A_2=\begin{bmatrix} A_{112} & A_{122} \\ A_{212} & A_{222} \end{bmatrix}$$

$$A_1^*=\begin{bmatrix} A_{111}^* & A_{121}^* \\ A_{211}^* & A_{221}^* \end{bmatrix}, \quad A_2^*=\begin{bmatrix} A_{112}^* & A_{122}^* \\ A_{212}^* & A_{222}^* \end{bmatrix},$$

where the subscripts p, q of A_{pqr} indicates that the element is located at the p-th row and q-th column of matrix A, and where the subscript r indicates the mode number. To find the residue matrix A_1, multiplying Eq. g by $(s-\lambda_1)$ and letting the Laplace variable be equal to λ_1 in the result, we have

$$A_1=H(s)(s-\lambda_1)\Big|_{s=\lambda_1}, \tag{h}$$

from which we obtain

$$A_{111}=\frac{2\lambda_1^2+2\lambda_1+4}{8(\lambda_1-\lambda_2)(\lambda_1-\lambda_1^*)(\lambda_1-\lambda_2^*)}=0.0078-j0.0883$$

$$A_{211}=\frac{\lambda_1+2}{8(\lambda_1-\lambda_2)(\lambda_1-\lambda_1^*)(\lambda_1-\lambda_2^*)}=0.0052-j0.0873$$

In the same manner, we have

$$A_{112}=\frac{2\lambda_2^2+2\lambda_2+4}{8(\lambda_2-\lambda_1)(\lambda_2-\lambda_1^*)(\lambda_2-\lambda_2^*)}=0.0078-j0.0233$$

$$A_{212}=\frac{\lambda_2+2}{8(\lambda_2-\lambda_1)(\lambda_2-\lambda_1^*)(\lambda_2-\lambda_2^*)}=0.0052-j0.0552$$

$$A_{111}^*=0.0078+j0.0883$$

$$A_{211}^*=0.0052+j0.0873$$

$$A^*_{112} = 0.0078 + j0.0233$$

$$A^*_{212} = 0.0052 + j0.0552$$

It has been shown that any column of the adjoint matrix $adjZ(s)$ is proportional to the modal vector (see Section 5.3). Since the matrix $Z(s)$ is symmetric, so also is the adjoint matrix $adjZ(s)$ symmetric. Thus, the residue matrix can be written in the form

$$A_r = Q_r \mathbf{u}_r \mathbf{u}_r^T, \tag{i}$$

where Q_r is the modal scaling factor for mode r, and where $\mathbf{u}_r$ is the modal vector for mode r.

For the two-degree-of-freedom system, let the elements of $\mathbf{u}_r$ be ψ_1 and ψ_2. That is,

$$\mathbf{u}_r = \begin{bmatrix} \psi_1 \\ \psi_2 \end{bmatrix}_r \tag{j}$$

Then Eq. i can be written as

$$A_r = Q_r \begin{bmatrix} \psi_1\psi_1 & \psi_1\psi_2 \\ \psi_2\psi_1 & \psi_2\psi_2 \end{bmatrix} \tag{k}$$

Choosing $Q_1 = 1$ (it may be chosen arbitrarily), we have

$$\psi_1\psi_1 = A_{111} = 0.0078 - j0.0883,$$

from which we obtain

$$\psi_1 = 0.2196 - j0.2011$$

Next, since

$$\psi_2\psi_1 = A_{211} = 0.0052 - j0.0837,$$

we have

$$\psi_2 = A_{211} / \psi_1 = 0.1770 - j0.2190$$

Therefore, the first modal vector can be written as

$$\mathbf{u}_1 = \begin{bmatrix} \psi_1 \\ \psi_2 \end{bmatrix} = \begin{bmatrix} 0.2196 - j0.2011 \\ 0.1770 - j0.2190 \end{bmatrix}, \ Q_1 = 1$$

In the same manner, we obtain

$$\mathbf{u}_2 = \begin{bmatrix} 0.0920 - j0.1275 \\ 0.2655 + j0.2321 \end{bmatrix}, \; Q_2 = 1$$

$$\mathbf{u}_1^* = \begin{bmatrix} 0.2196 + j0.2011 \\ 0.1770 + j0.2190 \end{bmatrix}, \; Q_1^* = 1$$

$$\mathbf{u}_2^* = \begin{bmatrix} 0.0920 + j0.1275 \\ 0.2655 - j0.2321 \end{bmatrix}, \; Q_2^* = 1$$

With the above results, the transfer function matrix can be written as

$$H(s) = \sum_{r=1}^{N} \left[\frac{Q_r \mathbf{u}_r \mathbf{u}_r^T}{s - \lambda_r} + \frac{Q_r^* \mathbf{u}_r^* \mathbf{u}_r^{*T}}{s - \lambda_r^*} \right] \tag{1}$$

When $s = j\omega$, Eq. 1 becomes the same as Eq. 7.6-34.

The above results indicate that for a system with nonproportional damping, the modal vectors are complex (complex modes). In this case, no matter what modal scaling factors are chosen, these complex modes cannot be converted into real modes. In contrast, for proportional damping systems, we can chose pure imaginary scaling factors to obtain real modes, as illustrated in Example 7.6-1.

7.7 Determination of Residues and Poles in State Space

We saw in the previous section that once the residue matrix and the poles of a system are known, the impulse response functions of the system can be determined. For a multi-degree-of-freedom system, it is generally difficult to manually calculate the residue matrix and poles. Therefore, it is desirable to develop a general method, which is suitable for numerical and computer driven solutions.

To this end, the equation of motion, Eq. 7.6-1, is transformed into state space form. As in Section 6.2, let us introduce the following identity:

$$M\dot{\mathbf{x}} - M\dot{\mathbf{x}} \equiv 0 \tag{7.7-1}$$

Combining Eq. 7.6-1 and 7.7-1 yields the following equation:

$$\begin{bmatrix} C & M \\ M & 0 \end{bmatrix} \begin{bmatrix} \dot{\mathbf{x}} \\ \ddot{\mathbf{x}} \end{bmatrix} + \begin{bmatrix} K & 0 \\ 0 & -M \end{bmatrix} \begin{bmatrix} \mathbf{x} \\ \dot{\mathbf{x}} \end{bmatrix} = \begin{bmatrix} \mathbf{f} \\ 0 \end{bmatrix} \tag{7.7-2}$$

Next, introduce the state vector $\mathbf{y}$ defined by the $2n \times 1$ vector:

$$\mathbf{y} = \begin{bmatrix} \mathbf{y}_1 \\ \mathbf{y}_2 \end{bmatrix} = \begin{bmatrix} \mathbf{x} \\ \dot{\mathbf{x}} \end{bmatrix} \tag{7.7-3}$$

Then Eq. 7.7-2 can be recognized as a first-order vector differential equation in the form

$$A\dot{\mathbf{y}} + B\mathbf{y} = \bar{\mathbf{f}}, \tag{7.7-4}$$

where

$$A = \begin{bmatrix} C & M \\ M & 0 \end{bmatrix}, \quad B = \begin{bmatrix} K & 0 \\ 0 & -M \end{bmatrix}, \quad \text{and} \quad \bar{\mathbf{f}} = \begin{bmatrix} \mathbf{f} \\ 0 \end{bmatrix} \tag{7.7-5}$$

Taking the Laplace transform in Eq. 7.7-4, and assuming that all initial conditions are zero, we have

$$(As + B)Y(s) = F(s) \tag{7.7-6}$$

The receptance transfer function matrix $\bar{H}(s)$ in state space can be defined as

$$Y(s) = \bar{H}(s)F(s) \tag{7.7-7}$$

Therefore, the transfer function matrix can be expressed as

$$\bar{H}(s) = [As + B]^{-1} \tag{7.7-8}$$

To find the relationship between the transfer function matrix and modal parameters, it is useful to review the following eigenvalue problem (see Section 6.2):

$$B\mathbf{z} = -\lambda A\mathbf{z} \tag{7.7-9}$$

This is the generalized eigenvalue problem in state space. There are $2n$ eigenvalues and $2n$ eigenvectors. The $2n$ eigenvalues λ_i appear in complex conjugate pairs. The corresponding $2n$ eigenvectors also appear in complex conjugate pairs. When the $2n$ pairs of complex conjugate eigenvectors are assembled into a $2n \times 2n$ square matrix with each eigenvector represented by a column, we have the complex modal matrix P given by

$$P = \begin{bmatrix} \mathbf{z}_1 & \cdots & \mathbf{z}_n & \mathbf{z}_1^* & \cdots & \mathbf{z}_n^* \end{bmatrix} = \begin{bmatrix} S & S^* \\ S\Lambda & S^*\Lambda^* \end{bmatrix}, \tag{7.7-10}$$

where

$$\Lambda = diag(\lambda_1, \cdots, \lambda_n) \tag{7.7-11}$$

$$\Lambda^* = diag(\lambda_1^*, \cdots, \lambda_n^*) \tag{7.7-12}$$

$$S = \begin{bmatrix} \mathbf{u}_1 & \cdots & \mathbf{u}_n \end{bmatrix} \tag{7.7-13}$$

$$S^* = \begin{bmatrix} \mathbf{u}_1^* & \cdots & \mathbf{u}_n^* \end{bmatrix} \tag{7.7-14}$$

Since PP^{-1} is an identity matrix, we have

$$\bar{H}(s) = PP^{-1}\left(As+B\right)^{-1} P^{-T} P^{T} \tag{7.7-15}$$

$$\bar{H}(s) = P\left[P^{-1}\left(As+B\right)^{-1} P^{-T}\right]P^{T} \tag{7.7-16}$$

$$\bar{H}(s) = P\left[P^{T}\left(As+B\right)\ P\right]^{-1} P^{T} \tag{7.7-17}$$

By virtue of orthogonality, P^TAP and P^TBP are diagonal. That is,

$$P^{T}AP = \text{diag}(a_1,\cdots,a_n,a_1^*,\cdots,a_n^*) \tag{7.7-18}$$

$$P^{T}BP = \text{diag}(b_1,\cdots,b_n,b_1^*,\cdots,b_n^*) \tag{7.7-19}$$

Thus, we obtain

$$P^{T}\left(As+B\right)P = \text{diag}(a_r s+b_r) \quad (r=1,2,\cdots,2n) \tag{7.7-20}$$

Since this is a diagonal matrix, the inverse is simply

$$\left[P^{T}\left(As+B\right)P\right]^{-1} = \text{diag}\left(\frac{1}{a_r(s+b_r/a_r)}\right) \tag{7.7-21}$$

Since $b_r/a_r = -\lambda_r$ (see Eq. 6.2-40), Eq. 7.7-21 can be written as

$$\left[P^{T}\left(As+B\right)P\right]^{-1} = \text{diag}\left(\frac{1}{a_r(s-\lambda_r)}\right) \tag{7.7-22}$$

It follows from Eq. 7.7-17 that the transfer function matrix in state space can has the form

$$\bar{H}(s) = P\,\text{diag}\left(\frac{1}{a_r(s-\lambda_r)}\right)P^{T} \tag{7.7-23}$$

Therefore, the transfer function between displacement **x** and excitation force **f** is given by

$$H(s) = \sum_{r=1}^{2n}\frac{\mathbf{u}_r\mathbf{u}_r^T}{a_r(s-\lambda_r)} = \sum_{r}^{n}\left[\frac{A_r}{s-\lambda_r}+\frac{A_r^*}{s-\lambda_r^*}\right] \tag{7.7-24}$$

The residue matrix is given by

$$A_r = \frac{\mathbf{u}_r\mathbf{u}_r^T}{a_r} \quad (r=1,2,\cdots,n) \tag{7.7-25a}$$

$$A_r^* = \frac{\mathbf{u}_r^* \mathbf{u}_r^{*T}}{a_r^*} \quad (r = 1, 2, \cdots, n), \tag{7.7-25b}$$

where $\mathbf{u}_r$ is the modal vector corresponding to the coordinate $\mathbf{x}$ (the vector composed of the first n elements of the r-th column in modal matrix P), and a_r is the r-th generalized mass defined by $P^T AP$ (see Eq. 7.7-18).

Once the residue matrix and poles are obtained, the impulse response functions can be determined from Eq. 7.6-33. The following example illustrates how to use Matlab to calculate the residue matrix, poles, and impulse response functions.

Example 7.7-1 Rework Example 7.6-1 numerically and compare the results with the closed form solutions.

Solution: The equation of motion in state space is given by

$$A\dot{\mathbf{y}} + B\mathbf{y} = \overline{\mathbf{f}},$$

where A and B are

$$A = \begin{bmatrix} C & M \\ M & 0 \end{bmatrix}, \ B = \begin{bmatrix} K & 0 \\ 0 & -M \end{bmatrix}$$

By using Matlab, the modal matrix and poles can be found by a simple command:

```
[P,D]=eig(-B,A);
```

Next, the generalized mass a_r is defined by $[\text{diag}(a_r)] = P^T AP$. The command in Matlab is

```
ma=P.'*A*P;
```

The residue matrix is then given by

$$A_r = \frac{\mathbf{u}_r \mathbf{u}_r^T}{a_r}$$

Finally, the impulse response functions are obtained by using Eq. 7.6-33. The driving-point impulse function $h_{11}(t)$ and the cross-point impulse response function $h_{12}(t)$ are graphed in Fig. 7.7-1 and 7.7-2, respectively. The solid lines are numerical solutions; the closed-form solutions given in Example 7.6-1 are

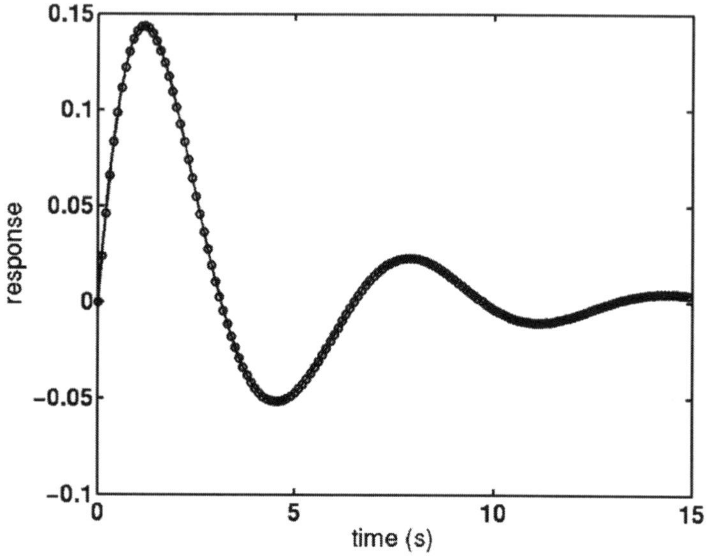

Fig. 7.7-1 The impulse response function $h_{11}(t)$. Solid line: numerical solution; circles: the closed-form solution from Example 7.6-1.

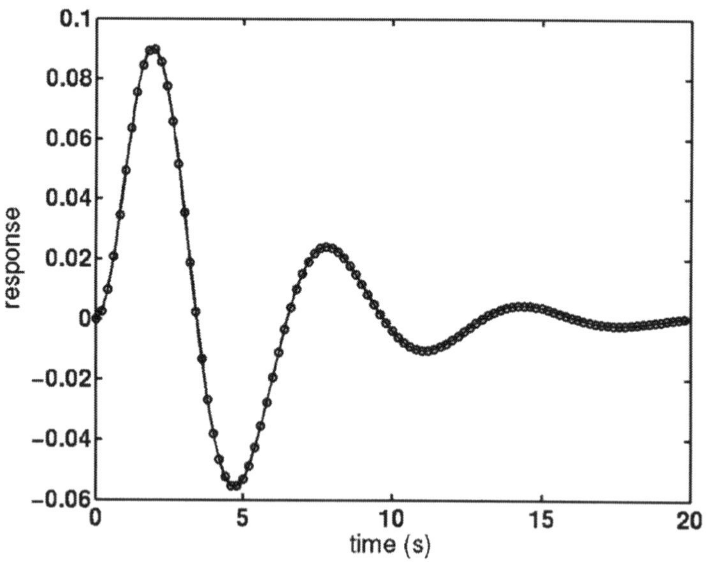

Fig. 7.7-2 The impulse response function $h_{12}(t)$. Solid line: numerical solution; circles: the closed-form solution from Example 7.6-1.

represented with small circles. The solutions from the two methods are, as expected, identical. The Matlab code for this calculation is

```
clear all
m1=4; m2=2;
k1=4; k2=2; k3=k2;
c1=2;c2=1;c3=1;
mass=[m1,0;0,m2];
stiff=[k1+k2,-k2;-k2,k2+k3];
damp=[c1+c2,-c2;-c2,c2+c3];
% form state space formula
a=[damp,mass;mass,zeros(2)];
b=[stiff,zeros(2);zeros(2),-mass];
% find poles and modal vector
[y,d]=eig(b,-a);
y,d
lambda=diag(d); % ........ complex frequency
% modal mass "ma"
ma=y.'*a*y; %.............y.'= transpose;
% residue matrices
A1=y(1:2,1)*y(1:2,1).'./ma(1,1);
A2=y(1:2,2)*y(1:2,2).'./ma(2,2);
A3=y(1:2,3)*y(1:2,3).'./ma(3,3);
A4=y(1:2,4)*y(1:2,4).'./ma(4,4);
% Impulse response functions
t=linspace(0,20,501);
xt11=A1(1,1)*exp(lambda(1).*t)+A2(1,1)*exp(lambda(2).*
  t)...,
+A3(1,1)*exp(lambda(3).*t)+A4(1,1)*exp(lambda(4).*t);
xt12=A1(1,2)*exp(lambda(1).*t)+A2(1,2)*exp(lambda(2).*
  t)...,
+A3(1,2)*exp(lambda(3).*t)+A4(1,2)*exp(lambda(4).*t);
h=plot(t,xt11,'k');
%h=plot(t,xt12,'k');
set(h,'LineWidth',2);
hx=get(gca, 'Xlabel');
hy=get(gca, 'Ylabel');
set(hx, 'String', 'time (s)');
set(hy, 'String', 'response');
hold;
for ii=1:100
t1(ii)=0.2*(ii-1);
xtta(ii)=2*sqrt(15)/45*exp(-1/4*t1(ii))*sin(sqrt(15)/4*t1
  (ii))...,
```

```
+2*sqrt(15)/135*exp(-5/8*t1(ii))*sin(3/8*sqrt(15)*t1(ii));
xttb(ii)=2*sqrt(15)/45*exp(-1/4*t1(ii))*sin(sqrt(15)/4*t1
   (ii))...,
-4*sqrt(15)/135*exp(-5/8*t1(ii))*sin(3/8*sqrt(15)*t1(ii));
end
plot(t,xt11,'k', t1,xtta,'ok');
%plot(t,xt12,'k', t1,xttb,'ok');
```

Problems

7.1 The transfer function of a single-degree-of-freedom system is given by

$$H(s)=\frac{1}{m(s^2+2\varsigma\omega_n s+\omega_n^2)}$$

Write it in the partial fraction form.

7.2 Show that

$$h(t)=\frac{e^{-\varsigma\omega_n t}}{m\omega_d}\sin\omega_d t \quad \text{and} \quad H(s)=\frac{1}{m(s^2+2\varsigma\omega_n s+\omega_n^2)}$$

form a Laplace transform pair.

7.3 Given m=20 kg, c=30 N·s/m, k=17000 N/m for a single-degree-of-freedom system, find the system poles and residues.

Ans. $\lambda_1=-0.75+j29.1451;\ \lambda_1^*=-0.75-j29.1451$

$$A_1=-j0.8578\times10^{-3};\ A_2=A^*$$

7.4 Using the values of the poles and residues obtained in Problem 7.3, determine the impulse response function of the system and compare the result with the formula

$$h(t)=\frac{e^{-\varsigma\omega_n t}}{m\omega_d}\sin(\omega_d t)$$

Ans. $x(t)=1.7156\times10^{-3}e^{-0.75t}\sin(29.1451t)$

7.5 Verify Eqs. 7.4-18 to 7.4-20 and graph the Nyquist diagrams.

7.6 The equation of motion of a system with structural damping is given by

$$m\ddot{x}+(k+jk\eta)x=f$$

Show that the real part $H^R(\omega)$ and the imaginary part $H^I(\omega)$ of the frequency response function satisfies the equation of the circle

$$\left[H^R(\omega)\right]^2+\left[H^I(\omega)+\frac{1}{2k\eta}\right]^2=\left(\frac{1}{2k\eta}\right)^2$$

7.7 Show that for the Nyquist circle of Problem 7.6, the real part is zero at D_1, corresponding to the natural frequency of the system. Show that point D_2 is the corresponding imaginary component at resonance and that points D_1 and D_2 are located at the same point on the circle.

7.8 Let $adj(Z(s))$ be the adjoint matrix of the system impedance matrix $Z(s)=Ms^2+Cs+K$. Show that any column or any row of $adj(Z(s))$ can be regarded as the modal vector of the system. Thus, the transfer function matrix of the system can be written as

$$H(s)=[Z(s)]^{-1}=\frac{adj(Z(s))}{|Z(s)|}=\frac{SS^T}{D(s-\lambda_1)(s-\lambda_2)\cdots(s-\lambda_{2n})},\ D=\text{constant}$$

7.9 Write the transfer function matrix found in Problem 7.8 in partial fraction expansion form.

7.10 [7.7] Write the orthogonal relationships as

$$M=S^{-T}\text{diag}(m_r)S^{-1},\ C=S^{-T}\text{diag}(c_r)S^{-1},\quad K=S^{-T}\text{diag}(k_r)S^{-1},$$

where S is the modal matrix of the corresponding undamped system. Substituting these into $H(s)=\left(Ms+Cs+K\right)^{-1}$, show that the transfer function matrix can be written as

$$H(s)=S\,diag\left(\frac{1}{m_r s^2+c_r s+k_r}\right)S^T=\sum_{r=1}^{n}\frac{\mathbf{u}_r\mathbf{u}_r^T}{m_r s^2+c_r s+k_r},$$

where $\mathbf{u}_r$ is the modal vector for mode r.

7.11 Find the poles and residue matrix of the system shown in Fig. P7.11, where m_1=5 kg, m_2=10 kg, k_1=k_2=2 N/m, k_3=4 N/m, $c_1=c_2=c_3=0$.

Ans. $\lambda_1=j\sqrt{2/5},\ \lambda_1^*=-j\sqrt{2/5},\ \lambda_2=j,\ \lambda_2^*=-j$

$$A_1=-j\frac{\sqrt{2/5}}{12}\begin{bmatrix}1&1\\1&1\end{bmatrix},\ A_1^*=j\frac{\sqrt{2/5}}{12}\begin{bmatrix}1&1\\1&1\end{bmatrix}$$

$$A_2=-j\frac{1}{60}\begin{bmatrix}4&-2\\-2&1\end{bmatrix},\ A_2^*=j\frac{1}{60}\begin{bmatrix}4&-2\\-2&1\end{bmatrix}$$

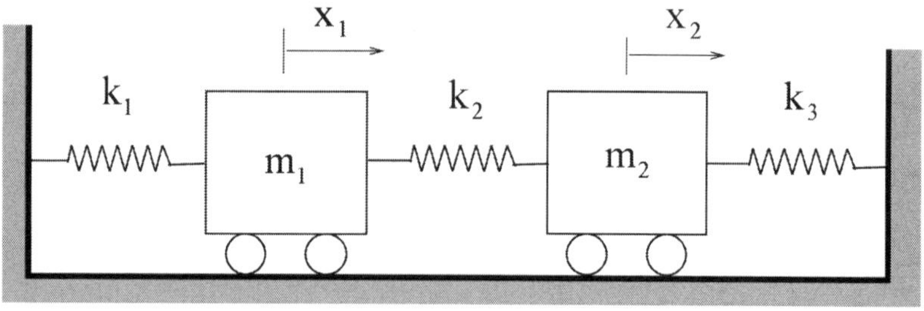

Fig. P7.11

7.12 Write a computer program to calculate the poles and residues of the system shown in Fig. P7.12, where m_1=4 kg, m_2=8 kg, m_3=5 kg, k_1=50000 N/m, k_2=30000 N/m, k_3=40000 N/m, $c_r=0.0002k_r$ $(r=1,2,3)$.

Ans. $\lambda_1=-0.13+j35.88$, $\lambda_2=\lambda_1^*$; $\lambda_3=-1.29+j113.64$, $\lambda_4=\lambda_3^*$,

$\lambda_5=-2.25+j150.14$, $\lambda_6=\lambda_5^*$

7.13 Show that the contribution of each mode to receptance of a structure at zero frequency has the form

$$H_{pq}^{(r)}(0)=\frac{1}{\omega_r^2 m_r}\left[\mathrm{Re}\left(\phi_p\phi_q\right)+\frac{\varsigma_r}{\sqrt{1-\varsigma_r^2}}\mathrm{Im}\left(\phi_p\phi_q\right)\right],$$

where ω_r is the natural frequency for mode r, m_r is the modal mass for mode r, ϕ_p is the value of mode shape function at point p, ϕ_q is the value of mode function at point q, and ς_r is the damping ratio of the system for mode r. Re(..) and Im(..) represent the real and imaginary parts, respectively. (Note: This equation can be used to estimate the contribution of each mode to static stiffness of the system.)

7.14 Validate the equation given in Problem 7.13 using the system in Example 7.6-1.

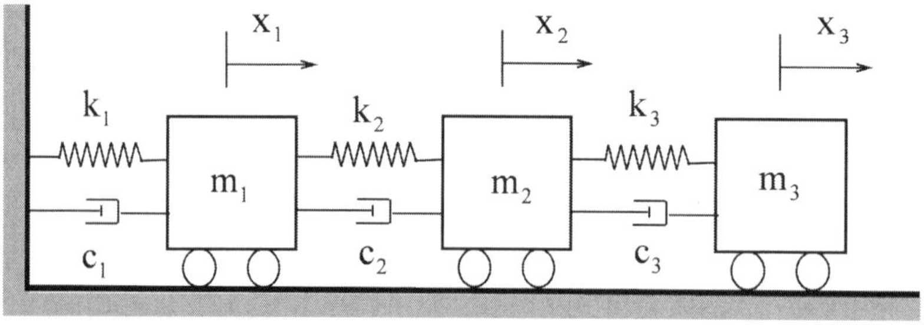

Fig. P7.12

References

7.1. Thomson, W. T., *Laplace Transformation*, 2nd ed., Prentice Hall, Englewood Cliffs, NJ, 1960.

7.2. Allemang, R. J., and Brown, D. L., "Experimental Modal Analysis," *Shock and Vibration Handbook*, C. M. Harris, ed., 3rd ed., McGraw-Hill, New York, 1987, pp. 21-1 to 21-34.

7.3. Allemang, R. J., *Vibrations: Analytical and Experimental Modal Analysis*, UC-SDRLL-CN-20-263-691, Department of Mechanical, Industrial, and Nuclear Engineering, University of Cincinnati, Cincinnati, OH, 1992.

7.4. Sneddon, I. A., *Fourier Transforms*, McGraw-Hill, New York, 1951.

7.5. Oran Brigham, E., *The Fast Fourier Transform and Its Applications*, Prentice Hall, Inc., Upper Saddle River, NJ, 1998.

7.6. McConnell, K. G., *Vibration Testing, Theory and Practice*, John Wiley & Sons, Inc., New York, 1995.

7.7. Inman, D. J., *Engineering Vibration*, Prentice Hall, Inc., Upper Saddle River, NJ, 1994.

Chapter Eight

Lumped Parameter Systems

8.1 Introduction

In a review of the foregoing chapters, we see that there are a number of relatively simple methods available for finding the natural frequencies and mode shapes of small, undamped systems. Nevertheless, when the number of degrees of freedom becomes large, the task of determining the frequencies and mode shapes becomes increasingly tedious and prone to error. Even with the use of high-speed computers and efficient, reliable software (e.g., Matlab), there are numerical approximation and round-off errors. But perhaps the greatest difficulty with numerical analysis of large degree-of-freedom systems is that intuitive understanding of the system characteristics is either lost or difficult to discern.

An alternative approach is to simplify the modeling of large systems and/or to simplify the analysis and the solution of the governing equations. In this chapter, we present some of these simplifying procedures. We also present and consider a few hand computation methods and some matrix methods.

We begin with a review of some fundamental concepts of structure analysis.

8.2 The Flexibility and Stiffness Matrices

Consider an elastic structure S. Let S be subjected to a series of forces, but let S be anchored so that it does not freely translate nor rotate due to the forces. Then, since S is elastic, the applied forces will simply deform S, leaving its overall position unchanged.

Next, let the magnitude of the applied forces be sufficiently small so that even though S is deformed, its deformed shape is not substantially different from its undeformed shape. That is, let the deformation of S be *small*. Specifically, if P is an arbitrary material point of S (i.e., P is a point of the space occupied by a physical particle of S), if u is the displacement of P due to the applied forces, and if L is a characteristic length of S, then u is small compared to L (i.e., $u << L$).

Under these conditions, we assume that S responds *linearly* to the applied forces. This means that if the magnitude of a force is increased by a factor κ, the magnitude of the resulting displacement is also increased by κ. Also, with two or more forces being applied, the order of application of the forces does not affect the final deformation state of S. In short, we assume the principle of superposition holds.

Consider again an arbitrary material point P of S. Let there be a force $\mathbf{f}$ applied at an arbitrary point $\hat{P}$ of S. Let $\mathbf{x}$ be the displacement at P due to the force at $\hat{P}$. Let $\mathbf{x}$ and $\mathbf{f}$ have components (x_1, x_2, x_3) and (f_1, f_2, f_3) relative to a Cartesian axis system. Then the x_i and the f_i $(i = 1,2,3)$ are related to each other as

$$
\begin{aligned}
x_1 &= \delta_{11} f_1 + \delta_{12} f_2 + \delta_{13} f_3 \\
x_2 &= \delta_{21} f_1 + \delta_{22} f_2 + \delta_{23} f_3 \\
x_3 &= \delta_{31} f_1 + \delta_{32} f_2 + \delta_{33} f_3
\end{aligned}
\tag{8.2-1}
$$

or simply as

$$\mathbf{x} = [\delta]\mathbf{f}, \tag{8.2-2}$$

where, by inspection of Eqs. 8.2-1 and 8.2-2, $[\delta]$ is the matrix

$$[\delta] = \begin{bmatrix} \delta_{11} & \delta_{12} & \delta_{13} \\ \delta_{21} & \delta_{22} & \delta_{23} \\ \delta_{31} & \delta_{32} & \delta_{33} \end{bmatrix} \tag{8.2-3}$$

$[\delta]$ is known as the *flexibility matrix*; δ_{ij} are called the *influence coefficients.*

With S being anchored, $[\delta]$ is non-singular, and its inverse $[\delta]^{-1}$ exists. Then, by pre-multiplying in Eq. 8.2-2 by $[\delta]^{-1}$, we have

$$[\delta]^{-1}\mathbf{x} = \mathbf{f} = [k]\mathbf{x}, \tag{8.2-4}$$

where $[k]$ is known as the *stiffness matrix.*

Eq. 8.2-4 shows that the flexibility and stiffness matrices are inverses of one another. That is,

$$[\delta]^{-1} = [k] \qquad \text{or} \qquad [\delta] = [k]^{-1} \tag{8.2-5}$$

Next, suppose that S is loaded at a series of points $\hat{P}_i$ $(i = 1, 2, \cdots, N)$ and that the displacement is measured at point P_j $(j = 1, 2, \cdots, M)$. Then the length of arrays $\mathbf{f}$ and $\mathbf{x}$ will expand to $3N$ and $3M$, respectively. In this case, $\mathbf{x}$ and $\mathbf{f}$ will again be related, as in Eq. 8.2-2, but now $[\delta]$ is a $3M \times 3N$ array.

For simplicity, suppose $M = N$ and that the points $\hat{P}_i$ and P_i are the same. That is, suppose S is loaded at a series of points and that the displacement of S is measured at the same points. Thus, with S being anchored, $[\delta]$ is again a non-singular $3M \times 3M$ square array and its inverse is a $3M \times 3M$ stiffness matrix $[k]$.

Due to the elasticity of S, and the linearity of the force-displacement relation, it is seen (see, for example, the next section) that $[\delta]$ and $[k]$ are symmetric. That is,

$$\delta_{ij} = \delta_{ji} \qquad \text{and} \qquad k_{ij} = k_{ji} \tag{8.2-6}$$

8.3 Maxwell's Reciprocity Theorem

Consider an elastic beam subjected to transverse loads f_1 and f_2 at points P_1 and P_2. Let δ_{ij} be the transverse displacement at point P_i due to unit load at point P_j $(i=1,2;\, j=1,2)$.

Imagine the beam to be initially unloaded, although supported and thus restrained against rigid-body motion. Let f_1 be applied first in a gradually increasing manner so that inertia forces (loading due to acceleration) are negligible. Thus, the work done by f_1 is simply $\frac{1}{2} f_1^2 \delta_{11}$.

Next, let f_2 be gradually applied. The work done on the beam while f_2 is applied is then $\frac{1}{2} f_2^2 \delta_{22} + f_1 \delta_{12} f_2$, with the second term occurring due to the displacement at P_1, now due to the constant load f_1 at P_1. Thus, the total work W done is

$$W = \frac{1}{2} f_1^2 \delta_{11} + \frac{1}{2} f_2^2 \delta_{22} + f_1 \delta_{12} f_2 \tag{8.3-1}$$

Similarly, if the forces were applied in reverse order, the total work done is

$$W = \frac{1}{2} f_2^2 \delta_{22} + \frac{1}{2} f_1^2 \delta_{11} + f_2 \delta_{21} f_1 \tag{8.3-2}$$

The work done for the two methods of loading must be the same, since the final states of the system are identical. Comparing the expressions for W, we see that

$$\delta_{12} = \delta_{21} \tag{8.3-3}$$

Eq. 8.3-3 is a symbolic statement of *Maxwell's reciprocity theorem*. In words, the displacement at P_1 due to a unit load at P_2 is equal to the displacement at P_2 due to a unit load at P_1.

Observe that in the above development of Eq. 8.3-3, using a beam, the loadings and the displacements are all in the same direction.

We can readily extend Maxwell's theorem to three-dimensional elastic structures [8.1]. Specifically, let S be an elastic structure and $\mathbf{f}_A$ be a unit magnitude force applied at a point A of S. Let $\mathbf{u}_{B/A}$ be the displacement at a point B of S, due to $\mathbf{f}_A$. Similarly, let $\mathbf{f}_B$ be a unit magnitude force applied at B and let $\mathbf{u}_{A/B}$ be the corresponding displacement at A due to $\mathbf{f}_B$. Then the projection of $\mathbf{u}_{B/A}$ along $\mathbf{f}_B$ is equal to the projection of $\mathbf{u}_{A/B}$ along $\mathbf{f}_A$. That is,

$$\mathbf{f}_A \cdot \mathbf{u}_{A/B} = \mathbf{f}_B \cdot \mathbf{u}_{B/A}, \qquad (8.3\text{-}4)$$

where the dot designates a scalar or *dot* product.

Maxwell's theorem can also be extended to concentrated couples and corresponding rotations [8.2].

For example, referring again to beams, we can state

1) The angle of rotation at point P_1 due to a unit force at point P_2 is numerically equal to the deflection at P_2 due to a unit couple at P_1.
2) The angle of rotation at P_1 due to a unit couple applied at P_2 is equal to the rotation at P_2 due to a unit couple applied at P_1.

All forms of the theorem can be applied to any component or structure subjected to bending, direct load, or torsional deformation provided that linear elastic conditions prevail and deflections are sufficiently small so as not to significantly affect the undeformed geometry.

Example 8.3-1 Determine the influence coefficients for the points (1) and (2) of the uniform beam, as shown in Fig. 8.3-1, expressed in terms of length L, Young's modulus E, and area moment of inertia of the beam.

Solution: When a unit force is applied at point 1, the displacements at points 1 and 2 can be determined by using Mohr's second theorem (see, for example, [8.2], p. 109). Referring to Fig. 8.3-2, we obtain

$$\delta_{11} = \frac{1}{EJ}\left[\frac{1}{2}\frac{L^2}{4}\left(\frac{2}{3}\frac{L}{2}\right)\right] = \frac{2L^3}{48EJ}$$

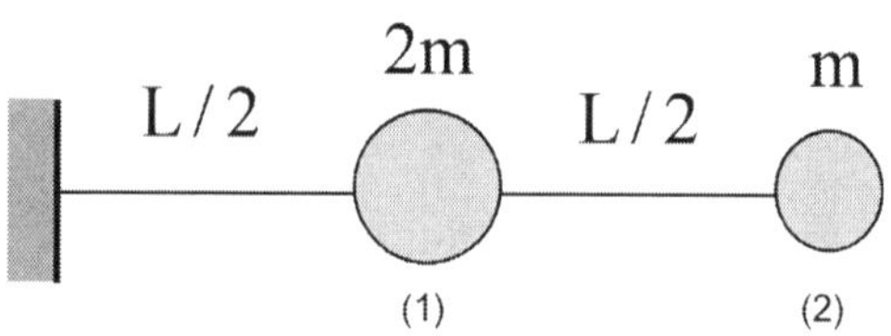

Fig. 8.3-1 A uniform beam with two masses.

$$\delta_{12} = \delta_{21} = \frac{1}{EJ}\left[\frac{1}{2}\frac{L^2}{4}\left(\frac{2}{3}\frac{L}{2}+\frac{L}{2}\right)\right] = \frac{5L^3}{48EJ}$$

Next, when a unit force is applied at point 2, we obtain

$$\delta_{22} = \frac{1}{EJ}\left[\frac{1}{2}L^2\left(\frac{2}{3}L\right)\right] = \frac{16L^3}{48EJ}$$

The flexibility matrix can now be written as

$$[\delta] = \frac{L^3}{48EJ}\begin{bmatrix} 2 & 5 \\ 5 & 16 \end{bmatrix}$$

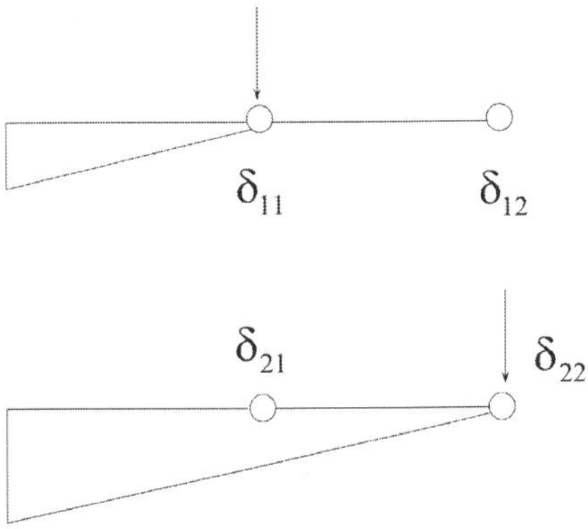

Fig. 8.3-2 The influence coefficients of the beam of Fig. 8.3-1.

Alternatively, the influence coefficients can be determined by calculating the deflections of the points from the beam deflection equation. For example, when a unit force is applied at point 1, we have

$$EJy'' = (1).(L/2 - x)$$

By integrating this and using the boundary condition, we obtain the deflection equation

$$EJy = \frac{1}{4}Lx^2 - \frac{1}{6}x^3$$

Letting $x = L/2$, we obtain δ_{11}. That is,

$$\delta_{11} = \frac{1}{EJ}\left(\frac{1}{16}L^3 - \frac{1}{48}L^3\right) = \frac{2L^3}{48EJ}$$

Similarly, when a unit force is applied at point 2, we have

$$EJy = \frac{L}{2}x^2 - \frac{1}{6}x^3$$

Letting $x = L/2$, we obtain $\delta_{21} = \delta_{12}$. Letting $x = L$ produces δ_{22}.

Example 8.3-2 The uniform light beam of Fig. 8.3-3 (A) vibrating in the plane as shown has three concentrated masses. Determine the influence coefficients.

Solution: It is known that the deflection at any point x, due to a unit force applied at a distance b from the end [Fig. 8.3-3 (B)], can be determined from the expression

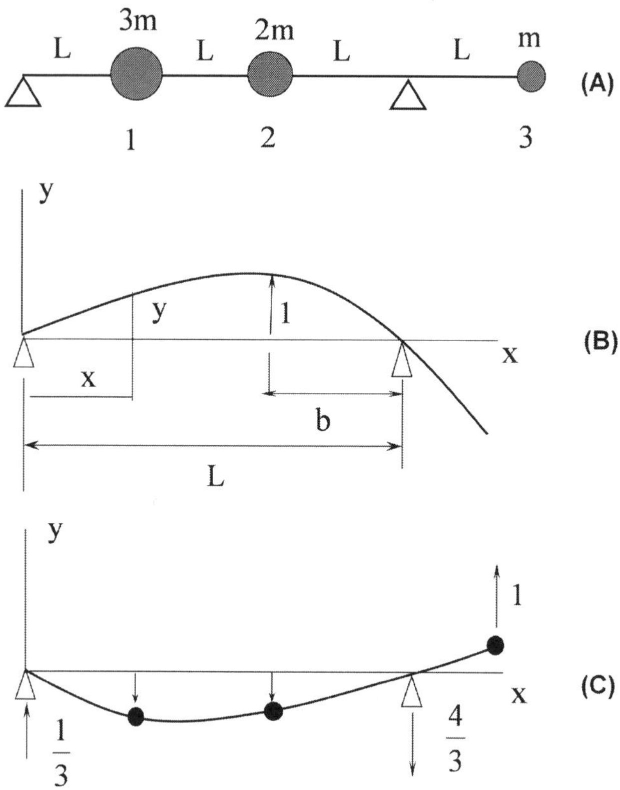

Fig. 8.3-3 A uniform beam with three concentrated masses.

$$y(x) = \frac{bx}{6EJL}\left(L^2 - x^2 - b^2\right), \tag{a}$$

which can be found from standard text and hand books on mechanics of materials. (See, for example, [8.3].)

Letting $x = L$ and $b = 2L$ in Eq. a yields the influence coefficient δ_{11} and δ_{22} as

$$\delta_{11} = \delta_{22} = \frac{(2L)(L)}{6EJ(3L)}\left[(3L)^2 - L^2 - (2L)^2\right] = \frac{8L^3}{18EJ} \tag{b}$$

Next, letting $x = L$ and $b = L$ in Eq. a yields δ_{12} and δ_{21} as

$$\delta_{12} = \delta_{21} = \frac{(L)(L)}{6EJ(3L)}\left[(3L)^2 - L^2 - L^2\right] = \frac{7L^3}{18EJ} \tag{c}$$

When a unit force is applied at point 3, the beam deflection can be obtained by integrating the moment equation

$$EJy'' = \frac{1}{3}x - \frac{4}{3}\left[x - 3L\right], \tag{d}$$

where $[x\text{–}3L]$ is called a *Macaulay bracket* (see, for example, [8.2], p. 102) having the values $x - 3L$ when $x \geq 3L$, and 0 when $x < 3L$. When integrated, $[x - 3L]$ must be integrated relative to $(x - 3L)$, not x. Hence, by integrating in Eq. d, we obtain the deflection equation

$$EJy = \frac{1}{18}x^3 - \frac{2}{9}\left[x - 3L\right]^3 - \frac{1}{2}L^2 x, \tag{e}$$

where again $[x\text{–}3L]$ is a Macaulay bracket.

Letting $x = L$ in Eq. e yields the influence coefficients δ_{13} and δ_{31} as

$$\delta_{13} = \delta_{31} = y\big|_{x=L} = -\frac{8L^3}{18EJ} \tag{f}$$

Letting $x = 2L$, we obtain the coefficients δ_{23} and δ_{32} as

$$\delta_{23} = \delta_{32} = y\big|_{x=2L} = -\frac{10L^3}{18EJ} \tag{g}$$

Finally, letting $x = 4L$ in Eq. e, we obtain the coefficient δ_{33} as

$$\delta_{33} = y\big|_{x=4L} = \frac{24L^3}{18EJ} \tag{h}$$

Therefore, the flexibility matrix is seen to be

$$\left[\delta\right] = \begin{bmatrix} \delta_{11} & \delta_{12} & \delta_{13} \\ \delta_{21} & \delta_{22} & \delta_{23} \\ \delta_{31} & \delta_{32} & \delta_{33} \end{bmatrix} = \frac{L^3}{18EJ}\begin{bmatrix} 8 & 7 & -8 \\ 7 & 8 & -10 \\ -8 & -10 & 24 \end{bmatrix} \tag{i}$$

8.4 Eigenvalue Analysis

Consider the matrix form of the undamped vibration equation

$$M\ddot{\mathbf{x}} + K\mathbf{x} = 0 \tag{8.4-1}$$

In Section 6.2, we discussed several eigenvalue and eigenvector procedures for analyzing solutions of Eq. 8.4-1. In this section, we will consider another eigenvalue analysis of the equation.

To this end, suppose we pre-multiply the terms of Eq. 8.4-1 by the flexibility matrix $[\delta]$. Since the flexibility and stiffness matrices are inverses of each other, we then obtain

$$D\ddot{\mathbf{x}}+\mathbf{x}=0, \tag{8.4-2}$$

where D, defined as $[\delta]M$, is sometimes called the *dynamic matrix.*

By assuming harmonic motion such that $\ddot{\mathbf{x}}$ is $-\omega^2\mathbf{x}$, Eq. 8.4-2 takes the form

$$D\mathbf{x}=\frac{1}{\omega^2}\mathbf{x} \tag{8.4-3}$$

Eq. 8.4-3 is an eigenvalue equation whose characteristic equation may be written as

$$\left|D-\frac{1}{\omega^2}I\right|=0 \tag{8.4-4}$$

When the mass matrix M is diagonal, we have

$$\begin{vmatrix} \delta_{11}m_{11}-\dfrac{1}{\omega^2} & \delta_{12}m_{22} & \cdots & \delta_{1n}m_{nn} \\ \delta_{21}m_{11} & \delta_{22}m_{22}-\dfrac{1}{\omega^2} & \cdots & \delta_{2n}m_{nn} \\ \cdot & \cdot & \cdot & \cdot \\ \delta_{n1}m_{11} & \delta_{n2}m_{22} & \cdots & \delta_{nn}m_{nn}-\dfrac{1}{\omega^2} \end{vmatrix}=0 \tag{8.4-5}$$

Expansion of Eq. 8.4-5 leads to

$$\left(\frac{1}{\omega^2}\right)^n-\left(\delta_{11}m_{11}+\delta_{22}m_{22}+\cdots+\delta_{nn}m_{nn}\right)\left(\frac{1}{\omega^2}\right)^{n-1}+\cdots=0 \tag{8.4-6}$$

If the roots of this equation are $1/\omega_1^2, 1/\omega_2^2, \ldots, 1/\omega_n^2$, the equation can be factored into the form

$$\left(\frac{1}{\omega^2}-\frac{1}{\omega_1^2}\right)\left(\frac{1}{\omega^2}-\frac{1}{\omega_2^2}\right)\cdots\left(\frac{1}{\omega^2}-\frac{1}{\omega_n^2}\right)=0 \tag{8.4-7}$$

According to the algebraic eigenvalue theorem, the sum of all the roots is equal to the sum of the diagonal terms of D (called the *trace* of the matrix). Thus, we have

$$\frac{1}{\omega_1^2}+\frac{1}{\omega_2^2}+\cdots+\frac{1}{\omega_n^2}=\delta_{11}m_{11}+\delta_{22}m_{22}+\cdots+\delta_{nn}m_{nn} \tag{8.4-8}$$

Example 8.4-1 Determine the influence coefficients for the spring-mass system of Fig. 8.4-1 and find the natural frequencies.

Solution: First, determine the influence coefficients. When a unit force is applied on mass 1, we have

$$\delta_{11}=\delta_{12}=\frac{1}{k}$$

For mass 2, there are two springs connected in series. Thus, applying a unit force to mass 2, we have

$$\delta_{22}=\frac{1}{k}+\frac{1}{k}=\frac{2}{k}$$

The flexibility influence matrix may thus be written as

$$[\delta]=\frac{1}{k}\begin{bmatrix}1 & 1\\ 1 & 2\end{bmatrix}$$

The mass matrix is

$$M=\begin{bmatrix}m & 0\\ 0 & m\end{bmatrix}$$

The characteristic equation is then

$$\left|[\delta]M-\frac{1}{\omega^2}I\right|=0$$

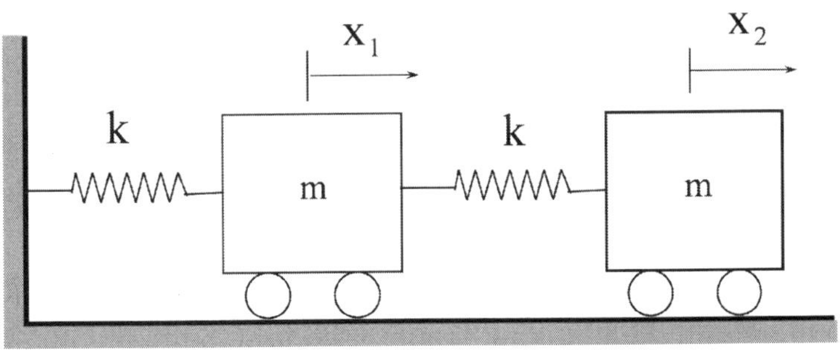

Fig. 8.4-1 Spring-mass system for Example 8.4-1.

Upon expansion, this leads to

$$\frac{1}{\omega^4}-\frac{3m}{k}\frac{1}{\omega^2}+\frac{m^2}{k^2}=0$$

Solving for $1/\omega^2$, we obtain

$$\frac{1}{\omega^2}=\frac{3\pm\sqrt{5}}{2}\frac{m}{k}$$

The natural frequencies are then given by

$$\omega_1=\sqrt{\frac{\left(3-\sqrt{5}\right)k}{2m}},\quad \omega_2=\sqrt{\frac{\left(3+\sqrt{5}\right)k}{2m}}$$

It is evident that

$$\frac{1}{\omega_1^2}+\frac{1}{\omega_2^2}=\delta_{11}m_{11}+\delta_{22}m_{22}=\frac{3m}{k}$$

Example 8.4-2 Determine the natural frequencies of the system shown in Example 8.3-2.

Solution: The flexibility matrix found in Example 8.3-2 is

$$[\delta]=\frac{L^3}{18EJ}\begin{bmatrix}8 & 7 & -8\\ 7 & 8 & -10\\ -8 & -10 & 24\end{bmatrix} \tag{a}$$

From the given data, the mass matrix is

$$M=m\begin{bmatrix}3 & & \\ & 2 & \\ & & 1\end{bmatrix} \tag{b}$$

The characteristic equation (see Eq. 8.4-4) is

$$\left|[\delta]M-\frac{1}{\omega^2}I\right|=0 \tag{c}$$

Expansion of Eq. c leads to

$$\lambda^3+a\lambda^2+b\lambda+c=0, \tag{d}$$

where

$$\lambda = 1/\omega^2$$

$$a = -(\delta_{11}m_{11} + \delta_{22}m_{22} + \delta_{33}m_{33})$$

$$b = m_{11}m_{22}(\delta_{11}\delta_{22} - \delta_{12}\delta_{21}) + m_{11}m_{33}(\delta_{11}\delta_{33} - \delta_{13}\delta_{31}) + m_{22}m_{33}(\delta_{22}\delta_{33} - \delta_{23}\delta_{32})$$

$$c = m_{11}m_{22}m_{33}(\delta_{11}\delta_{23}\delta_{32} + \delta_{22}\delta_{13}\delta_{31} + \delta_{33}\delta_{12}\delta_{21} - \delta_{11}\delta_{22}\delta_{33} - \delta_{21}\delta_{32}\delta_{13} - \delta_{31}\delta_{23}\delta_{12})$$

Substituting the values, we obtain

$$\lambda^3 - 64z\lambda^2 + 658z^2\lambda - 1008z^3 = 0, \quad \text{where} \quad z = \frac{mL^3}{18EJ} \tag{e}$$

Solving for λ, we obtain

$$\lambda_1 = 2.8686\frac{mL^3}{EJ}; \quad \lambda_2 = 0.5848\frac{mL^3}{EJ}; \quad \lambda_3 = 0.1031\frac{mL^3}{EJ} \tag{f}$$

Finally, the natural frequencies are seen to be

$$\omega_1 = 0.590\sqrt{\frac{EJ}{mL^3}}; \quad \omega_2 = 1.308\sqrt{\frac{EJ}{mL^3}}; \quad \omega_3 = 3.114\sqrt{\frac{EJ}{mL^3}} \tag{g}$$

8.5 Dunkerley's Equation

S. Dunkerley [8.4] in the late nineteenth century developed a method for approximating the fundamental frequency of multi-rotor systems. It provides good results if damping is negligible and if the frequencies of the higher harmonics are considerably greater than that of the fundamental.

Consider again Eq. 8.4-8. The influence coefficient δ_{ii} is the deflection at point i due to a unit force applied also at point i. The reciprocal of δ_{ii} is the stiffness coefficient k_{ii}. Thus, the term $\delta_{ii}m_i$ can be written as

$$\delta_{ii}m_i = \frac{m_i}{k_{ii}} = \frac{1}{\omega_{ii}^2}, \tag{8.5-1}$$

where ω_{ii} is the natural frequency of an equivalent mass-spring system with mass m_i acting alone at point i. Thus, we can write Eq. 8.4-8 in the following equivalent forms

$$\begin{aligned}\frac{1}{\omega_1^2} + \frac{1}{\omega_2^2} + \cdots + \frac{1}{\omega_n^2} &= \delta_{11}m_{11} + \delta_{22}m_{22} + \cdots + \delta_{nn}m_{n:} \\ &= \frac{m_{11}}{k_{11}} + \frac{m_{22}}{k_{22}} + \cdots + \frac{m_{nn}}{k_{nn}} \\ &= \frac{1}{\omega_{11}^2} + \frac{1}{\omega_{22}^2} + \cdots + \frac{1}{\omega_{nn}^2}\end{aligned} \tag{8.5-2}$$

By neglecting the natural frequencies of higher-order modes on the left side, we have

$$\frac{1}{\omega_1^2} \approx \frac{1}{\omega_{11}^2} + \frac{1}{\omega_{22}^2} + \cdots + \frac{1}{\omega_{nn}^2} = \sum_{i=1}^{n} \frac{1}{\omega_{ii}^2} \tag{8.5-3}$$

Eq. 8.5-3 is called *Dunkerley's equation*. Observe in this expression that the estimated fundamental frequency is always lower than the exact value. That is, Dunkerley's equation produces a lower bound to the fundamental frequency.

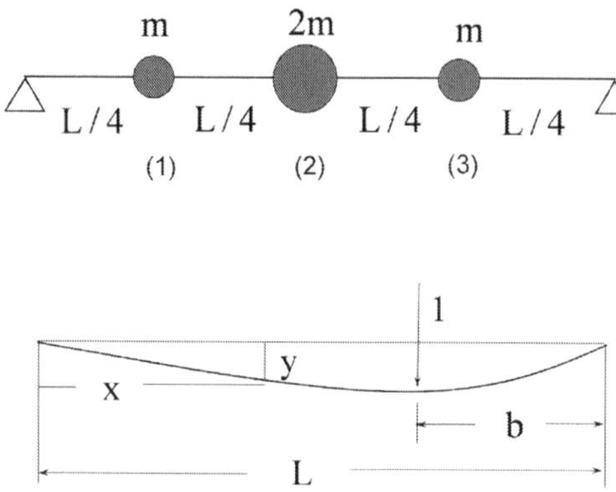

Fig. 8.5-1 Simple supported beam with three lumped masses.

Example 8.5-1 Fig. 8.5-1 depicts a simple supported beam with three concentrated masses. Estimate the fundamental natural frequency by using Dunkerley's equation.

Solution: Referring to Fig. 8.5-1, the deflection at any point x, due to a unit force applied at a distance b from the right end, can be determined from the equation (see, for example, [8.3])

$$y(x) = \frac{bx}{6EJL}\left(L^2 - x^2 - b^2\right) \tag{a}$$

Letting $x = L/4$ and $b = 3L/4$ in Eq. a yields the influence coefficient δ_{11}. Similarly, letting $x = L/2$ and $b = L/2$ in Eq. a leads to δ_{22}. Hence, we obtain

$$\delta_{11} = \delta_{33} = \frac{3L^3}{256EJ}, \quad \delta_{22} = \frac{L^3}{48EJ} \tag{b}$$

The mass matrix is

$$M = \begin{bmatrix} m & 0 & 0 \\ 0 & 2m & 0 \\ 0 & 0 & m \end{bmatrix} \tag{c}$$

Thus, we have $m_{11} = m_{33} = m$ and $m_{22} = 2m$. Then

$$\frac{1}{\omega_{11}^2} = \frac{1}{\omega_{33}^2} = \delta_{11} m_{11} = \frac{3mL^3}{256EJ} \tag{d}$$

$$\frac{1}{\omega_{22}^2} = \delta_{22} m_{22} = \frac{mL^3}{24EJ} \tag{e}$$

Substituting into Dunkerly's equation (Eq. 8.5-3), we have

$$\frac{1}{\omega_1^2} \approx \frac{3mL^3}{256EI} + \frac{mL^3}{24EJ} + \frac{3mL^3}{256EJ} = \frac{25mL^2}{384EJ} \tag{f}$$

Therefore,

$$\omega_1 \approx 3.919\sqrt{\frac{EJ}{mL^3}} \tag{g}$$

8.6 Rayleigh Principle

In Chapter Two, we used Rayleigh principle to study a one-degree-of-freedom system. This method can be extended to multi-degree-of-freedom systems if a reasonable estimate of the deformation is known. The method is based upon two fundamental assumptions: (1) the system is conservative, and (2) in the principal mode, the maximum potential energy is equal to the maximum kinetic energy.

As before, let M and K be the mass and stiffness matrices, respectively, and let $\mathbf{x}$ be the assumed displacement vector for the vibration. For a harmonic motion, we can express the maximum kinetic and potential energies as

$$T_{\max} = \frac{1}{2}\omega^2 \mathbf{x}^T M\mathbf{x} \tag{8.6-1}$$

and

$$U_{\max} = \frac{1}{2}\mathbf{x}^T K\mathbf{x} \tag{8.6-2}$$

By equating these two expressions and solving for ω^2, we obtain the Rayleigh's quotient

$$\omega^2 = \frac{\mathbf{x}^T K\mathbf{x}}{\mathbf{x}^T M\mathbf{x}} \tag{8.6-3}$$

This quotient approaches the lowest natural frequency (or fundamental frequency) from the higher side. In other words, Rayleigh's quotient gives the upper bound to the fundamental frequency.

To explore and illustrate this, suppose we express a shaft displacement curve in terms of the normal modes $\mathbf{u}_i$ as

$$\mathbf{x} = \mathbf{u}_1 + C_2\mathbf{u}_2 + C_3\mathbf{u}_3 + \cdots \tag{8.6-4}$$

Then $\mathbf{x}^T K\mathbf{x}$ becomes

$$\begin{aligned}\mathbf{x}^T K\mathbf{x} &= \mathbf{u}_1^T K\mathbf{u}_1 + C_2^2\mathbf{u}_2^T K\mathbf{u}_2 + C_3^2\mathbf{u}_3^2 K\mathbf{u}_3 + \cdots \\ &= \omega_1^2 + C_2^2\omega_2^2 + C_3^2\omega_3^2 + \cdots\end{aligned} \tag{8.6-5}$$

and $\mathbf{x}^T M\mathbf{x}$ is

$$\mathbf{x}^T M\mathbf{x} = 1 + C_2^2 + C_3^2 + \cdots \tag{8.6-6}$$

Thus, Eq. 8.6-3 becomes

$$\omega^2 = \frac{\omega_1^2 + C_2^2\omega_2^2 + C_3^2\omega_3^2 + \cdots}{1 + C_2^2 + C_3^2 + \cdots} = \omega_1^2 \left[\frac{1 + C_2^2 \dfrac{\omega_2^2}{\omega_1^2} + C_3^2 \dfrac{\omega_3^2}{\omega_1^2} + \cdots}{1 + C_2^2 + C_3^2 + \cdots} \right] > \omega_1^2 \tag{8.6-7}$$

It is evident that ω^2 is greater than ω_1^2 because $\omega_i^2 / \omega_1^2 > 1$ $(i = 2, 3, \cdots)$.

This analysis shows that if the exact fundamental deflection (or mode shape) $\mathbf{u}_1$ is assumed, the frequency calculated will be exact, since $C_2, C_3, \ldots$, will then be zero. If the assumed mode shape is not exact, it is equivalent to the application of additional constraints on the vibrating system. Hence, the calculated frequency is higher than the actual value. That is, the Rayleigh method gives a high value for the estimated frequency. Nevertheless, if a reasonable assumed mode shape is known, we will generally obtain a satisfactory result.

Another form of Rayleigh's quotient, providing a better estimate of the fundamental frequency, can be obtained by starting with the flexibility equation of motion (see Eq. 8.4-2)

$$\mathbf{x} = -D\ddot{\mathbf{x}} = \omega^2 D\mathbf{x} \tag{8.6-8}$$

Pre-multiplying by $\mathbf{x}^T M$, we obtain

$$\mathbf{x}^T M\mathbf{x} = \omega^2 \mathbf{x}^T MD\mathbf{x} \tag{8.6-9}$$

When $\mathbf{x}$ is expressed in terms of the normal modes, as before, the orthogonality conditions will again eliminate all terms for which $i \neq j$. Specifically, we have

$$\begin{aligned} \mathbf{x}^T MD\mathbf{x} &= \left(\mathbf{u}_1^T + C_2\mathbf{u}_2^T + C_3\mathbf{u}_3^T \cdots\right) M \left(\frac{\mathbf{u}_1}{\omega_1^2} + C_2 \frac{\mathbf{u}_2}{\omega_2^2} + C_3 \frac{\mathbf{u}_3}{\omega_3^2} + \cdots \right) \\ &= \frac{1}{\omega_1^2} + \frac{C_2^2}{\omega_2^2} + \frac{C_3^2}{\omega_3^2} + \cdots \end{aligned} \tag{8.6-10}$$

Consider now ω^2 obtained from Eq. 8.6-9 as

$$\omega^2 = \frac{\mathbf{x}^T M\mathbf{x}}{\mathbf{x}^T MD\mathbf{x}} \tag{8.6-11}$$

When compared with Eq. 8.6-3, we see that we now have another form of Raleigh's quotient.

By substituting from Eqs. 8.6-6 and 8.6-10 into Eq. 8.6-11, we have

$$\omega^2 = \frac{1+C_2^2+C_3^2+\cdots}{\frac{1}{\omega_1^2}+\frac{C_2^2}{\omega_2^2}+\frac{C_3^2}{\omega_3^2}+\cdots} = \omega_1^2\left[\frac{1+C_2^2+C_3^2+\cdots}{1+\frac{\omega_1^2}{\omega_2^2}C_2^2+\frac{\omega_1^2}{\omega_3^2}C_3^2+\cdots}\right] > \omega_1^2 \qquad (8.6\text{-}12)$$

A comparison of Eqs. 8.6-12 with 8.6-7 shows that with $\omega_1 < \omega_i^2$ $(i=2,3,\cdots)$, Eq. 8.6-11 provides a better estimate of ω^2 than does Eq. 8.6-3.

We can use Rayleigh's quotient to estimate the fundamental frequency of a vibrating beam. (We will consider vibration of continuous systems in detail in Chapter Nine.)

To briefly explore this, consider a flexible beam with mass m per unit length and maximum displacement $y(x)$ for an assumed first-mode displacement shape. For harmonic motion, the maximum transverse vibration speed at a position x along the beam is simply ωy. Thus, the maximum kinetic energy T_{max} of the beam is

$$T_{max} = \frac{1}{2}\omega^2\int y^2 dm \qquad (8.6\text{-}13)$$

The maximum potential energy U_{max} of the beam may be expressed as

$$U_{max} = \frac{1}{2}\int EJ\left(\frac{d^2y}{dx^2}\right)^2 dx \qquad (8.6\text{-}14)$$

Equating the kinetic and potential energies, the fundamental frequency of the beam is then given by

$$\omega^2 = \frac{\int EJ\left(d^2y/dx^2\right)^2 dx}{\int y^2 dm} \qquad (8.6\text{-}15)$$

If the beam is represented by a series of lumped masses $m_1, m_2, \ldots, m_n$, the maximum potential energy is equal to the work done as these loads are gradually added. As a first approximation, the static deflection $y_1, y_2, \ldots, y_n$ of corresponding points may be used. Thus, the maximum kinetic and potential energies are

$$T_{max} = \frac{1}{2}\omega^2\sum_{i=1}^{n} m_i y_i^2 \quad \text{and} \quad U_{max} = \frac{1}{2}\sum_{i=1}^{n} m_i g y_i \qquad (8.6\text{-}16)$$

Equating these two expressions yields the frequency equation

$$\omega^2 = \frac{g\sum m_i y_i}{\sum m_i y_i^2} \tag{8.6-17}$$

8.7 Method of Matrix Iteration

The method developed by Duncan and Collar in 1934 [8.5] is an iterative procedure providing one frequency and modal vector for each iteration sequence.

To develop the method, consider again the equation of motion Eq. 8.4-3, rewritten here as

$$\mathbf{x} = \omega^2 D\mathbf{x} \tag{8.7-1}$$

To determine the fundamental frequency ω_1 and the corresponding fundamental modal vector, we assume an arbitrary modal vector $\mathbf{x}_0$. Then, if $\mathbf{u}_i$ $(i = 1, 2, \cdots, n)$ are the actual modal vectors (unknown at this stage of the analysis), we can express $\mathbf{x}_0$ in terms of the $\mathbf{u}_i$ as

$$\mathbf{x}_0 = \sum_{i=1}^{n} C_i \mathbf{u}_i, \tag{8.7-2}$$

where the C_i $(i = 1, 2, \cdots, n)$ are constants.

Next, we form a sequence of vectors by iteration as

$$\mathbf{x}_0, \mathbf{x}_1 = D\mathbf{x}_0, \mathbf{x}_2 = D\mathbf{x}_1 = D^2\mathbf{x}_0, \ldots \tag{8.7-3}$$

By substituting from Eq. 8.7-3 into Eq. 8.7-2, and by recalling that $D\mathbf{u}_i = \mathbf{u}_i / \lambda_i$ where, as before, λ_i is ω_i^2, we obtain the series of expressions

$$\mathbf{x}_1 = D\mathbf{x}_0 = \sum_{i=1}^{n} C_i D\mathbf{u}_i = \sum_{i=1}^{n} \frac{1}{\lambda_i} C_i \mathbf{u}_i$$

$$\mathbf{x}_2 = D^2\mathbf{x}_0 = \sum_{i=1}^{n} C_i D^2\mathbf{u}_i = \sum_{i=1}^{n} \frac{1}{\lambda_i^2} C_i \mathbf{u}_i \tag{8.7-4}$$

$$\cdots\cdot$$

$$\mathbf{x}_s = D^s\mathbf{x}_0 = \sum_{i=1}^{n} C_i D^s\mathbf{u}_i = \sum_{i=1}^{n} \frac{1}{\lambda_i^s} C_i \mathbf{u}_i$$

From our previous analyses, we assume that the λ_i are distinct and that they are monotonically increasing in value. That is,

$$\lambda_1 < \lambda_2 < \cdots < \lambda_n \tag{8.7-5}$$

Therefore, if s is sufficiently large, we have

$$(1/\lambda_1)^s >> (1/\lambda_i)^s \; (i=2,3,\cdots,n) \tag{8.7-6}$$

Thus, we obtain

$$\mathbf{x}_s \approx \frac{1}{\lambda_1^s} C_1 \mathbf{u}_1 \tag{8.7-7}$$

If now we perform one more iteration, we have

$$\mathbf{x}_{s+1} = \frac{1}{\lambda_1^{s+1}} C_1 \mathbf{u}_1 = \frac{1}{\lambda_1} \mathbf{x}_s \tag{8.7-8}$$

In view of Eq. 8.7-8, we can develop a procedure for obtaining the fundamental frequency and the corresponding first modal vector. We begin with assuming an arbitrary modal vector $\mathbf{x}_0$. Then, by substituting $\mathbf{x}_0$ into the right side of Eq. 8.7-1, we obtain a column of numbers which can be *normalized* by adjusting the first element to the same value as the first element of $\mathbf{x}_0$ (see the following example). This procedure is then repeated until the iterated column $\mathbf{x}_{s+1}$ approaches $\mathbf{x}_s$.

Example 8.7-1 Using the method of matrix iteration, determine the natural fundamental frequency of the system in Example 8.5-1.

Solution: Using Eq. a of Example 8.5-1, we obtain the flexibility matrix $[\delta]$ as

$$[\delta] = \frac{L^3}{768EJ} \begin{bmatrix} 9 & 11 & 7 \\ 11 & 16 & 11 \\ 7 & 11 & 9 \end{bmatrix} \tag{a}$$

The mass matrix is

$$M = m \begin{bmatrix} 1 & 0 & 0 \\ 0 & 2 & 0 \\ 0 & 0 & 1 \end{bmatrix} \tag{b}$$

Thus, the dynamic matrix is

$$D = [\delta]M = \frac{mL^3}{768EJ} \begin{bmatrix} 9 & 22 & 7 \\ 11 & 32 & 11 \\ 7 & 22 & 9 \end{bmatrix} \tag{c}$$

The eigenvalue problem is then

$$\mathbf{x} = \omega^2 D \mathbf{x} \tag{d}$$

Now, we can use the iteration method to find the fundamental frequency. Specifically, we start with $\mathbf{x}_0 = \begin{bmatrix} 1 & 2 & 1 \end{bmatrix}^T$ for the right column of Eq. d. We then obtain

$$\mathbf{x} = \omega^2 D\mathbf{x}_0 = \frac{\omega^2 mL^3}{768EJ}\begin{bmatrix} 60 \\ 86 \\ 60 \end{bmatrix} = \frac{60\omega^2 mL^3}{768EJ}\begin{bmatrix} 1.00000 \\ 1.433333 \\ 1.00000 \end{bmatrix},$$

where we have *normalized* the column in the last equality.

Next, we let $\mathbf{x}_1 = \begin{bmatrix} 1.00 & 1.433333 & 1.00 \end{bmatrix}^T$ and iterate again, obtaining

$$\mathbf{x} = \omega^2 D\mathbf{x}_1 = \frac{\omega^2 mL^3}{768EJ}\begin{bmatrix} 47.5333 \\ 67.8667 \\ 47.5333 \end{bmatrix} = \frac{47.5333\omega^2 mL^2}{768EJ}\begin{bmatrix} 1.0000 \\ 1.42777 \\ 1.00000 \end{bmatrix},$$

where we have again *normalized* the column in the last equality.

By repeating this procedure a few more times, the columns converge as

$$\begin{bmatrix} 1.0000 \\ 1.42770 \\ 1.0000 \end{bmatrix} = \frac{47.409\omega^2 mL^3}{768EJ}\begin{bmatrix} 1.0000 \\ 1.42770 \\ 1.0000 \end{bmatrix}$$

The mode shape is thus

$$\mathbf{u}_1 = \begin{bmatrix} 1.0000 \\ 1.42770 \\ 1.0000 \end{bmatrix}, \tag{e}$$

as indicated in Fig. 8.7-1.

Finally, the fundamental frequency can be found from

$$\frac{47.409\omega^2 mL^3}{768EJ} = 1 \tag{f}$$

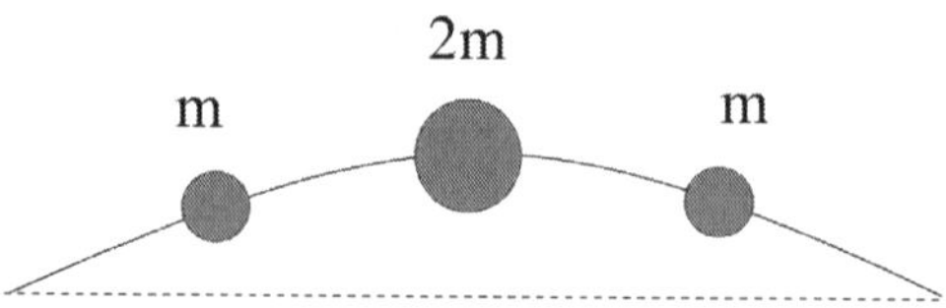

Fig. 8.7-1 First mode of Example 8.7-1.

That is,

$$\omega_1 = \sqrt{\frac{768EI}{47.409mL^3}} = 4.0248\sqrt{\frac{EJ}{mL^3}} \tag{g}$$

This result is seen to be more accurate than that obtained using Dunkerley's equation.

8.8 Determining of Higher-Order Modes and Frequencies

When the equations of motion are expressed in terms of the flexibility influence coefficients, the foregoing iterative procedure converges to the lowest frequency mode. Once this *lowest* mode is obtained we can, in effect, delete it from the system, and then use the iteration method again to obtain the next lowest mode.

To develop this concept, consider a vibrating system with modes $\mathbf{u}_1, \mathbf{u}_2, \ldots, \mathbf{u}_n$ (to be determined). Let a deformation $\mathbf{x}$ of the system be expressed as

$$\mathbf{x} = C_1\mathbf{u}_1 + C_2\mathbf{u}_2 + \cdots C_n\mathbf{u}_n \tag{8.8-1}$$

By pre-multiplying by $\mathbf{u}_1^T M$ all the terms on the right side, except the first one, are eliminated due to the orthogonal relationship. Then we have

$$\mathbf{u}_1^T M\mathbf{x} = C_1 m_1 \tag{8.8-2}$$

or

$$C_1 = \frac{\mathbf{u}_1^T M\mathbf{x}}{m_1}, \tag{8.8-3}$$

where m_1 is the first modal mass, $m_1 = \mathbf{u}_1^T M\mathbf{u}_1$.

Now to eliminate the first mode, we introduce a modified deformation $\mathbf{x}$ as

$$\mathbf{x}^* = \mathbf{x} - C_1\mathbf{u}_1 = \mathbf{x} - \mathbf{u}_1\left(\frac{\mathbf{u}_1^T M\mathbf{x}}{m_1}\right) = \left[I - \frac{\mathbf{u}_1\mathbf{u}_1^T M}{m_1}\right]\mathbf{x} = S_1\mathbf{x}, \tag{8.8-4}$$

where I is the $n \times n$ unit matrix, and where S_1, called the *sweeping matrix*, is defined by inspection as

$$S_1 = I - \frac{\mathbf{u}_1\mathbf{u}_1^T M}{m_1} \tag{8.8-5}$$

The first mode is then swept out of the assumed deformation by the sweeping matrix.

By substituting $\mathbf{x}^{*} = S_1\mathbf{x}$ into the right side of Eq. 8.7-1, we obtain the iterative expression

$$\mathbf{x} = \omega^2 D S_1 \mathbf{x} \tag{8.8-6}$$

The iteration procedure used for Eq. 8.8-6 will converge to the second mode.

For the third and higher modes, the sweeping procedures are repeated. For example, the following sweeping matrix can be used for the third mode:

$$S_2 = I - \frac{\mathbf{u}_1\mathbf{u}_1^T M}{m_1} - \frac{\mathbf{u}_2\mathbf{u}_2^T M}{m_2} \tag{8.8-7}$$

Eq. 8.8-7 can be written as

$$S_2 = S_1 - \frac{\mathbf{u}_2\mathbf{u}_2^T M}{m_2} \tag{8.8-8}$$

In general, we have [8.6]

$$S_n = S_{n-1} - \frac{\mathbf{u}_n\mathbf{u}_n^T M}{m_n} \tag{8.8-9}$$

Example 8.8-1 Determine the second and third modes of the system in Example 8.7-1.

Solution: Using Eqs. b and e in Example 8.7-1, the first modal mass is obtained as

$$m_1 = \mathbf{u}_1^T M \mathbf{u}_1 = 6.07665m \tag{a}$$

The sweeping matrix is then given by

$$S_1 = I - \frac{\mathbf{u}_1\mathbf{u}_1^T M}{m_1} = \begin{bmatrix} 0.83544 & -0.469897 & -0.16456 \\ -0.23494 & 0.329128 & -0.23495 \\ -0.16456 & -0.469897 & 0.835436 \end{bmatrix} \tag{b}$$

Starting with $\mathbf{x} = \begin{bmatrix} -1 & 0 & 1 \end{bmatrix}^T$ for the right column of Eq. 8.8-6, we obtain

$$\begin{bmatrix} -1 \\ 0 \\ 1 \end{bmatrix} = \frac{\omega_2^2 mL^3}{768EJ} \begin{bmatrix} 2.0000 \\ 0 \\ -2.0000 \end{bmatrix} \begin{bmatrix} -1 \\ 0 \\ 1 \end{bmatrix} = 2.60417 \times 10^{-3} \frac{\omega_2^2 mL^3}{EJ} \begin{bmatrix} -1 \\ 0 \\ 1 \end{bmatrix} \tag{c}$$

This indicates that $\mathbf{x} = \begin{bmatrix} -1 & 0 & 1 \end{bmatrix}^T$ is the second mode $\mathbf{u}_2$. Hence, the second frequency is

$$\omega_2 = \sqrt{\frac{EJ}{2.60417 \times 10^3 \, mL^3}} = 19.5919\sqrt{\frac{EJ}{mL^3}} \tag{d}$$

If we start the iteration process with $\mathbf{x} = \begin{bmatrix} 1 & 1 & 1 \end{bmatrix}^T$, it will also converge to the second mode, but it will take longer.

Now we can calculate the third mode. From the second mode and mass matrix, the second modal mass is found to be

$$m_2 = \mathbf{u}_2^T M \mathbf{u}_2 = 2m \tag{e}$$

We now use the sweeping matrix

$$S_2 = S_1 - \frac{\mathbf{u}_2 \mathbf{u}_2^T M}{m_2} = \begin{bmatrix} 0.33544 & -0.46990 & -0.33544 \\ -0.23495 & 0.32912 & -0.23495 \\ -0.33544 & -0.46990 & 0.33544 \end{bmatrix}$$

Starting with $\mathbf{x} = \begin{bmatrix} 1 & -1 & 1 \end{bmatrix}^T$ for the right column of equation $\mathbf{x} = \omega^2 D S_2 \mathbf{x}$, it converges to the third mode as

$$\begin{bmatrix} 1.0000 \\ -0.740043 \\ 1.0000 \end{bmatrix} = \frac{\omega_3^2 mL^3}{768EJ} \begin{bmatrix} 0.590600 \\ -0.413672 \\ 0.590600 \end{bmatrix} = 7.69011 \times 10^{-4} \frac{\omega_3^2 mL^3}{EI} \begin{bmatrix} 1.0000 \\ -0.740043 \\ 1.0000 \end{bmatrix}$$

Thus, the third modal vector is

$$\mathbf{u}_3 = \begin{bmatrix} 1.0000 \\ -0.740043 \\ 1.0000 \end{bmatrix} \tag{f}$$

The natural frequency for the third mode is given by

$$\omega_3 = \sqrt{\frac{EJ}{7.69011 \times 10^{-4} \, mL^3}} = 36.0606\sqrt{\frac{EJ}{mL^3}} \tag{g}$$

8.9 Transfer Matrix Method

Using the transfer matrix method, a large system is divided into smaller subsystems. The analysis is conducted from one end of the system to the other, satisfying interface conditions between the subsystems.

The method employs *state vectors, state variables, point matrices*, and *field matrices*, defined as follows [8.7, 8.8]:

> A *state vector* is a column array of displacements and interactive forces at an internal location of the system.
>
> A *state variable* is an element of a state vector.
>
> A *point matrix* has elements defining the dynamic properties of a subsystem. Point matrices relate state vectors on opposite boundaries of a subsystem.
>
> A *field matrix* has elements defining the elastic properties of the subsystems. Field matrices relate state vectors across a boundary of adjoining subsystems.

In the following paragraph, we attempt to illustrate these concepts with rather simple systems.

(1) A Spring-Mass System

Fig. 8.9-1 shows a portion of an inline spring-mass system [8.8]. Envision an x-coordinate axis along the system axis, with the positive direction being to the right.

Consider a typical subsystem (or segment) consisting of a massless spring with modulus k and a particle with mass m. For the n-th segment, the spring is to the left of the particle, with modulus k_n, and the particle mass m_n. Finally, let superscripts L and R designate the left and right sides of the particle.

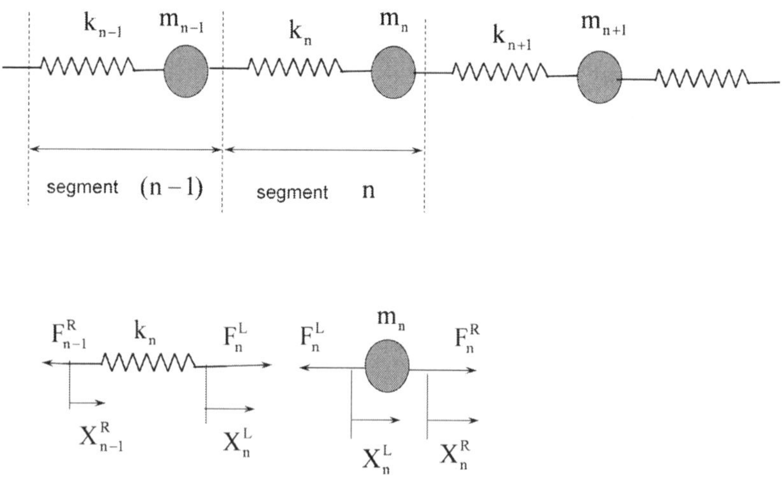

Fig. 8.9-1 A spring-mass system [8.8]. (This graph is reproduced with the permission of the F. S. Tse family and co-authors.)

Consider a typical particle P_n with mass m_n and displacement x_n. Then the free-body diagram of Fig. 8.9-1 produces the equation

$$m_n \ddot{x}_n = F_n^R - F_n^L \tag{8.9-1}$$

For harmonic motion with frequency ω, Eq. 8.9-1 becomes

$$F_n^R = -\omega^2 m_n X_n + F_n^L \tag{8.9-2}$$

Since the displacements on both sides of P_n are the same, we have

$$X_n = X_n^R = X_n^L \tag{8.9-3}$$

Eqs. 8.9-2 and 8.9-3 can be assembled into a matrix equation as

$$\begin{bmatrix} X \\ F \end{bmatrix}_n^R = \begin{bmatrix} 1 & 0 \\ -\omega^2 m & 1 \end{bmatrix}_n \begin{bmatrix} X \\ F \end{bmatrix}_n^L, \tag{8.9-4}$$

where $\begin{bmatrix} X \\ F \end{bmatrix}$ is called the *state vector* and the square matrix is the *point matrix.*

Next, consider the spring k_n. With the spring being massless, the opposite end forces are equal. The free-body diagram of Fig. 8.9-1 then leads to the equations

$$F_n^L = F_{n-1}^R \qquad \text{and} \qquad F_{n-1}^R = k_n\left(X_n^L - X_{n-1}^R\right) \tag{8.9-5}$$

Putting Eq. 8.9-5 into matrix form, we have

$$\begin{bmatrix} X \\ F \end{bmatrix}_n^L = \begin{bmatrix} 1 & 1/k \\ 0 & 1 \end{bmatrix}_n \begin{bmatrix} X \\ F \end{bmatrix}_{n-1}^R, \tag{8.9-6}$$

where the square matrix is the *field matrix.*

By substituting from Eq. 8.9-6 into Eq. 8.9-4, we obtain

$$\begin{bmatrix} X \\ F \end{bmatrix}_n^R = \begin{bmatrix} 1 & 0 \\ -\omega^2 m & 1 \end{bmatrix}_n \begin{bmatrix} 1 & 1/k \\ 0 & 1 \end{bmatrix}_n \begin{bmatrix} X \\ F \end{bmatrix}_{n-1}^R$$

or

$$\begin{bmatrix} X \\ F \end{bmatrix}_n^R = \begin{bmatrix} 1 & 1/k \\ -\omega^2 m & \left(1-\omega^2 m/k\right) \end{bmatrix}_n \begin{bmatrix} X \\ F \end{bmatrix}_{n-1}^R \tag{8.9-7}$$

or simply

$$Z_n^R = H_n Z_{n-1}^R, \tag{8.9-8}$$

where the square matrix H_n is the *transfer matrix*, through which the state vector at the end of the segment $n-1$ is transferred to the end of the segment n.

Using the recurrence formula in Eq. 8.9-8, the state vector Z_n^R at a typical location n can be related to the state vector Z_0^R at the boundary of the problem. That is,

$$Z_n^R = H_n H_{n-1} \cdots H_2 H_1 Z_0^R \tag{8.9-9}$$

For the case where the boundary condition at the left side of location 0 is known, using the point matrix in Eq. 8.9-4, the state vector Z_n^R can be related to the state vector Z_0^L; that is,

$$Z_n^R = H_n H_{n-1} \cdots H_2 H_1 S_0 Z_0^L, \tag{8.9-10}$$

where S_0 is the point matrix of location 0; that is,

$$Z_0^R = S_0 Z_0^L \tag{8.9-11}$$

With known values of the state vector elements at location 0 and a chosen value of ω^2, it is possible to progressively compute the state vectors to the last location n. Depending on the boundary conditions, either X_n or F_n can be graphed as a function of ω^2. The natural frequencies of the system are then established when the boundary conditions are satisfied.

Problems of this type, where only displacement is associated with each mass, are called Holzer-type problems [8.8]. Holzer developed a tabular method to find the natural frequencies of the system [8.9]. This method has general applications, including systems with rectilinear and angular motions, damped or undamped systems, semidefinite systems, branched systems, or systems with fixed ends.

In the development so far, the locations were numbered in increasing order from left to right, with the transfer matrix also progressing to the right. In some problems, it is convenient to proceed with the transfer matrix in the opposite direction, in which case we need only to invert Eqs. 8.9-7. We then obtain the relationship

$$\begin{bmatrix} X \\ F \end{bmatrix}_{n-1}^R = \begin{bmatrix} \left(1-\omega^2 m/k\right) & -1/k \\ \omega^2 m & 1 \end{bmatrix}_n \begin{bmatrix} X \\ F \end{bmatrix}_n^R \tag{8.9-12}$$

This equation can be easily verified with free-body diagrams.

Example 8.9-1 Determine the natural frequencies of the system shown in Fig. 8.9-2.

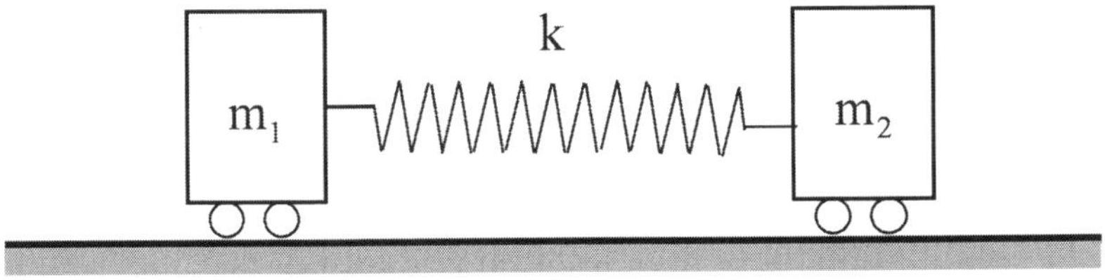

Fig. 8.9-2 A spring-mass system for Example 8.9-1.

Solution: Since there are no external forces applied to the two masses, we have

$$F_1^L = F_2^R = 0 \tag{a}$$

Then, from Eq. 8.9-7, we have

$$\begin{bmatrix} X \\ F \end{bmatrix}_2^R = \begin{bmatrix} 1 & 1/k \\ -\omega^2 m_2 & 1-\omega^2 m_2/k \end{bmatrix} \begin{bmatrix} X \\ F \end{bmatrix}_1^R \tag{b}$$

From Eq. 8.9-4, we have

$$\begin{bmatrix} X \\ F \end{bmatrix}_1^R = \begin{bmatrix} 1 & 0 \\ -\omega^2 m_1 & 1 \end{bmatrix} \begin{bmatrix} X \\ F \end{bmatrix}_1^L \tag{c}$$

Substituting from Eq. c into Eq. b, and noting the boundary conditions, we have

$$\begin{bmatrix} X \\ F \end{bmatrix}_2^R = \begin{bmatrix} 1-\omega^2 m_1/k & 1/k \\ -\omega^2 m_2 - \left(1-\omega^2 m_2/k\right)\left(\omega^2 m_1\right) & 1-\omega^2 m_2/k \end{bmatrix} \begin{bmatrix} X \\ 0 \end{bmatrix}_1^L \tag{d}$$

From this result we obtain

$$X_2 = X_2^R = \left(1-\omega^2 m_1/k\right) X_1^L = \left(1-\omega^2 m_1/k\right) X_1 \tag{e}$$

$$F_2^R = \left[-\omega^2 m_2 - \left(1-\omega^2 m_2/k\right)\left(\omega^2 m_1\right)\right] X_1 = 0 \tag{f}$$

If $X_1 \neq 0$, we then have

$$\left[-\omega^2 m_2 - \left(1-\omega^2 m_2 / k\right)\left(\omega^2 m_1\right)\right] = 0 \tag{g}$$

From this expression, the two natural frequencies are found to be

$$\omega_1^2 = 0,\ \omega_1 = 0 \tag{h}$$

and

$$\omega_2^2 = \frac{k(m_1+m_2)}{m_1 m_2},\ \omega_2 = \sqrt{\frac{k(m_1+m_2)}{m_1 m_2}} \tag{i}$$

Substituting $\omega_1 = 0$ into Eq. e, we obtain the first mode shape

$$X_1 / X_2 = 1 \tag{j}$$

This is a rigid body mode. Next, substituting ω_2 into Eq. e, we obtain

$$X_1 / X_2 = -m_2 / m_1 \tag{k}$$

From this result, we see that the two masses are always moving in opposition to each other.

Example 8.9-2 Determine the natural frequencies of the system shown in Fig. 8.9-3 by using the transfer matrix method.

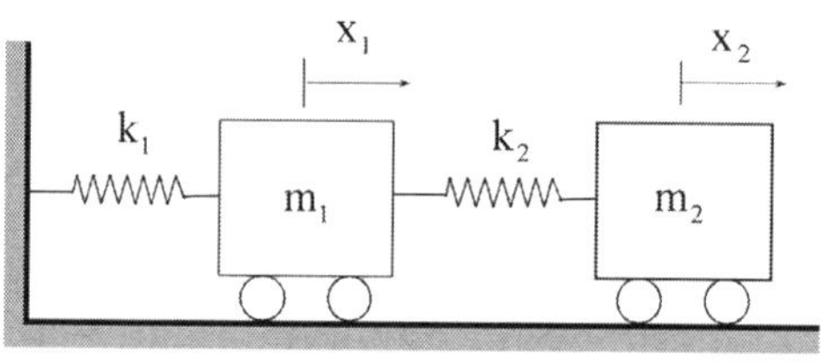

Fig. 8.9-3 A two-degree-of-freedom mass-spring system.

Solution: The recurrence formulas for the computations are

$$Z_1^R = H_1 Z_0^R \tag{a}$$

$$Z_2^R = H_2 Z_1^R = H_2 H_1 Z_0^R, \tag{b}$$

where the state vectors and transfer matrices are

$$Z_2^R = \begin{bmatrix} X \\ F \end{bmatrix}_2^R = \begin{bmatrix} X \\ 0 \end{bmatrix}_2^R,\ Z_0^R = \begin{bmatrix} X \\ F \end{bmatrix}_0^R = \begin{bmatrix} 0 \\ F \end{bmatrix}_0^R$$

$$H_2 = \begin{bmatrix} 1 & 1/k_2 \\ -\omega^2 m_2 & \left(1-\omega^2 m_2 / k_2\right) \end{bmatrix},\ H_1 = \begin{bmatrix} 1 & 1/k_1 \\ -\omega^2 m_1 & \left(1-\omega^2 m_1 / k_1\right) \end{bmatrix}$$

Substituting into Eq. b, we have

$$\begin{bmatrix} X \\ 0 \end{bmatrix}_2^R = \begin{bmatrix} \left(1-\omega^2 m_1/k_2\right) & 1/k_1 + 1/k_2\left(1-\omega^2 m_1/k_1\right) \\ -\omega^2 m_2 - \omega^2 m_1\left(1-\omega^2 m_2/k_1\right) & -\omega^2 m_2/k_1 + \left(1-\omega^2 m_1/k_1\right)\left(1-\omega^2 m_2/k_2\right) \end{bmatrix} \begin{bmatrix} 0 \\ F \end{bmatrix}_0^R$$

The second equation yields

$$F_2^R = 0 = \left[-\omega^2 m_2/k_1 + \left(1-\omega^2 m_1/k_1\right)\left(1-\omega^2 m_2/k_2\right)\right] F_0$$

Since $F_0 \neq 0$, due to oscillation of the system, we have

$$\left[-\omega^2 m_2/k_1 + \left(1-\omega^2 m_1/k_1\right)\left(1-\omega^2 m_2/k_2\right)\right] = 0$$

This equation can be written as

$$\omega^4 - \left[\frac{k_1}{m_1} + \frac{k_2}{m_2}\left(1+\frac{m_2}{m_1}\right)\right]\omega^2 + \frac{k_1 k_2}{m_1 m_2} = 0, \tag{c}$$

from which the natural frequencies are found to be

$$\omega_n = \sqrt{\frac{1}{2}\left[\frac{k_1}{m_1} + \frac{k_2}{m_2}\left(1+\frac{m_2}{m_1}\right) \pm \sqrt{\left[\frac{k_1}{m} + \frac{k_2}{m_2}\left(1+\frac{m_2}{m_1}\right)\right]^2 - \frac{4k_1k_2}{m_1 m_2}}\right]} \tag{d}$$

This solution can also be obtained by using Eq. 8.9-12 to proceed with the transfer matrix in the opposite direction. That is,

$$\begin{bmatrix} X \\ F \end{bmatrix}_1^R = \begin{bmatrix} \left(1-\omega^2 m_2/k_2\right) & -1/k_2 \\ \omega^2 m_2 & 1 \end{bmatrix} \begin{bmatrix} X \\ 0 \end{bmatrix}_2^R \tag{e}$$

$$\begin{bmatrix} 0 \\ F \end{bmatrix}_0^R = \begin{bmatrix} \left(1-\omega^2 m_1/k_1\right) & -1/k_1 \\ \omega^2 m_1 & 1 \end{bmatrix} \begin{bmatrix} X \\ F \end{bmatrix}_1^R \tag{f}$$

Substituting from Eq. e into Eq. f, we obtain

$$\begin{bmatrix} 0 \\ F \end{bmatrix}_0^R = \begin{bmatrix} \left(1-\omega^2 m_1/k_1\right) & -1/k_1 \\ \omega^2 m_1 & 1 \end{bmatrix} \begin{bmatrix} \left(1-\omega^2 m_2/k_2\right) & -1/k_2 \\ \omega^2 m_2 & 1 \end{bmatrix} \begin{bmatrix} X \\ 0 \end{bmatrix}_2^R \tag{g}$$

The first equation of (g) yields the same frequency equation (c).

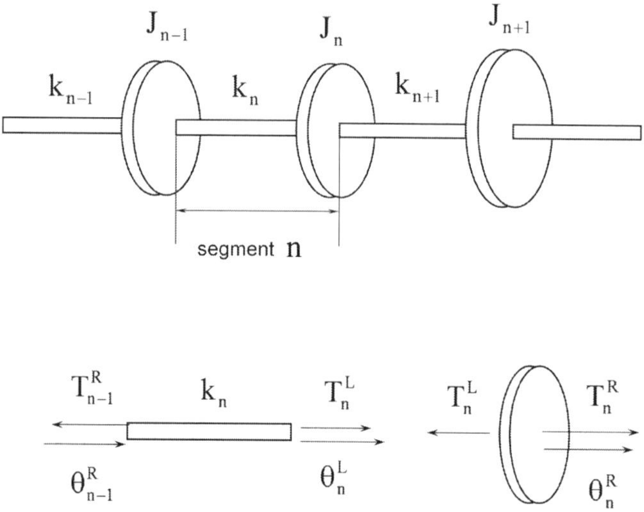

Fig. 8.9-4 A torsional system segment.

For these sample examples, the natural frequencies can be calculated by hand. For large systems, the numerical calculation can be performed as follows: (1) Assume a trial frequency ω and let $F_0 = 1$, arbitrarily; (2) Calculate Z_1^R from Eq. a and Z_2^R from Eq. b for a different value of ω; (3) Check whether the constraint condition ($F_2^R = 0$) is satisfied. If not, a new value of ω is assumed and the process repeated. The frequencies that result in $F_2^R = 0$ converge to the natural frequencies.

(2) Torsional Systems

Consider next a torsional, or rotating, system, as in Fig. 8.9-4 [8.7, 8.8]. To study such system, it is helpful to first establish a sign convention. In Fig. 8.9-5, a coordinate axis is placed along the torsional axis with the positive direction to the right. A *positive face* at a position, say, location n, has its outer normal in the positive direction. Next, positive angular displacements and positive moments are defined, following the right-hand screw rule. Similarly,

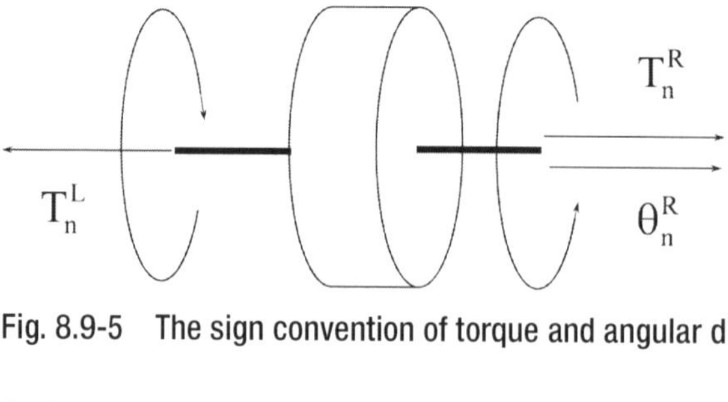

Fig. 8.9-5 The sign convention of torque and angular displacement.

a *negative face* at position n has its outer normal in the negative coordinate direction. Finally, positive moments are also defined following the right-hand screw rule. As shown in Fig. 8.9-5, the quantities are positive.

With these definitions, the development of the transfer matrix of the torsional system is analogous to that of the linear spring-mass system, with $\begin{bmatrix} \theta \\ T \end{bmatrix}$ being the state vector.

Fig. 8.9-4 depicts a portion of a generic torsional system. Consider a typical segment n. Assuming the shaft to be light, compared with the large diameter disks or masses, then the dynamical equation may be written in matrix form as

$$\begin{bmatrix} \theta \\ T \end{bmatrix}_n^R = \begin{bmatrix} 1 & 0 \\ -\omega^2 J & 1 \end{bmatrix}_n \begin{bmatrix} \theta \\ T \end{bmatrix}_n^L \tag{8.9-13}$$

and

$$\begin{bmatrix} \theta \\ T \end{bmatrix}_n^L = \begin{bmatrix} 1 & 1/k \\ 0 & 1 \end{bmatrix}_n \begin{bmatrix} \theta \\ T \end{bmatrix}_{n-1}^R, \tag{8.9-14}$$

where the J in Eq. 8.9-13 is the polar (or axial) moment of inertia of the n-th disk. By combining Eqs. 8.9-13 and 8.9-14, we have

$$\begin{bmatrix} \theta \\ T \end{bmatrix}_n^R = \begin{bmatrix} 1 & 1/k \\ -\omega^2 J & \left(1-\omega^2 J/k\right) \end{bmatrix}_n \begin{bmatrix} \theta \\ T \end{bmatrix}_{n-1}^R \tag{8.9-15}$$

Observe that each of Eqs. 8.9-13, 8.9-14, and 8.9-15 is, respectively, directly analogous to Eqs. 8.9-4, 8.9-6, and 8.9-7 of the inline spring-mass system.

As before, in some problems, it is convenient to proceed with the transfer matrix forward in the opposite direction. In this case, we need only invert Eq. 8.9-15. We then obtain

$$\begin{bmatrix} \theta \\ T \end{bmatrix}_{n-1}^R = \begin{bmatrix} \left(1-\omega^2 J/k\right) & -1/k \\ \omega^2 J & 1 \end{bmatrix}_n \begin{bmatrix} \theta \\ T \end{bmatrix}_n^R \tag{8.9-16}$$

Example 8.9-3 Find the natural frequencies of the system depicted in Fig. 8.9-6. We illustrate here an analysis using a Matlab computer code. By adjusting the data, the analysis procedure is applicable to similarly configured torsional systems.

For numerical specificity, let the disk axial inertial moments and the shaft stiffnesses be

J_2=40 kg.m^2, J_2=20 kg.m^2, J_2=15 kg.m^2

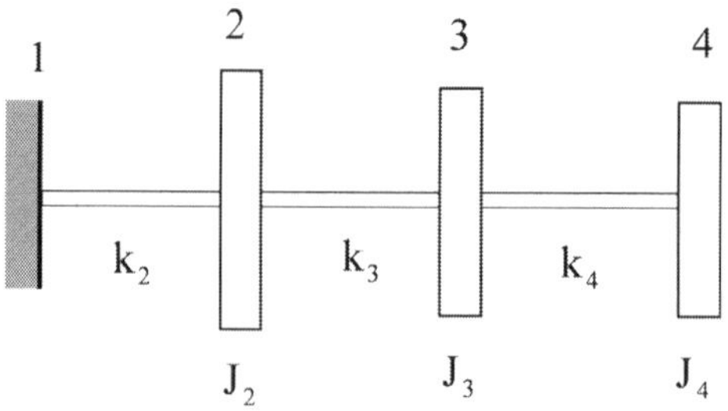

Fig. 8.9-6 A torsional system.

$k_2 = 3\times10^6$ N.m/rad,
$k_3 = k_4 = 2\times10^6$ N.m/rad.

Solution: From Eq. 8.9-16, we have

$$Z_1^R = H_2 H_3 H_4 Z_4^R, \tag{a}$$

where Z_1^R and Z_4^R are the state vectors at positions 1 and 4, respectively, and are

$$Z_1^R = \begin{bmatrix} \theta \\ T \end{bmatrix}_1^R \quad \text{and} \quad Z_4^R = \begin{bmatrix} \theta \\ T \end{bmatrix}_4^R \tag{b}$$

The transfer matrices H_n are defined as

$$H_n = \begin{bmatrix} 1-\omega^2 J_n / k_n & -1/k_n \\ \omega^2 J_n & 1 \end{bmatrix} \quad (n=2,3,4) \tag{c}$$

Since the left end of the system is fixed, the displacement θ_1 must be zero. Correspondingly, since there is no external torque applied at the right end, T_4^R is zero. We begin at disk 4, with the state vector

$$Z_4^R = \begin{bmatrix} 1 \\ 0 \end{bmatrix} \tag{d}$$

The state vector at position 1 now can be calculated from Eq. a for different values of ω. Fig. 8.9-7 provides a graph of θ_1 against ω. The natural frequencies of the system correspond to frequencies where θ_1=0, and it is seen from Fig. 8.9-7 that they are $\omega_1 = 160$rad/s, $\omega_2 = 356$rad/s, and $\omega_3 = 552$rad/s.

The Matlab code for this calculation is

```
clear all
J2=40; J3=20; J4=15;
K2=3E6;K3=2E6;K4=2E6;
V4=[1;0];
for i=1:120
w(i)=(i-1)*5;
H4=[1-w(i)^2*J4/K4,-1/K4;w(i)^2*J4,1];
H3=[1-w(i)^2*J3/K3, -1/K3;w(i)^2*J3,1];
H2=[1-w(i)^2*J2/K2,-1/K2;w(i)^2*J2,1];
V1=H2*H3*H4*V4;
theta1(i)=V1(1);
end
h=plot(w,theta1);
```

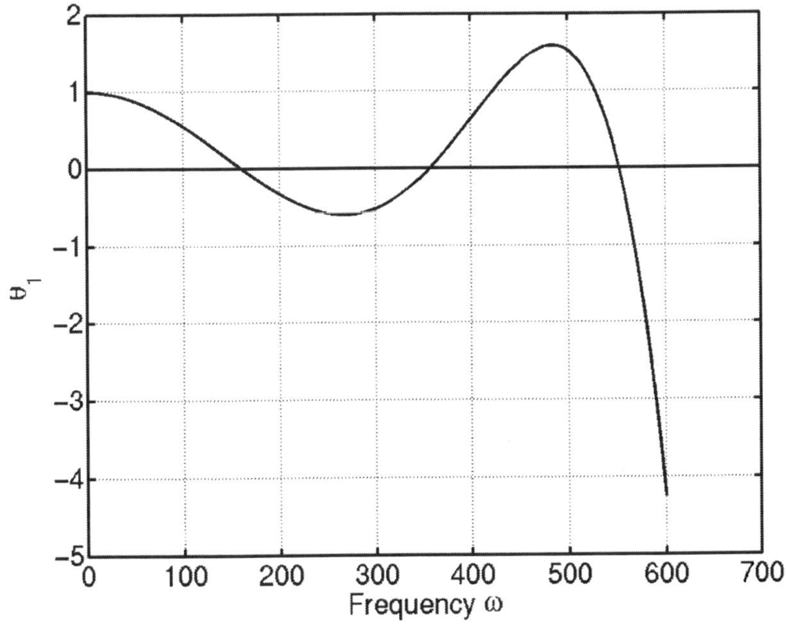

Fig. 8.9-7 Graph of θ_1 against frequency ω. (The frequencies that result in $\theta_1 = 0$ are the natural frequencies. They are $\omega_1 = 160$, $\omega_2 = 356$, and $\omega_3 = 552$.)

(3) A Geared System

Consider next the geared torsional system of Fig. 8.9-8 [8.7]. If the speed ratio of shaft 2 to shaft 1 is n, the speed ω_2 of shaft 2 is then $n\omega_1$. Assuming that the inertia effects of the gears are negligible, the kinetic energy of the system is

$$T = \frac{1}{2} J_1 \omega_1^2 + \frac{1}{2} J_2 n^2 \omega_1^2 \tag{8.9-17}$$

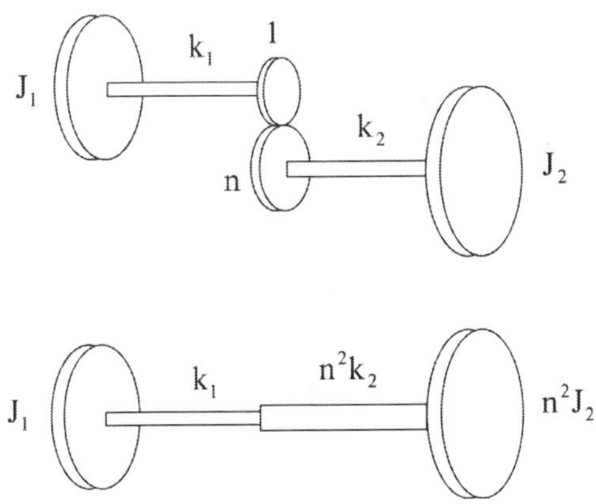

Fig. 8.9-8 Equivalent torsional system.

Thus, the equivalent inertia of disk 2 referred to shaft 1 is $n^2 J_2$.

We can determine the equivalent stiffness of shaft 2 relative to shaft 1 by clamping disks 1 and 2 and then applying a torque to gear 1, rotating it through an angle θ_1. Gear 2 will then rotate through an angle θ_2, where θ_2 is

$n\theta_1$. Shaft 2 will be twisted through the same angle. The potential energy U of the system is then

$$U = \frac{1}{2}k_1\theta_1^2 + \frac{1}{2}k_2 n^2\theta_1^2 \tag{8.9-18}$$

The equivalent stiffness of shaft 2 referred to shaft 1 is thus n^2k_2.

The above results indicate that the geared system can be reduced to an equivalent single shaft system referred to a reference shaft with a simple rule [8.7]: *multiply all stiffnesses and inertias of the geared shaft by n^2 where n is the speed ratio of the geared shaft relative to the reference shaft.*

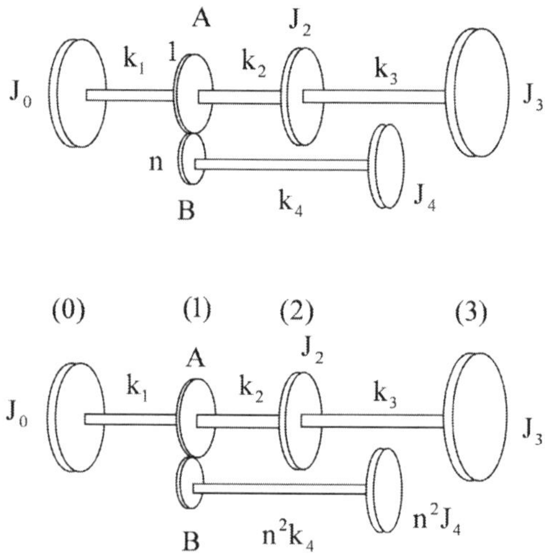

Fig. 8.9-9 Branched system (above) and equivalent reduced system (below).

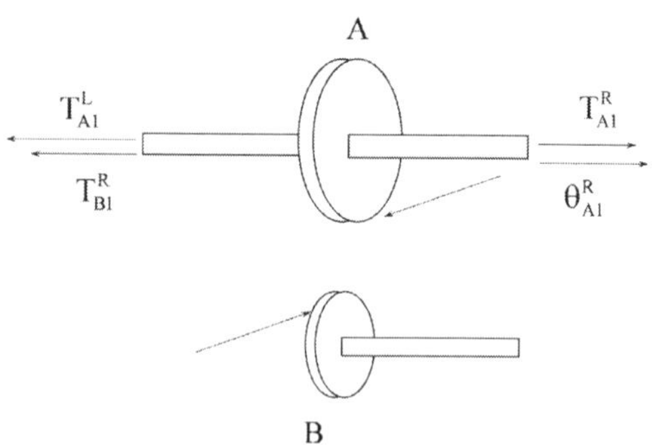

Fig. 8.9-10 Torques and displacements on the meshing gears.

(4) Branched Systems

A branched system is a vibrating system connected to a vibrating subsystem. Fig. 8.9-9 shows an example [8.7]. We can conduct an analysis by modeling the dynamic properties of the branch and then shifting them to the branch point.

To illustrate the method, consider the system of Fig. 8.9-9. Let us assume that the speed ratio of shaft B to shaft A is n. The system may then be replaced by a system having one-to-one gears at the branch point, with the stiffness and inertia of the branch being multiplied by n^2 and then assigned to gear B. The system of Fig. 8.9-9 then becomes equivalent to the system of Fig. 8.9-6 of Example 8.9-3. The only difference is the modified torque at gear A.

Specifically, we transmit the dynamic effects of the branch to that of the main system through the forces between gears B and A at the gear tooth mesh, as represented in Fig. 8.9-10. We can then proceed in an orderly analysis from position (0) to position (3).

Consider first the branch: Using Eq. 8.9-16, we have

$$\begin{bmatrix} \theta_B \\ T_B \end{bmatrix}_1^R = \begin{bmatrix} \left(1 - \dfrac{\omega^2 n^2 J_4}{n^2 k_4}\right) & -\dfrac{1}{n^2 k_4} \\ \omega^2 n^2 J_4 & 1 \end{bmatrix} \begin{bmatrix} \theta_B \\ 0 \end{bmatrix}_4^R$$

$$= \begin{bmatrix} 1 - \dfrac{\omega^2 n^2 J_4}{n^2 k_4} \\ \omega^2 n^2 J_4 \end{bmatrix} \theta_{B4}^R \qquad (8.9\text{-}19)$$

Since $\theta_{B1}^R = -\theta_{A1}^L = -\theta_{A1}^R$, we obtain

$$\theta_{B1}^R = \left(1 - \omega^2 J_4 / k_4\right) \theta_{B4}^R = -\theta_{A1}^L \qquad (8.9\text{-}20)$$

$$T_{B1}^R = \omega^2 n^2 J_4 \theta_{B4}^R = \frac{-\omega^2 n^2 J_4}{1 - \omega^2 J_4 / k_4} \theta_{A1}^L \qquad (8.9\text{-}21)$$

A balance of moments on gear A then leads to

$$T_{A1}^R = T_{A1}^L + T_{B1}^R = -\frac{-\omega^2 n^2 J_4}{1 - \omega^2 J_4 / k_4} \theta_{A1}^L + T_{A1}^L \qquad (8.9\text{-}22)$$

Thus, the state vectors of shaft A across the gears are related as

$$\begin{bmatrix} \theta_A \\ T_A \end{bmatrix}_1^R = \begin{bmatrix} 1 & 0 \\ \dfrac{-\omega^2 n^2 J_4}{1 - \omega^2 J_4 / k_4} & 1 \end{bmatrix} \begin{bmatrix} \theta_A \\ T_A \end{bmatrix}_1^L \qquad (8.9\text{-}23)$$

The square matrix is the point transfer matrix at position (1).

The field transfer matrix is the square matrix of the following equation:

$$\begin{bmatrix} \theta_A \\ T_A \end{bmatrix}_1^L = \begin{bmatrix} 1 & 1/k_1 \\ 0 & 1 \end{bmatrix} \begin{bmatrix} \theta \\ T \end{bmatrix}_0^R \qquad (8.9\text{-}24)$$

Substituting from Eq. 8.9-24 into Eq. 8.9-23, we obtain the transfer matrix H_1

$$\begin{bmatrix} \theta_A \\ T_A \end{bmatrix}_1^R = \begin{bmatrix} 1 & 1/k_1 \\ \dfrac{-\omega^2 n^2 J_4}{1 - \omega^2 J_4 / k_4} & 1 - \dfrac{\omega^2 n^2 J_4 / k_1}{1 - \omega^2 J_4 / k_4} \end{bmatrix} \begin{bmatrix} \theta \\ T \end{bmatrix}_0^R \equiv H_1 \begin{bmatrix} \theta \\ T \end{bmatrix}_0^R \qquad (8.9\text{-}25)$$

We can now proceed in the analysis from position (1) to position (3) in the usual manner.

8.10 Beams Modeled as Lumped Masses

Consider a vibrating beam. If the beam has a uniform cross section and a constant mass density, the mass will be distributed uniformly along the length. We may thus think of a beam as being a *continuous system*. (As noted earlier, we will study vibration of continuous systems in detail in Chapter Nine.)

Theoretically, a continuous system has an infinite number of degrees of freedom, and thus it also has an infinite number of natural frequencies and mode shapes. We can obtain a good approximation to the lower values of these frequencies and their corresponding mode shapes by modeling the beam as a *lumped mass* system. Specifically, we replace the uniformly distributed mass of the beam by a series of uniformly spaced concentrated (or *lumped*) masses along the span. We then consider the structure between the lumped masses as being massless, and modeled as springs representing the flexibility of the beam.

In 1944, N. O. Myklestad introduced a tabular method for obtaining the natural frequencies and vibration modes of lumped mass systems [8.10]. We will use this method to obtain approximate values for the frequencies and modes of vibrating beams. We begin with flexural vibrations and then move on to torsional vibrations.

(1) Flexural Vibration

As with the finite element method, we begin by dividing the beam into a series of segments. Fig. 8.10-1 shows a typical segment consisting of a massless span and a point mass.

We illustrate the analyses using the transfer matrix technique: We represent the flexural properties of the segment with the field matrix of segment span,

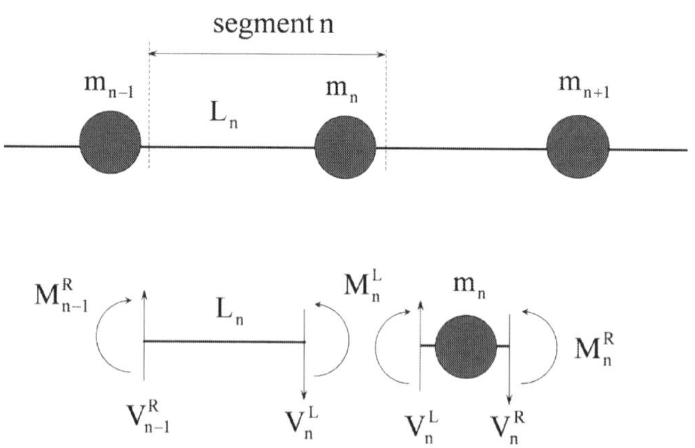

Fig. 8.10-1 Derivation of transfer matrix of a beam with lumped masses.

and the inertia properties by the point transfer matrix. In this way, the procedure is the same as that described in Section 8.9, except for the state variables of the different systems.

To develop the *field transfer matrix*, consider a typical segment n. Fig. 8.10-1 shows free-body diagrams of the segment mass and span. From these diagrams we obtain the shear and moment equations [8.7, 8.8]

$$V_n^L = V_{n-1}^R \qquad \text{and} \qquad M_n^L = M_{n-1}^R + L_n V_{n-1}^R, \tag{8.10-1}$$

where M and V are the moment and the shear force, respectively. The quantities as indicated in Fig. 8.10-1 are positive. The subscript and superscript notations are the same as in Section 8.9.

We can describe the elastic deformation of the segment by referring to Fig. 8.10-2, where we see that the change in slope θ is due to the moment and shear forces. Specifically, we have [8.7, 8.8]

$$\theta_n^L - \theta_{n-1}^R = \left(\frac{L}{EJ}\right)_n M_n^L - \left(\frac{L^2}{2EJ}\right) V_n^L \tag{8.10-2}$$

Substituting from Eq. 8.10-1 into Eq. 8.10-2 and rearranging, we obtain

$$\theta_n^L = \theta_{n-1}^R + \left(\frac{L}{EJ}\right)_n M_{n-1}^R + \left(\frac{L^2}{2EJ}\right)_n V_{n-1}^R \tag{8.10-3}$$

Similarly, the change in displacement (or *deflection*) is

$$y_n^L - y_{n-1}^R = L_n \theta_{n-1}^R + \left(\frac{L^2}{2EJ}\right)_n M_n^L - \left(\frac{L^3}{3EJ}\right)_n V_n^L \tag{8.10-4}$$

The first term on the right side is the deflection due to the initial slope of the span, the second term is due to the moment, and the third term is due to the shear force. The shear deformation of the beam is assumed to be negligible. Substituting from Eq. 8.10-1 into Eq. 8.10-4 and rearranging, we obtain

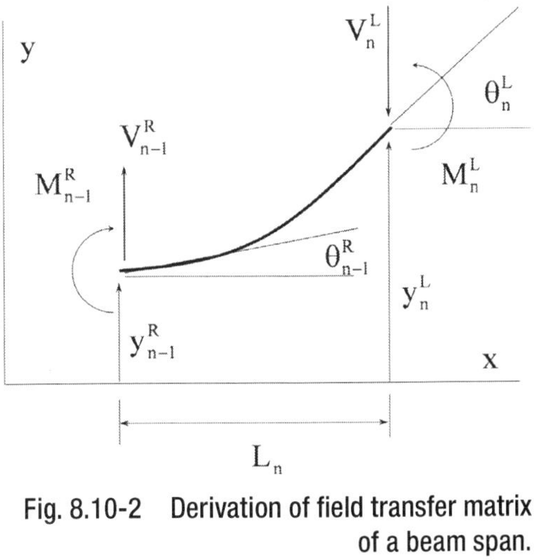

Fig. 8.10-2 Derivation of field transfer matrix of a beam span.

$$y_n^L = y_{n-1}^R + \mathrm{L}_n \theta_{n-1}^R + \left(\frac{L^2}{2EJ} \right)_n M_{n-1}^R + \left(\frac{L^3}{6EJ} \right)_n V_{n-1}^R \tag{8.10-5}$$

The field transfer matrix is obtained by writing Eqs. 8.10-1, 8.10-3, and 8.10-5 in the matrix form

$$\begin{bmatrix} y \\ \theta \\ M \\ V \end{bmatrix}_n^L = \begin{bmatrix} 1 & \mathrm{L} & \dfrac{L^2}{2EJ} & \dfrac{L^3}{6EJ} \\ 0 & 1 & \dfrac{1}{EJ} & \dfrac{L^2}{2EJ} \\ 0 & 0 & 1 & L \\ 0 & 0 & 0 & 1 \end{bmatrix}_n \begin{bmatrix} y \\ \theta \\ M \\ V \end{bmatrix}_{n-1}^R \tag{8.10-6}$$

To derive the *point transfer matrix*, consider the free-body diagram of mass m_n. Neglecting the applied force, the applied torque, and the mass moment of inertia, the equations for the shear and moment are

$$V_n^L - V_n^R = -\omega^2 m_n y_n^R \qquad \text{and} \qquad M_n^R = M_n^L \tag{8.10-7}$$

For the point mass, we have

$$\theta_n^L = \theta_n^R \qquad \text{and} \qquad y_n^L = y_n^R \tag{8.10-8}$$

The point transfer matrix is obtained from Eqs. 8.10-7 and 8.10-8 as

$$\begin{bmatrix} y \\ \theta \\ M \\ V \end{bmatrix}_n^R = \begin{bmatrix} 1 & 0 & 0 & 0 \\ 0 & 1 & 0 & 0 \\ 0 & 0 & 1 & 0 \\ \omega^2 m & 0 & 0 & 1 \end{bmatrix}_n \begin{bmatrix} y \\ \theta \\ M \\ V \end{bmatrix}_n^L \tag{8.10-9}$$

The *transfer matrix* for the segment n is obtained by substituting the state vector Z_n^L from Eq. 8.10-6 in Eq. 8.10-9, leading to

$$\begin{bmatrix} y \\ \theta \\ M \\ V \end{bmatrix}_n^R = \begin{bmatrix} 1 & 0 & 0 & 0 \\ 0 & 1 & 0 & 0 \\ 0 & 0 & 1 & 0 \\ \omega^2 m & 0 & 0 & 1 \end{bmatrix}_n \begin{bmatrix} 1 & \mathrm{L} & \dfrac{L^2}{2EJ} & \dfrac{L^3}{6EJ} \\ 0 & 1 & \dfrac{L}{EJ} & \dfrac{L^2}{2EJ} \\ 0 & 0 & 1 & L \\ 0 & 0 & 0 & 1 \end{bmatrix}_n \begin{bmatrix} y \\ \theta \\ M \\ V \end{bmatrix}_{n-1}^R$$

or

$$\begin{bmatrix} y \\ \theta \\ M \\ V \end{bmatrix}_n^R = \begin{bmatrix} 1 & L & \dfrac{L^2}{2EJ} & \dfrac{L^3}{6EJ} \\ 0 & 1 & \dfrac{L}{EJ} & \dfrac{L^2}{2EJ} \\ 0 & 0 & 1 & L \\ \omega^2 m & \omega^2 mL & \dfrac{\omega^2 mL^2}{2EJ} & 1+\dfrac{\omega^2 mL^3}{6EJ} \end{bmatrix}_n \begin{bmatrix} y \\ \theta \\ M \\ V \end{bmatrix}_{n-1}^R \tag{8.10-10}$$

Eq. 8.10-10 is a recurrence relation. It can be written simply as

$$Z_n^R = H_n Z_{n-1}^R \tag{8.10-11}$$

By repeated application, the state vector Z_n^R at a typical position n can be related to the state vector Z_0^R at position 0 by the equation

$$Z_n^R = H_n H_{n-1} \cdots H_1 Z_0^R \tag{8.10-12}$$

Generally, two boundary conditions at each end are known. Thus, the frequencies that satisfy these conditions are the natural frequencies of the beam.

For the case where the boundary condition at the left side of position 0 is known, using the point matrix in Eq. 8.10-9, the state vector Z_n^R can be related to the state vector Z_0^L, that is,

$$Z_n^R = H_n H_{n-1} \cdots H_1 S_0 Z_0^L, \tag{8.10-13}$$

where S_0 is the point matrix of position 0, that is,

$$Z_0^R = S_0 Z_0^L \tag{8.10-14}$$

In this development, the positions were numbered in increasing order from left to right, with the transfer matrix also progressing to the right. In some problems, however, it is convenient to proceed with the transfer matrix in the opposite direction. In this case, we need only invert Eq. 8.10-10. We then obtain

$$\begin{bmatrix} y \\ \theta \\ M \\ V \end{bmatrix}_{n-1}^R = \begin{bmatrix} 1+\dfrac{\omega^2 mL^3}{6EJ} & -L & \dfrac{L^2}{2EJ} & -\dfrac{L^3}{6EJ} \\ -\dfrac{\omega^2 mL^2}{2EJ} & 1 & -\dfrac{L}{EJ} & \dfrac{L^2}{2EJ} \\ \omega^2 mL & 0 & 1 & -L \\ -\omega^2 m & 0 & 0 & 1 \end{bmatrix}_n \begin{bmatrix} y \\ \theta \\ M \\ V \end{bmatrix}_n^R \tag{8.10-15}$$

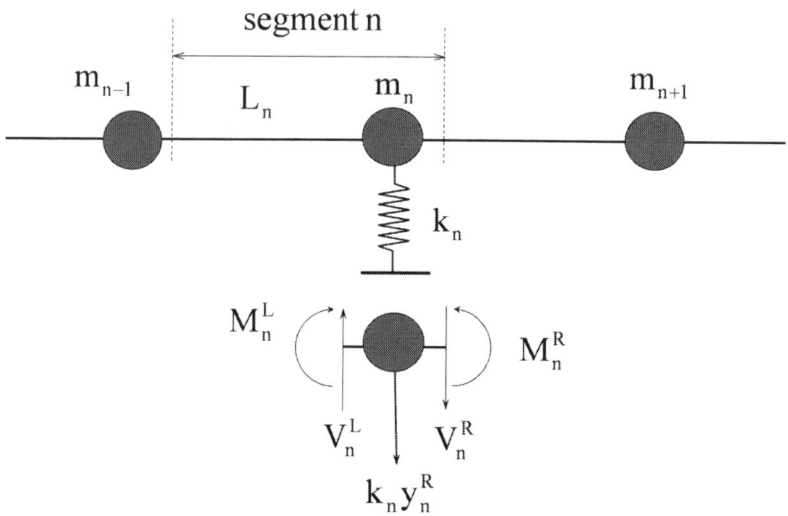

Fig. 8.10-3 Derivation of point matrix of a beam with spring support.

Example 8.10-1 Fig. 8.10-3 shows a point mass m_n supported by a spring with stiffness k_n. Develop the transfer matrix for segment n.

Solution: The field transfer matrix is the same as Eq. 8.10-6. To derive the point matrix, consider a free-body diagram of m_n. Neglecting the mass moment of inertia, the equations for the shear and moment are

$$V_n^L - V_n^R - k_n y_n^R = -\omega^2 m_n y_n^R \qquad \text{and} \qquad M_n^R = M_n^L \tag{a}$$

For the point mass, we have

$$\theta_n^L = \theta_n^R \qquad \text{and} \qquad y_n^L = y_n^R \tag{b}$$

The point transfer matrix is then obtained from Eqs. a and b as

$$\begin{bmatrix} y \\ \theta \\ M \\ V \end{bmatrix}_n^R = \begin{bmatrix} 1 & 0 & 0 & 0 \\ 0 & 1 & 0 & 0 \\ 0 & 0 & 1 & 0 \\ \omega^2 m - k & 0 & 0 & 1 \end{bmatrix}_n \begin{bmatrix} y \\ \theta \\ M \\ V \end{bmatrix}_n^L \tag{c}$$

The transfer matrix for the segment n is obtained by substituting the state vector Z_n^L from Eq. 8.10-6 in Eq. c. The result is

$$\begin{bmatrix} y \\ \theta \\ M \\ V \end{bmatrix}_n^R = \begin{bmatrix} 1 & L & \dfrac{L^2}{2EJ} & \dfrac{L^3}{6EJ} \\ 0 & 1 & \dfrac{L}{EJ} & \dfrac{L^3}{2EJ} \\ 0 & 0 & 1 & L \\ \omega^2 m-k & (\omega^2 m-k)L & \dfrac{(\omega^2 m-k)L^2}{2EJ} & 1+\dfrac{(\omega^2 m\ \ k)L^3}{6EJ} \end{bmatrix}_n \begin{bmatrix} y \\ \theta \\ M \\ V \end{bmatrix}_{n-1}^R$$

Example 8.10-2 Apply the transfer matrix method to find the natural frequencies of the lumped system shown in Fig. 8.10-4. Assume $m_1 = m_2 = 20$ kg and that the flexural stiffness EJ of the beam is 3000 N.m^2.

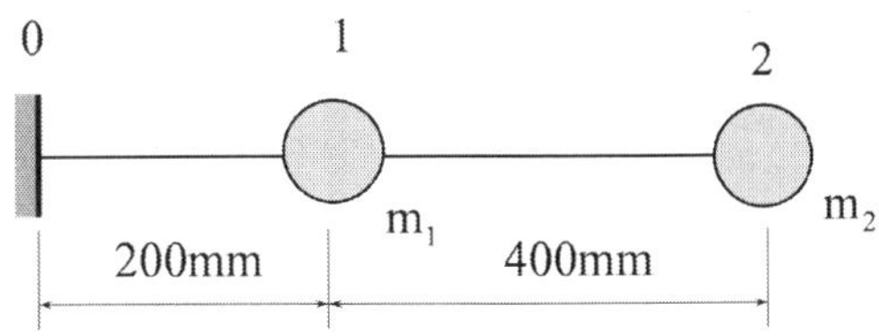

Fig. 8.10-4 Lumped-mass representation of a beam.

Solution: The recurrence formulas for the computation are

$$Z_1^R = H_1 Z_0^R \qquad \text{and} \qquad Z_2^R = H_2 Z_1^R, \tag{a}$$

where

$$Z_0^R = \begin{bmatrix} y \\ \theta \\ M \\ V \end{bmatrix}_0^R = \begin{bmatrix} 0 \\ 0 \\ M \\ V \end{bmatrix}_0^R \qquad \text{and} \qquad Z_2^R = \begin{bmatrix} y \\ \theta \\ M \\ V \end{bmatrix}_2^R = \begin{bmatrix} y \\ \theta \\ 0 \\ 0 \end{bmatrix}_2^R$$

Applying Eq. 8.10-10, we obtain

$$\begin{bmatrix} y \\ \theta \\ M \\ V \end{bmatrix}_1^R = \begin{bmatrix} 1 & 0.2 & 6.667(10^{-6}) & 0.444(10^{-6}) \\ 0 & 1 & 66.667(10^{-6}) & 6.667(10^{-6}) \\ 0 & 0 & 1 & 0.2 \\ 20\omega^2 & 4\omega^2 & 133.333(10^{-6})\omega^2 & 1+8.889(10^{-6})\omega^2 \end{bmatrix} \begin{bmatrix} y \\ \theta \\ M \\ V \end{bmatrix}_0^R$$

$$\begin{bmatrix} y \\ \theta \\ M \\ V \end{bmatrix}_2^R = \begin{bmatrix} 1 & 0.4 & 26.667(10^{-6}) & 3.556(10^{-6}) \\ 0 & 1 & 133.333(10^{-6}) & 26.667(10^{-6}) \\ 0 & 0 & 1 & 0.4 \\ 20\omega^2 & 8\omega^2 & 533.333(10^{-6})\omega^2 & 1+71.111(10^{-6})\omega^2 \end{bmatrix} \begin{bmatrix} y \\ \theta \\ M \\ V \end{bmatrix}_1^R$$

From $Z_2^R = H_2 H_1 Z_0^R = HZ_0^R$, we have

$$\begin{bmatrix} y \\ \theta \\ M \\ V \end{bmatrix}_2^R = \begin{bmatrix} H_{11} & H_{12} & H_{13} & H_{14} \\ H_{21} & H_{22} & H_{23} & H_{24} \\ H_{31} & H_{32} & H_{33} & H_{34} \\ H_{41} & H_{42} & H_{43} & H_{44} \end{bmatrix} \begin{bmatrix} 0 \\ 0 \\ M \\ V \end{bmatrix}_0^R \tag{b}$$

Since M_2^R and V_2^R must be zero at the free end of the beam, from Eq. b we have the expression

$$\begin{bmatrix} M \\ V \end{bmatrix}_2^R = \begin{bmatrix} H_{33} & H_{34} \\ H_{43} & H_{44} \end{bmatrix} \begin{bmatrix} M \\ V \end{bmatrix}_0^R = \begin{bmatrix} 0 \\ 0 \end{bmatrix} \tag{c}$$

For a nontrivial solution of the homogeneous equations, the determinant of the coefficient matrix must vanish, that is,

$$\begin{vmatrix} H_{33} & H_{34} \\ H_{43} & H_{44} \end{vmatrix} = 0 \tag{d}$$

From the given data, the corresponding equation is

$$\begin{vmatrix} 1+53.333(10^{-6})\omega^2 & 0.6+3.5556(10^{-6})\omega^2 \\ 1.333(10^{-3})\omega^2+9.481(10^{-9})\omega^4 & 1+248.88(10^{-6})\omega^2+0.6321(10^{-9})\omega^4 \end{vmatrix} = 0$$

or

$$\omega^4 - 143.1(10^3)\omega^2 + 287.6(10^6) = 0 \tag{e}$$

The natural frequencies are found as $\omega_1 = 45.2$ rad/s and $\omega_2 = 375.6$ rad/s. The analysis is routine and a computer can be used to make the calculations.

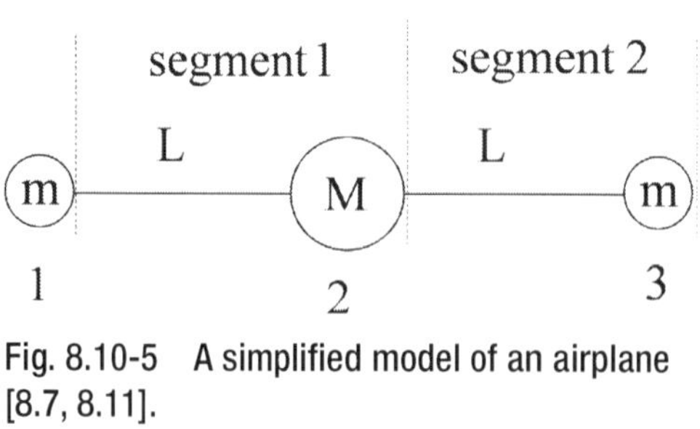

Fig. 8.10-5 A simplified model of an airplane [8.7, 8.11].

Example 8.10-3 Apply the transfer matrix method to determine the natural frequency of the simplified model of an airplane, as shown in Fig. 8.10-5 (see [8.7, 8.11]). Assume that (1) $M/m = \mu$, (2) the beam with length L is uniform, and (3) the flexural stiffness of the beam EJ is known.

Solution: Let S_1 be the point matrix of the mass at position 1. Let H_2 and H_3 be the transfer matrices of segments 1 and 2, respectively. Then we have

$$Z_1^R = S_1 Z_1^L \tag{a}$$

$$Z_2^R = H_2 Z_1^R = H_2 S_1 Z_1^L \tag{b}$$

$$Z_3^R = H_3 Z_2^R = H_3 H_2 S_1 Z_1^L, \tag{c}$$

where

$$Z_n = \begin{bmatrix} y \\ \theta \\ M \\ V \end{bmatrix}_n \quad (n=1,2,3) \tag{d}$$

$$S_1 = \begin{bmatrix} 1 & 0 & 0 & 0 \\ 0 & 1 & 0 & 0 \\ 0 & 0 & 1 & 0 \\ \omega^2 m & 0 & 0 & 1 \end{bmatrix} \tag{e}$$

$$H_2 = \begin{bmatrix} 1 & L & \dfrac{L^2}{2EJ} & \dfrac{L^3}{6EJ} \\ 0 & 1 & \dfrac{L}{EJ} & \dfrac{L^2}{2EJ} \\ 0 & 0 & 1 & L \\ \omega^2 M & \omega^2 ML & \dfrac{\omega^2 ML^2}{2EJ} & 1+\dfrac{\omega^2 ML^3}{6EJ} \end{bmatrix} \tag{f}$$

and

$$H_3 = \begin{bmatrix} 1 & L & \dfrac{L^2}{2EJ} & \dfrac{L^3}{6EJ} \\ 0 & 1 & \dfrac{L}{EJ} & \dfrac{L^2}{2EJ} \\ 0 & 0 & 1 & L \\ \omega^2 m & \omega^2 mL & \dfrac{\omega^2 mL^2}{2EJ} & 1+\dfrac{\omega^2 mL^3}{6EJ} \end{bmatrix} \tag{g}$$

From these expressions, we obtain

$$H_2S_1 = \begin{bmatrix} 1+\dfrac{\omega^2 mL^3}{6EJ} & L & \dfrac{L^2}{2EJ} & \dfrac{L^3}{6EJ} \\ \dfrac{\omega^2 mL^2}{2EJ} & 1 & \dfrac{L}{EJ} & \dfrac{L^2}{2EJ} \\ \omega^2 mL & 0 & 1 & L \\ \omega^2(m+M)+\dfrac{\omega^4 mML^3}{6EJ} & \omega^2 ML & \dfrac{\omega^2 ML^2}{2EJ} & 1+\dfrac{\omega^2 ML^3}{6EJ} \end{bmatrix} \tag{h}$$

From $Z_3^R = H_3H_2S_1Z_1^L = HZ_1^L$, we have

$$\begin{bmatrix} y \\ \theta \\ M \\ V \end{bmatrix}_2^R = \begin{bmatrix} H_{11} & H_{12} & H_{13} & H_{14} \\ H_{21} & H_{22} & H_{23} & H_{24} \\ H_{31} & H_{32} & H_{33} & H_{34} \\ H_{41} & H_{42} & H_{43} & H_{44} \end{bmatrix} \begin{bmatrix} y \\ \theta \\ 0 \\ 0 \end{bmatrix}_1^L \tag{i}$$

Since M_3^R and V_3^R must be zero at the right end of the beam, we have the expression

$$\begin{bmatrix} M \\ V \end{bmatrix}_2^R = \begin{bmatrix} H_{33} & H_{34} \\ H_{43} & H_{44} \end{bmatrix} \begin{bmatrix} M \\ V \end{bmatrix}_0^R = \begin{bmatrix} 0 \\ 0 \end{bmatrix} \tag{j}$$

For a nontrivial solution of the homogeneous equations, the determinant of the coefficient matrix must vanish, that is,

$$\begin{vmatrix} H_{33} & H_{34} \\ H_{43} & H_{44} \end{vmatrix} = 0 \tag{k}$$

From the given expressions, the corresponding equation is

$$\begin{vmatrix} \omega^2(2m+M)L+\dfrac{\omega^4 mML^4}{6EJ} & \omega^2 ML^2 \\ \omega^2(2m+M)+\omega^4\dfrac{m(4m+M)L^3}{3EJ}+\dfrac{\omega^6 m^2 ML^6}{36(EJ)^2} & \omega^2(2m+M)L+\dfrac{\omega^4 mML^4}{6EJ} \end{vmatrix} = 0$$

or

$$\omega^2\left[\omega^2 - \frac{6EJ}{L^3}\frac{(1+\mu/2)}{M}\right] = 0 \tag{l}$$

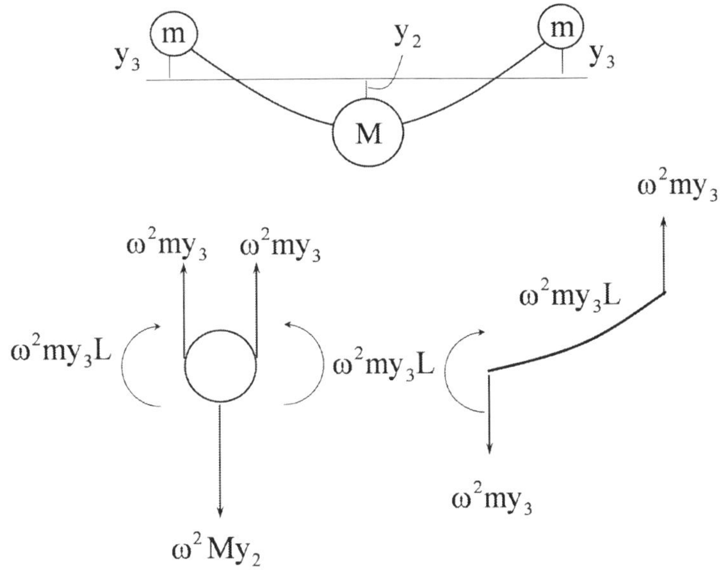

Fig. 8.10-6 Free-body diagrams for Example 8.10-3.

The natural frequencies are

$$\omega_1 = 0, \ \omega_2 = \sqrt{\frac{6EJ}{ML^3}\left(1+\frac{\mu}{2}\right)} \tag{m}$$

Note that the solution can be obtained rapidly from the free-body diagram shown in Fig. 8.10-6. Specifically, the equation of force balance on mass M is

$$\omega^2 My_2 = 2\omega^2 my_3 \qquad \text{or} \qquad y_3 = \frac{\mu}{2} y_2 \tag{n}$$

The deflection equation of the beam may be written as

$$y_2 + y_3 = \frac{\omega^2 my_3 L^3}{3EJ} \tag{o}$$

Substituting from Eq. n into Eq. o, we obtain the same result as in Eq. m.

(2) Rotating Beams

In this section, we examine rotating beams with vibration perpendicular to the plane of rotation [8.7, 8.8, 8.12]. Examples include propellers and turbine blades. To treat such problems, it is necessary to review the concept of centrifugal force.

Fig. 8.10-7 shows a model of a turbine blade, rotating at a constant angular velocity Ω about the Y-axis [8.12]. Consider the transfer matrix of a typical

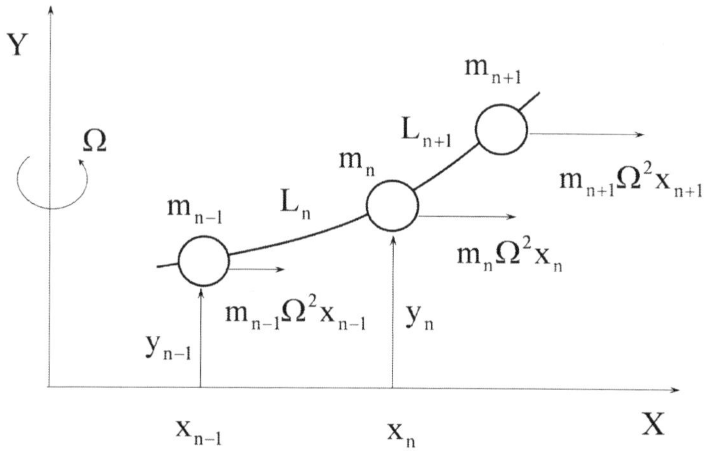

Fig. 8.10-7 A rotating beam.

segment of the blade. The analysis is similar to that of the beam analysis, except that the equations are modified by the centrifugal force. The equations to follow are arranged for the calculations to proceed from right to left, similar to Eq. 8.10-15.

The centrifugal force due to point mass m_n is $\Omega^2 m_n x_n$, normal to the axis of rotation, as shown in Fig. 8.10-7. The force equation at m_n in the X-direction is then

$$F_n^L = F_n^R + m_n \Omega^2 x_n, \tag{8.10-16}$$

where

$$F_n^L = \sum_{i=n}^{N} m_i \Omega^2 x_i \tag{8.10-17}$$

The moment balance equation becomes

$$M_{n-1}^R = M_n^L - V_n^L L_n - F_n (y_n^L - y_{n-1}^R) \tag{8.10-18}$$

The component of the force F_n^L normal to the segment is $F_n^L \theta_n^L$, where θ_n^L is the slope at station n. (Note, θ is small.) The deflection and slope are also affected by this force as a shear load

$$y_n^L = y_{n-1}^R + L_n \theta_{n-1}^R - F_n^L \theta_n^L \left(\frac{L^3}{3EJ}\right)_n + M_n^L \left(\frac{L^3}{2EJ}\right)_n - V_n^L \left(\frac{L^3}{3EJ}\right)_n \tag{8.10-19}$$

and

$$\theta_n^L = \theta_{n-1}^R + M_n\left(\frac{L}{EJ}\right)_n - V_n\left(\frac{L^2}{2EJ}\right)_n - F_n\theta_n^L\left(\frac{L^2}{2EJ}\right)_n \tag{8.10-20}$$

These equations can now be rearranged for evaluation to proceed from right to left as follows:

$$y_{n-1}^L = y_n^L - \theta_n^L\left[L + \frac{F^L L^3}{6EJ}\right]_n + M_n^L\left(\frac{L^2}{2EJ}\right)_n - V_n^L\left(\frac{L^3}{6EJ}\right)_n$$

$$\theta_{n-1}^R = \theta_n^L\left[1 + \frac{F^L}{2EJ}\right] - M_n^L\left(\frac{L}{EJ}\right)_n + V_n^L\left(\frac{L^2}{2EJ}\right)_n$$

$$M_{n-1}^R = M_n^L\left[1 + \frac{F^L L^2}{2EJ}\right]_n - V_n^L\left[\mathrm{L} + \frac{F^L L^3}{6EJ}\right]_n - \theta_n F_n^L\left[1 + \frac{F^L L^3}{6EJ}\right]_n$$

and

$$V_{n-1}^R = V_n^L$$

In the *field matrix*, these equations appears as

$$\begin{bmatrix} y \\ \theta \\ M \\ V \end{bmatrix}_{n-1}^R = \begin{bmatrix} 1 & \left(L + \dfrac{F^L L^3}{6EJ}\right) & \dfrac{L^2}{2EJ} & -\dfrac{L^3}{6EJ} \\ 0 & \left(1 + \dfrac{F^L L^2}{2EJ}\right) & -\dfrac{L}{EJ} & \dfrac{L^2}{2EJ} \\ 0 & -F^L\left(L + \dfrac{F^L L^3}{6EJ}\right) & 1 + \dfrac{F^L L^2}{2EJ} & -\left(L + \dfrac{F^L L^3}{6EJ}\right) \\ 0 & 0 & 0 & 1 \end{bmatrix}_n \begin{bmatrix} y \\ \theta \\ M \\ V \end{bmatrix}_n^L \tag{8.10-21}$$

The *point matrix* for m_n can be obtained from Eqs. 8.10-7 and 8.10-8 as

$$\begin{bmatrix} y \\ \theta \\ M \\ V \end{bmatrix}_n^L = \begin{bmatrix} 1 & 0 & 0 & 0 \\ 0 & 1 & 0 & 0 \\ 0 & 0 & 1 & 0 \\ -\omega^2 m & 0 & 0 & 1 \end{bmatrix}_n \begin{bmatrix} y \\ \theta \\ M \\ V \end{bmatrix}_n^R \tag{8.10-22}$$

Substituting from Eq. 8.10-22 into Eq. 8.10-21, we obtain the *transfer matrix* of the segment as

$$\begin{bmatrix} y \\ \theta \\ M \\ V \end{bmatrix}_{n-1}^{R} = \begin{bmatrix} \left(1+\dfrac{m\omega^2 L^3}{6EJ}\right) & -\left(L+\dfrac{F^L L^3}{6EJ}\right) & \dfrac{L^2}{2EJ} & -\dfrac{L^3}{6EJ} \\ -\dfrac{m\omega^2 L^2}{2EJ} & \left(1+\dfrac{F^L L^2}{2EJ}\right) & -\dfrac{L}{EJ} & \dfrac{L^2}{2EJ} \\ m\omega^2\left(L+\dfrac{F^L L^3}{6EJ}\right) & -F^L\left(L+\dfrac{F^L L^3}{6EJ}\right) & 1+\dfrac{F^L L^2}{2EJ} & -\left(L+\dfrac{F^L L^3}{6EJ}\right) \\ -m\omega^2 & 0 & 0 & 1 \end{bmatrix}_n \begin{bmatrix} y \\ \theta \\ M \\ V \end{bmatrix}_n^{R} \tag{8.10-23}$$

Observe that when $\Omega = 0$, then $F_n^L = 0$; in this case, Eq. 8.10-23 becomes Eq. 8.10-15.

Problems

8.1 Neglecting the mass of the shafts or beams, determine the influence coefficients of the systems shown in each part of Fig. P8.1 [8.8].

Ans.

$$[\delta]_{(A)} = \frac{L^3}{6EJ}\begin{bmatrix} 2 & 5 & 8 \\ 5 & 16 & 28 \\ 8 & 28 & 54 \end{bmatrix}; \quad [\delta]_{(B)} = \frac{L^3}{18EJ}\begin{bmatrix} 8 & 7 \\ 7 & 8 \end{bmatrix};$$

$$[\delta]_{(C)} = \frac{L^3}{768EJ}\begin{bmatrix} 16 & 11 \\ 11 & 9 \end{bmatrix}; \quad [\delta]_{(D)} = \frac{L^3}{768EJ}\begin{bmatrix} 9 & 11 & 7 \\ 11 & 16 & 11 \\ 7 & 11 & 9 \end{bmatrix}$$

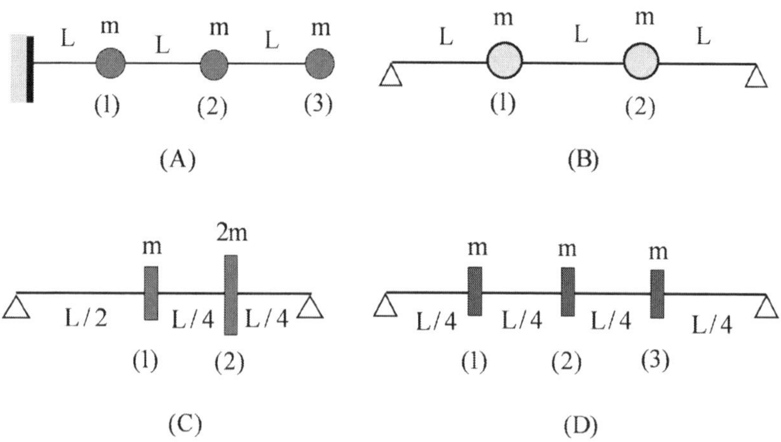

Fig. P8.1

8.2 Determine the fundamental frequencies of the system shown in each part of Fig. P8.1 by using (1) Dunkerley's equation, (2) Rayleigh's method, and (3) the characteristic equation, and compare the results estimated by the different methods.

Ans.

$$\text{(A) (1) } 0.289\sqrt{\frac{EJ}{mL^3}};\ \text{(2) } 0.294\sqrt{\frac{EJ}{mL^3}};\ \text{(3) } 0.293\sqrt{\frac{EJ}{mL^3}}$$

$$\text{(B) (1) } 1.061\sqrt{\frac{EJ}{mL^3}};\ \text{(2) } 1.095\sqrt{\frac{EJ}{mL^3}};\ \text{(3) } 1.095\sqrt{\frac{EJ}{mL^3}}$$

$$\text{(C) (1) } 4.753\sqrt{\frac{EJ}{L^3}};\ \text{(2) } 4.855\sqrt{\frac{EJ}{L^3}};\ \text{(3) } 4.856\sqrt{\frac{EJ}{L^3}}$$

$$\text{(D) (1) } 4.753\sqrt{\frac{EJ}{mL^3}};\ \text{(2) } 4.933\sqrt{\frac{EJ}{mL^3}};\ \text{(3) } 4.934\sqrt{\frac{EJ}{mL^3}}$$

8.3 Determine the natural frequencies and mode shapes of the system shown in Fig. P8.3 by using the method of matrix iteration.

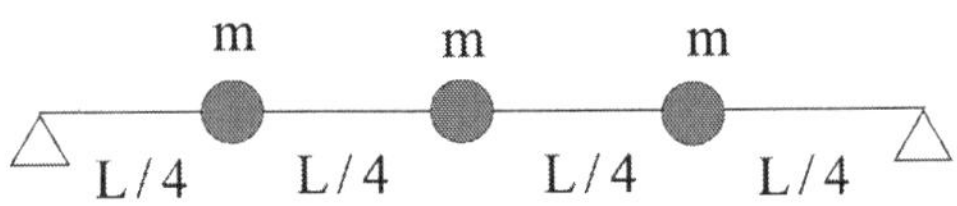

Fig. P8.3

Ans.

$$\omega_1 = 4.93\sqrt{\frac{EJ}{mL^3}},\quad \omega_2 = 19.60\sqrt{\frac{EJ}{mL^3}},\quad \omega_3 = 41.61\sqrt{\frac{EJ}{mL^3}}$$

$$\mathbf{u}_1 = \begin{bmatrix} 1 \\ 1.4142 \\ 1 \end{bmatrix},\quad \mathbf{u}_2 = \begin{bmatrix} 1 \\ 0 \\ -1 \end{bmatrix},\quad \mathbf{u}_3 = \begin{bmatrix} 1 \\ -1.4152 \\ 1 \end{bmatrix}$$

8.4 Referring to Example 8.9-3, write a computer program to find the natural frequencies of the torsional system shown in Fig. P8.4. Assume that J_2=50 kg.m^2, J_3=30 kg.m^2, J_4=25 kg.m^2, k_2=3×10^6 N.m/rad, k_3=k_4=2×10^6 N.m/rad.

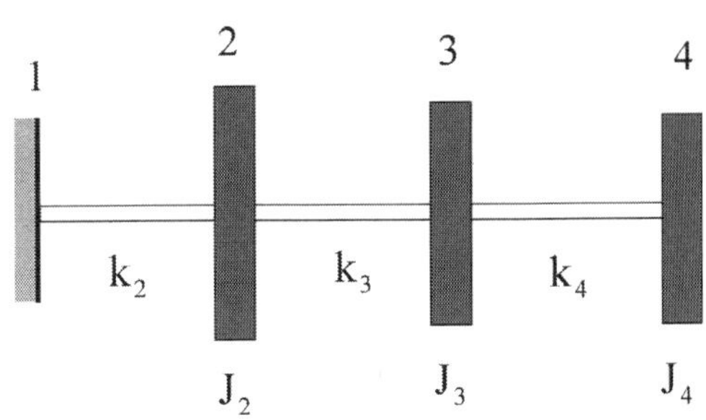

Fig. P8.4

Ans.

$\omega_1 = 130$ rad/s, $\omega_2 = 306$ rad/s, $\omega_3 = 450$ rad/s

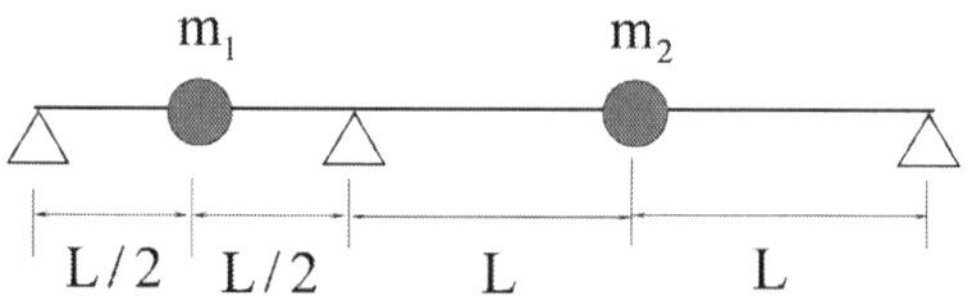

Fig. P8.5

8.5 Fig. P8.5 shows a uniform beam on three supports carrying two lumped masses. Determine the influence coefficients and the characteristic equation.

Ans.

$$\delta_{12}=\delta_{21}=-\frac{36L^3}{2304EJ},\ \delta_{22}=\frac{240L^3}{2304EJ}$$

$$\lambda^2-(39m_1+240m_2)z\lambda+8064m_1m_2z=0,\text{ where }\lambda=\frac{1}{\omega^2},\ z=\frac{L^3}{2304EJ}$$

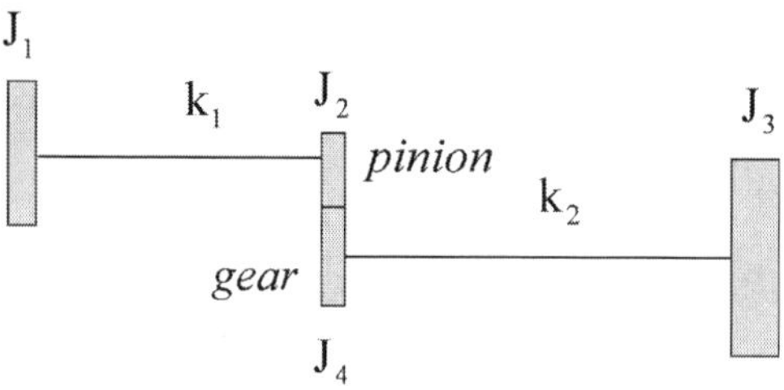

Fig. P8.6

8.6 [8.8] Two shaft systems in Fig. P8.6 are connected by a pair of gears. Assume that the speed ratio of the gear to the pinion is *n*. (1) Neglecting the inertial effect of the gears, determine the frequencies of the system. (2) Repeat part (1) but include the inertial effects of the gears.

Ans.

$$(1)\ \omega=\sqrt{\frac{k_1k_2(J_1+n^2J_3)}{(k_1+n^2k_2)J_1J_3}}$$

$$(2)\ \omega^4-\left[k_1\left(\frac{1}{J_1}+\frac{1}{J_g}\right)+n^2k_2\left(\frac{1}{J_g}+\frac{1}{n^2J_3}\right)\right]\omega^2+k_1k_2\frac{J_1+J_g+n^2J_3}{J_1J_gJ_3}=0$$

$$J_g=J_2+n^2J_4$$

8.7 Derive the transfer matrix for a typical segment of a spring-mass system when damping is included.

Ans. $$\begin{bmatrix}x\\F\end{bmatrix}_n^R=\begin{bmatrix}1 & \dfrac{1}{k+j\omega c}\\ -\omega^2m & \left(1-\dfrac{\omega^2m}{k+j\omega c}\right)\end{bmatrix}_n\begin{bmatrix}x\\F\end{bmatrix}_{n-1}^R$$

8.8 Determine the influence coefficients in vertical vibration of the system as shown in Fig. P8.8, where the bending stiffness EJ and torsional stiffness GJ_p are given.

Ans. $\delta_{11} = L^3/3EJ$, $\delta_{12} = \delta_{21} = L^3/3EJ$,

$\delta_{22} = 2L^3/3EJ + L^3/GJ_p$

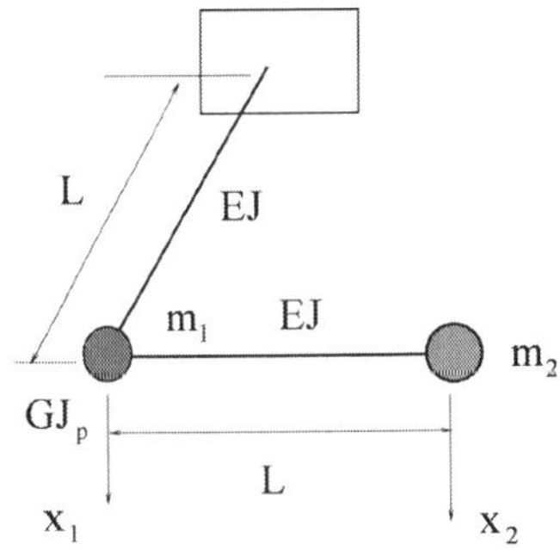

Fig. P8.8

8.9 Determine the influence coefficients and natural frequencies of vibration in the xy-plane of the mass m shown in Fig. P8.9.

Ans. $\omega_{1,2}^2 = \frac{6}{7}\left(5 \pm 3\sqrt{2}\right)\frac{EJ}{mL^3}$

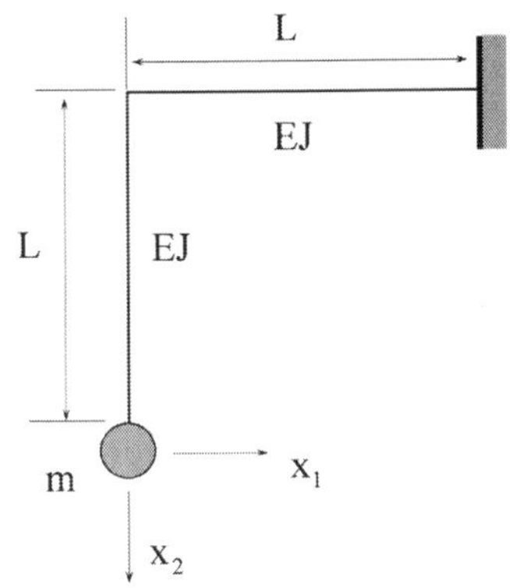

Fig. P8.9

8.10 Determine the influence coefficients and natural frequencies of the torsional system shown in Fig. P8.10.

Ans.

$$[\delta] = \frac{1}{k_r}\begin{bmatrix} 1 & 1 & 1 \\ 1 & 2 & 2 \\ 1 & 2 & 3 \end{bmatrix}$$

$$\lambda^3 - 6(I/k_r)\lambda^2 + 5(I/k_r)^2\lambda - (I/k_r)^3 = 0$$

$$\lambda_1 = (1/\omega_1)^2 = 5.05I/k_r$$
$$\lambda_2 = (1/\omega_2)^2 = 0.643I/k_r$$
$$\lambda_3 = (1/\omega_3)^2 = 0.308I/k_r$$

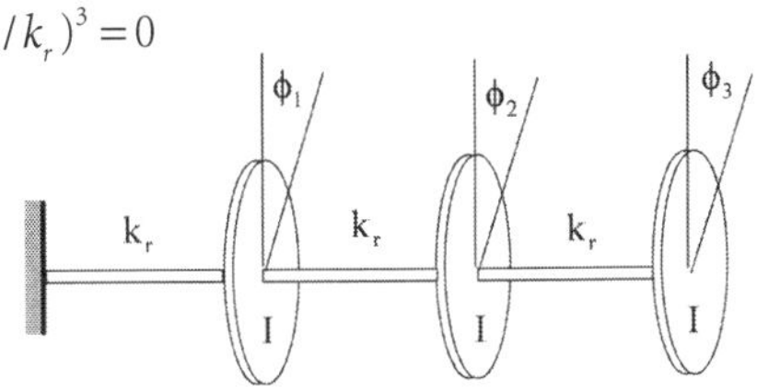

Fig. P8.10

8.11 Determine the influence coefficients and natural frequencies of vibration in the xy-plane of the system shown in Fig. P8.11.

Ans.

$$[\delta] = \frac{L^3}{48EJ}\begin{bmatrix} 64 & 24 & 6 \\ 24 & 16 & 5 \\ 6 & 5 & 2 \end{bmatrix}$$

$$\omega_3 = 0.802\sqrt{\frac{EJ}{mL^3}} \quad \omega_3 = 2.621\sqrt{\frac{EJ}{mL^3}} \quad \omega_3 = 12.504\sqrt{\frac{EJ}{mL^3}}$$

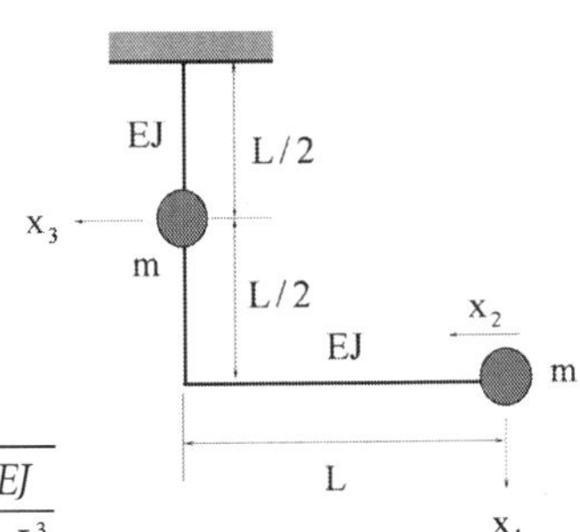

Fig. P8.11

8.12 Find the fundamental frequency of a cantilever beam under the action of its weight w per unit length.

Ans. $\omega = 3.53\sqrt{\frac{gEJ}{wL^4}}$

References

8.1. Pilkey, W. D., and Wonderlich, W., *Mechanics of Structures, Variational and Computational Methods*, CRC Press, Boca Raton, FL, 1994.

8.2. Hearn, E. J., *Mechanics of Materials*, 2nd ed., Pergamon Press, New York, 1985.

8.3. Young , W. C., and Budymas, R. G., *Roark's Formulas for Stress and Strain*, 7th ed., McGraw-Hill, New York, 2002.

8.4. Dunkerley, S., "On the Whirling and Vibration of Shafts," *Philosophical Transactions of the Royal Society of London*, Series A, Vol. 185, 1894, pp. 279–360.

8.5. Duncan, W. J., and Collar, A. R., "A Method for the Solution of Oscillation Problems by Matrices," *Philosophical Magazine*, Series 7, Vol. 17, 1934, p. 865.

8.6. Greig, R. R., Jr., *Structural Dynamics: An Introduction to Computer Methods*, Wiley, New York, 1981.

8.7. Thomson, W. T., and Dahleh, M. D., *Theory of Vibration with Applications*, 5th ed., Prentice Hall, Inc., Upper Saddle River, NJ, 1998.

8.8. Tse, F. S., Morse, I. E., and Hinkle, R. T., *Mechanical Vibration, Theory and Applications*, Allyn and Bacon, Inc., Boston, 1979.

8.9. Holzer, H., *Die Berechnung der Drehschwingungen und ihre Anwendung im Maschinenbau*, Julius Springer, Berlin, 1921 (in German).

8.10. Myklestad, N. O., "A New Method of Calculating Natural Modes of Uncoupled Bending Vibration of Airplane Wings and Other Types of Beams," *Journal of Aeronautical Science*, Vol. 11, No. 2, 1944, pp. 153–162.

8.11. Inman, D. J., *Engineering Vibration*, Prentice Hall, Inc., Upper Saddle River, NJ, 1994.

8.12. Myklestad, N. O., *Fundamentals of Vibration Analysis*, McGraw-Hill Book Co., 1956.

Chapter Nine

Continuous Systems

9.1 Introduction

Imagine an arbitrarily shaped solid, elastic body B, where *solid* simply means that the mass of B is distributed continuously throughout the region occupied by B. *Elastic* means that if B is subjected to a loading, consisting of forces and/or moments, B will deform so that its shape is changed and/or the region occupied by B will change. Then, if the loading is removed, B will return to its original unloaded shape or region.

For brevity we will call such a body a *continuous system.*

Observe that even though the mass of the body, or system, is continuously distributed, the distribution is not necessarily uniform. Similarly, the elastic properties (e.g., the stiffness) is also not necessarily uniform throughout the body. In our consideration, however, we will restrict our attention to bodies with *uniformly* distributed mass and elastic properties in all directions. That is, we will consider continuous systems whose material is homogeneous and isotropic.

We may also envision a continuous system as being an infinite set of closely packed particles, with each particle having the same infinitesimal mass and with each particle connected to its neighboring particles by linear springs and possibly dampers.

As such, a continuous system has an *infinite* number of degrees of freedom, and, consequently, an infinite number of natural frequencies and mode shapes.

As a contrast, a discrete system is a set of distinct and separated concentrated masses, or relatively small bodies, connected to one another by springs and possibly dampers; and, correspondingly, with a *finite* number of natural frequencies and mode shapes.

Perhaps the simplest example of a continuous system is an elastic string. Elastic beams and twisting shafts are other simple examples.

The vibration of a continuous system is described in terms of space coordinates and time. Consequently, the governing equations are partial differential equations. The solution of these equations will thus require knowledge of both initial conditions and boundary (or support) conditions.

In this chapter, we will illustrate these concepts through a series of relatively simple systems, beginning with a vibrating string.

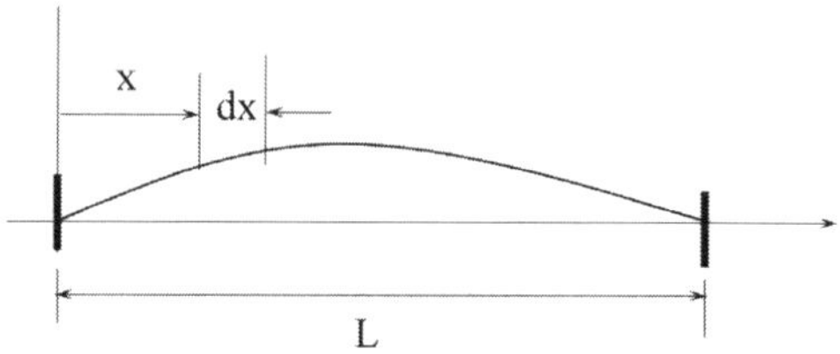

(A) Lateral vibration

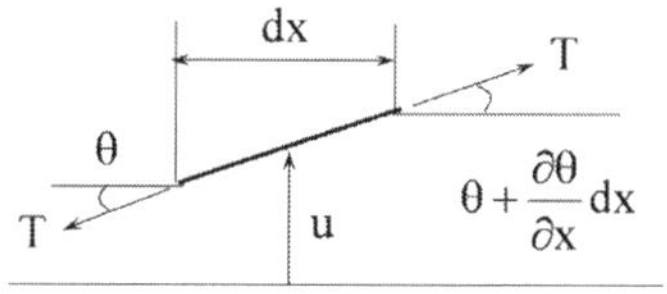

(B) Element

Fig. 9.2-1 A model of lateral vibration of a flexible string.

9.2 Lateral Vibration of a String

Fig. 9.2-1 (A) depicts the lateral displacement of an elastic string. Let u be the lateral displacement and let the string have a uniform mass ρ per unit length. Let the string be in tension T. Since the displacement varies depending upon the distance x along the string, and since the displacement also depends upon the time t, the displacement is a function of both x and t. That is,

$$u = u(x,t) \tag{9.2-1}$$

Consider an element of the string as in Fig. 9.2-1 (B). By assuming the lateral displacement u to be small, the change in tension due to the displacement is negligible. Then, by applying Newton's law, we readily obtain the equation of motion as

$$\rho dx \frac{\partial^2 u}{\partial t^2} = T\left(\theta + \frac{\partial \theta}{\partial x} dx\right) - T\theta, \tag{9.2-2}$$

where θ is the slope of the string, given by $\partial u / \partial x$. Eq. 9.2-2 reduces to

$$\frac{\partial^2 u}{\partial t^2} = c^2 \frac{\partial^2 u}{\partial x^2}, \tag{9.2-3}$$

where c is $\sqrt{T/\rho}$.

(1) d'Alembert's Solution

Eq. 9.2-3 is known as the *one-dimensional wave equation*, and c is seen to be the speed of wave propagation.

To briefly explore this, we know (see [9.1]) that the general solution of Eq. 9.2-3 has the form

$$u = f(x+ct) + g(x-ct) \tag{9.2-4}$$

Suppose next that initially (t=0) the string displacement $u(x,0)$ and the displacement rate $\partial u(x,0)/\partial t$ are

$$u(x,0) = \phi(x) \quad \text{and} \quad \frac{\partial u(x,0)}{\partial t} = \psi(x) \tag{9.2-5}$$

Then, by substituting into Eq. 9.2-4, we have

$$\phi(x)=f(x)+g(x) \quad \text{and} \quad \psi(x)=cf'(x)-cg'(x), \tag{9.2-6}$$

where, as is customary, the prime designates a total derivative. That is, $f'(\eta)\equiv df/d\eta$.

By differentiating in the first expression of Eq. 9.2-6 and then appending the result to the second expression, we obtain the pair of equations

$$f'(x)+g'(x)=\phi'(x) \tag{9.2-7}$$

and

$$cf'(x)-cg'(x)=\psi(x) \tag{9.2-8}$$

By solving for $f'(x)$, we have

$$f'(x)=\frac{1}{2c}\psi(x)+\frac{1}{2}\phi'(x) \tag{9.2-9}$$

By integrating, we obtain

$$f(x)=\frac{1}{2c}\int_0^x \psi(\eta)d\eta+\frac{1}{2}\phi(x) \tag{9.2-10}$$

Similarly, by solving for $g(x)$, we obtain

$$g(x)=-\frac{1}{2c}\int_0^x \psi(\eta)d\eta+\frac{1}{2}\phi(x) \tag{9.2-11}$$

Finally, by substituting from the results of Eqs. 9.2-10 and 9.2-11 into Eq. 9.2-4, we find the string deflection to be

$$u(x,t)=\frac{1}{2}[\phi(x+ct)+\phi(x-ct)]+\frac{1}{2c}\int_{x-ct}^{x+ct} \psi(\eta)d\eta \tag{9.2-12}$$

Eq. 9.2-12 is commonly known as *d'Alembert's solution to the wave equation.*

To illustrate this solution, suppose an infinitely long string is given an initial finite length displacement $\phi(x)$ and then released from rest $[\psi(x)=0]$. The string displacement will then move to the left and right along the string at speed c with amplitude $(1/2)\phi(x)$, as illustrated in Fig. 9.2-2.

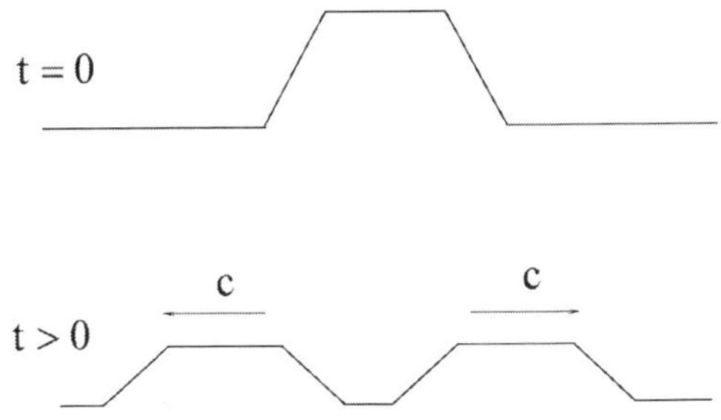

Fig. 9.2-2 Wave propagation along an infinite string.

These results confirm that the speed of wave movement c $\left(\text{defined as} \sqrt{T/\rho}\right)$ increases/decreases with the square root of the string tension/mass density.

(2) Separation of Variables

A popular method for solving partial differential equations is to separate the independent variables and thereby convert the partial differential equations into ordinary differential equations.

Consider again Eq. 9.2-3:

$$\frac{\partial^2 u}{\partial t^2} = c^2 \frac{\partial^2 u}{\partial x^2} \tag{9.2-13}$$

In the separation of variables method, we assume a solution in the form

$$u(x,t) = \phi(x) q(t), \tag{9.2-14}$$

where $\phi(x)$ is a function of x alone, and $q(t)$ is a function of t alone. Substituting from Eq. 9.2-14 into Eq. 9.2-13 and simplifying, we obtain

$$c^2 \frac{1}{\phi} \frac{d^2\phi}{dx^2} = \frac{1}{q} \frac{d^2 q}{dt^2} \tag{9.2-15}$$

Since the left side of Eq. 9.2-15 is independent of t and the right side is independent of x, and since the equality holds for all values of x and t, it follows that each side must be a constant. Let the constant be $-\omega^2$. We then obtain two ordinary differential equations:

$$\frac{d^2\phi}{dx^2} + \left(\frac{\omega}{c}\right)^2 \phi = 0 \tag{9.2-16}$$

$$\frac{d^2 q}{dt^2} + \omega^2 q = 0 \tag{9.2-17}$$

The general solutions of Eqs. 9.2-16 and 9.2-17 are

$$\phi = A \sin\frac{\omega x}{c} + B \cos\frac{\omega x}{c} \tag{9.2-18}$$

$$q = C \sin\omega t + D \cos\omega t \tag{9.2-19}$$

Thus, from Eq. 9.2-14, the formal solution of Eq. 9.2-13 is

$$u(x,t) = \left(A \sin\frac{\omega x}{c} + B \cos\frac{\omega x}{c}\right)\left(C \sin\omega t + D \cos\omega t\right) \tag{9.2-20}$$

The coefficients A, B, C, and D of Eq. 9.2-20 are to be determined by boundary conditions and initial conditions. The boundary condition $u(0,t)=0$ requires that $B=0$. The boundary condition $u(L,t)=0$ gives

$$\sin\frac{\omega L}{c} = 0 \quad \text{or} \quad \frac{\omega L}{c} = n\pi \quad (n=1,2,\cdots) \tag{9.2-21}$$

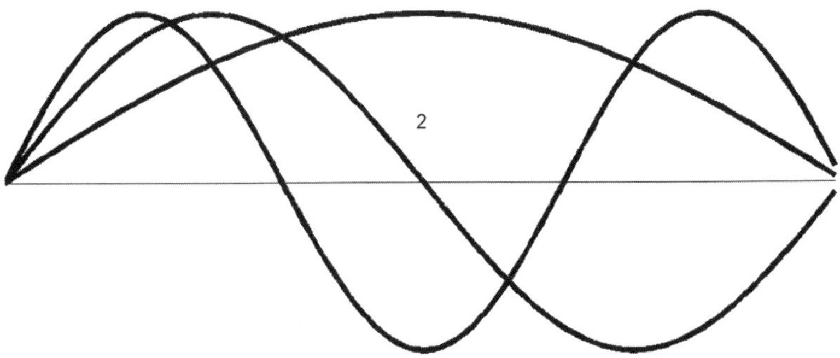

Fig. 9.2-3 The first three modes of vibration of a string fixed at its ends.

Eq. 9.2-21 is sometimes called the *frequency equation* or the *characteristic equation*. The various values of the integer n provide various values of frequencies ω, designated now as ω_n. For example, the frequency for the n-th mode is

$$\omega_n = n\pi c / L \tag{9.2-22}$$

Consequently, from Eq. 9.2-18, the n-th mode shape function $\phi_n(x)$ is given by

$$\phi_n(x) = A_n \sin\frac{\omega_n x}{c} = A_n \sin\frac{n\pi x}{L}, \tag{9.2-23}$$

where A_n is a constant.

Fig. 9.2-3 shows the first three mode shapes. (Observe and recall that the string is fixed at both ends.)

With the vibration amplitude being small and thus with the resulting governing equation (Eq. 9.2-13) being linear, its general solution may be obtained by the superposition of the mode shapes. That is,

$$u(x,t) = \sum_{n=1}^{\infty} \sin\frac{n\pi x}{L}\left(C_n \sin\omega_n t + D_n \cos\omega_n t\right), \quad \omega_n = \frac{n\pi c}{L}, \tag{9.2-24}$$

where the constants C_n and D_n are determined by initial conditions.

The following example illustrates the procedure.

Example 9.2-1 A uniform string of length L is fixed at its ends and stretched under tension T. The string is placed into the shape, as shown in Fig. 9.2-4, and released. The objective is to find the subsequent motion $u(x,t)$.

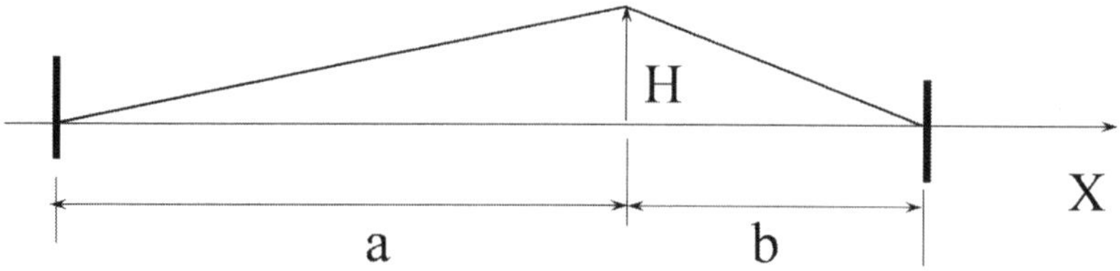

Fig. 9.2-4 Initial displacement of a string.

Solution: The initial displacement can be expressed as

$$f(x)=\begin{cases} Hx/a, & 0\le x\le a \\ H(L-x)/b, & a\le x\le L \end{cases} \tag{a}$$

At t=0, the displacement and velocity are then given by

$$u(x,0)=\sum_{n=1}^{\infty} D_n \sin\frac{n\pi x}{L}=f(x) \tag{b}$$

$$\frac{\partial u(x,0)}{\partial t}=\sum_{n=1}^{\infty} \omega_n C_n \sin\frac{n\pi x}{L}=0 \tag{c}$$

Multiplying each expression of Eq. b by $\sin(k\pi x/L)$ and integrating from x=0 to x=L, all of the terms on the right side are zero except the term with n=k. Thus, we have

$$D_n\int_0^L \sin^2\frac{n\pi x}{L}dx=\int_0^L f(x)\sin\frac{n\pi x}{L}dx, \tag{d}$$

from which we obtain

$$\begin{aligned} D_n &=\frac{2}{L}\int_0^L f(x)\sin\frac{n\pi x}{L}dx \\ &=\frac{2}{L}\left[\int_0^a \frac{H}{a}x\sin\frac{n\pi x}{L}dx+\int_a^L \frac{H}{b}(L-x)\sin\frac{n\pi x}{L}dx\right] \\ &=\frac{2HL^2}{ab\pi^2 n^2}\sin\frac{n\pi a}{L} \quad (n=1,2,\cdots) \end{aligned} \tag{e}$$

Next, for zero initial velocity, $C_n = 0$ in Eq. c. Hence, Eq. 9.2-24 becomes

$$u(x,t) = \frac{2HL^2}{ab\pi^2} \sum_{n=1}^{\infty} \frac{1}{n^2} \sin\frac{n\pi a}{L} \sin\frac{n\pi x}{L} \cos\frac{n\pi ct}{L} \tag{f}$$

Particularly, if $a=b=L/2$, Eq. f then becomes [9.2]

$$u(x,t) = \frac{8H}{\pi^2} \sum_{n=1}^{\infty} \frac{(-1)^{n-1}}{(2n-1)^2} \sin\frac{(2n-1)\pi x}{L} \cos\frac{(2n-1)\pi ct}{L} \tag{g}$$

9.3 Longitudinal Vibration of Rods

Consider a thin homogeneous rod, as represented in Fig. 9.3-1 (A) [9.2, 9.3]. Consider a differential element with length dx, as shown. Let $u(x,t)$ be the axial, or longitudinal, displacement of the element.

Fig. 9.3-1 (B) provides a force representation for the element where P is the axial force. From Hooke's law, as a stress-strain relation, we have

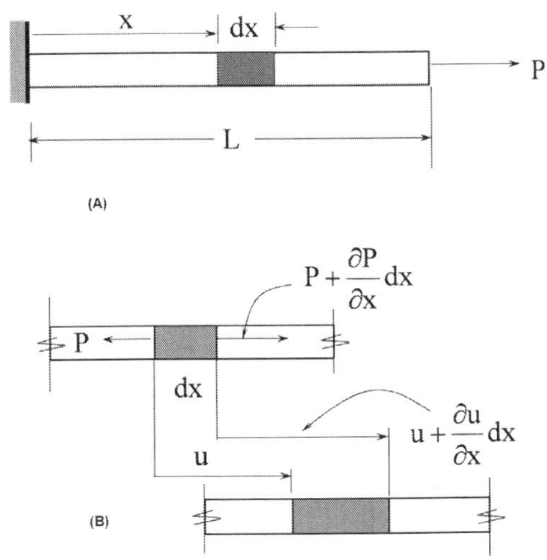

Fig. 9.3-1 Displacement of a longitudinal vibration rod.

$$P = AE\frac{\partial u}{\partial x}, \tag{9.3-1}$$

where A is the cross-section area of the rod, E is the Young's modulus of elasticity, and $\partial u / \partial x$ is the strain.

Applying Newton's law with Fig. 9.3-1 (B), we have

$$(\rho dx)\frac{\partial^2 u}{\partial t^2} = \left(P + \frac{\partial P}{\partial x}dx\right) - P, \tag{9.3-2}$$

where ρ is the mass of the rod per unit length. Substituting from Eq. 9.3-1 into Eq. 9.3-2 and simplifying, we obtain the equation of motion

$$\rho\frac{\partial^2 u}{\partial t^2} = \frac{\partial}{\partial x}\left(AE\frac{\partial u}{\partial x}\right) \tag{9.3-3}$$

If AE is a constant, we have

$$\frac{\partial^2 u}{\partial t^2} = c^2 \frac{\partial^2 u}{\partial x^2}, \tag{9.3-4}$$

where $c = \sqrt{AE/\rho}$ is the velocity of propagation of the displacement, or the stress wave, in the rod.

Eq. 9.3-4 is identical in form to Eq. 9.2-3 (the *wave equation* for the vibrating string). Thus, by the separation of variables method, we seek a solution in the form

$$u(x,t) = \phi(x) q(t) \tag{9.3-5}$$

By substituting into Eq. 9.3-4, we obtain two ordinary differential equations similar to Eqs. 9.2-16 and 9.2-17. Then, as in Eqs. 9.2-18 and 9.2-19, we have solutions $\phi(x)$ and $q(t)$ in the form

$$\phi(x) = A\sin\frac{\omega x}{c} + B\cos\frac{\omega x}{c} \tag{9.3-6}$$

$$q(t) = C\sin\omega t + D\cos\omega t, \tag{9.3-7}$$

Where, as before, the constants A, B, C, and D are to be determined by boundary conditions and initial conditions.

Table 9.3-1 lists typical boundary (or end) conditions for the longitudinal vibration (see [9.2]).

Table 9.3-1 Common Boundary Conditions for Longitudinal Vibration of a Rod [9.2]

	$\left.\frac{\partial u(x,t)}{\partial x}\right\vert_{x=0} = \left.\frac{\partial u(x,t)}{\partial x}\right\vert_{x=L} = 0$
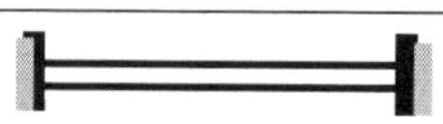	$u(0,t) = u(L,t) = 0$
	$u(0,t) = 0 \quad ku(L,t) = -EA\left.\frac{\partial u(x,t)}{\partial x}\right\vert_{x=L}$
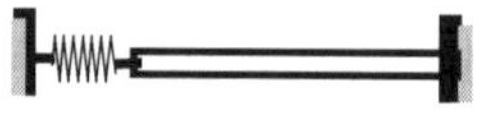	$ku(0,t) = EA\left.\frac{\partial u(x,t)}{\partial x}\right\vert_{x=0} \quad u(L,t) = 0$
	$m\left.\frac{\partial^2 u(x,t)}{\partial t^2}\right\vert_{x=0} = EA\left.\frac{\partial u(x,t)}{\partial x}\right\vert_{x=0} \quad u(L,t) = 0$
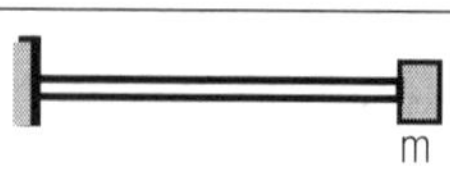	$u(0,t) = 0 \quad m\left.\frac{\partial^2 u(x,t)}{\partial t^2}\right\vert_{x=L} = -EA\left.\frac{\partial u(x,t)}{\partial x}\right\vert_{x=L}$

Example 9.3-1 A uniform bar of length L is fixed at both ends. The bar, initially at rest, is loaded by a constant axial force P_0 at the middle point. If the force is suddenly removed, find the longitudinal vibration.

Solution: From Eqs. 9.3-5 to 9.3-7, the solution has the form

$$u(x,t)=\left(B_1\sin\frac{\omega x}{c}+B_2\cos\frac{\omega x}{c}\right)\left(C\sin\omega t+D\cos\omega t\right) \tag{a}$$

From Table 9.3-1, the boundary conditions are

$$u(0,t)=0 \quad \text{and} \quad u(L,t)=0 \tag{b}$$

The first of Eq. b leads to $B_2=0$. From the second of Eq. b, we obtain

$$\sin\frac{\omega L}{c}=0 \quad \text{or} \quad \omega_n=\frac{n\pi c}{L} \quad (n=1,2,\cdots) \tag{c}$$

The solution thus becomes

$$u(x,t)=\sum_{n=1}^{\infty}\sin\frac{n\pi x}{L}\left(C_n\sin\frac{n\pi c}{L}t+D_n\cos\frac{n\pi c}{L}t\right) \tag{d}$$

With the initial velocity being zero, we have

$$\frac{\partial u(x,0)}{\partial t}=0 \tag{e}$$

This leads to $C_n=0$. Thus, the solution has the form

$$u(x,t)=\sum_{n=1}^{\infty}D_n\sin\frac{n\pi x}{L}\cos\frac{n\pi c}{L}t \tag{f}$$

Next, from Hooke's law, the stress-strain relation is

$$\frac{P_0}{2A}=E\frac{u(x,0)}{x} \quad \text{or} \quad u(x,0)=\frac{P_0 x}{2EA}, \quad (0<x<L/2) \tag{g}$$

$$\frac{P_0}{2A}=E\frac{u(x,0)}{L-x} \quad \text{or} \quad u(x,0)=\frac{P_0}{2EA}(L-x), \quad (L/2<x<L) \tag{h}$$

Applying these initial conditions in Eq. f, we obtain

$$\begin{aligned} D_n &= \frac{2}{L}\int_0^{L/2}\frac{P_0 x}{2AE}\sin\frac{n\pi x}{L}dx+\frac{2}{L}\int_{L/2}^{L}\frac{P_0}{2EA}(L-x)\sin\frac{n\pi x}{L}dx \\ &= \frac{2P_0L}{AE\pi^2}\frac{1}{n^2}\sin\frac{n\pi}{2} \end{aligned} \qquad (n=1,2,\cdots) \tag{i}$$

Therefore, the longitudinal vibration of the bar is

$$u(x,t)=\frac{2P_0L}{AE\pi^2}\sum_{n=1}^{\infty}\frac{(-1)^{n-1}}{(2n-1)^2}\sin\frac{(2n-1)\pi x}{L}\cos\frac{(2n-1)\pi ct}{L} \tag{j}$$

Example 9.3-2 [9.4]

Fig. 9.3-2 shows a test setup for obtaining the frequency response function (FRF) of a structure in an area where the impact hammer cannot reach. That is, a uniform bar *B* is attached to a structure *S*, where the FRF is difficult to be measured by hammer impact testing. A transducer *T* is mounted on the top of the bar. With a hammer *H*, the driving point FRF of the top of the bar can easily be measured. Once this FRF is obtained, the FRF of the structure on the bottom of the bar can accurately be calculated based on the driving point FRF of the top of the bar and the physical properties of the bar as explained below.

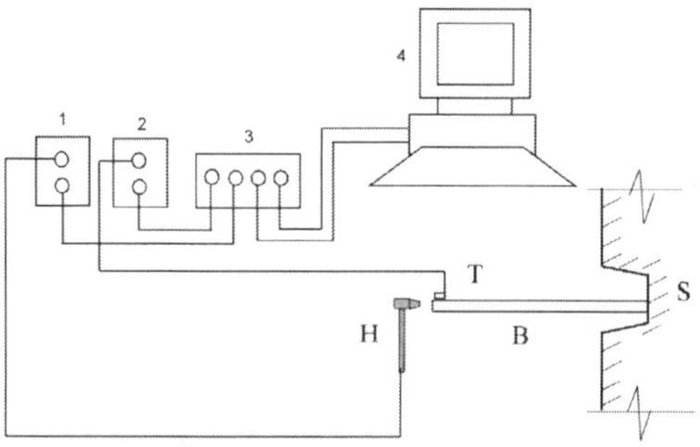

Fig. 9.3-2 Experiment setup for FRF measurement. *S*: structure of which FRF will be measured; *B*: bar attached to the structure; *H*: hammer; *T*: transducer to measure the response of the top of the bar; 1 and 2: signal conditioning; 3: antialiasing filters; 4: Fourier analyzer.

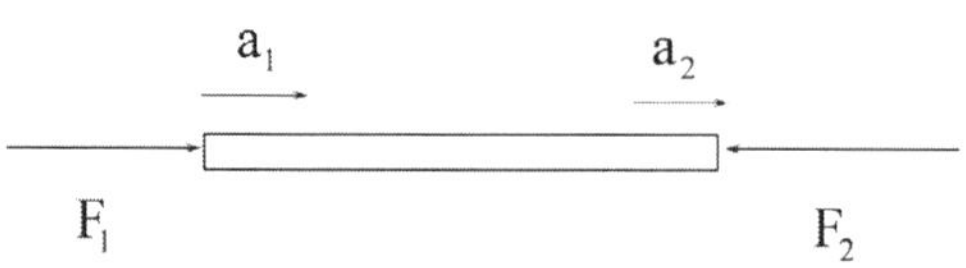

Fig. 9.3-3 Free-body diagram of the bar in Fig. 9.3-2.

Fig. 9.3-3 shows a free-body diagram of the bar.

The solution of the wave equation for the bar has the form:

$$u(x,t)=\left[C_1\sin\frac{\omega x}{c}+C_2\cos\frac{\omega x}{c}\right]\sin(\omega t+\varphi), \tag{a}$$

where C_1, C_2, and φ are constants that are dependent on the boundary conditions and initial conditions. For this case, the boundary conditions for the left end of the bar are

$$a_1=-\omega^2 u\big|_{x=0}=-\omega^2 C_2\sin(\omega t+\varphi) \tag{b}$$

$$F_1=-AE\frac{\partial u}{\partial x}\bigg|_{x=0}=-AE\frac{\omega}{c}C_1\sin(\omega t+\varphi), \tag{c}$$

where a_1 is the acceleration of the bar at the left end, and where A is the rod cross-section area.

The boundary conditions for the right end of the rod are

$$\begin{aligned} a_2&=-\omega^2 u\big|_{x=L} \\ &=-\omega^2\left[C_1\sin\left(\frac{\omega L}{c}\right)+C_2\cos\left(\frac{\omega L}{c}\right)\right]\sin(\omega t+\phi) \end{aligned} \tag{d}$$

and

$$F_2 = -AE\frac{\partial u}{\partial x}\bigg|_{x=L} \tag{e}$$
$$= -AE\frac{\omega}{c}\left[C_1\cos\left(\frac{\omega L}{c}\right) - C_2\sin\left(\frac{\omega L}{c}\right)\right]\sin(\omega t + \phi)$$

At this point, it is convenient to introduce a dimensionless parameter α defined as

$$\alpha = \frac{\omega L}{c} = \frac{\omega}{\sqrt{K_r / m_r}}, \tag{f}$$

where K_f is the rod stiffness given by

$$K_r = \frac{AE}{L} \tag{g}$$

and m_r is the rod mass given by

$$m_r = \rho L \tag{h}$$

From the definition of Eq. f, we can interpret α as a frequency ratio ω / ω_r, where ω_r is $\sqrt{K_r / m_r}$, a reference frequency.

From Eqs. b to e, the final equations relating the state vectors at the left and right ends are found to be

$$a_2 = a_1\cos\alpha + F_1\frac{\alpha\sin\alpha}{m_r} \tag{i}$$

$$F_2 = -a_1\frac{m_r\sin\alpha}{\alpha} + F_1\cos\alpha \tag{j}$$

Let us now define the FRFs as

$$H_{22} = \frac{a_2}{F_2} \quad \text{and} \quad H_{11} = \frac{a_1}{F_1} \tag{k}$$

By substituting from Eqs. i and j, H_{22} is seen to be

$$H_{22} = \frac{\dfrac{\alpha\sin\alpha}{m_r} + H_{11}\cos\alpha}{\cos\alpha - H_{11}\dfrac{m_r\sin\alpha}{\alpha}} \tag{l}$$

By inspection of Eq. l, we see that the expression is an algorithm for determining the accelerance a/F of the structure as it would be measured in the recessed space (see Fig. 9.3-2). Specifically, since it is physically difficult to take measurements in the recessed space, we can simply attach a slender bar to the desired measurement location and then measure the accelerance H_{11} at the free end of the bar. Then, using Eq. l, once H_{11} is known, the accelerance H_{22} of the structure is readily determined.

9.4 Torsional Vibration of Rods

Consider next the twisting oscillation of a circular rod, as represented in Fig. 9.4-1 [9.2, 9.3]. Let θ be the angle of twist. Then, θ will depend upon the axial position x along the rod, and upon the time t. That is, $\theta = \theta(x,t)$.

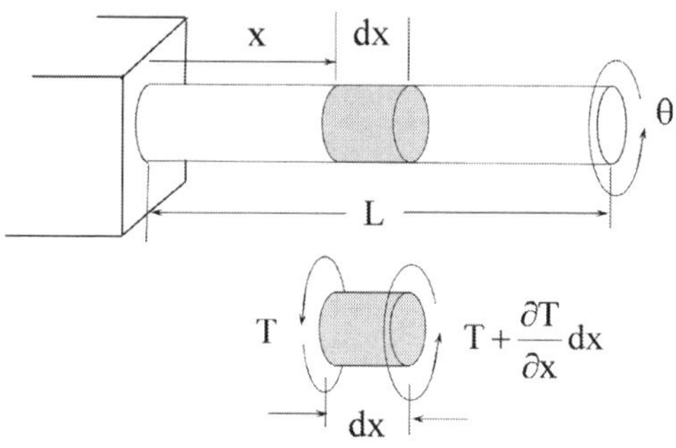

Fig. 9.4-1 Torsional vibration of a long rod.

Consider a differential element with length dx at location x along the rod, as represented in Fig. 9.4-1. From Hooke's law, the axial moment T at x is

$$T = J_p G \frac{\partial \theta}{\partial x}, \tag{9.4-1}$$

where J_p is the second polar moment of area of the cross section and G is the shear modulus. The product $J_p G$ is known as the *torsional stiffness* of the rod.

Applying Newton's law with a free-body diagram of the element, we have

$$\rho J_p \frac{\partial^2 \theta}{\partial t^2} dx = \left(T + \frac{\partial T}{\partial x} dx \right) - T, \tag{9.4-2}$$

where ρ is the mass density (mass per unit volume).

By substituting for T from Eq. 9.4-1 into Eq. 9.4-2, we have the equation of motion

$$\rho J_p \frac{\partial^2 \theta}{\partial t^2} = \frac{\partial}{\partial x} \left(J_p G \frac{\partial \theta}{\partial x} \right) \tag{9.4-3}$$

Then, if $J_p G$ is a constant, as with a rod with a uniform circular cross section, we again obtain the wave equation

$$\frac{\partial^2 \theta}{\partial t^2} = c^2 \frac{\partial^2 \theta}{\partial x^2}, \tag{9.4-4}$$

where now c is $\sqrt{G/\rho}$. (Observe that with the uniform circular rod, J_p disappears.)

Table 9.4-1 Common Boundary Conditions for Torsional Vibration of a Rod [9.2*].

<table>
<tr><td></td><td>$\left.\frac{\partial\theta(x,t)}{\partial x}\right|_{x=0} = \left.\frac{\partial\theta(x,t)}{\partial x}\right|_{x=L} = 0$</td></tr>
<tr><td></td><td>$\theta(0,t) = \theta(L,t) = 0$</td></tr>
<tr><td>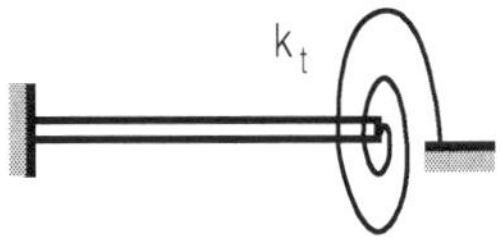</td><td>$\theta(0,t) = 0 \quad k\theta(L,t) = -J_pG\left.\frac{\partial\theta(x,t)}{\partial x}\right|_{x=L}$</td></tr>
<tr><td>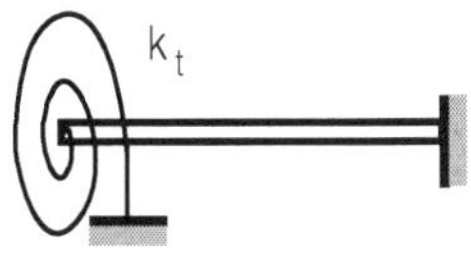</td><td>$k\theta(0,t) = J_pG\left.\frac{\partial\theta(x,t)}{\partial x}\right|_{x=0} \quad \theta(L,t) = 0$</td></tr>
<tr><td>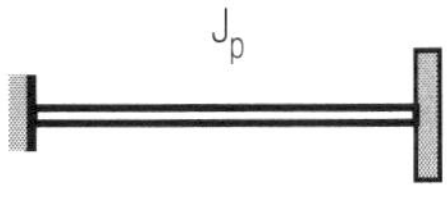</td><td>$\theta(0,t) = 0$
$J_0\left.\frac{\partial^2\theta(x,t)}{\partial t^2}\right|_{x=L} = -J_pG\left.\frac{\partial\theta(x,t)}{\partial x}\right|_{x=L}$</td></tr>
<tr><td>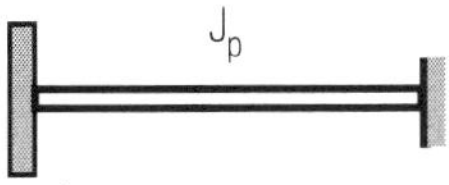</td><td>$J_0\left.\frac{\partial^2\theta(x,t)}{\partial t^2}\right|_{x=0} = J_pG\left.\frac{\partial\theta(x,t)}{\partial x}\right|_{x=0}$
$\theta(L,t) = 0$</td></tr>
<tr><td colspan="2">*J_0 is the mass moment of inertia of the disk, and J_p is the second polar moment of area of the cross-section of the rod.</td></tr>
</table>

Eq. 9.4-4 is identical in form to Eq. 9.2-3 (the *wave equation* for the vibrating string). Then, by comparison, following the same solution procedure, we obtain θ in the form

$$\theta(x,t) = \left(A\sin\omega\sqrt{\frac{\rho}{G}}x + B\cos\omega\sqrt{\frac{\rho}{G}}x\right)\left(C\sin\omega t + D\cos\omega t\right), \qquad (9.4\text{-}5)$$

where, as before, the coefficients A, B, C, and D are to be determined by the boundary and initial conditions. Table 9.4-1 lists the common boundary conditions for torsional vibration of a rod (see [9.2]).

Example 9.4-1 Fig. 9.4-1 depicts a long uniform rod with one end fixed and the other free. Suppose a torque T is applied at the free end and then released. Find the subsequent torsional motion.

Solution: From Eq. 9.4-5, the general solution has the form

$$\theta(x,t) = \left(A\sin\omega\sqrt{\frac{\rho}{G}}x + B\cos\omega\sqrt{\frac{\rho}{G}}x\right)\left(C\sin\omega t + D\cos\omega t\right), \qquad \text{(a)}$$

where ρ is mass per unit volume, and G is the shear modulus.

The boundary conditions are

$$\theta(0,t)=0 \tag{b}$$

$$\left.\frac{\partial\theta}{\partial x}\right|_{x=L}=0 \tag{c}$$

Eq. b leads to B=0. Eq. c yields the frequency equation

$$\cos\omega_n\sqrt{\frac{\rho}{G}}L=0, \quad \text{or} \quad \omega_n\sqrt{\frac{\rho}{G}}L=\frac{(2n-1)\pi}{2} \quad (n=1,2,\cdots) \tag{d}$$

For $\omega_n=(2n-1)\pi\sqrt{G/\rho}/(2L)$, Eq. a yields

$$\theta(x,t)=\sum_{n=1}^{\partial}\sin\frac{(2n-1)\pi x}{2L}\left(C_n\sin\omega_n t+D_n\cos\omega_n t\right) \tag{e}$$

The constants C_n and D_n may now be determined from the initial conditions

$$\theta(x,0)=\frac{Tx}{J_pG} \quad \text{and} \quad \frac{\partial\theta(x,0)}{\partial t}=0 \tag{f}$$

The first of these conditions demonstrates the uniform initial twist at t=0. The second condition states that the initial velocity is zero, since the rod is released at t=0. Hence, C_n=0. As a result, the solution has the form

$$\theta(x,t)=\sum_{n=1}^{\partial}D_n\sin\frac{(2n-1)\pi x}{2L}\cos\frac{(2n-1)\pi ct}{2L} \tag{g}$$

The first condition of Eq. f yields

$$\sum_{n=1}^{\partial}D_n\sin\frac{(2n-1)\pi x}{2L}=\frac{Tx}{J_pG} \tag{h}$$

Multiplying in Eq. h by $\sin((2n-1)\pi x/2L)$ and integrating from x=0 to x=L, we obtain

$$D_n=\frac{8TL}{(2n-1)^2\pi^2J_pG}(-1)^{(n-1)} \quad (n=1,2,\cdots) \tag{i}$$

The torsional vibration of the rod is then given by

$$\theta(x,t)=\frac{8TL}{\pi^2J_pG}\sum_{n=1}^{\partial}\frac{(-1)^{(n-1)}}{(2n-1)^2}\sin\frac{(2n-1)\pi x}{2L}\cos\frac{(2n-1)\pi ct}{2L} \tag{j}$$

9.5 Lateral Vibration of Beams

Fig. 9.5-1 (a) shows a long, thin flexible beam which can undergo displacement in the *x-y* plane. For simplicity, we begin our analysis with a beam whose displacement u is small compared with the beam length and is due to bending along the beam. The corresponding model is known as the *Euler-Bernoulli beam* [9.2].

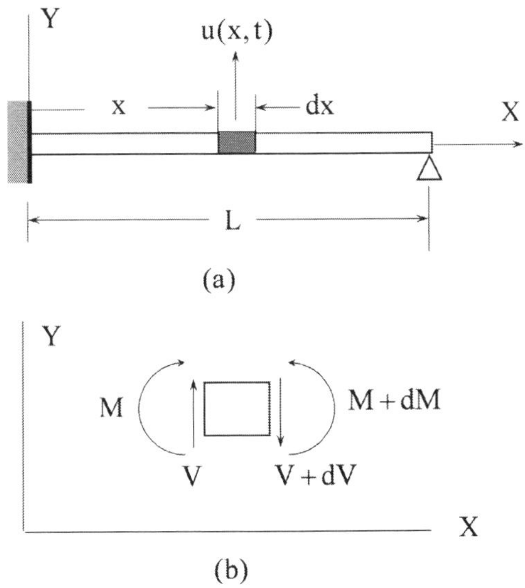

Fig. 9.5-1 Lateral vibration of a beam (see [9.2]). (This graph is reproduced with permission of the F. S. Tse family and co-authors.)

The displacement u of the beam will depend upon both the position x along the beam and time t. Thus, $u=u(x,t)$.

Fig. 9.5-1 (b) shows a free-body diagram of a differential element of the beam. The figure itself shows the positive directions for the bending moment M and the shear force V.

From a balance of forces and Newton's law, we have

$$\rho dx\frac{\partial^2 u}{\partial t^2}=-\left(V+dV\right)+V$$

or simply

$$\rho\frac{\partial^2 u}{\partial t^2}=-\frac{\partial V}{\partial x}, \qquad (9.5\text{-}1)$$

where ρ is the beam mass per unit length.

Next, from a balance of moments in Fig. 9.5-1 (b), we have

$$\frac{\partial M}{\partial x}dx-Vdx=0$$

or simply

$$\frac{\partial M}{\partial x}=V, \qquad (9.5\text{-}2)$$

where M is the bending moment with positive direction defined as in Fig. 9.5-1 (b).

From elementary strength of materials analyses [9.5], we have the relation

$$EJ\frac{\partial^2 u}{\partial x^2}=M, \tag{9.5-3}$$

where EJ is the flexural rigidity of the beam. Combining Eqs. 9.5-1 to 9.5-3, the differential equation for lateral vibration of the beam becomes

$$\rho\frac{\partial^2 u}{\partial t^2}=-\frac{\partial^2}{\partial x^2}\left(EJ\frac{\partial^2 u}{\partial x^2}\right) \tag{9.5-4}$$

In the special case where the flexural rigidity EJ is constant along the length of the beam, the above equation may be written as

$$\frac{\partial^2 u}{\partial t^2}+a^2\frac{\partial^4 u}{\partial x^4}=0, \tag{9.5-5}$$

where

$$a^2=\frac{EJ}{\rho} \tag{9.5-6}$$

As with the wave equation of the foregoing sections, we can solve Eq. 9.5-5 using the method of separation of variables. That is, we seek a solution in the form

$$u(x,t)=\phi(x)q(t) \tag{9.5-7}$$

Substituting into Eq. 9.5-5 and simplifying, we obtain

$$-a^2\frac{1}{\phi}\frac{d^4\phi}{dx^4}=\frac{1}{q}\frac{d^2q}{dt^2} \tag{9.5-8}$$

Since the left side is a function of x and the right side is a function of t, and since x and t are independent variables, the two sides are equal only if both are equal to the same constant. Thus, letting each side of the equation be equal to $-\omega^2$, where ω is a constant, we obtain

$$\frac{d^4\phi}{dx^4}-\beta^4\phi=0 \quad \text{and} \quad \frac{d^2q}{dt^2}+\omega^2 q=0, \tag{9.5-9}$$

where

$$\beta^4=\frac{\omega^2}{a^2}=\frac{\rho\omega^2}{EJ} \tag{9.5-10}$$

The first of Eqs. 9.5-9 forms an eigenvalue problem for a continuous beam. Therefore, $\phi(x)$ is called an *eigenfunction*. The general solution of Eq. 9.5-9 can be shown to be

$$\phi(x)=A_1\sin\beta x+A_2\cos\beta x+A_3\sinh\beta x+A_4\cosh\beta x \tag{9.5-11}$$

$$q(t)=A_5\sin\omega t+A_6\cos\omega t, \tag{9.5-12}$$

where $A_1,\ldots,A_4$ are constants to be determined by boundary conditions, and the constants A_5 and A_6 are determined by the initial conditions.

It is convenient to write the eigenfunction in the form [9.6]

$$\begin{aligned}\phi(x) &= C_1(\cos\beta x+\cosh\beta x)+C_2(\cos\beta x-\cosh\beta x)\\ &\quad +C_3(\sin\beta x+\sinh\beta x)+C_4(\sin\beta x-\sinh\beta x)\end{aligned} \tag{9.5-13}$$

In applying the boundary conditions, the following relations are used where primes indicate derivatives with respect to x:

$$\begin{aligned}\phi'(x) &= \beta\left[C_1(-\sin\beta x+\sinh\beta x)+C_2(-\sin\beta x-\sinh\beta x)\right]\\ &\quad +\beta\left[C_3(\cos\beta x+\cosh\beta x)+C_4(\cos\beta x-\cosh\beta x)\right]\end{aligned} \tag{9.5-14}$$

$$\begin{aligned}\phi''(x) &= \beta^2\left[C_1(-\cos\beta x+\cosh\beta x)+C_2(-\cos\beta x-\cosh\beta x)\right]\\ &\quad +\beta^2\left[C_3(-\sin\beta x+\sinh\beta x)+C_4(-\sin\beta x-\sinh\beta x)\right]\end{aligned} \tag{9.5-15}$$

$$\begin{aligned}\phi'''(x) &= \beta^3\left[C_1(\sin\beta x+\sinh\beta x)+C_2(\sin\beta x-\sinh\beta x)\right]\\ &\quad +\beta^3\left[C_3(-\cos\beta x+\cosh\beta x)+C_4(-\cos\beta x-\cosh\beta x)\right]\end{aligned} \tag{9.5-16}$$

For the usual boundary conditions, two of the constants are zero, and there remain two equations containing two constants. These can be combined to give the *frequency equation*. Using the frequency equation, one of the unknown constants can be found in terms of the other. Therefore, the modes of natural vibration and their frequencies will be established. Table 9.5-1 lists the common boundary conditions (see [9.2]).

By superimposing all possible normal vibrations, the general expression for the free lateral vibration of beam becomes

$$u(x,t)=\sum_{n=1}^{\infty}\phi_n(A_n\sin\omega_n t+B_n\cos\omega_n t), \tag{9.5-17}$$

where the constants A_n and B_n are to be determined by initial conditions.

Example 9.5-1 Derive the frequency equation and eigenfunction for the lateral vibration of a uniform beam with the left end fixed and the right end simply supported, as in Fig. 9.5-1 (a).

Solution: Let the lateral displacement be $u(x,t)=\phi(x)q(t)$. From Eq. 9.5-13, $\phi(x)$ has the form

$$\begin{aligned}\phi(x) &= C_1(\cos\beta x+\cosh\beta x)+C_2(\cos\beta x-\cosh\beta x)\\ &\quad +C_3(\sin\beta x+\sinh\beta x)+C_4(\sin\beta x-\sinh\beta x)\end{aligned} \tag{a}$$

The boundary conditions for the fixed end are

$$(1)\ \ u(0,t)=0 \quad \text{and} \quad (2)\ \ \left.\frac{\partial u}{\partial x}\right|_{x=0}=0$$

Table 9.5-1 Common Boundary Conditions for Lateral Vibration of Beams [9.2]

	$M\|_{x=0}=0,\ V\|_{x=0}=0,\ M_{x=L}=0,\ V\|_{X=L}=0$
	$u(0,t)=(L,t)=0 \quad \left.\frac{\partial u}{\partial x}\right\|_{x=0}=\left.\frac{\partial u}{\partial x}\right\|_{x=L}=0$
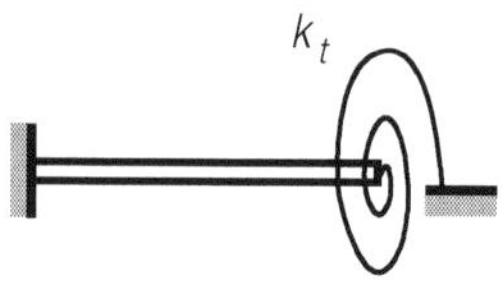	$u(0,t)=0 \quad \left.\frac{\partial u}{\partial x}\right\|_{x=0}=0$ $k_t\frac{\partial u}{\partial x}=-EJ\left.\frac{\partial^2 u}{\partial x^2}\right\|_{x=L} \quad \left.\frac{\partial^3 u}{\partial x^3}\right\|_{x=L}=0$
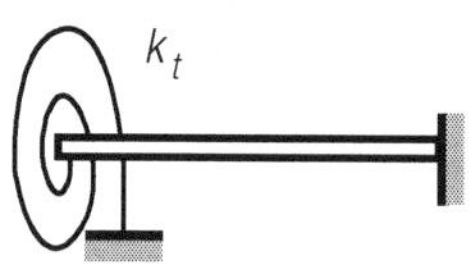	$k_t\frac{\partial u}{\partial x}=-EJ\left.\frac{\partial^2 u}{\partial x^2}\right\|_{x=0} \quad \left.\frac{\partial^3 u}{\partial x^3}\right\|_{x=0}=0$ $u(L,t)=0 \quad \left.\frac{\partial u}{\partial x}\right\|_{x=L}=0$
	$u(0,t)=0 \quad \left.\frac{\partial u}{\partial x}\right\|_{x=0}=0$ $\left.\frac{\partial^2 u}{\partial x^2}\right\|_{x=L}=0 \quad EJ\left.\frac{\partial^3 u}{\partial x^3}\right\|_{x=L} \quad \left.\frac{\partial^2 u}{\partial t^2}\right\|_{x=L}$
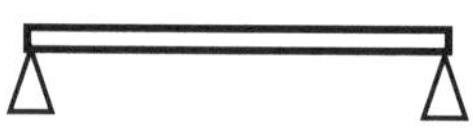	$u(0,t)=0 \quad \left.\frac{\partial^2 u}{\partial x^2}\right\|_{x=0}=0$ $u(L,t)=0 \quad \left.\frac{\partial^2 u}{\partial x^2}\right\|_{x=L}=0$

Those for the simply supported end are

$$(3)\ u(L,t)=0 \quad \text{and} \quad (4)\ \left.\frac{\partial^2 u}{\partial x^2}\right|_{x=L}=0$$

Condition (1) requires that $C_1=0$. Condition (2) requires that $C_3=0$. Thus, we have

$$\phi(x)=C_2(\cos\beta x-\cosh\beta x)+C_4(\sin\beta x-\sinh\beta x)$$

Conditions (3) and (4) lead to

$$(\cos\beta L-\cosh\beta L)C_2+(\sin\beta L-\sinh\beta L)C_4=0 \tag{b}$$

$$-(\cos\beta L+\cosh\beta L)C_2-(\sin\beta L+\sinh\beta L)C_4=0 \tag{c}$$

A solution for the constants C_2 and C_4 different from zero can be obtained only in the case when the determinant of their coefficients is equal to zero. That is,

$$\begin{vmatrix} (\cos\beta L-\cosh\beta L) & (\sin\beta L-\sinh\beta L) \\ -(\cos\beta L+\cosh\beta L) & -(\sin\beta L+\sinh\beta L) \end{vmatrix}=0$$

Expanding the determinant yields the frequency equation

$$\cos\beta L\sinh\beta L-\sin\beta L\cosh\beta L=0$$

or

$$\tan\beta L=\tanh\beta L \tag{d}$$

The first three roots of the frequency equation are

$$\beta_1 L=3.9266, \quad \beta_2 L=7.0686, \quad \beta_3 L=10.2102$$

Next, we can determine the ratio C_2/C_4 from Eqs. b and c. That is,

$$\frac{C_2}{C_4}=-\frac{\sin\beta L-\sinh\beta L}{\cos\beta L-\cosh\beta L}=-\frac{\sin\beta L+\sinh\beta L}{\cos\beta L+\cosh\beta L}$$

Defining A as C_2/C_4, an eigenfunction $\phi(x)$ can be expressed as

$$\phi(x)=A(\cos\beta x-\cosh\beta x)+(\sin\beta x-\sinh\beta x) \tag{e}$$

Example 9.5-2 Derive the frequency equation for lateral vibration of a uniform beam of length L with hinged-hinged ends.

Solution: The general solution for $\phi(x)$ is

$$\begin{aligned}\phi(x)=&C_1(\cos\beta x+\cosh\beta x)+C_2(\cos\beta x-\cosh\beta x)\\ &+C_3(\sin\beta x+\sinh\beta x)+C_4(\sin\beta x-\sinh\beta x)\end{aligned} \tag{a}$$

For the case of hinged ends, the end conditions are

$$(1)\quad \phi(0)=0; \quad (2)\quad \left.\frac{d^2\phi}{dx^2}\right|_{x=0}=0 \tag{b}$$

$$(3)\quad \phi(L)=0; \quad (4)\quad \left.\frac{d^2\phi}{dx^2}\right|_{x=L}=0 \tag{c}$$

From conditions b, it can be concluded that $C_1=C_2=0$. From conditions c, we obtain

$$C_3(\sin\beta L+\sinh\beta L)+C_4(\sin\beta L-\sinh\beta L)=0$$

$$C_3(-\sin\beta L+\sinh\beta L)+C_4(-\sin\beta L-\sinh\beta L)=0$$

These two equations yield $C_3=C_4$ and

$$\sin\beta L=0 \tag{d}$$

Eq. d is the frequency equation for the case under consideration. The consecutive roots are

$$\beta_n L=n\pi \quad (n=1,2,\cdots) \tag{e}$$

The circular frequencies of the consecutive modes of vibration may be obtained from Eq. 9.5-10. Specifically,

$$\omega_n=a\beta_n^2=a\left(\frac{n\pi}{L}\right)^2=\sqrt{\frac{EJ}{\rho}}\left(\frac{n\pi}{L}\right)^2 \tag{f}$$

In the case of rotating circular shafts of uniform cross section, the frequencies calculated by Eq. f represent the critical frequencies. When the frequency of rotation of the shaft approaches one of the frequencies of Eq. f, a considerable lateral vibration of the shaft should be expected.

The shape of the deflection curve for various modes of vibration is determined by the normal function of Eq. a. In the case we are considering, $C_1=C_2=0$ and $C_3=C_4$. Thus, the eigenfunction can be expressed as

$$\phi(x)=D\sin\beta x,\ (D=C_3+C_4) \tag{g}$$

Finally, the general solution can be written as

$$u(x,t)=\sum_{n=1}^{\infty}\sin\frac{n\pi x}{L}(A_n\sin\omega_n t+B_n\cos\omega_n t) \tag{h}$$

The constants A_n and B_n of this solution can be determined by initial conditions.

Example 9.5-3 A uniform beam of length L, with hinged ends, has zero initial deflection but has initial transverse velocity V everywhere except at two ends. Find the subsequent motion of the beam.

Solution: From Eq. h of Example 9.5-2, the general solution is

$$u(x,t)=\sum_{n=1}^{\infty}\sin\frac{n\pi x}{L}(A_n \sin\omega_n t+B_n \cos\omega_n t), \tag{a}$$

where

$$\omega_n - a\beta_n^2 = \sqrt{\frac{EJ}{\rho}}\left(\frac{n\pi}{L}\right)^2$$

The initial deflection of the beam is zero:

$$u(x,0)=0 \quad (0\leq x\leq L) \tag{b}$$

The initial velocity is given by

$$\frac{\partial u(x,0)}{\partial t}=V \quad (0<x<L) \tag{c}$$

Substituting into Eq. a, we obtain

$$0=\sum_{n=1}^{\infty}B_n \sin\frac{n\pi x}{L} \tag{d}$$

$$V=\sum_{n=1}^{\infty}A_n\omega_n \sin\frac{n\pi x}{L} \tag{e}$$

Multiplying Eqs. d and e by $\sin(n\pi x/L)$ and integrating both sides of these equations from x=0 to x=L, we obtain

$$A_n=\frac{2}{L}\frac{1}{\omega_n}\int_0^L V\sin\frac{n\pi x}{L}dx=\frac{2V}{n\pi\omega_n}(1-\cos n\pi) \tag{f}$$

$$B_n=0 \tag{g}$$

Substituting these results into Eq. a, we obtain

$$u(x,t)=\frac{4V}{\pi}\sum_{n=1,3,5\ldots}^{\infty}\frac{1}{n\omega_n}\sin\frac{n\pi x}{L}\sin\omega_n t \tag{h}$$

Example 9.5-4 Derive the frequency equation and eigenfunction for the lateral vibration of a uniform beam with the left end fixed and the right end free.

Solution: Let the lateral displacement of the beam be $u(x,t)=\phi(x)q(t)$. From Eq. 9.5-13, we have

$$\begin{aligned}\phi(x)=&C_1(\cos\beta x+\cosh\beta x)+C_2(\cos\beta x-\cosh\beta x)\\&+C_3(\sin\beta x+\sinh\beta x)+C_4(\sin\beta x-\sinh\beta x)\end{aligned}, \tag{a}$$

where $\beta^4 = \rho\omega^2 / EJ$. The boundary conditions are

$$(1)\quad u(0,t)=0 \qquad (2)\quad \frac{\partial u(0,t)}{\partial x}=0$$

$$(3)\quad \frac{\partial^2 u(L,t)}{\partial^2 x}=0 \qquad (4)\quad \frac{\partial^3 u(L,t)}{\partial x^3}=0$$

From conditions (1) and (2) we have C_1=C_3=0. From conditions (3) and (4) we obtain

$$(\cos\beta L+\cosh\beta L)C_2+(\sin\beta L+\sinh\beta L)C_4=0 \tag{b}$$

$$(\sin\beta L-\sinh\beta L)C_2-(\cos\beta L+\cosh\beta L)C_4=0 \tag{c}$$

For a nontrivial solution for C_2 and C_4, the determinant of the coefficients must be zero. In this manner, we obtain the frequency equation as

$$\cos\beta L\cosh\beta L=-1 \tag{d}$$

The first two roots of this equation are

$$\beta_1 L=1.8751, \qquad \beta_2 L=4.6941 \tag{e}$$

Introducing A_n as

$$A_n=\frac{C_2}{C_4}=\frac{\cos\beta_n L+\cosh\beta_n L}{\sin\beta_n L-\sinh\beta_n L}=-\frac{\sin\beta L+\sinh\beta L}{\cos\beta L+\cosh\beta L} \tag{f}$$

The eigenfunction $\phi_n(x)$ can be expressed as

$$\phi_n(x)=A_n(\cos\beta_n x-\cosh\beta_n x)+(\sin\beta_n x-\sinh\beta_n x) \tag{g}$$

By using Eq. f for the first two modes, we have

$$A_1=-1.3622, \quad A_2=-0.9819 \tag{h}$$

Thus, from Eq. g, the first two eigenfunctions are given by

$$\phi_1(x)=-1.3622\left(\cos\frac{1.8751x}{L}-\cosh\frac{1.8751x}{L}\right)+\left(\sin\frac{1.8751x}{L}-\sinh\frac{1.8751x}{L}\right)$$

$$\phi_2(x)=-0.9819\left(\cos\frac{4.6941x}{L}-\cosh\frac{4.6941x}{L}\right)+\left(\sin\frac{4.6941x}{L}-\sinh\frac{4.6941x}{L}\right)$$

Table 9.5-2 lists the frequency equations, eigenfunctions, and the roots of the frequency equations of a single-span uniform beam for different end conditions (see [9.2]). Note that, for semidefinite systems, zero modes are not included in the Table.

Table 9.5-2 Summary of Equations of Lateral Vibration of a Uniform Beam [9.2] (This table is reproduced with permission of the F. S. Tse family and co-authors.)

End conditions	Equations	$\omega_n = \frac{(\beta_n L)^2}{L^2}\sqrt{\frac{EJ}{\rho}}$
	$\cos\beta L \cosh\beta L = 1$ $\phi(x) = A(\cos\beta x - \cosh\beta x) + (\sin\beta x - \sinh\beta x)$ $A = -\frac{\sin\beta L - \sinh\beta L}{\cos\beta L - \cosh\beta L} = \frac{\cos\beta L - \cosh\beta L}{\sin\beta L + \sin\beta L}$	$\beta_1 L = 4.7300$ $\beta_2 L = 7.8532$ $\beta_3 L = 10.9956$
	$\cos\beta L \cosh\beta L = 1$ $\phi(x) = A(\cos\beta x - \cosh\beta x) + (\sin\beta x - \sinh\beta x)$ $A = -\frac{\sin\beta L - \sinh\beta L}{\cos\beta L - \cosh\beta L} = \frac{\cos\beta L - \cosh\beta L}{\sin\beta L + \sinh\beta L}$	$\beta_1 L = 4.7300$ $\beta_2 L = 7.8532$ $\beta_3 L = 10.9956$
	$\tan\beta L = \tanh\beta L$ $\phi(x) = A(\cos\beta x - \cosh\beta x) + (\sin\beta x - \sinh\beta x)$ $A = -\frac{\sin\beta L - \sinh\beta L}{\cos\beta L - \cosh\beta L} = -\frac{\sin\beta L + \sinh\beta L}{\cos\beta L + \cosh\beta L}$	$\beta_1 L = 3.9266$ $\beta_2 L = 7.0686$ $\beta_3 L = 10.2102$
	$\tan\beta L = \tanh\beta L$ $\phi(x) = A\sin\beta x + \sin\beta x$ $A = -\frac{\sinh\beta L}{\sin\beta L} = -\frac{\cosh\beta L}{\cos\beta L}$	$\beta_1 L = 3.9266$ $\beta_2 L = 7.0686$ $\beta_3 L = 10.2102$
	$\sin\beta L = 0$ $\phi(x) = A\sin\beta x$	$\beta_n L = n\pi$
	$\cos\beta L \cosh\beta L = -1$ $\phi(x) = A(\cos\beta x - \cosh\beta x) + (\sin\beta x - \sinh\beta x)$ $A = -\frac{\sin\beta L + \sinh\beta L}{\cos\beta L + \cosh\beta L} = \frac{\cos\beta L + \cosh\beta L}{\sin\beta L - \sinh\beta L}$	$\beta_1 L = 1.8751$ $\beta_2 L = 4.6941$ $\beta_3 L = 7.8548$

9.6 Orthogonal Relations

As with discrete systems, the eigenfunctions satisfy orthogonal relations [9.7]. We can develop these by following the same procedure as with the discrete systems.

To this end, let $\phi_m(x)$ and $\phi_n(x)$ be two of eigenfunctions. From the first expression of Eq. 9.5-9 we have

$$\frac{d^4\phi_m}{dx^4} = \beta_m^4\phi_m \tag{9.6-1}$$

$$\frac{d^4\phi_n}{dx^4} = \beta_n^4\phi_n, \tag{9.6-2}$$

where the β_n are

$$\beta_m^4 = \frac{\omega_m^2}{a^2} = \frac{\rho\omega_m^2}{EJ}$$

Multiplying in Eq. 9.6-1 by ϕ_n and then integrating by parts, we obtain

$$\left[\phi_n \frac{d^3\phi_m}{dx^3}\right]_0^L - \left[\frac{d\phi_n}{dx}\frac{d^2\phi_m}{dx^2}\right]_0^L + \int_0^L \frac{d^2\phi_m}{dx^2}\frac{d^2\phi_n}{dx^2}dx = \beta_m^4 \int_0^L \phi_m\phi_n dx \qquad (9.6\text{-}3)$$

Similarly, by multiplying in Eq. 9.6-2 by ϕ_m and then integrating by parts, we obtain

$$\left[\phi_m \frac{d^3\phi_n}{dx^3}\right]_0^L - \left[\frac{d\phi_m}{dx}\frac{d^2\phi_n}{dx^2}\right]_0^L + \int_0^L \frac{d^2\phi_m}{dx^2}\frac{d^2\phi_n}{dx^2}dx = \beta_n^4 \int_0^L \phi_m\phi_n dx \qquad (9.6\text{-}4)$$

Finally, by subtracting Eq. 9.6-4 from Eq. 9.6-3, we obtain the relation [9.7]

$$\left(\beta_m^4 - \beta_n^4\right)\int_0^L \phi_m\phi_n dx = \left[\left(\phi_n \frac{d^3\phi_m}{dx^3} - \phi_m \frac{d^3\phi_n}{dx^3}\right) + \left(\frac{d\phi_m}{dx}\frac{d^2\phi_n}{dx^2} - \frac{d\phi_n}{dx}\frac{d^2\phi_m}{dx^2}\right)\right]_0^L \qquad (9.6\text{-}5)$$

The integration limits on the right side of Eq. 9.6-5 are the boundary conditions for $x=0$ and $x=L$. For free-free, hinged-hinged, and fixed-fixed beams, the right side of Eq. 9.6-5 is identically zero (see Table 9.5-1). Thus, if $m \neq n$, we have orthogonal relation

$$\int_0^L \phi_m\phi_n dx = 0 \quad (m \neq n) \qquad (9.6\text{-}6)$$

Substituting from Eq. 9.6-6 into Eq. 9.6-3 or 9.6-4, we obtain a second orthogonal relation

$$\int_0^L \frac{d^2\phi_m}{dx^2}\frac{d^2\phi_n}{dx^2}dx = 0 \quad (m \neq n) \qquad (9.6\text{-}7)$$

Observe that the relation of Eq. 9.6-5 can also be used to evaluate integrals such as $\int_0^L \phi_n^2 dx$ and $\int_0^L \left(\frac{d^2\phi_n}{dx^2}\right)^2 dx$ [9.7]. These integrals appear in expressions for the kinetic and potential energies of vibrating beams. Observe further, however, that if we let $m=n$ in Eq. 9.6-5, we do not obtain the anticipated result since both sides of the equation become zero. We can get around this difficulty by letting ϕ_n be a function nearly equal to ϕ_m. Specifically, let $\delta\beta_n$ be a small increment to β_n so that ϕ_m and ϕ_n become equal as $\delta\beta_n$ becomes zero. That is,

$$\phi_m = \phi_n + \frac{d\phi_n}{d\beta_n}\delta\beta_n \tag{9.6-8}$$

$$\beta_m^4 = \left(\beta_n + \delta\beta_n\right)^4 = \beta_n^4 + 4\beta_n^3\delta\beta_n \quad \text{or} \quad \beta_m^4 - \beta_n^4 = 4\beta_n^3\delta\beta_n \tag{9.6-9}$$

By substituting into Eq. 9.6-5 and neglecting small quantities of higher order, we obtain

$$4\beta_n^3\int_0^L \phi_n^2 dx = \left[\phi_n \frac{d}{d\beta_n}\frac{d^3\phi_n}{dx^3} - \frac{d\phi_n}{d\beta_n}\frac{d^3\phi_n}{dx^3} + \frac{d}{d\beta_n}\left(\frac{d\phi_n}{dx}\right)\frac{d^2\phi_n}{dx^2} - \frac{d\phi_n}{dx}\frac{d}{d\beta_n}\left(\frac{d^2\phi_n}{dx^2}\right)\right]_0^L$$

$$\tag{9.6-10}$$

For convenience, let ϕ_n', ϕ_n'', ϕ_n''' represent consecutive derivatives of ϕ_n with respect to $\beta_n x$; then we have

$$\frac{d\phi_n}{dx} = \beta_n\phi_n' \quad \text{and} \quad \frac{d\phi_n}{d\beta_n} = x\phi_n' \tag{9.6-11}$$

With this notation, Eq. 9.6-2 becomes

$$\phi_n'''' = \phi_n \tag{9.6-12}$$

and Eq. 9.6-10 has the form [9.7]

$$4\beta_n\int_0^L \phi_n^2 dx = \left[3\phi_n\phi_n''' + \beta_n x\phi_n^2 - 2\beta_n x\phi_n'\phi_n''' - \phi_n'\phi_n'' + \beta_n x\left(\phi_n''\right)^2\right]_0^L \tag{9.6-13}$$

For free, hinged, and fixed ends, the products $\phi_n\phi_n''$ and $\phi_n'\phi_n''$ are equal to zero; hence we have

$$\int_0^L \phi_n^2 dx = \frac{L}{4}\left|\phi_n^2 - 2\phi_n'\phi_n''' + \left(\phi_n''\right)^2\right|_{x=L} \tag{9.6-14}$$

Eq. 9.6-14 can now readily be used for the various kinds of end conditions. For example,

(1) for a free right end, since $\left(\phi_n''\right)\big|_{x=L} = 0$ and $\left(\phi_n'''\right)\big|_{x=L} = 0$, we obtain

$$\int_0^L \phi_n^2 dx = \frac{L}{4}\left(\phi_n^2\right)_{x=L} \tag{9.6-15}$$

(2) for a fixed right end, since $\phi_n = 0$ and $\phi_n{}' = 0$, we obtain

$$\int_0^L \phi_n^2 dx = \frac{L}{4}\left(\phi_n{}''\right)_{x=L}^2 \tag{9.6-16}$$

(3) for a hinged right end, since $\phi_n = 0$ and $\phi_n{}'' = 0$, we obtain

$$\int_0^L \phi_n^2 dx = -\frac{L}{2}\left(\phi_n{}'\phi_n{}'''\right)_{x=L} \tag{9.6-17}$$

To calculate $\int_0^L \left(\frac{d^2\phi_n}{dx^2}\right)^2 dx$, we can use Eq. 9.6-2. Multiplying by ϕ_n and integrating along the length of the beam, we have

$$\int_0^L \frac{d^4\phi_n}{dx^4}\phi_n dx = \beta_n^4 \int_0^L \phi_n^2 dx \tag{9.6-18}$$

Integrating the left side by parts, we obtain

$$\int_0^L \left(\frac{d^2\phi_n}{dx^2}\right)^2 dx = \beta_n^4 \int_0^L \phi_n^2 dx \tag{9.6-19}$$

This result together with Eq. 9.6-14 evaluates the integral.

Eqs. 9.6-14 and 9.6-19 are useful in investigating forced vibrations of beams with end conditions other than hinged. The advantage of these results is that calculation of the integrals is obtained by using the values of the eigenfunctions at the end conditions.

As an example, consider the eigenfunction for a uniform beam with hinged-hinged ends. That is, $\phi_n(x) = \sin\beta_n x$, where $\beta_n = n\pi / L$. By direct integration, we have

$$\int_0^L \phi_n^2 dx = \int_0^L \sin^2\beta_n x dx = \frac{L}{2} \tag{9.6-20}$$

This result can also be obtained by using Eq. 9.6-14. Since $\left(\phi_n{}'\phi_n{}'''\right) = -\cos^2\beta_n x$, thus,

$$\int_0^L \phi_n^2 dx = -\frac{L}{2}\left(\phi_n{}'\phi_n{}'''\right)_{x=L} = \frac{L}{2}\left(\cos^2 n\pi\right) = \frac{L}{2} \tag{9.6-21}$$

The results, Eqs. 9.6-20 and 9.6-21, are identical. For end conditions other than hinged, it is more efficient to use Eq. 9.6-14 than direct integration.

Example 9.6-1 A uniform cantilever beam of length L has a mass m attached at $x=L$, as shown in Fig. 9.6-1. Find the eigenfunctions for lateral vibration and show that these eigenfunctions arc orthogonal.

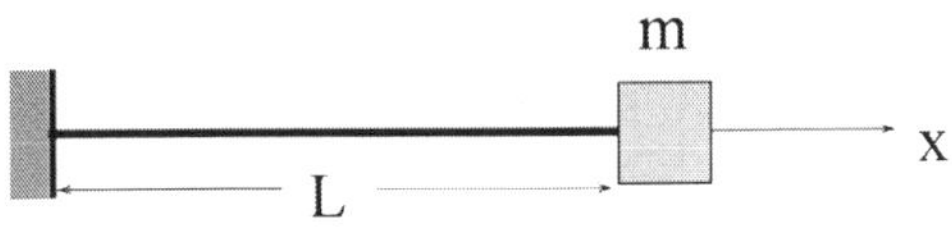

Fig. 9.6-1 A uniform cantilever beam with a mass attached at its free end.

Solution: The eigenfunction has the form (see Eq. 9.5-13)

$$\phi(x)=C_1(\cos\beta x+\cosh\beta x)+C_2(\cos\beta x-\cosh\beta x) \\ +C_3(\sin\beta x+\sinh\beta x)+C_4(\sin\beta x-\sinh\beta x)$$

The boundary conditions at the fixed end are

$$\phi(0)=0 \quad \text{and} \quad \phi'(0)=0: \tag{a}$$

These two boundary conditions lead to

$$C_1=C_3=0 \tag{b}$$

Thus, the eigenfunction can be written as

$$\phi(x)=C_2(\cos\beta x-\cosh\beta x)+C_4(\sin\beta x-\sinh\beta x)$$

The boundary condition at the free end is

$$\phi''(L)=0 \tag{d}$$

Eq. d leads to

$$C_2(-\cos\beta L-\cosh\beta L)+C_4(-\sin\beta L-\sinh\beta L)=0 \tag{e}$$

Letting $A=C_2/C_4$, from Eq. e, we have

$$A=\frac{C_2}{C_4}=-\frac{\sin\beta L+\sinh\beta L}{\cos\beta L+\cosh\beta L} \tag{f}$$

Thus, the eigenfunction can be written as

$$\phi(x)=A(\cos\beta x-\cosh\beta x)+(\sin\beta x-\sinh\beta x) \tag{g}$$

Finally, from Eq. 9.6-5, we have

$$\left(\beta_m^4-\beta_n^4\right)\int_0^L\phi_m\phi_n dx=\left[\left(\phi_n\frac{d^3\phi_m}{dx^3}-\phi_m\frac{d^3\phi_n}{dx^3}\right)+\left(\frac{d\phi_m}{dx}\frac{d^2\phi_n}{dx^2}-\frac{d\phi_n}{dx}\frac{d^2\phi_m}{dx^2}\right)\right]_0^L \tag{h}$$

To verify the orthogonal relations, we need to show that the right side of Eq. h is identically equal to zero if $m \neq n$. Indeed, this can be shown as follows:

When $x = 0$, since $\phi_m(0)=\phi_n(0)=0$ and $\phi'_m(0)=\phi'_n=0$, the right side of Eq. h is zero.

Next, when $x=L$, from the boundary condition at the free end, we have

$$\frac{d^3\phi_n}{dx^3}=-\frac{m\omega^2}{EJ}\phi_n \tag{i}$$

Thus, the first term of the right side of Eq. h becomes

$$\phi_n(L)\left(-\frac{m\omega^2}{EJ}\phi_m(L)\right)-\phi_m(L)\left(-\frac{m\omega^2}{EJ}\phi_n(L)\right)=0 \tag{j}$$

Finally, when $x=L$, we have $\phi''(L)\equiv 0$. That is, when $x=L$, the second term of Eq. h is zero.

Taken together, if $m \neq n$, the right side of Eq. h is identically equal to zero. Thus, we have orthogonal relations

$$\int_0^L \phi_m\phi_n dx=0 \quad (m \neq n) \tag{k}$$

9.7 Effects of Rotary Inertia, Shear, and Axial Loading

In our foregoing discussion of beam vibration, we restricted our attention to *Euler-Bernoulli beams*—that is, beams whose cross-section dimensions are small compared with the beam length and whose displacements are solely due to bending moments. In this section, we relax these restrictions, allowing for larger dimension cross sections and for axial loading.

(1) The Effect of Rotary Inertia and Shearing Deformation

When the cross-section dimensions of a beam are sufficiently large so that shear forces (forces tangent to the cross section) produce rotation in the cross section comparable to rotation produced by bending moments, the beam is sometimes called a *Timoshenko beam* [9.2]. With such beams, the cross section may significantly resist rotation during vibration due to the moment of inertia of a cross-sectional beam element [9.2, 9.7].

To explore this, consider a beam vibrating laterally in the X-Y plane—that is, where the X-axis is along the beam axis and the displacement is in the Y direction. Consider an element of the beam, also in the X-Y plane, as represented in Fig. 9.7-1. The positive directions for shear forces, bending moments, and displacement are those shown in the figure.

The slope of the deflection curve of the beam is $\partial u/\partial x$, which depends not only on the rotation of cross sections of the beam but also on the shear. Let θ be the slope of the deflection curve when the shear force is neglected and let φ be the angle of shear at the neutral axis in the same section. Then we have

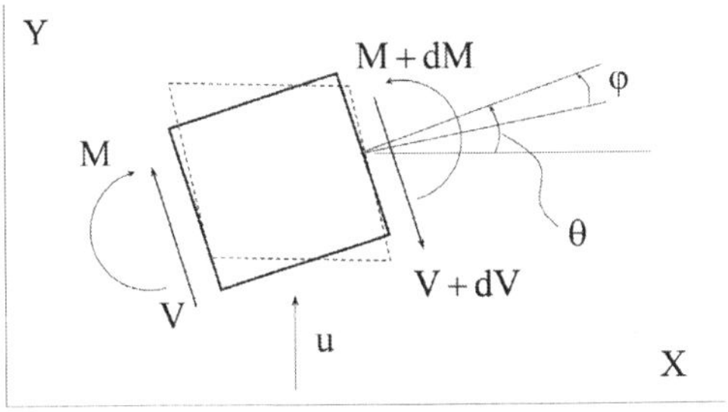

Fig. 9.7-1 Beam element of a beam vibrating in the *X–Y* plane [9.2, 9.7].

$$\frac{\partial u}{\partial x}=\theta-\varphi \tag{9.7-1}$$

From the elementary beam bending theory, we have the following relations between the bending moment and shear force [9.8]

$$M=EJ\frac{\partial \theta}{\partial x} \quad \text{and} \quad V=\varphi\,\kappa GA=\left(\theta-\frac{\partial u}{\partial x}\right)\kappa GA, \tag{9.7-2}$$

where A is the cross-section area, κ is called a shape factor, G is the shear modulus, and κGA is called an *effective shear modulus* (see, for example, [9.8], p. 343). The differential equation of rotation of an element then becomes

$$\left(J_p dx\right)\frac{\partial^2 \theta}{\partial t^2}=\frac{\partial M}{\partial x}dx-Vdx, \tag{9.7-3}$$

where J_p is the rotary moment of inertia per unit length of the beam in the X-Y plane. Substituting from Eq. 9.7-2 into Eq. 9.7-3, we obtain

$$EJ\frac{\partial^2 \theta}{\partial x^2}+\left(\frac{\partial u}{\partial x}-\theta\right)\kappa GA-J_p\frac{\partial^2 \theta}{\partial t^2}=0 \tag{9.7-4}$$

The differential equation for the translatory motion of the same element in the vertical direction is

$$\frac{\partial V}{\partial x}dx=-(\rho dx)\frac{\partial^2 u}{\partial t^2}, \tag{9.7-5}$$

where ρ is mass per unit length of the beam. Substituting from the second equation of 9.7-2 into Eq. 9.7-5, we obtain

$$\left(\frac{\partial \theta}{\partial x}-\frac{\partial^2 u}{\partial x^2}\right)\kappa GA+\rho\frac{\partial^2 u}{\partial t^2}=0 \tag{9.7-6}$$

Eliminating θ between Eqs. 9.7-4 and 9.7-6, the following differential equation for the lateral vibration of a uniform beam is obtained [9.7]

$$EJ\frac{\partial^4 u}{\partial x^4}+\rho\frac{\partial^2 u}{\partial t^2}-\left(J_p+\frac{EJ\rho}{\kappa GA}\right)\frac{\partial^4 u}{\partial x^2\partial t^2}+\frac{J_p\rho}{\kappa GA}\frac{\partial^4 u}{\partial t^4}=0 \tag{9.7-7}$$

If the effect of shear force on the slope of the deflection curve is neglected ($\varphi=0$), it follows from Eq. 9.7-2 that the shear modulus κGA must be infinite (otherwise, V would be zero). Therefore, letting κGA approach infinity in Eq. 9.7-7, we obtain

$$EJ\frac{\partial^4 u}{\partial x^4}+\rho\frac{\partial^2 u}{\partial t^2}-J_p\frac{\partial^4 u}{\partial x^2\partial t^2}=0 \tag{9.7-8}$$

Eq. 9.7-8 is the differential equation for the lateral vibration of a uniform beam without the effect of the shear force but with the effect of rotary inertia on the slope of the deflection curve.

Example 9.7-1 [9.7] Determine the effect of rotary inertia on the natural frequency of a uniform beam with hinged-hinged ends.

Solution: From Example 9.5-2, we know that the natural frequencies for a uniform beam with hinged-hinged ends without the rotary inertia effect are

$$\omega_n=\sqrt{\frac{EJ}{\rho}}\left(\frac{n\pi}{L}\right)^2 \tag{a}$$

The equation of motion including the effect of rotary inertia is Eq. 9.7-8. Since we have a hinged-hinged beam, the mode shapes must be sine functions, as shown in Example 9.5-2. The solution of Eq. 9.7-8 in this case is thus of the form

$$u(x,t)=C\sin\frac{n\pi x}{L}\sin(\omega_c t+\psi), \tag{b}$$

where ω_c is a natural frequency corrected for the rotary inertia effect. Substituting from Eq. b into Eq. 9.7-8, we obtain

$$\omega_c^2\left[1+\left(\frac{n\pi r}{L}\right)^2\right]=\frac{EJ}{\rho}\left(\frac{n\pi}{L}\right)^4, \tag{c}$$

where $r^2=J/A$. (Recall that $J_p=\rho J/A$.) From Eq. a, we see that the right side of this equation is ω_n^2. Thus, we have

$$\omega_c=\frac{\omega_n}{\sqrt{1+\left(\frac{n\pi r}{L}\right)^2}}\approx\left[1-\frac{1}{2}\left(\frac{n\pi r}{L}\right)^2\right]\omega_n, \tag{d}$$

where the approximation is obtained by using the first term of the binomial expression of $\left[1+(n\pi r/L)^2\right]^{(-1/2)}$.

Observe that the effect of rotary inertia is to decrease the natural frequency, and the effect is more pronounced for the higher modes.

(2) The Effect of Axial Force on Lateral Vibration

A beam in lateral vibration may also be subjected to an axial force. To evaluate the effect of an axial force, consider the free-body diagram of an element of a long beam in lateral vibration, as in Fig. 9.7-2 [9.2, 9.7]. The force T is assumed to be constant for small deflections of the beam, and other higher-order effects are assumed to be negligible. From Newton's law, the equation of motion of the element is seen then to be

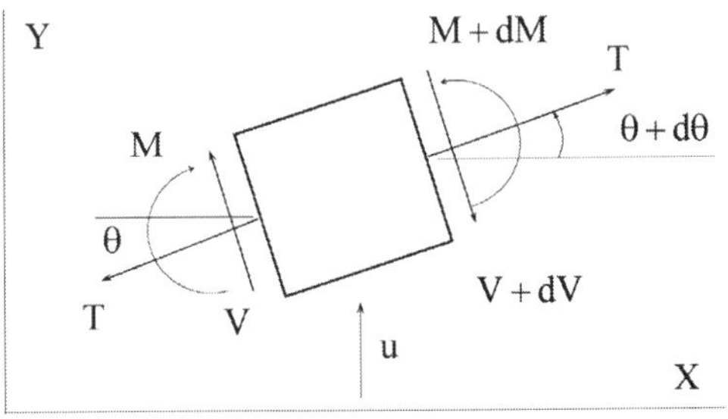

Fig. 9.7-2 Lateral vibration of beam with axial tension [9.2, 9.7].

$$\rho\frac{\partial^2 u}{\partial t^2}dx = -\left(V+\frac{\partial V}{\partial x}dx\right)+V+T\left(\theta+\frac{\partial\theta}{\partial x}dx\right)-T\theta$$

or

$$\rho\frac{\partial^2 u}{\partial t^2} = -\frac{\partial V}{\partial x}+T\frac{\partial^2 u}{\partial x^2}, \tag{9.7-9}$$

where ρ is mass per unit length. By using the relations

$$V=\frac{\partial M}{\partial x} \quad \text{and} \quad M=EJ\frac{\partial^2 u}{\partial x^2},$$

Eq. 9.7-9 becomes

$$\rho\frac{\partial^2 u}{\partial t^2}+\frac{\partial^2}{\partial x^2}\left(EJ\frac{\partial^2 u}{\partial x^2}\right)-T\frac{\partial^2 u}{\partial x^2}=0 \tag{9.7-10}$$

If EJ is constant, we obtain

$$\frac{\partial^2 u}{\partial t^2}+a^2\frac{\partial^4 u}{\partial x^4}-c^2\frac{\partial^2 u}{\partial x^2}=0, \tag{9.7-11}$$

where

$$a^2=\frac{EJ}{\rho} \quad \text{and} \quad c^2=\frac{T}{\rho} \tag{9.7-12}$$

Eq. 9.7-10 is the governing equation including the effect of axial tension. If the beam is under compression, the sign of T is reversed.

Example 9.7-3 [9.2, 9.7] Find the natural frequency of lateral vibration of a uniform hinged-hinged beam subjected to an axial force T.

Solution: From Example 9.5-2, the general solution of the problem has the form

$$u(x,t)=C\sin\frac{n\pi x}{L}\sin(\omega t+\psi) \tag{a}$$

Substituting into Eq. 9.7-11 and simplifying, we obtain

$$\omega^2=a^2\left(\frac{n\pi}{L}\right)^4+c^2\left(\frac{n\pi}{L}\right)^2 \tag{b}$$

Note that from Example 9.5-2 and Eq. 9.7-12, a^2 and c^2 are

$$a^2=\frac{EJ}{\rho},\quad c^2=\frac{T}{\rho} \tag{c}$$

With these results, it follows from Eq. b that

$$\omega=\left(\frac{n\pi}{L}\right)\sqrt{\left(\frac{n\pi}{L}\right)^2\frac{EJ}{\rho}+\frac{T}{\rho}} \tag{d}$$

Observe that if T=0, the natural frequency is identical to that of a simply supported beam, as in Example 9.5-2. If EJ=0, the natural frequency is that of lateral vibration of a string, as in Eq. 9.2-21. The tension will stiffen the beam and thereby increase its natural frequency. If, instead of tension, a compressive force is applied, the frequency of vibration is

$$\omega=\left(\frac{n\pi}{L}\right)\sqrt{\left(\frac{n\pi}{L}\right)^2\frac{EJ}{\rho}-\frac{T}{\rho}} \tag{e}$$

If T approaches $\pi^2 EJ/L$, the frequency of the fundamental type vibration approaches zero, because at this value of T, the compressive force attains its critical value, and the beam buckles.

9.8 Forced Vibration

To study forced vibration of continuous systems, it is helpful to introduce and use Lagrange's equations. Recall that Lagrange's equations use potential and kinetic energy functions. In this section, we develop these functions for the forced lateral vibration of a beam, and then substitute them into Lagrange's equations to obtain the equations of motion.

To begin, recall also that by using the separation of variables method, we found the general solution of wave equation has the form (see Sections 9.2 and 9.3)

$$u(x,t)=\sum_{n=1}^{\infty}\phi_n(x)q_n(t) \tag{9.8-1}$$

Next, recall further that the potential energy U for lateral deformation of a beam may be expressed as [9.9]

$$U=\frac{EJ}{2}\int_0^L\left(\frac{\partial^2 u}{\partial x^2}\right)^2 dx, \tag{9.8-2}$$

where E is the modulus of elasticity, J is the second moment of cross-section area, L is the beam length, and u is the lateral displacement.

By substituting from Eq. 9.8-1 into Eq. 9.8-2, we have

$$U=\frac{EJ}{2}\sum_{n=1}^{\infty}q_n^2\int_0^{\infty}\left(\frac{d^2\phi}{dx^2}\right)^2 dx=\frac{1}{2}\sum_{n=1}^{\infty}q_n^2 k_{nn}, \tag{9.8-3}$$

where

$$k_{nn}=EJ\int_0^L\left(\frac{d^2\phi_n}{dx^2}\right)^2 dx \tag{9.8-4}$$

k_{nn} is called the *generalized stiffness* for the n-th mode. The terms containing products of different q_n disappear from the expressions of Eqs. 9.8-3 and 9.8-4 due to orthogonal relations (see Eq. 9.6-7).

Similarly, the kinetic energy of the vibrating beam is

$$T=\frac{\rho}{2}\sum_{n=1}^{\infty}\dot{q}_n^2\int_0^L\phi_n^2 dx=\frac{1}{2}\sum_{n=1}^{\infty}\dot{q}_n^2 m_{nn}, \tag{9.8-5}$$

where ρ is the mass per unit length and m_{nn} is defined as

$$m_{nn}=\rho\int_0^L\phi_n^2 dx \tag{9.8-6}$$

m_{nn} is called the *generalized mass* for the n-th mode. Again, the terms containing products of different q_n disappear from the expressions of Eqs. 9.8-5 and 9.8-6 due to orthogonal relations (see Eq. 9.6-6).

It is convenient to use Eqs. 9.6-14 and 9.6-19 to calculate the potential and kinetic energies of a beam in lateral vibration.

Lagrange's equations of motion for small oscillation of a beam are then

$$\frac{d}{dt}\left(\frac{\partial T}{\partial \dot{q}_n}\right)+\frac{\partial U}{\partial q_n}=Q_n \quad (n=1,2,3,\cdots), \tag{9.8-7}$$

where Q_n is the generalized force corresponding to the coordinate q_n. Substituting from Eqs. 9.8-3 and 9.8-5 into Eq. 9.8-7, the equations of motion become

$$m_{nn}\ddot{q}_n+k_{nn}q_n=Q_n \quad (n=1,2,3,\cdots) \tag{9.8-8}$$

Example 9.8-1 A force $P=P_0\sin\omega t$ is applied laterally at the free end of a uniform cantilever beam, as represented in Fig. 9.8-1. (1) Calculate the first three generalized masses m_{nn} and stiffnesses k_{nn} $(n=1,2,3)$, and show that $k_{nn}/m_{nn}=\omega_n^2$.
(2) Find the general expression for the forced vibration response.

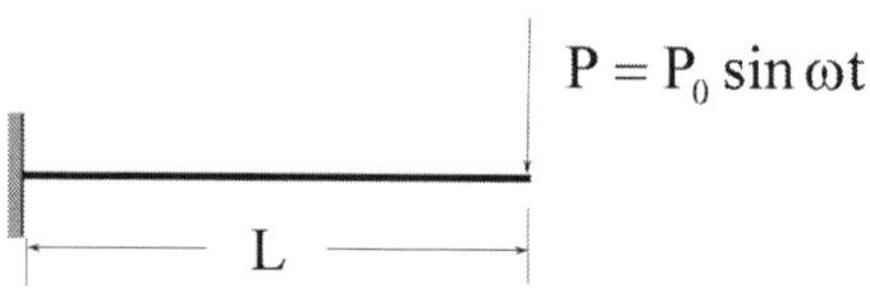

Fig. 9.8-1 Forced vibration of a uniform cantilever beam.

Solution: (1) From Example 9.5-4, an eigenfunction $\varphi_n(x)$ can be expressed as

$$\phi_n(x)=A_n(\cos\beta_n x-\cosh\beta_n x)+(\sin\beta_n x-\sinh\beta_n x) \tag{a}$$

where

$$A_n=\frac{\cos\beta_n L+\cosh\beta_n L}{\sin\beta_n L-\sinh\beta_n L}=-\frac{\sin\beta L+\sinh\beta L}{\cos\beta L+\cosh\beta L} \tag{b}$$

The frequency equation is

$$\cos\beta L\cosh\beta L=-1 \tag{c}$$

The first three roots of this equation are

$$\beta_1 L=1.8751, \quad \beta_2 L=4.6941, \quad \beta_3 L=7.8548 \tag{d}$$

The $\phi_n(x)$ can be used to evaluate the generalized masses m_{nn} as (see Eq. 9.8-6)

$$m_{nn}=\rho\int_0^L \phi_n^2 dx \tag{e}$$

From Eq. 9.6-14, we have

$$\int_0^L \phi_n^2 dx=\frac{L}{4}\left(\phi_n^2\right)_{x=L} \tag{f}$$

For the first three modes, we obtain (see Eqs. a and b)

$$\phi_1(L)=2.7245, \quad \phi_2(L)=-1.9637, \quad \phi_3(L)=2.0016 \tag{g}$$

Thus, we obtain

$$m_{11} = \frac{\rho L}{4}(7.4229), \quad m_{22} = \frac{\rho L}{4}(3.8563), \quad m_{33} = \frac{\rho L}{4}(4.0062) \tag{h}$$

Next, the generalized stiffnesses can be calculated as (see Eq. 9.8-4)

$$k_{nn} = EJ\int_0^L \left(\frac{d^2\phi_n}{dx^2}\right)^2 dx = EJ\beta_n^4 \int_0^L \phi_n^2 dx = \frac{EJL\beta_n^4}{4}\phi_n^2(L), \tag{i}$$

from which we have

$$k_{11} = \frac{EJL\beta_1^4}{4}(7.4240), \quad k_{22} = \frac{EJL\beta_2^4}{4}(3.8563), \quad k_{33} = \frac{EJL\beta_3^4}{4}(4.0062) \tag{j}$$

From the above results, it can be seen that

$$\frac{k_{nn}}{m_{nn}} = \frac{EJ}{\rho}\beta_n^4 = a^2\beta_n^4 = \omega_n^2 \tag{k}$$

(2) The generalized force corresponding to the generalized coordinate q_n is

$$Q_n = \phi_n(L)P_0 \sin\omega t \qquad (n=1,2,3,\cdots) \tag{l}$$

Substituting Q_n into Lagrange's equations, we have

$$m_{nn}\ddot{q}_n + k_{nn}q_n = \phi_n(L)P_0 \sin\omega t \qquad (n=1,2,3,\cdots) \tag{m}$$

Considering only this part of the forced vibrations, we obtain

$$q_n = \frac{\phi_n(L)P_0}{m_{nn}}\frac{1}{\omega_n^2 - \omega^2}\left[\sin\omega t - \frac{\omega}{\omega_n}\sin\omega_n t\right] \qquad (n=1,2,3,\cdots) \tag{n}$$

The general expression for the vibration produced by the force is then given by

$$u(x,t) = P_0 \sum_{n=1}^{\infty} \frac{\phi_n(L)\phi_n(x)}{m_{nn}(\omega_n^2 - \omega^2)}\left[\sin\omega t - \frac{\omega}{\omega_n}\sin\omega_n t\right] \tag{o}$$

The steady-state forced vibration is

$$u(x,t) = P_0 \sum_{n=1}^{\infty} \frac{\phi_n(L)\phi_n(x)}{m_{nn}(\omega_n^2 - \omega^2)}\sin\omega t \tag{p}$$

The steady-state forced vibration at the free end is found to be

$$u(L,t) = \frac{4P_0L^3}{EJ}\sin\omega t\left(0.08091\alpha_1 + 0.00206\alpha_2 + 0.000206\alpha_3 + \cdots\right), \tag{q}$$

where

$$\alpha_n = \frac{1}{1-\omega^2/\omega_n^2} \tag{r}$$

9.9 Ritz Method

We have seen earlier that for large, complex discrete systems, the Rayleigh method is useful for obtaining an approximation for the fundamental frequency. In applying the Rayleigh method for continuous systems, we need to first make an assumption concerning the deformation shape of the vibration. If, however, the assumed deformation shape is inaccurate, it then simulates a system having additional constraint. The calculated frequency will then be larger than the true fundamental frequency.

A better approximation for the fundamental frequency, and even approximation of the higher-order frequencies, can be obtained using the Ritz method, which is an extension of the Rayleigh method.

In using the Ritz method, we approximate the vibration mode $\psi(x)$ as a finite series of linearly independent functions $\phi_n(x)$, each satisfying the boundary conditions. That is, the mode $\psi(x)$ is represented as

$$\psi(x) = \sum_{i=1}^{n} a_i \phi_i(x), \tag{9.9-1}$$

where the a_i ($i=1,2,3,\cdots,n$) are parameters to be determined.

Approximating a mode shape by a finite series of functions, as in Eq. 9.9-1, is equivalent to modeling the continuous system by a discrete system, in this case, with n degrees of freedom. With the representation of Eq. 9.9-1 being discrete, it is (as with the Rayleigh method) analogous to placing constraints on the continuous system. Since constraints make a system stiffer, the discrete approximation of Eq. 9.9-1 will produce higher valued frequencies than occur with the original continuous system. The Ritz method, therefore, proceeds by adjusting the a_i so as to minimize the frequency values.

To illustrate the method, consider the lateral vibration of a uniform beam. Let the displacement $u(x,t)$ of the beam axis be of the form

$$u(x,t) = \psi(x)q(t), \tag{9.9-2}$$

where $q(t)$ is a harmonic function. Then the potential energy U of the beam due to the bending occurring from the vibration is

$$U = \frac{1}{2}q^2 \int_0^L EI\left(\frac{d^2\psi}{dx^2}\right)^2 dx, \tag{9.9-3}$$

where, as before, EJ is the flexural rigidity of the beam. Without loss of generality, assuming $q_{max}=1$, the maximum potential energy is then

$$U_{max}=\frac{1}{2}\int_0^L EJ\left(\frac{d^2\psi}{dx^2}\right)^2 dx=\frac{1}{2}\int_0^L EJ\left(\sum_{i=1}^n a_i\phi_i''\right)\left(\sum_{j=1}^n a_j\phi_j''\right)dx \quad \left(\phi''=d^2\phi/dx^2\right) \tag{9.9-4}$$

In like manner, the kinetic energy T of the beam becomes

$$T=\frac{1}{2}\dot{q}^2\int_0^L \rho\psi^2 dx, \tag{9.9-5}$$

where ρ is mass per unit length of the beam. Noting that $\dot{q}_{max}=\omega q_{max}$, and assuming $q_{max}=1$, the maximum kinetic energy is then

$$T_{max}=\frac{1}{2}\omega^2\int_0^L \rho\psi^2 dx=\frac{1}{2}\omega^2\int_0^L \rho\left(\sum_{i=1}^n a_i\phi_i\right)\left(\sum_{j=1}^n a_j\phi_j\right)dx\equiv\omega^2 T^*_{max}, \tag{9.9-6}$$

where T^*_{max} is called the *reference kinetic energy* and is defined as

$$T^*_{max}=\frac{1}{2}\int_0^L \rho\left(\sum_{i=1}^n a_i\phi_i\right)\left(\sum_{j=1}^n a_j\phi_j\right)dx \tag{9.9.7}$$

Consider a conservative system, with a continuous exchange between kinetic and potential energies; the maximum values of the energies will be equal. That is,

$$U_{max}=T_{max} \tag{9.9-8}$$

By substituting from Eqs. 9.9-4 and 9.9-6 into Eq. 9.9-8 and solving for ω^2, we obtain

$$\omega^2=\frac{U_{max}}{T^*_{max}} \tag{9.9-9}$$

Following the Ritz method, we now attempt to choose the a_i $(i=1,2,3,\cdots,n)$ so as to minimize the approximate frequency or, equivalently, to minimize the square ω^2 of the approximate frequency. That is, we choose the a_i so that

$$\frac{\partial}{\partial a_i}\left(\frac{U_{max}}{T^*_{max}}\right)=0 \quad (i=1,2,3,\cdots,n) \tag{9.9-10}$$

Performing the indicated differentiation, we obtain

$$T^*_{max}\frac{\partial U_{max}}{\partial a_i}-U_{max}\frac{\partial T^*_{max}}{\partial a_i}=0 \quad (i=1,2,3,\cdots,n) \tag{9.9-11}$$

Recall from Eq. 9.9-9 that $U_{max} = \omega^2 T^*_{max}$. Thus, from Eq. 9.9-11, we obtain

$$\frac{\partial U_{max}}{\partial a_i} - \omega^2 \frac{\partial T^*_{max}}{\partial a_i} = 0 \quad (i=1,2,3,\cdots, n) \tag{9.9-12}$$

Substituting from Eqs. 9.9-4 and 9.9-7 into Eq. 9.9-12, we obtain a homogeneous set of linear algebraic equations for the a_i as

$$\sum_{j=1}^{n} (k_{ij} - \omega^2 m_{ij}) a_j = 0 \quad (i=1,2,\cdots,n), \tag{9.9-13}$$

where k_{ij} and m_{ij} are the generalized stiffnesses and generalized masses, respectively, defined as

$$k_{ij} = \int_0^L EJ\phi_i''\phi_j''dx \tag{9.9-14}$$

$$m_{ij} = \int_0^L \rho\phi_i\phi_j dx \tag{9.9-15}$$

Eq. 9.9-13 may be written in matrix form as

$$\left[K - \omega^2 M\right]\{a\} = 0 \tag{9.9-16}$$

For a nontrivial solution, the determinant of the coefficients must be zero. That is,

$$\left|K - \omega^2 M\right| = 0 \tag{9.9-17}$$

Expansion of the determinant in Eq. 9.9-17 provides a polynomial equation for n values of ω^2

In a similar way, we see that for longitudinal beam vibration, the frequencies satisfy an equation identical in form to Eq. 9.9-17, except that the generalized stiffnesses and masses are

$$k_{ij} = \int_0^L AE\phi_i'\phi_j'dx \tag{9.9-18}$$

$$m_{ij} = \int_0^L \rho\phi_i\phi_j dx, \tag{9.9-19}$$

where, as before, A is the cross-section area.

For torsional beam vibration, the equations are the same except that the generalized stiffnesses and masses are

$$k_{ij} = \int_0^L GJ_p \phi_i' \phi_j' dx \tag{9.9-20}$$

$$m_{ij} = \int_0^L \rho \phi_i \phi_j dx, \tag{9.9-21}$$

where, as before, G is the shear modulus and J_p is the second polar moment of cross-section area.

Example 9.9-1 Find the fundamental frequency of lateral vibration of a uniform cantilever beam, as shown in Fig. 9.9-1, by Ritz method.

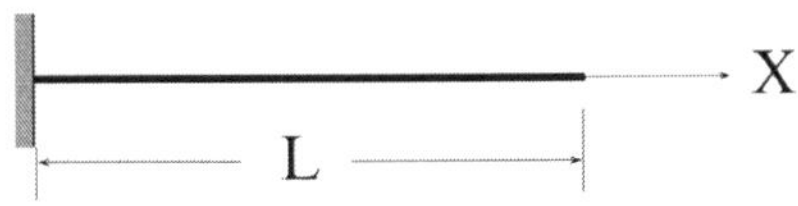

Fig. 9.9-1 A uniform cantilever beam.

Solution: Let the beam displacement be approximated by $\psi(x)$, a function represented in turn by two functions $\phi_1(x)$ and $\phi_2(x)$ given by

$$\psi(x) = a_1\phi_1(x) + a_2\phi_2(x) = a_1\left(\frac{x}{L}\right)^2 + a_2\left[2\left(\frac{x}{L}\right)^2 - \left(\frac{x}{L}\right)^3\right]$$

It is easy to see that each term of $\psi(x)$ satisfies the geometric boundary conditions

$$\psi(0) = 0 \quad \text{and} \quad \psi'(0) = 0 \tag{a}$$

From Eqs. 9.9-14 and 9.9-15, we find the k_{ij} and m_{ij} $(i,j=1,2)$ to be

$$k_{11} = \int_0^L EJ\phi''^2_1 dx = 4EJ/L^3$$

$$k_{12} = k_{21} = \int_0^L EJ\phi_1''\phi_2'' dx = 2EJ/L^3$$

$$k_{22} = \int_0^L EJ\phi''^2_2 dx = 4EJ/L^3$$

$$m_{11} = \int_0^L \rho\phi_1^2 dx = \rho L/5$$

$$m_{12} = m_{21} = \int_0^L \rho\phi_1\phi_2 dx = 30\rho L/7$$

$$m_{22} = \int_0^L \rho\phi_2^2 dx = 29\rho L/105,$$

where ρ is the mass per unit length of the beam.

We can assemble these results into matrices K and M as

$$K = \frac{EJ}{L^3}\begin{bmatrix} 4 & 2 \\ 2 & 4 \end{bmatrix} \quad \text{and} \quad M = \rho L \begin{bmatrix} 1/5 & 30/7 \\ 30/7 & 29/105 \end{bmatrix}$$

The frequency equation is

$$\left| K - \omega^2 M \right| = 0 \tag{b}$$

Expansion of Eq. b leads to

$$0.000794\omega^4 - 0.9714(EJ / \rho L^4)\omega^2 + 12(EJ\rho L^4)^2 = 0, \tag{c}$$

from which we obtain the fundamental frequency as

$$\omega_1 = \frac{3.533}{L^2}\sqrt{\frac{EJ}{\rho}} \tag{d}$$

The exact solution is $\omega_1 = \dfrac{3.516}{L^2}\sqrt{\dfrac{EJ}{\rho}}$ (see Table 9.5-2) (e)

Comparing Eqs. d and e, we see that the error of the approximation is less than 0.5%. Observe that, although the function $\psi(x) = a_1 x^2 + a_2 x^3$ does not satisfy all the boundary conditions (only the geometric boundary conditions), the result obtained with the Ritz method is still satisfactory.

Problems

9.1 Find the natural frequencies and mode shapes of longitudinal vibration of a rod with both ends free (a *free-free* rod).

Ans. $\omega_n = \dfrac{n\pi}{L}\sqrt{\dfrac{AE}{\rho}}, \quad \phi(x) = \cos\dfrac{n\pi x}{L}, \quad (n=1,2,\cdots)$

9.2 Derive mode shape functions and frequency equations for various configurations of a uniform beam of length L in longitudinal vibration, as shown in Table 9.3-1.

Ans. (1) $\cos(n\pi x / L), \quad \omega_n = n\pi c / L,$

(2) $\sin(n\pi x / L), \quad \omega_n = n\pi c / L$

(3) $\sin(\beta_n x / L), \quad \beta_n \cos\beta_n = -kL / EA, \quad \omega_n = \beta_n c / L$

(4) $\sin(\beta_n x / L) + \beta_n (EA / k)\cos(\beta_n x / L), \quad \tan\beta_n + \beta_n (EA / kL) = 0, \quad \omega_n = \beta_n c / L$

(5) $\sin(\beta_n x / L) - (m_{\text{rod}} / m\beta_n)\cos(\beta_n x / L), \quad \beta_n \tan\beta_n = m_{\text{rod}} / m, \quad \omega_n = \beta_n c / L$

(6) $\sin(\beta_n x / L), \quad \beta_n \tan\beta_n = m_{\text{rod}} / m, \quad \omega_n = \beta_n c / L$

9.3 Assume a uniform beam with hinged-hinged ends is loaded by a uniformly distributed load of intensity w and released with zero initial velocity. Find the subsequent motion of the beam.

Hint: $u(x,0) = \dfrac{w}{24EJ}\left(L^3 x - 2Lx^3 + x^4\right)$

Ans. $u = \dfrac{4wL^4}{\pi^5 EJ} \sum_{n=1,3,5,...}^{\infty} \dfrac{1}{n^5} \sin\dfrac{n\pi x}{L} \cos\omega_n t, \quad \omega_n = \sqrt{\dfrac{EJ}{\rho}} \left(\dfrac{n\pi}{L}\right)^2$

9.4 A pulsating force $P = P_0 \sin\omega t$ is applied at $x=x_1$ to a uniform beam with hinged-hinged ends. Find the forced vibration of the beam.

Ans. $u = \dfrac{2P_0}{\rho L} \sin\omega t \sum_{n=1}^{\infty} \dfrac{\sin\dfrac{n\pi x_1}{2} \sin\dfrac{n\pi x}{L}}{\omega_n^2 - \omega^2}, \quad \omega_n = \sqrt{\dfrac{EJ}{\rho}} \left(\dfrac{n\pi}{L}\right)^2$

9.5 A uniform beam with hinged-hinged ends is loaded by a uniformly distributed load of intensity $w = w_0 \sin\omega t$. Find the forced vibration of the beam.

Ans. $u = \dfrac{4w_0}{\rho\pi} \sin\omega t \sum_{n=1}^{\infty} \dfrac{\sin\dfrac{n\pi x}{L}}{n\left(\omega_n^2 - \omega^2\right)}, \quad \omega_n = \sqrt{\dfrac{EJ}{\rho}} \left(\dfrac{n\pi}{L}\right)^2$

9.6 A pulsating moment $M = M_0 \sin\omega t$ is applied at $x=0$ to a uniform beam with hinged-hinged ends. Find the forced vibration of the beam.

Ans. $u = \dfrac{2M_0 L^2}{\pi^3 EJ} \sin\omega t \sum_{n=1}^{\infty} \dfrac{1}{n^3 (1 - \omega^2 / \omega_n^2)} \sin\dfrac{n\pi x}{L}, \quad \omega_n = \sqrt{\dfrac{EJ}{\rho}} \left(\dfrac{n\pi}{L}\right)^2$

9.7 If an axial constant force P_0 is suddenly applied at the free end of a cantilever beam, find the forced vibration of the beam.

Ans. $u = \dfrac{8P_0 L}{AE\pi^2} \sum_{n=1,3,5,..}^{\infty} \dfrac{1}{n^2} \sin\dfrac{n\pi}{2} \sin\dfrac{n\pi x}{2L} \left(1 - \cos\dfrac{n\pi ct}{2L}\right), \quad c = \sqrt{AE/\rho}$

9.8 A constant vertical force P_0 is moving with a constant velocity v along the length of a uniform beam with hinged-hinged ends. Let the force be at the left support at the initial time. Find the forced vibration of the beam.

Ans. $u = \dfrac{2P_0}{\rho L} \sum_{n=1}^{\infty} \dfrac{1}{\omega_n^2 - \omega^2} \sin\dfrac{n\pi x}{L} \sin\dfrac{n\pi vt}{L}, \quad \omega_n = \dfrac{EJ}{\rho} \left(\dfrac{n\pi}{L}\right)^2$

9.9 The contribution of each mode to the dynamic behavior of an undamped structure can be evaluated as follows: from Eq. 7.6-16 in Chapter Seven, the FRF matrix can be expressed as

$$H(\omega) = \sum_{n=1}^{\infty} \frac{\mathbf{u}_n \mathbf{u}_n^T}{m_n(\omega_n^2 - \omega^2)},$$

where $\mathbf{u}_n$ is the n-th mode vector, and m_n is the generalized mass for mode n. When $\omega = 0$, $H(0)$ becomes

$$H(0) = \sum_{n=1}^{\infty} \frac{\mathbf{u}_n \mathbf{u}_n^T}{m_n \omega_n^2}$$

For a uniform beam with hinged-hinged ends, a mode function is $\phi_n(x)=\sin\frac{n\pi x}{L}$.

(1) Taking the first mode, find the approximate stiffness at the middle of the beam.

(2) Repeat this procedure by taking the first and third modes.

Ans. (1) $\frac{1}{k}=\frac{2L^3}{EJ\pi^4}=\frac{L^3}{48.7EJ}$; (2) $\frac{1}{k}=\frac{L^3}{48.1EJ}$

9.10 A pulsating force $P=P_0\sin\frac{\pi x}{L}\sin\omega t$ is applied to a uniform beam with hinged-hinged ends. Find the forced vibration of the beam.

Ans. $u(x,t)=\frac{P_0L^4}{\pi^4EJ(1-\omega^2/\omega_1^2)}\sin\frac{\pi x}{L}\sin\omega t,\ \omega_1=\frac{\pi^2}{L^2}\sqrt{\frac{EJ}{\rho}}$

9.11 Find the frequency equation and eigenfunction for lateral vibration of a uniform beam with fixed-fixed ends.

Ans. $\cos\beta L\cosh\beta L=1,\quad \beta^4=\omega^2\rho/EJ$

$$\phi(x)=A(\cos\beta x-\cosh\beta x)+(\sin\beta x-\sinh\beta x)$$

$$A=-\frac{\sin\beta L-\sinh\beta L}{\cos\beta L-\cosh\beta L}=\frac{\cos\beta L-\cosh\beta L}{\sin\beta L+\sinh\beta L}$$

9.12 Find the frequency equation and eigenfunction for lateral vibration of a uniform beam with free-free ends.

Ans. $\cos\beta L\cosh\beta L=1,\quad \beta^4=\omega^2\rho/EJ$

$$\phi(x)=A(\cos\beta x+\cosh\beta x)+(\sin\beta x+\sinh\beta x)$$

$$A=-\frac{\sin\beta L-\sinh\beta L}{\cos\beta L-\cosh\beta L}=\frac{\cos\beta L-\cosh\beta L}{\sin\beta L+\sinh\beta L}$$

9.13 Find the fundamental frequency of longitudinal vibration of a uniform cantilever beam. Assume that $\psi(x)=a_1(x/L)+a_2(x/L)^2$

Ans. $\omega_1=1.58\sqrt{\frac{AE}{\rho L^2}}$; Exact solution: $\omega_1=\frac{\pi}{2}\sqrt{\frac{AE}{\rho L^2}}$

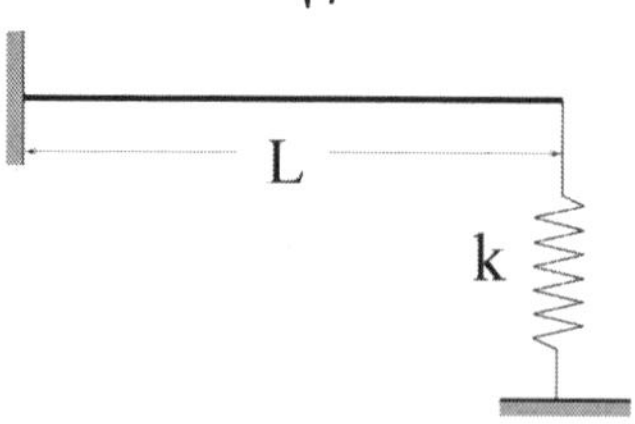

Fig. P9.14

9.14 A uniform beam of length L has a spring attached at $x=L$, as shown in Fig. P9.14. Find the frequency equation for lateral vibration.

Ans. $\frac{kL^3}{EJ(\beta L)^3}=\frac{1+\cos\beta L\cosh\beta L}{\cos\beta L\sinh\beta L-\sin\beta L\cosh\beta L}$

9.15 Show that Eq. 9.9-13 can be obtained from Hamilton's principle

$$\delta \int_0^{2\pi/\omega} (T-U)dt = 0,$$

where T and U are the kinetic energy and potential energy, respectively, of the beam.

9.16 Show that the eigenfunctions for the lateral vibrations of each of the beams in Fig.P9.14 are orthogonal.

References

9.1. Hildebrand, F. B., *Advanced Calculus for Applications*, 2nd ed., Prentice Hall, Inc., Upper Saddle River, NJ, 1976.

9.2. Tse, F. S., Morse, I. E., and Hinkle, R. T., *Mechanical Vibration, Theory and Applications*, Allyn and Bacon, Inc., Boston, 1979.

9.3. Thomson, W. T., and Dahleh, M. D., *Theory of Vibration with Applications*, 5th ed., Prentice Hall, Inc., Upper Saddle River, NJ, 1998.

9.4. Liu, C. Q., et al., "A New Method for Obtaining FRF of a Structure in Area Where Impact Hammer Cannot Reach," SAE paper 2007-01-2385, SAE International, Warrendale, PA, 2007.

9.5. Hearn, E. J., *Mechanics of Materials*, 2nd ed., Pergamon Press, New York, 1985.

9.6. Stokey, W. F., "Vibration of Systems Having Distributed Mass and Elasticity," in *Shock and Vibration Handbook*, C. M. Harris, ed., 3rd ed., McGraw-Hill, New York, 1987, pp. 7–13.

9.7. Timoshenko, S. P., Young, D. H., and Weaver, W., Jr., *Vibration Problems in Engineering*, 4th ed., John Wiley & Sons, New York, 1967.

9.8. Crandall, S. H., et al., *Dynamics of Mechanical and Electromechanical Systems*, Krieger Publishing Co., Malabar, FL, 1982.

9.9. Barr, A. H., *Mechanical Analysis and Design*, Elsevier, New York, 1981.

Chapter Ten

Engine Mounting Systems

10.1 Introduction

The design of engine mounting systems is one of the most important applications of vibration analysis. Engine mounting is a fundamental aspect in the design and assembly of motor vehicles. Engine mounting is so important that we are devoting an entire chapter to the underlying concepts, principles, and procedures.

In addition to the engine itself, an engine mounting system consists of three or four rubber or hydraulic mounts and a foundation (typically a vehicle frame). The mounts provide the connection between the engine and the foundation. With a vehicle frame (and also with a fixed foundation) and with the engine being the vehicle's largest concentrated mass, an improper design and positioning of the mounts can allow extremely harmful vibration to occur.

The effectiveness of an engine mounting system depends not only upon the performance characteristics of the mounts, but also upon the behavior of the engine and the foundation themselves. For example, the pulsating torques of a multicylinder engine is a major source of vibration in motor vehicles. Looking to the future in multicylinder engine technology where, for fuel economy, some of the cylinders will be selectively idled, the need for well-designed mounting systems will be even more important.

The design of an engine mounting system has the following major tasks: (1) location of the engine mass center [or center of gravity (*CG*)] and the orientation of the engine, (2) location and orientation of individual mounts, and (3) selection of stiffness characteristics for each mount. In engine mounting system design, a major objective is to decouple the engine rigid body modes. There are, however, only a few special configurations where the modes can be completely decoupled. Moreover, theoretical results are not always directly applicable in a given mounting configuration.

It is, therefore, virtually imperative to design engine mounting systems using available experimental data. The complexity of design problems requires accurate modeling to understand the dynamical issues. With reliable models, we can perform analytical and numerical simulations while an engine design, the mounting system, and the foundation are still in the conceptual stage. In this way, alternative engine mounting systems can be evaluated.

A significant portion of machine design time can also be reduced if generic high performance mounting systems can be incorporated, enabling computer-aided design and manufacture (CAD/CAM).

The objective of this chapter is to present a design optimization tool to assist engineers and designers in producing engine mounting systems with the greatest potential for decoupling the modes without violating packaging and/or manufacturing constraints.

10.2 Inertia Properties of an Engine

With engines being massive structures, their inertia properties play a principal role in their vibration and in the oscillatory forces transmitted to the mounts and through the mounts to the foundation (typically a vehicle frame). The inertia properties of an engine determine its resistance to motion. An engine's resistance to translation is determined by its mass, and its resistance to rotation is determined by its moments and products of inertia.

The mass of an engine is simply its weight divided by the gravity acceleration g. The moments and products of inertia are measures of the mass distribution of the engine. The mass distribution, in turn, depends upon the density of the engine materials and the geometric properties of the engine.

To make dynamic analyses of an engine, it is helpful to introduce a couple of geometric parameters: the *mass center* (or *center of gravity*) and *principal axes of inertia*. The center of gravity is an interior point, where, in a gravity field (e.g., the earth), the engine would have no tendency to rotate when supported at that point. Alternatively, the center of gravity (*CG*) is that point where the engine could be supported in *balanced equilibrium* (no tendency to move) by a single vertical force passing through the *CG*.

The principal inertia axes are perhaps less intuitive, but they are simply axes through an arbitrary reference point (typically the center of gravity, *CG*) in directions for which the inertia matrix is diagonal. The *inertia matrix* is a 3×3 symmetric array whose diagonal elements are moments of inertia and whose off-diagonal elements are products of inertia. The moments and products of inertia are determined by integration of the mass density multiplied by product combinations of coordinates, relative to the reference point, throughout the geometric region occupied by the engine.

References [10.1] and [10.2] provide details of analyses and computations based upon these concepts.

The irregularity of the geometry of most engines, however, is such that analytical determination of the *CG* location, the moments and products of inertia, and the principal axes of inertia is cumbersome and impractical. Instead, engineers and designers often resort to experimental techniques. We discuss some of these techniques in the following paragraphs.

We can readily measure an engine's weight. Determining the mass center (or *CG*) location and the elements of inertia matrix, however, requires ingenuity

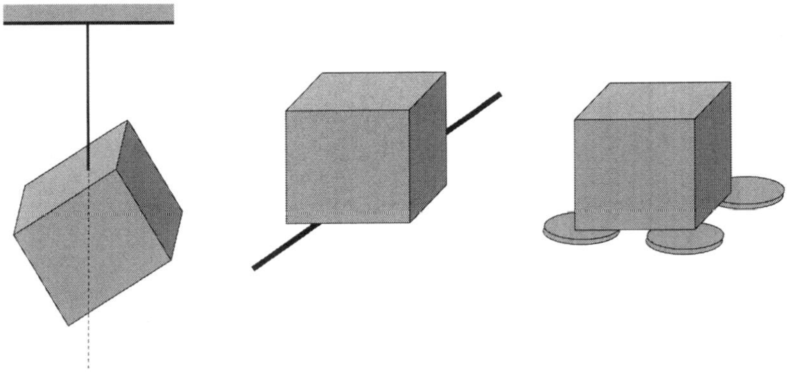

(a) (b) (c)

Fig. 10.2-1 Methods for finding the CG location.

in the experimental procedures. Suppose, for example, that an engine is suspended by a wire, as in Fig. 10.2-1 (a). Then, if the attachment point is moved to another point, and then again to still another point, the lines along the wire supports for the three attachment configurations will theoretically intersect at the *CG*, inside the engine. But in practice, the *CG* location is difficult to find using this method.

Another approach is to attempt to balance the engine on an edge, or line, as in Fig. 10.2-1 (b). The difficulty with this method, however, is maintaining stability of the rather heavy engine. The instability can be reduced by replacing the edge support by a narrow flat, but then the accuracy of the method is compromised.

Still another method is to place the engine on three scales, as in Fig. 10.2-1 (c). The *CG* location is then determined from moment equilibrium equations taken about the scale support points. There are practical limitations to this method as well in that the locations of the scale support points are difficult to precisely determine.

Recently, experimentalists have been attempting to determine inertia properties by using frequency response functions (FRFs). A drawback of this approach, however, is that at least six response locations are needed and all six rigid body modes need to be established. It is also difficult to excite all six modes.

In the following paragraphs, we present a new experimental method for finding not only the *CG* location but also the elements of the inertia matrix. This method is developed using a traditional torsion pendulum method, where the engine is suspended by a rod acting as a torsion spring.

Fig. 10.2-2 An illustrative disk used for determining the torsional stiffness of a suspension rod.

The first step in this method is to determine the equivalent spring modulus of the torsion rod.

To this end, we hang a precisely measured circular disk from the rod, as in Fig. 10.2-2. By knowing the weight of the disk and its dimensions, we can immediately determine its axial moment of inertia I_d. Then, by oscillating the disk about its axis and by measuring the period T_d of the torsional oscillation, we can determine the torsional stiffness k_r of the rod from the expression

$$k_r = \frac{4\pi^2 I_d}{T_d^2} \tag{10.2-1}$$

An accurate value for the oscillation period can be obtained by recording the time T_{total} for, say, n oscillations and then dividing by n. That is,

$$T_d = \frac{T_{\text{total}}}{n} \tag{10.2-2}$$

The second step is to conduct a test using the engine itself.

Specifically, the engine is hung and suspended by the torsion rod, as represented in Figs. 10.2-3 and 10.2-4. Then, by torsionally oscillating the

Fig. 10.2-3 Experimental setup for measuring moment of inertia of an engine. (Courtesy of Raven Engineering, Inc)

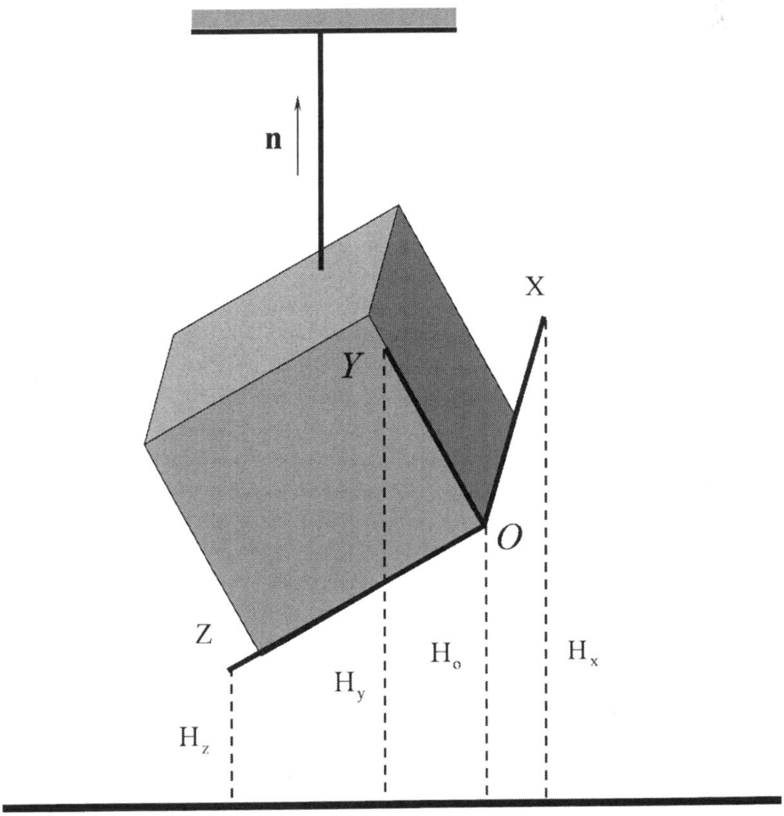

Fig. 10.2-4 An engine is suspended on a rod and a reference frame is mounted on it.

engine, we can determine the moment of inertia of the engine I_L about a line passing through the rod axis and the engine CG. That is,

$$I_L = \frac{k_r T^2}{4\pi^2}, \tag{10.2-3}$$

where T is the period of torsional vibration of the engine about the suspension rod.

In the remaining paragraphs of this section, we present details of the procedure and the associated computations.

(1) Reference Frame and Direction Cosines

Prior to hanging the test engine, we can establish a three-dimensional reference coordinate frame on the engine. Then, when the engine is suspended on the rod, we can accurately measure the following parameters: the heights, H_x, H_y, H_z, of points on the axes of the reference frame, and the height, H_O, of the origin O of the reference frame, all relative to a horizontal reference plane (see Fig. 10.2-4). Then, the direction cosines of the suspension rod relative to the reference frame can be calculated as

$$\cos\alpha = \frac{H_x - H_O}{a_x} \tag{10.2-4}$$

$$\cos\beta = \frac{H_y - H_O}{a_y} \tag{10.2-5}$$

$$\cos\gamma = \frac{H_z - H_O}{a_z}, \tag{10.2-6}$$

where a_x, a_y, a_z are the distances from the origin of the selected points on the X, Y, and Z axes of the reference frame. The angles α, β, γ can also be directly measured by using a digital projector. But it is more accurate to simply measure the distances.

(2) Inertia Matrix

Next, we can establish the relationship between the moment of inertia of the engine with respect to the suspension rod and the moments of inertia with respect to the reference frame. To this end, let $P(x, y, z)$ be any point in the engine, and let a unit vector **n** along the suspension rod L with respect to the reference frame be $\mathbf{n} = (\cos\alpha, \cos\beta, \cos\gamma)$. The moment of inertia of the engine about the suspension rod, I_L, may be expressed as

$$\begin{aligned} I_L &= I_{xx}\cos^2\alpha + I_{yy}\cos^2\beta + I_{zz}\cos^2\gamma \\ &\quad + I_{xy}2\cos\alpha\cos\beta + I_{yz}2\cos\beta\cos\gamma + I_{zx}\cos\gamma\cos\alpha, \end{aligned} \tag{10.2-7}$$

where

$$I_{xx} = \int (y^2 + z^2)dm, \quad I_{yy} = \int (z^2 + x^2)dm$$

$$I_{zz} = \int (x^2 + y^2)dm, \quad I_{xy} = -\int xydm$$

$$I_{yz} = -\int yzdm, \quad I_{zx} = -\int zxdm$$

In theory, we need six equations to solve for six unknowns: $I_{ij}\ (i, j = x, y, z)$. To avoid singularities, however, it is useful to consider a *redundant* set of equations. From experience, a solution with error less than 2% can be obtained by finding a least squares solution through testing 12 different hanging positions. That is, by writing Eq. 10.2-7 12 times for 12 different hanging positions, we have a set of equations for the six unknowns as

$$\left[H\right]\begin{bmatrix} I_{xx} \\ I_{yy} \\ I_{zz} \\ I_{xy} \\ I_{yz} \\ I_{zx} \end{bmatrix} = \begin{bmatrix} I_L(1) \\ I_L(2) \\ \bullet \\ \bullet \\ \bullet \\ I_L(12) \end{bmatrix}, \tag{10.2-8}$$

where the coefficient matrix $\left[H\right]$ is defined as

$$\left[H\right] = \begin{bmatrix} c_\alpha^2(1) & c_\beta^2(1) & c_\gamma^2(1) & 2c_\alpha(1)c_\beta(1) & 2c_\beta(1)c_\gamma(1) & 2c_\gamma(1)c_\alpha(1) \\ \\ \\ c_\alpha^2(12) & c_\beta^2(12) & c_\gamma^2(12) & 2c_\alpha(12)c_\beta(12) & 2c_\beta(12)c_\gamma(12) & 2c_\gamma(12)c_\alpha(12) \end{bmatrix}, \tag{10.2-9}$$

where c_α, c_β, c_γ are abbreviations for $\cos\alpha$, $\cos\beta$, and $\cos\gamma$, respectively.

Using Eq. 10.2-8, we can find the least squares solution for the inertia matrix elements. Thus, we have

$$\left[I\right] = \begin{bmatrix} I_{xx} & I_{xy} & I_{xz} \\ I_{yx} & I_{yy} & I_{yz} \\ I_{zx} & I_{zy} & I_{zz} \end{bmatrix} \tag{10.2-10}$$

The principal moments of inertia and the principal axes can now be determined by solving the eigenvalue problem

$$[I]\{x\}=\lambda\{x\}, \qquad (10.2\text{-}11)$$

where

$[I]$ = inertia matrix, as in Eq. 10.2-10;

$\{x\}$ = eigenvector, representing the principal axis;

λ = eigenvalue, representing principal moment of inertia.

From Eq. 10.2-11, one can find three eigenvalues and three corresponding eigenvectors, and, hence, the three principal moments of inertia and the corresponding principal axes.

The above inertia matrix is relative to a coordinate system whose origin is located at the mass center and whose axes are parallel to those of the test reference frame, respectively. The elements of the inertia matrix can be transformed to other coordinate systems. Two transformations are required as follows:

Transformation to Inclined Axes. The inertia matrix may be transformed to inclined axes *x, y, z* with origin at the *CG*. Let $\hat{\mathbf{n}}_x$, $\hat{\mathbf{n}}_y$, $\hat{\mathbf{n}}_z$ be mutually perpendicular unit vectors parallel to axes *x, y, z*, respectively. Let $\mathbf{N}_X$, $\mathbf{N}_Y$, $\mathbf{N}_Z$ be mutually perpendicular unit vectors parallel to the axes *X, Y, Z*, respectively. The orthogonal transformation matrix, *S*, between the two coordinate systems can be expressed in terms of the direction cosines for the respective sets of axes. For example, λ_{Xx} is the cosine of the angle between the *X* and *x* axes. Thus, we have

$$\begin{bmatrix}\mathbf{N}_X\\ \mathbf{N}_Y\\ \mathbf{N}_Z\end{bmatrix}=\begin{bmatrix}\lambda_{Xx} & \lambda_{Xy} & \lambda_{Xz}\\ \lambda_{Yx} & \lambda_{Yy} & \lambda_{Yz}\\ \lambda_{Zx} & \lambda_{Zy} & \lambda_{Zz}\end{bmatrix}\begin{bmatrix}\hat{\mathbf{n}}_x\\ \hat{\mathbf{n}}_y\\ \hat{\mathbf{n}}_z\end{bmatrix}\equiv S\begin{bmatrix}\hat{\mathbf{n}}_x\\ \hat{\mathbf{n}}_y\\ \hat{\mathbf{n}}_z\end{bmatrix} \qquad (10.2\text{-}12)$$

Then the inertia matrix $\hat{I}$ relative to the axes *x, y, z* is

$$\hat{I}=S^T IS \qquad (10.2\text{-}13)$$

Transformation to Parallel Axes. Referring to Fig. 10.2-5, the axes *x, y, z* are chosen parallel to *X, Y, Z*, with their origin at the *CG* of the engine. Let the perpendicular distance between *x* and *X* be a_x, between *y* and *Y* be a_y, and between *z* and *Z* be a_z. The moments and products of inertia of the engine of mass m with respect to *x, y, z* axes are then [10.3]

$$I_{xx}=I_{XX}+ma_x^2, \quad I_{yy}=I_{YY}+ma_y^2, \quad I_{zz}=I_{ZZ}+ma_z^2 \qquad (10.2\text{-}14)$$

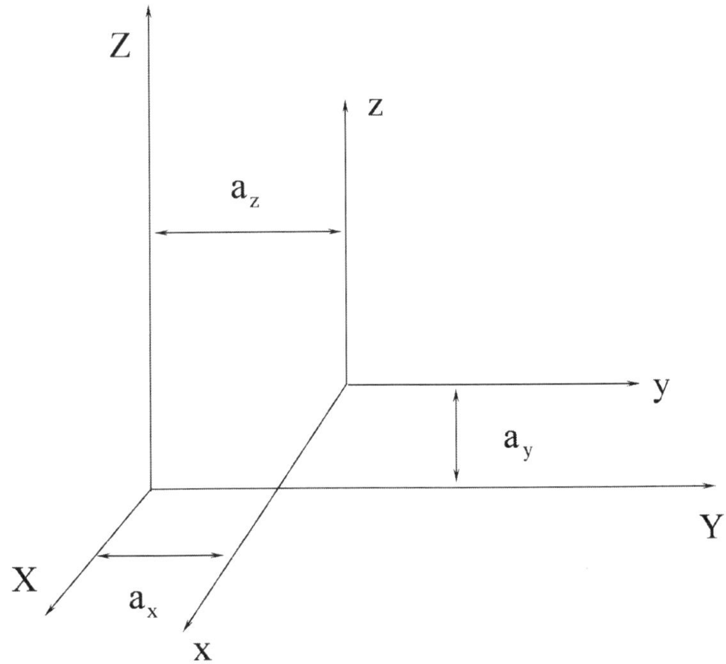

Fig. 10.2-5 Two-coordinate system with parallel axes.

Similarly, the corresponding products of inertia are

$$I_{xy} = I_{XY} + ma_x a_y, \quad I_{xz} = I_{XZ} + ma_x a_z, \quad I_{yz} = I_{YZ} + ma_y a_z \tag{2.10-15}$$

(3) Mass Center

As noted, the term *center of gravity* (*CG*) is often used to designate the mass center. Generally, there is no important difference between the center of gravity and the mass center, but technically they are different [10.2]. The center of gravity of a body is defined as that point through which all the gravitational forces could be replaced by a single force. The mass center is defined as that point for which the first moment of mass is zero. For small bodies near the earth's surface, however, this distinction is generally unimportant.

We are now in the position to derive the formula for determination of the mass center (or *CG*) of the test engine. Let x_c, y_c, z_c be the coordinates of the *CG* relative to the origin *O* of the reference frame. The horizontal distance

d from the CG to the origin O of the reference frame can be calculated as follows (see Fig. 10.2-6):

$$d^2 = \left|\mathbf{n} \times \vec{\overline{OC}}\right|^2$$
$$= (y_c \cos\gamma - z_c \cos\beta)^2 + (z_c \cos\alpha - x_c \cos\gamma)^2 + (x_c \cos\beta - y_c \cos\alpha)^2$$

or

$$d^2 = (y_c^2 + z_c^2)\cos^2\alpha + (z_c^2 + x_c^2)\cos^2\beta + (x_c^2 + y_c^2)\cos^2\gamma \quad (10.2.16)$$
$$-x_c y_c 2\cos\alpha\cos\beta - y_c z_c 2\cos\beta\cos\gamma - z_c x_c 2\cos\gamma\cos\alpha$$

For each suspension position, we have an equation similar to Eq. 10.2-16. For 12 different suspension points, 12 equations can be obtained. These equations can be written in matrix form as

$$[H]\begin{bmatrix} q_1 \\ q_2 \\ q_3 \\ q_4 \\ q_5 \\ q_6 \end{bmatrix} = \begin{bmatrix} d_1^2 \\ d_2^2 \\ \bullet \\ \bullet \\ \bullet \\ d_{12}^2 \end{bmatrix}, \quad (10.2\text{-}17)$$

where the coefficient matrix $[H]$ is defined in Eq. 10.2-9, and where new auxiliary variables $q_i\ (i = 1, 2, \cdots, 6)$ are introduced as follows:

$$q_1 = y_c^2 + z_c^2, \quad q_2 = z_c^2 + x_c^2, \quad q_3 = x_c^2 + y_c^2,$$
$$q_4 = -x_c y_c, \quad q_5 = -y_c z_c, \quad q_6 = -z_c x_c \quad (10.2\text{-}18)$$

Once the auxiliary variables are evaluated from Eq. 10.2-17, the coordinates of the mass center of the engine can be obtained by three methods:

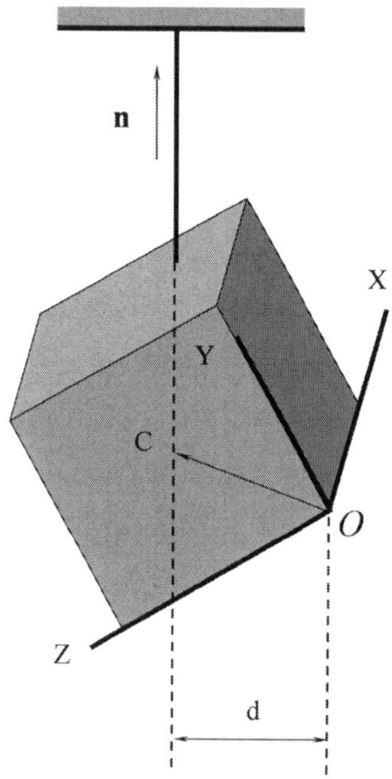

Fig. 10.2-6 The horizontal distance d of the mass center to the origin of the reference frame can be measured.

Method 1:

$$x_c = \pm\sqrt{\frac{q_4 q_6}{-q_5}} \quad (10.2\text{-}19a)$$

$$y_c = \pm\sqrt{\frac{q_5 q_4}{-q_6}} \quad (10.2\text{-}19b)$$

$$z_c = \pm\sqrt{\frac{q_6 q_5}{-q_4}} \quad (10.2\text{-}19c)$$

Method 2:

$$x_c = \pm\sqrt{\frac{q_1 + q_2 - q_3}{2}} \tag{10.2-20a}$$

$$y_c = \pm\sqrt{\frac{q_2 + q_3 - q_1}{2}} \tag{10.2-20b}$$

$$z_c = \pm\sqrt{\frac{q_3 + q_1 - q_2}{2}} \tag{10.2-20c}$$

Method 3:

Solve the following linear algebraic equations:

$$x_c + y_c = \pm\sqrt{q_3 - 2q_4} \equiv \pm b_1 \tag{10.2-21a}$$

$$y_c + z_c = \pm\sqrt{q_1 - 2q_5} \equiv \pm b_2 \tag{10.2-21b}$$

$$z_c + x_c = \pm\sqrt{q_2 - 2q_6} \equiv \pm b_3 \tag{10.2-21c}$$

In the first method, the sign of one of x_c, y_c, and z_c must be freely assigned whereas the sign of the others must match the sign of the products. However, the solution is possible only if $q_i \neq 0$ $(i = 4,5,6)$; otherwise, the equation is singular. In the second and third methods, eight different solutions are always found, but there is no analytical determination of the correct sign. In the first method, only q_4, q_5, and q_6 are used. In the second method, only q_1, q_2, and q_3 are used. In the third method, all of the q_1 $(i = 1,2,\cdots,6)$ are used. Therefore, the third method is recommended. The solution (including magnitude and sign) can be determined from the eight possible solutions by inspecting the configurations of the test engine and the coordinate reference frame attached on it.

10.3 Orientation Angles and Transformation Matrix

As we mentioned before, the engine mounting system includes an engine, three to four engine rubber or hydraulic mounts, and a foundation (typically a vehicle frame), and usually the mounts are inclined with respect to the engine. A typical engine mount is represented by three mutually perpendicular springs, and it is assumed that the top surface is attached to the engine and the bottom surface to the vehicle. To define the orientation of an engine mount with respect to a reference coordinate system, we introduce the engine mount coordinate system as follows: let the origin of each mount coordinate system be located at the center of each mount, and let the three principal axes of the mount be mutually perpendicular.

The orientation of a typical engine mount in a reference coordinate system may be defined in terms of angles between the respective axes. Specifically, let $\mathbf{n}_1$, $\mathbf{n}_2$, $\mathbf{n}_3$ be mutually perpendicular unit vectors parallel to the mount principle axes, respectively, and let $\mathbf{N}_1$, $\mathbf{N}_2$, $\mathbf{N}_3$ be mutually perpendicular unit vectors parallel, respectively, to the axes *X*, *Y*, *Z* of a reference coordinate system. Then the angles describing the inclination of the axes of the engine mount coordinate system relative to those of the reference coordinate system by their cosines are determined by the products $\mathbf{N}_i \cdot \mathbf{n}_j$ $(i=1,2,3)$. We can then form a matrix *S* defined as

$$S=[S_{ij}]=[\mathbf{N}_i \cdot \mathbf{n}_j] \tag{10.3-1}$$

It is readily seen that *S* is non-singular and orthogonal. That is, the inverse is the transpose

$$S^{-1}=S^T \tag{10.3-2}$$

S is called the *transformation matrix* or *direction cosine matrix* between the two coordinate systems. *S* may be conveniently constructed and expressed in terms of orientation angles as follows: Let the axes of the engine mount coordinate system be initially aligned with those of the reference coordinate system. Then let the engine mount coordinate system be brought into its general orientation relative to the reference coordinate system by three successive dextral rotations of the engine mount coordinate system about $\mathbf{n}_1$, $\mathbf{n}_2$, and $\mathbf{n}_3$ through angles α, β, and γ. Then *S* may be expressed as

$$S=ABC=\begin{bmatrix} 1 & 0 & 0 \\ 0 & c_\alpha & -s_\alpha \\ 0 & s_\alpha & c_\alpha \end{bmatrix}\begin{bmatrix} c_\beta & 0 & s_\beta \\ 0 & 1 & 0 \\ -s_\beta & 0 & c_\beta \end{bmatrix}\begin{bmatrix} c_\gamma & -s_\gamma & 0 \\ s_\gamma & c_\gamma & 0 \\ 0 & 0 & 1 \end{bmatrix}, \tag{10.3-3}$$

where s_α and c_α are abbreviations for $\sin\alpha$ and $\cos\alpha$, etc., and where the matrices *A*, *B*, and *C* (defined by inspection) represent the transformation matrix between the two coordinate systems after the first, second, and third rotations of the engine mount coordinate system, respectively.

After the multiplication in Eq. 10.3-3 is carried out, *S* may be expressed as

$$S=\begin{bmatrix} c_\beta c_\gamma & -c_\beta s_\gamma & s_\beta \\ (c_\alpha s_\gamma + s_\alpha s_\beta c_\gamma) & (c_\alpha c_\gamma - s_\alpha s_\beta s_\gamma) & -s_\alpha c_\beta \\ (s_\alpha s_\gamma - c_\alpha s_\beta c_\gamma) & (s_\alpha c_\gamma + c_\alpha s_\beta s_\gamma) & c_\alpha c_\beta \end{bmatrix} \tag{10.3-4}$$

The development of *S* in the foregoing paragraph is based on a *1-2-3* rotation sequence. That is, successive right-handed rotations of the engine mount coordinate system about its 1-axis through the angle α, about its 2-axis through the angle β, and, finally, about its 3-axis through the angle γ brings it into a general orientation. This type of rotation (each rotation occurs about an axis whose index differs from that of all previous axes) was used first by Tait in 1869 and later by Bryan in 1911. Thus, the angles sometimes are called Tait-Bryan angles. In the German literature, they sometimes are referred to as Cardan angles. In vehicle dynamics, they are called the *roll angle*, *pitch angle*, and *yaw angle*, although the order in which these are performed depends on the user. The roll axis is the forward axis of the vehicle, yaw is about the vertical axis, and pitch is about the lateral axis [10.4].

An alternative rotation sequence is the *3-1-3* rotations. That is, let the axes of the engine mount coordinate system be initially aligned with those of the reference coordinate system. Then let the engine mount coordinate system be brought into its general orientation relative to the reference coordinate system by three successive dextral rotations: rotation about the 3-axis through an angle ϕ, rotation about the 1-axis through an angle θ, and, finally, rotation about the 3-axis again through an angle ψ. In this case, the rotation angles are known as Euler angles. In classical rigid body dynamics, ϕ is called the *precession angle*, θ is called the *nutation angle*, and ψ is called the *spin angle*. The direction cosine matrix for this case is

$$S=\begin{bmatrix} c_\phi & -s_\phi & 0 \\ s_\phi & c_\phi & 0 \\ 0 & 0 & 1 \end{bmatrix}\begin{bmatrix} 1 & 0 & 0 \\ 0 & c_\theta & -s_\theta \\ 0 & s_\theta & c_\theta \end{bmatrix}\begin{bmatrix} c_\psi & -s_\psi & 0 \\ s_\psi & c_\psi & 0 \\ 0 & 0 & 1 \end{bmatrix} \tag{10.3-5}$$

When written out in full, this matrix becomes

$$S=\begin{bmatrix} (c_\psi c_\phi - s_\psi c_\theta s_\phi) & (-s_\psi c_\phi - c_\psi c_\theta s_\phi) & s_\theta s_\phi \\ (c_\psi s_\phi + s_\psi c_\theta c_\phi) & (-s_\psi s_\phi + c_\psi c_\theta c_\phi) & -s_\theta c_\phi \\ s_\psi s_\theta & c_\psi s_\theta & c_\theta \end{bmatrix}, \tag{10.3-6}$$

where s_ϕ and c_ϕ are abbreviations for $\sin\phi$ and $\cos\phi$, etc.

It is important to note that although Euler angles themselves are the same and the symbols used to denote them are the same as well, the sequences of carrying out the Euler rotation may be different. As a result, the final direction cosine matrices are not the same. Indeed, the direction cosine matrix of two coordinate systems may be defined by 24 rotation sequences (see [10.2, 10.5] and Appendix C). Determining the direction cosine matrix is an error-prone process. Therefore, it is helpful to develop algorithms to routinely construct the direction cosine matrix.

To this end, consider a pair of unit vector sets $\mathbf{N}_i$ and $\hat{\mathbf{N}}_i$ $(i=1,2,3)$ as in Fig. 10.3-1, where the unit vectors $\mathbf{N}_1$ and $\hat{\mathbf{N}}_1$ are coincident and where the unit vector pairs $\mathbf{N}_2$, $\hat{\mathbf{N}}_2$ and $\mathbf{N}_3$, $\hat{\mathbf{N}}_3$ are inclined relative to each other by an angle α, as shown. Observe that for this configuration, it is relatively easy to express the unit vectors in terms of one another, leading to the equations

$$\begin{aligned}\mathbf{N}_1 &= \hat{\mathbf{N}}_1 \\ \mathbf{N}_2 &= c_\alpha \hat{\mathbf{N}}_2 - s_\alpha \hat{\mathbf{N}}_3 \\ \mathbf{N}_3 &= s_\alpha \hat{\mathbf{N}}_2 + c_\alpha \hat{\mathbf{N}}_3\end{aligned} \quad \text{and} \quad \begin{aligned}\hat{\mathbf{N}}_1 &= \mathbf{N}_1 \\ \hat{\mathbf{N}}_2 &= c_\alpha \mathbf{N}_2 + s_\alpha \mathbf{N}_3 \\ \hat{\mathbf{N}}_3 &= -s_\alpha \mathbf{N}_2 + c_\alpha \mathbf{N}_3,\end{aligned} \qquad (10.3\text{-}7a,b)$$

where, as before, s_α and c_α are abbreviations for $\sin\alpha$ and $\cos\alpha$.

In matrix form, Eq. 10.3-7 may be written as

$$\begin{bmatrix}\mathbf{N}_1 \\ \mathbf{N}_2 \\ \mathbf{N}_3\end{bmatrix} = \begin{bmatrix}1 & 0 & 0 \\ 0 & c_\alpha & -s_\alpha \\ 0 & s_\alpha & c_\alpha\end{bmatrix}\begin{bmatrix}\hat{\mathbf{N}}_1 \\ \hat{\mathbf{N}}_2 \\ \hat{\mathbf{N}}_3\end{bmatrix} \quad \text{and} \quad \begin{bmatrix}\hat{\mathbf{N}}_1 \\ \hat{\mathbf{N}}_2 \\ \hat{\mathbf{N}}_3\end{bmatrix} = \begin{bmatrix}1 & 0 & 0 \\ 0 & c_\alpha & s_\alpha \\ 0 & -s_\alpha & c_\alpha\end{bmatrix}\begin{bmatrix}\mathbf{N}_1 \\ \mathbf{N}_2 \\ \mathbf{N}_3\end{bmatrix} \qquad (10.3\text{-}8a,b)$$

or as

$$\mathbf{N} = A\hat{\mathbf{N}} \quad \text{and} \quad \hat{\mathbf{N}} = A^T\mathbf{N}, \qquad (10.3\text{-}9a,b)$$

where $\mathbf{N}$ and $\hat{\mathbf{N}}$ are the column arrays of Eq. 10.3-8, and where A is the direction cosine matrix, which by inspection is seen to be orthogonal. That is,

$$A^T = A^{-1} \qquad (10.3\text{-}10)$$

Eq. 10.3-7 and, as a result, Eqs. 10.3-8 and 10.3-9 are readily developed by observing that when the inclination angle α is zero, the unit vector sets are mutually aligned. This, in turn, means that when α is not zero, the unit vectors with like subscripts ($\mathbf{N}_2$, $\hat{\mathbf{N}}_2$ and $\mathbf{N}_3$, $\hat{\mathbf{N}}_3$) are related by $\cos\alpha$. [Recall that $\cos 0 = 1$, $\cos(\pi/2) = 0$.] Similarly, unit vectors with unlike subscripts are related by $\pm\sin\alpha$. [Recall that $\sin 0 = 0$, and $\sin(\pi/2) = 1$.] Then, the only decision in Eq. 10.3-7 is where to place the minus signs. In the plane of the vectors $\mathbf{N}_2$, $\hat{\mathbf{N}}_2$, $\mathbf{N}_3$, and $\hat{\mathbf{N}}_3$, as in Fig. 10.3-2, we see that two of the vectors, namely, $\hat{\mathbf{N}}_2$ and $\mathbf{N}_3$, are *inside* the vectors $\mathbf{N}_2$ and $\hat{\mathbf{N}}_3$. Alternatively, $\mathbf{N}_2$ and $\hat{\mathbf{N}}_3$

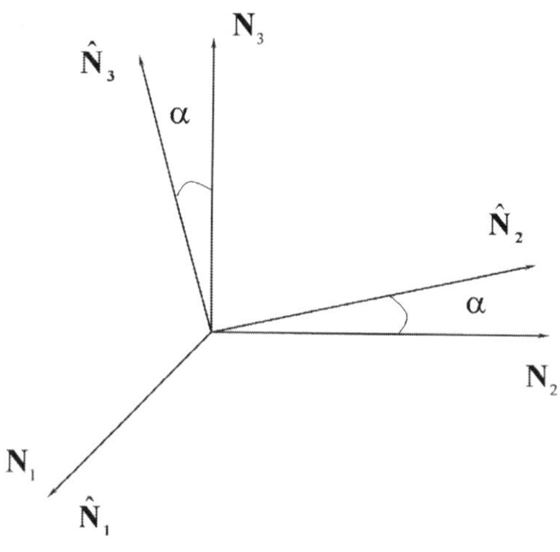

Fig. 10.3-1 Unit vector sets $\mathbf{N}_i$ and $\hat{\mathbf{N}}_i$ $(i = 1, 2, 3)$.

are *outside* of $\mathbf{N}_2$ and $\mathbf{N}_3$. By inspection, we see that the outside vectors require the negative signs for the $\sin\alpha$ terms.

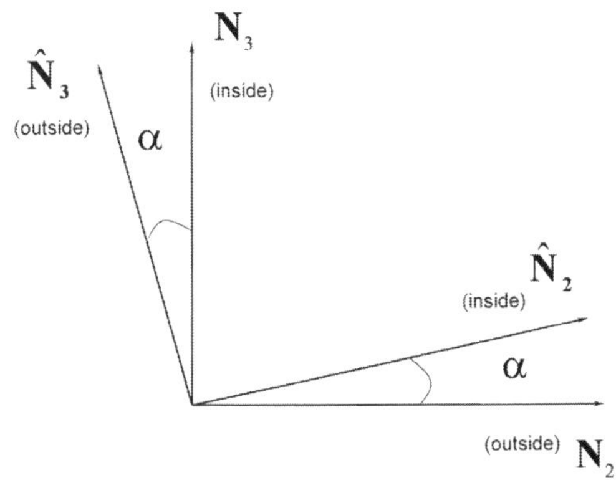

Fig. 10.3-2 Plane of unit vector shows inside and outside vectors.

The information contained in Eqs. 10.3-7 to 10.3-9 and in the description of the foregoing paragraph may be condensed into a simple diagram called a *configuration graph* [10.1, 10.2], as in Fig. 10.3-3. In this diagram, there are six dots, or nodes, each representing one of the six unit vectors of Fig. 10.3-1 and as indicated in Fig. 10.3-3. The horizontal line then designates equality ($\mathbf{N}_1 = \hat{\mathbf{N}}_1$). The inclined line connects nodes associated with the *inside* unit vectors ($\hat{\mathbf{N}}_2$ and $\mathbf{N}_3$). Consequently, the unconnected nodes are associated with the *outside* unit vectors ($\mathbf{N}_2$ and $\hat{\mathbf{N}}_3$). Alternatively, the inclined line can be easily obtained as follows: First, move the left end of the horizontal line forward two nodes in a cyclic permutation of 1,2,3. (That is, move the left end of the horizontal line from node 1 to node 3, in this case.) Next, move the right end of the horizontal line forward one node in a cyclic permutation of 1,2,3. (That is, move the right end of the horizontal line from node 1 to node 2, in this case.) These then arrived at nodes which are the end points of the sloped line. These observations enable us to immediately use the diagram of Fig. 10.3-3 to produce Eqs. 10.3-7 and 10.3-8. Indeed, we could even develop an algorithm for equation development, as in Fig. 10.3-4 [10.2].

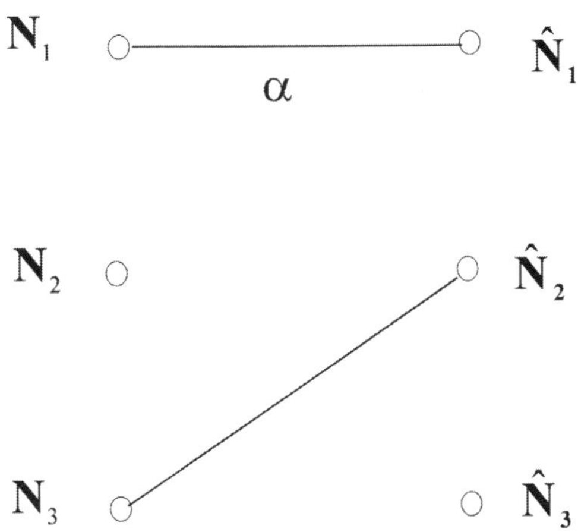

Fig. 10.3-3 Configuration graph for the unit vector sets of Fig. 10.3-1.

To illustrate the use of configuration graphs, consider a pair of mutually perpendicular unit vector sets $\hat{\mathbf{N}}_i$ and $\hat{\mathbf{n}}_i$ $(i=1,2,3)$ inclined relative to each other by the angle β, as in Fig. 10.3-5. By inspection in the figure, we see that the vectors $\hat{\mathbf{N}}_2$ and $\hat{\mathbf{n}}_2$ are equal. Thus, the horizontal line connects the unit vectors $\hat{\mathbf{N}}_2$ and $\hat{\mathbf{n}}_2$. The inclined line can be simply obtained as

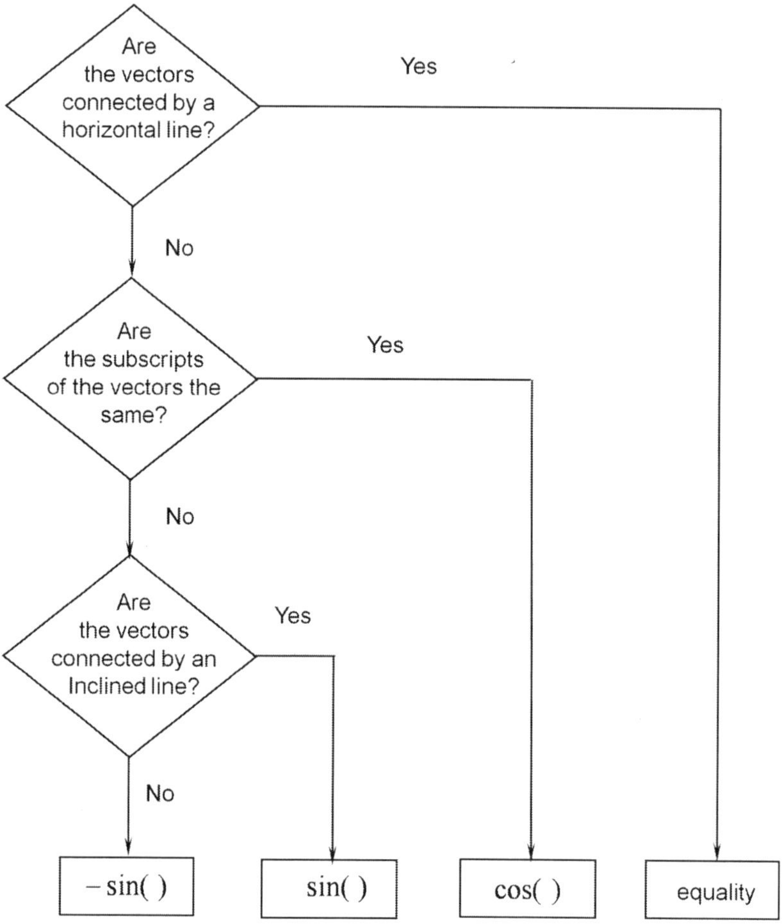

Fig. 10.3-4 Equation coefficients and signs obtained from a configuration graph [10.2].

follows: First, move the left end of the horizontal line forward two steps in a cyclic permutation of 1,2,3; that is, move it from node 2 to node 3 and finally to node 1. Next, move the right end of the horizontal line forward one step; that is, move it from node 2 to node 3.

Following the procedure of Fig. 10.3-4, we immediately obtain the expressions

$$\begin{aligned} \hat{\mathbf{N}}_1 &= c_\beta \hat{\mathbf{n}}_1 + s_\beta \hat{\mathbf{n}}_3 \\ \hat{\mathbf{N}}_2 &= \hat{\mathbf{n}}_2 \\ \hat{\mathbf{N}}_3 &= -s_\beta \hat{\mathbf{n}}_1 + c_\beta \hat{\mathbf{n}}_3 \end{aligned} \quad \text{and} \quad \begin{aligned} \hat{\mathbf{n}}_1 &= c_\beta \hat{\mathbf{N}}_1 - s_\beta \hat{\mathbf{N}}_3 \\ \hat{\mathbf{n}}_2 &= \hat{\mathbf{N}}_2 \\ \hat{\mathbf{n}}_3 &= s_\beta \hat{\mathbf{N}}_1 + c_\beta \hat{\mathbf{N}}_3 \end{aligned} \qquad (10.3\text{-}11\text{a,b})$$

Or, in matrix form,

$$\begin{bmatrix} \hat{\mathbf{N}}_1 \\ \hat{\mathbf{N}}_2 \\ \hat{\mathbf{N}}_3 \end{bmatrix} = \begin{bmatrix} c_\beta & 0 & s_\beta \\ 0 & 1 & 0 \\ -s_\beta & 0 & c_\beta \end{bmatrix} \begin{bmatrix} \hat{\mathbf{n}}_1 \\ \hat{\mathbf{n}}_2 \\ \hat{\mathbf{n}}_3 \end{bmatrix}$$

and

$$\begin{bmatrix} \hat{\mathbf{n}}_1 \\ \hat{\mathbf{n}}_2 \\ \hat{\mathbf{n}}_3 \end{bmatrix} = \begin{bmatrix} c_\beta & 0 & -s_\beta \\ 0 & 1 & 0 \\ s_\beta & 0 & c_\beta \end{bmatrix} \begin{bmatrix} \hat{\mathbf{N}}_1 \\ \hat{\mathbf{N}}_2 \\ \hat{\mathbf{N}}_3 \end{bmatrix} \quad (10.3\text{-}12a,b)$$

or

$$\hat{\mathbf{N}} = B\hat{\mathbf{n}} \quad \text{and} \quad \hat{\mathbf{n}} = B^T \hat{\mathbf{N}}, \quad (10.3\text{-}13a,b)$$

where the matrices $\hat{\mathbf{N}}$, $\hat{\mathbf{n}}$, and B are defined by comparing Eqs. 10.3-12 and 10.3-13.

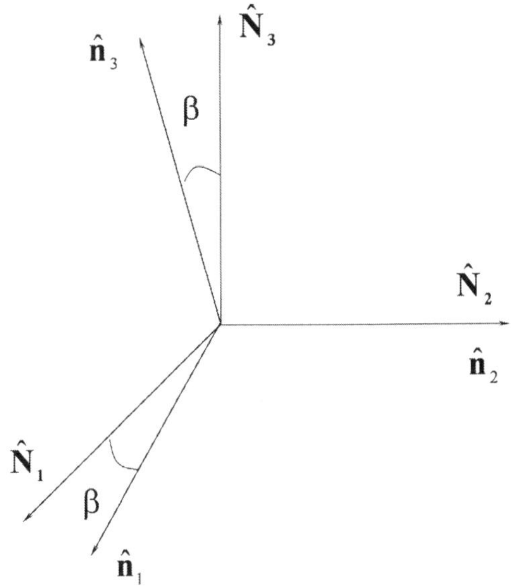

Fig. 10.3-5 Unit vector sets $\hat{\mathbf{N}}_i$ and $\hat{\mathbf{n}}_i$ $(i = 1,2,3)$.

Similarly, let unit vector sets $\hat{\mathbf{n}}_i$ and $\mathbf{n}_i$ $(i=1,2,3)$ be inclined relative to each other by the angle γ, as in Fig. 10.3-7. By inspection in this figure, we see that the vectors $\hat{\mathbf{n}}_3$ and $\mathbf{n}_3$ are equal. Then, the configuration graph relating $\hat{\mathbf{n}}_i$ and $\mathbf{n}_i$ is as shown in Fig. 10.3-8, and expressions relating the unit vector sets are

$$\begin{aligned} \hat{\mathbf{n}}_1 &= c_\gamma \mathbf{n}_1 - s_\gamma \mathbf{n}_2 \\ \hat{\mathbf{n}}_2 &= s_\gamma \mathbf{n}_1 + c_\gamma \mathbf{n}_2 \\ \hat{\mathbf{n}}_3 &= \mathbf{n}_3 \end{aligned}$$

and

$$\begin{aligned} \mathbf{n}_1 &= c_\gamma \hat{\mathbf{n}}_1 + s_\gamma \hat{\mathbf{n}}_2 \\ \mathbf{n}_2 &= -s_\gamma \hat{\mathbf{n}}_1 + c_\gamma \mathbf{n}_2 \\ \mathbf{n}_3 &= \hat{\mathbf{n}}_3 \end{aligned} \quad (10.3\text{-}14a,b)$$

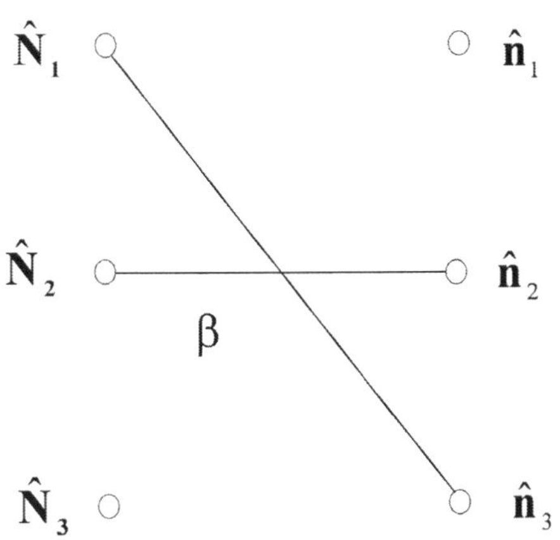

Fig. 10.3-6 Configuration graph for the unit vector sets of Fig. 10.3-5.

Or, in matrix form,

$$\begin{bmatrix} \hat{\mathbf{n}}_1 \\ \hat{\mathbf{n}}_2 \\ \hat{\mathbf{n}}_3 \end{bmatrix} = \begin{bmatrix} c_\gamma & -s_\gamma & 0 \\ s_\gamma & c_\gamma & 0 \\ 0 & 0 & 1 \end{bmatrix} \begin{bmatrix} \mathbf{n}_1 \\ \mathbf{n}_2 \\ \mathbf{n}_3 \end{bmatrix}$$

and

$$\begin{bmatrix} \mathbf{n}_1 \\ \mathbf{n}_2 \\ \mathbf{n}_3 \end{bmatrix} = \begin{bmatrix} c_\gamma & s_\gamma & 0 \\ -s_\gamma & c_\gamma & 0 \\ 0 & 0 & 1 \end{bmatrix} \begin{bmatrix} \hat{\mathbf{n}}_1 \\ \hat{\mathbf{n}}_2 \\ \hat{\mathbf{n}}_3 \end{bmatrix} \quad (10.3\text{-}15a,b)$$

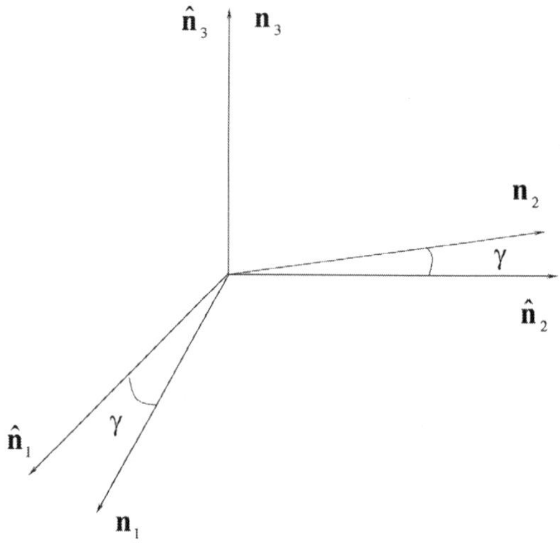

Fig. 10.3-7 Unit vector sets $\hat{\mathbf{n}}_i$ and $\mathbf{n}_i$ $(i=1,2,3)$.

or

$$\hat{\mathbf{n}} = C\mathbf{n} \quad \text{and} \quad \mathbf{n} = C^T\hat{\mathbf{n}}, \qquad (10.3\text{-}16a,b)$$

where the matrices $\mathbf{n}$, $\hat{\mathbf{n}}_i$, and C are defined by comparing Eqs. 10.3-15 and 10.3-16.

Several additional comments might be helpful as an aid to using the configuration graphs and these expressions. First, observe that in Figs. 10.3-1, 10.3-5, and 10.3-7, we have simulated rotations about the 1, 2, and 3-axes (through the angles α, β, and γ). By appropriately choosing α, β, and γ, and by linking these rotations together, we can bring a coordinate system (or a rigid body) into any desired orientation relative to a reference coordinate system. Specifically, if the $\mathbf{N}_i$ $(i=1,2,3)$ are fixed in a reference frame R, and if the $\mathbf{n}_i$ $(i=1,2,3)$ are fixed in the body (or the engine mount), then the rotation between the $\mathbf{N}_i$ and $\mathbf{n}_i$ are conveniently obtained by adjoining the respective configuration graphs of Figs. 10.3-3, 10.3-6, and 10.3-8. That is, by observing the common intermediate unit vector

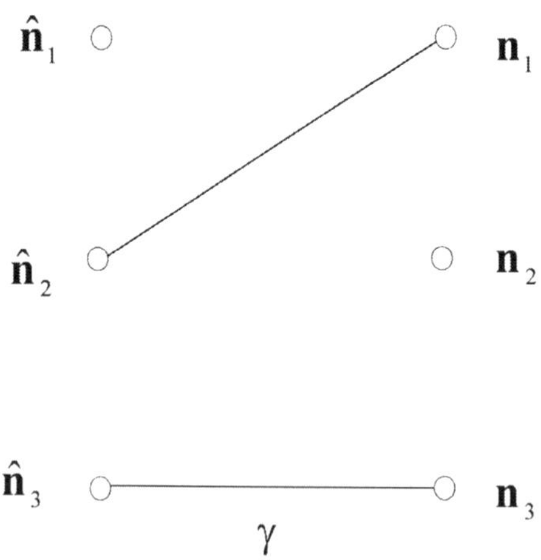

Fig. 10.3-8 Configuration graph for the unit vector sets of Fig. 10.3-7.

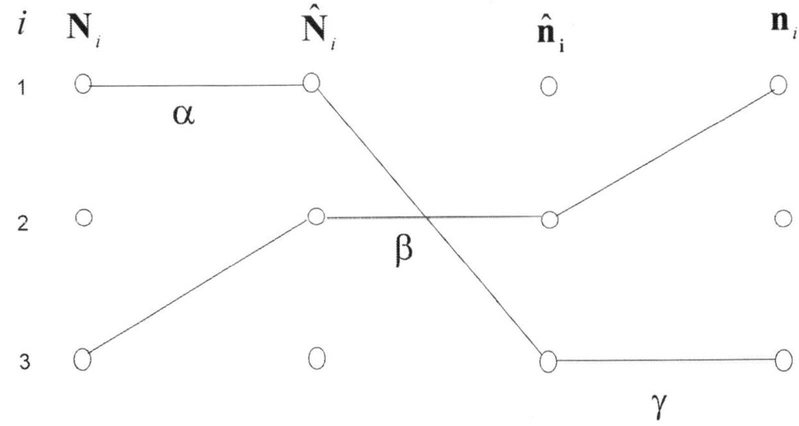

Fig. 10.3-9 Configuration graph to relate unit vectors in two-coordinate systems.

sets, namely, the $\hat{\mathbf{N}}_i$ and $\hat{\mathbf{n}}_i$ $(i=1,2,3)$, we obtain the general, or *three-rotation*, graph of Fig. 10.3-9.

By using the three-rotation graph, we can directly write the expressions relating the unit vectors. For example,

$$\begin{aligned} \mathbf{N}_1 &= \hat{\mathbf{N}}_1 \\ &= c_\beta \hat{\mathbf{n}}_1 + s_\beta \hat{\mathbf{n}}_3 \\ &= c_\beta \left(c_\gamma \mathbf{n}_1 - s_\gamma \mathbf{n}_2 \right) + s_\beta \left(\mathbf{n}_3 \right) \\ &= c_\beta c_\gamma \mathbf{n}_1 - c_\beta s_\gamma \mathbf{n}_2 + s_\beta \mathbf{n}_3 \end{aligned} \tag{10.3-17}$$

Similarly, the unit vector $\mathbf{N}_2$ and $\mathbf{N}_3$ can be expressed in terms of $\mathbf{n}_1$, $\mathbf{n}_2$, and $\mathbf{n}_3$.

Next, observe that by comparing the forms of the configuration graphs with the form of the associated direction cosine matrices, we can recognize a pattern enabling us to immediately obtain the direction cosine matrix, once the configuration graph is constructed. Specifically, consider the graphs of Figs. 10.3-3, 10.3-6, and 10.3-8 and the corresponding matrices A, B, and C of Eqs. 10.3-9, 10.3-13, and 10.3-16 placed atop one another as in Fig. 10.3-10. Then the position of the 1's and 0's in the matrices corresponds to the row of the horizontal line of the graphs. The cosines are then placed on the diagonals and the sines off-diagonal. The minus sign (which is always associated with one of the sine terms) is placed with the sine term in the *upper* row of the matrix if the slope of the inclined line in the graph is *positive* (that is upward). Alternatively, when the inclined line slope of the graph is negative (that is downward), the minus sign is placed with the sine term in the *lower* row of the matrix.

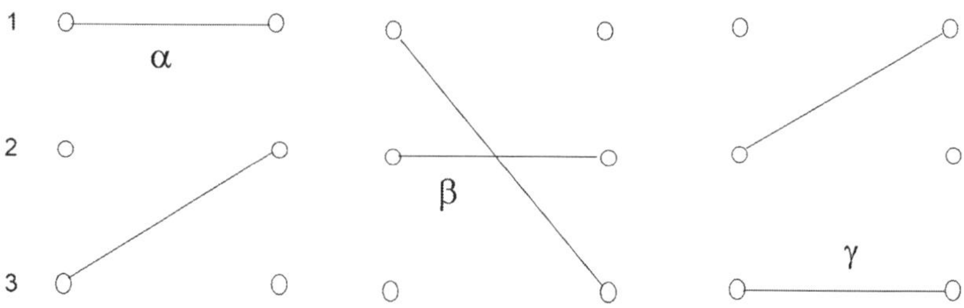

$$A = \begin{bmatrix} 1 & 0 & 0 \\ 0 & c_\alpha & -s_\alpha \\ 0 & s_\alpha & c_\alpha \end{bmatrix} \qquad B = \begin{bmatrix} c_\beta & 0 & s_\beta \\ 0 & 1 & 0 \\ -s_\beta & 0 & c_\beta \end{bmatrix} \qquad C = \begin{bmatrix} c_\gamma & -s_\gamma & 0 \\ s_\gamma & c_\gamma & 0 \\ 0 & 0 & 1 \end{bmatrix}$$

Fig. 10.3-10 Superposition and juxtaposition of configuration graphs and direction cosine matrices.

Finally, observe that in view of the direct relation between the configuration graphs and the direction cosine matrices as presented in Fig. 10.3-10, we see from the three rotation graph of Fig. 10.3-9 that the desired relations between $\mathbf{N}_i$ and $\mathbf{n}_i$ unit vectors may be obtained from the expression

$$S = ABC, \tag{10.3-18}$$

where, from Eq. 10.3-1, the elements S_{ij} of S are $\mathbf{N}_i \cdot \mathbf{n}_j$, and S itself is given by Eq. 10.3-4.

10.4 Equations of Motion

In this section, we derive the equations of motion of the engine mounting system. The engine is modeled as a rigid body with a given mass center and three principal inertia axes orthogonal to each other. The engine may translate in any of the three directions or rotate about any of the axes. Thus, the engine has six degrees of freedom. It is reasonable to model the engine as a rigid body because its natural frequencies of the flexible modes are much higher than those of the mounting system. Each mount is represented by three mutually perpendicular sets of linear springs in each of the three principal directions. Finally, the foundation (vehicle body) is modeled as a rigid body. The reasons to exclude the vibration of the foundation are: first, the displacement of the body side of the engine mount is small compared to the engine side so that the engine and mounts can be treated outside the context of a full vehicle; second, since mount optimization is usually needed in the early stages of the vehicle design process, it is desirable to optimize the mounting system before good definition of the remainder of the vehicle can be obtained.

To derive the equations of motion of the engine mounting system, let us set several coordinate systems. Fig. 10.4-1 shows a view of an engine at rest with an inertial set of axes *X*, *Y*, *Z* (fixed in the vehicle) and a coincident set of axes *x*, *y*, *z* fixed in the engine, both sets of axes passing through the mass center of the engine. A typical engine mount is represented by three mutually perpendicular springs and it is assumed that the top surface is attached to the engine and the bottom surface to the vehicle. The origin of each mount coordinate system is located at the center of the mount; the three principal axes of the mount, *p*, *q*, *r*, are perpendicular to each other. The point of attachment of each of the idealized engine mounts is located at the coordinate distance a_x, a_y, a_z of the elastic center of the mount (in Fig. 10.4-1 only one engine mount is shown).

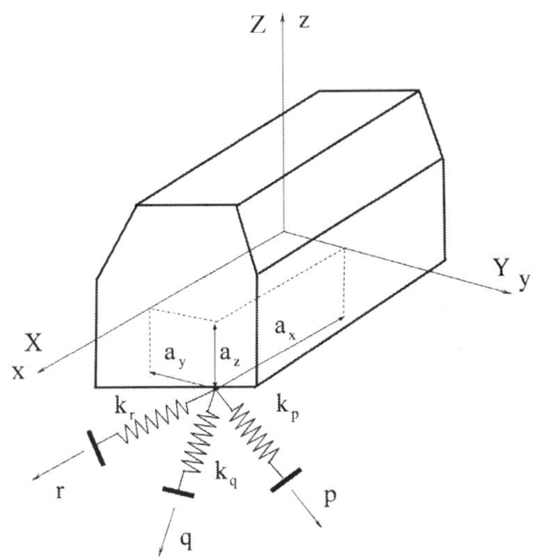

Fig. 10.4-1 Engine mounts and coordinate systems.

The translational displacements of the mass center of the engine are x_c, y_c, z_c in the *X*, *Y*, *Z* directions, respectively. The rotational displacements of the engine are characterized by the angles α, β, γ of the engine body axes about the *X*, *Y*, *Z* axes, respectively. Only small translations and rotations are considered. Hence, the rotations are commutative (i.e., the resulting position is independent of the order of the component rotations) and the angles of rotation about the engine body axes are equal to those about the global *X*, *Y*, *Z* axes.

The translational motion of the engine may be described by Newton's second law, and the rotational motion of the engine can be described by Euler's dynamic equations. Thus, the equations of motion of the engine, for small displacement and rotation, can be written as in the forms

$$m\ddot{x}_c = \sum F_X + \sum F_{RX} \tag{10.4-1a}$$

$$m\ddot{y}_c = \sum F_Y + \sum F_{RY} \tag{10.4-1b}$$

$$m\ddot{z}_c = \sum F_Z + \sum F_{RZ} \tag{10.4-1c}$$

$$I_{xx}\ddot{\alpha} + I_{xy}\ddot{\beta} + I_{xz}\ddot{\gamma} = \sum M_X + \sum M_{RX} \tag{10.4-2a}$$

$$I_{xy}\ddot{\alpha} + I_{yy}\ddot{\beta} + I_{yz}\ddot{\gamma} = \sum M_Y + \sum M_{RY} \tag{10.4-2b}$$

$$I_{zx}\ddot{\alpha} + I_{yz}\ddot{\beta} + I_{zz}\ddot{\gamma} = \sum M_Z + \sum M_{RZ}, \tag{10.4-2c}$$

where m is the mass of the engine; $\ddot{x}_c, \ddot{y}_c, \ddot{z}_c$ are the accelerations of the mass center of the engine in the inertial coordinate system; F_X, F_Y, F_Z are the reaction forces produced by an engine mount due to translational motion of the engine; F_{RX}, F_{RY}, F_{RZ} are the reaction forces produced by the engine mount due to rotational motion of the engine; M_X, M_Y, M_Z are the reaction torques produced by the engine mount due to translational motion of the engine; and M_{RX}, M_{RY}, M_{RZ} are the reaction torques produced by the engine mount due to rotational motion of the engine.

(1) Forces and Torques Produced by an Engine Mount due to Translational Motion of the Engine

Let $\mathbf{n}_p$, $\mathbf{n}_q$, $\mathbf{n}_r$ be mutually perpendicular unit vectors parallel to the axes p, q, r, respectively, of a typical engine mount. Let $\mathbf{N}_x$, $\mathbf{N}_y$, $\mathbf{N}_z$ be mutually perpendicular unit vectors parallel to global X, Y, Z axes, respectively. The two coordinate systems can be related by the direction cosines for the respective sets of axes. For example, λ_{xp} is the cosine of the angle between the $\mathbf{N}_x$ and the $\mathbf{n}_p$ axes. Thus, we have

$$\begin{bmatrix} \mathbf{n}_p \\ \mathbf{n}_q \\ \mathbf{n}_r \end{bmatrix} = \begin{bmatrix} \lambda_{xp} & \lambda_{yp} & \lambda_{zp} \\ \lambda_{xq} & \lambda_{yq} & \lambda_{zq} \\ \lambda_{xr} & \lambda_{yr} & \lambda_{zr} \end{bmatrix} \begin{bmatrix} \mathbf{N}_x \\ \mathbf{N}_y \\ \mathbf{N}_z \end{bmatrix} \tag{10.4-3}$$

Let Δ_x, Δ_y, Δ_z be the displacements of the engine in global X, Y, Z directions, respectively. Then, the displacements in the p, q, r directions of the engine mount coordinate system are given by

$$\begin{bmatrix} \Delta_p \\ \Delta_q \\ \Delta_r \end{bmatrix} = \begin{bmatrix} \lambda_{xp} & \lambda_{yp} & \lambda_{zp} \\ \lambda_{xq} & \lambda_{yq} & \lambda_{zq} \\ \lambda_{xr} & \lambda_{yr} & \lambda_{zr} \end{bmatrix} \begin{bmatrix} \Delta_x \\ \Delta_y \\ \Delta_z \end{bmatrix} \equiv C \begin{bmatrix} \Delta_x \\ \Delta_y \\ \Delta_z \end{bmatrix} \tag{10-4.4}$$

Hence, the force in the p, q, r directions can be calculated as

$$\begin{bmatrix} F_p \\ F_q \\ F_r \end{bmatrix} = \begin{bmatrix} -k_p\Delta_p \\ -k_q\Delta_q \\ -k_r\Delta_r \end{bmatrix} = -\begin{bmatrix} k_p & & \\ & k_q & \\ & & k_r \end{bmatrix} \begin{bmatrix} \lambda_{xp} & \lambda_{yp} & \lambda_{zp} \\ \lambda_{xq} & \lambda_{yq} & \lambda_{zq} \\ \lambda_{xr} & \lambda_{yr} & \lambda_{zr} \end{bmatrix} \begin{bmatrix} \Delta_x \\ \Delta_y \\ \Delta_z \end{bmatrix} \tag{10.4-5}$$

The forces can be expressed in the global X, Y, Z directions as

$$\begin{bmatrix} F_X \\ F_Y \\ F_Z \end{bmatrix} = \begin{bmatrix} \lambda_{px} & \lambda_{qx} & \lambda_{rx} \\ \lambda_{py} & \lambda_{qy} & \lambda_{ry} \\ \lambda_{pz} & \lambda_{qz} & \lambda_{rz} \end{bmatrix} \begin{bmatrix} F_p \\ F_q \\ F_r \end{bmatrix} = C^T \begin{bmatrix} F_p \\ F_q \\ F_r \end{bmatrix} \tag{10.4-6}$$

Substituting from Eq. 10.4-5 into Eq. 10.4-6, we have

$$\begin{bmatrix} F_X \\ F_Y \\ F_Z \end{bmatrix} = -\begin{bmatrix} k_{xx} & k_{xy} & k_{xz} \\ k_{yx} & k_{yy} & k_{yz} \\ k_{zx} & k_{zy} & k_{zz} \end{bmatrix} \begin{bmatrix} \Delta_x \\ \Delta_y \\ \Delta_z \end{bmatrix} = -C^T \begin{bmatrix} k_p & & \\ & k_q & \\ & & k_r \end{bmatrix} C \begin{bmatrix} \Delta_x \\ \Delta_y \\ \Delta_z \end{bmatrix} \tag{10.4-7}$$

That is,

$$\begin{bmatrix} k_{xx} & k_{xy} & k_{xz} \\ k_{yx} & k_{yy} & k_{yz} \\ k_{zx} & k_{zy} & k_{zz} \end{bmatrix} = C^T \begin{bmatrix} k_p & & \\ & k_q & \\ & & k_r \end{bmatrix} C, \tag{10.4-8}$$

from which we obtain [10.3]

$$k_{xx} = k_p\lambda_{xp}^2 + k_q\lambda_{xq}^2 + k_r\lambda_{xr}^2 \tag{10.4-9a}$$

$$k_{yy} = k_p\lambda_{yp}^2 + k_q\lambda_{yq}^2 + k_r\lambda_{yr}^2 \tag{10.4-9b}$$

$$k_{zz} = k_p\lambda_{zp}^2 + k_q\lambda_{zq}^2 + k_r\lambda_{zr}^2 \tag{10.4-9c}$$

$$k_{xy} = k_{yx} = k_p\lambda_{xp}\lambda_{yp} + k_q\lambda_{xq}\lambda_{yq} + k_r\lambda_{xr}\lambda_{yr} \tag{10.4-9d}$$

$$k_{xz} = k_{zx} = k_p\lambda_{xp}\lambda_{zp} + k_q\lambda_{xq}\lambda_{zq} + k_r\lambda_{xr}\lambda_{zr} \tag{10.4-9e}$$

$$k_{yz} = k_{zy} = k_p\lambda_{yp}\lambda_{zp} + k_q\lambda_{yq}\lambda_{zq} + k_r\lambda_{yr}\lambda_{zr} \tag{10.4-9f}$$

Next, consider the reaction torques produced by a typical engine mount. Let a_x, a_y, and a_z represent the coordinates of the mount. The torques produced by the mount about X, Y, Z axes can be calculated as

$$M_X\mathbf{N}_x + M_Y\mathbf{N}_y + M_Z\mathbf{N}_z = (a_x\mathbf{N}_x + a_y\mathbf{N}_y + a_z\mathbf{N}_z) \times (F_X\mathbf{N}_x + F_Y\mathbf{N}_y + F_Z\mathbf{N}_z)$$

This yields

$$M_X = a_y F_Z - a_z F_Y \tag{10.4-10a}$$

$$M_Y = a_z F_X - a_x F_Z \tag{10.4-10b}$$

$$M_Z = a_x F_Y - a_y F_X \tag{10.4-10c}$$

Substituting F_X, F_Y, F_Z from Eq. 10.4-7 into Eq. 10.4-10, we obtain

$$M_X = -(k_{zx} a_y - k_{yx} a_z)\Delta_x - (k_{yz} a_y - k_{yy} a_z)\Delta_y - (k_{zz} a_y - k_{yz} a_z)\Delta_z \tag{10.4-11a}$$

$$M_Y = -(k_{xx} a_z - k_{zx} a_x)\Delta_x - (k_{xy} a_z - k_{zy} a_x)\Delta_y - (k_{xz} a_z - k_{zz} a_x)\Delta_z \tag{10.4-11b}$$

$$M_Z = -(k_{yx} a_x - k_{xx} a_y)\Delta_x - (k_{yy} a_x - k_{xy} a_y)\Delta_y - (k_{yz} a_x - k_{xz} a_y)\Delta_z \tag{10.4-11c}$$

(2) Forces and Torques Produced by an Engine Mount due to Rotational Motion of the Engine

Let α, β, γ be the small rotation angles of the engine about the X, Y, Z axes, respectively. Then, the displacements Λ_x, Λ_y, Λ_z of the engine mount due to rotation of the engine may be expressed as

$$\Lambda_x \mathbf{N}_x + \Lambda_y \mathbf{N}_y + \Lambda_z \mathbf{N}_z = (\alpha \mathbf{N}_x + \beta \mathbf{N}_y + \gamma \mathbf{N}_z) \times (a_x \mathbf{N}_x + a_y \mathbf{N}_y + a_z \mathbf{N}_z),$$

from which we have

$$\Lambda_x = a_z \beta - a_y \gamma \tag{10.4-12a}$$

$$\Lambda_y = a_x \gamma - a_z \alpha \tag{10.4-12b}$$

$$\Lambda_z = a_y \alpha - a_x \beta \tag{10.4-12c}$$

By using Eqs. 10.4-7, the forces produced by displacements Λ_x, Λ_y, Λ_z are given by

$$F_{RX} = -(k_{xz} a_y - k_{xy} a_z)\alpha - (k_{xx} a_z - k_{xz} a_x)\beta - (k_{xy} a_x - k_{xx} a_y)\gamma \tag{10.4-13a}$$

$$F_{RY} = -(k_{zy} a_y - k_{yy} a_z)\alpha - (k_{xy} a_z - k_{yz} a_x)\beta - (k_{yy} a_x - k_{xy} a_y)\gamma \tag{10.4-13b}$$

$$F_{RZ} = -(k_{zz} a_y - k_{yz} a_z)\alpha - (k_{xz} a_z - k_{zz} a_x)\beta - (k_{yz} a_x - k_{xz} a_y)\gamma \tag{10.4-13c}$$

By using Eqs. 10.4-11, and substituting Λ_x, Λ_y, Λ_z for Δ_x, Δ_y, Δ_z, the torques caused by Λ_x, Λ_y, Λ_z are given by

$$\begin{aligned} M_{RX} = &-(k_{zz} a_y^2 - 2k_{yz} a_y a_z + k_{yy} a_z^2)\alpha \\ &-(k_{xz} a_z a_y - k_{zz} a_x a_y + k_{zy} a_x a_z - k_{xy} a_z^2)\beta \\ &-(k_{yz} a_x a_y - k_{xz} a_y^2 + k_{xy} a_y a_z - k_{yy} a_x a_z)\gamma \end{aligned} \tag{10.4-14a}$$

$$\begin{aligned} M_{RY} = &-(k_{xz}a_ya_z - k_{xy}a_z^2 - k_{zz}a_xa_y + k_{yz}a_xa_z)\alpha \\ &-(k_{xx}a_z^2 - 2k_{xz}a_za_x + k_{zz}a_x^2)\beta \\ &-(k_{xy}a_xa_z - k_{xx}a_ya_z - k_{yz}a_x^2 + k_{xz}a_xa_y)\gamma \end{aligned} \tag{10.4-14b}$$

$$\begin{aligned} M_{RZ} = &-(k_{zy}a_xa_y - k_{yy}a_xa_y - k_{xz}a_y^2 + k_{xy}a_ya_z)\alpha \\ &-(k_{xy}a_xa_z - k_{yz}a_x^2 - k_{xx}a_ya_z + k_{xz}a_xa_y)\beta \\ &-(k_{yy}a_x^2 - 2k_{xy}a_xa_y + k_{xx}a_y^2)\gamma \end{aligned} \tag{10.4-14c}$$

Summing over the number of engine mounts, the equations of motion of the engine may be written in the matrix form

$$[M]\{\ddot{q}\}+[K]\{q\}=0, \tag{10.4-15}$$

where

$$\{q\}=[x_c \quad y_c \quad z_c \quad \alpha \quad \beta \quad \gamma]^T \tag{10.4-16}$$

$$[M]=\begin{bmatrix} m &&&&& \\ & m &&&& \\ && m &&& \\ &&& I_{xx} & I_{xy} & I_{xz} \\ &&& I_{xy} & I_{yy} & I_{yz} \\ &&& I_{xz} & I_{yz} & I_{zz} \end{bmatrix} \tag{10.4-17}$$

and

$$[K]=[K_{ij}]_{6\times 6} \tag{10.4-18}$$

The elements of the matrix $[K]$ are given by [10.3]

$$K_{11}=\sum k_{xx}$$

$$K_{12}=K_{21}=\sum k_{xy}$$

$$K_{13}=K_{31}=\sum k_{xz}$$

$$K_{14}=K_{41}=\sum (k_{xz}a_y - k_{xy}a_z)$$

$$K_{15}=K_{51}=\sum (k_{xx}a_z - k_{xz}a_x)$$

$$\begin{aligned}
K_{16} &= K_{61} = \sum (k_{xy}a_x - k_{xx}a_y) \\
K_{22} &= \sum k_{yy} \\
K_{23} &= K_{32} = \sum k_{yz} \\
K_{24} &= K_{42} = \sum (k_{yz}a_y - k_{yy}a_z) \\
K_{25} &= K_{52} = \sum (k_{xy}a_z - k_{yz}a_x) \\
K_{26} &= K_{62} = \sum (k_{yy}a_x - k_{xy}a_y) \\
K_{33} &= \sum k_{zz} \qquad (10.4\text{-}19) \\
K_{34} &= K_{43} = \sum (k_{zz}a_y - k_{yz}a_z) \\
K_{35} &= K_{53} = \sum (k_{xz}a_z - k_{zz}a_x) \\
K_{36} &= K_{63} = \sum (k_{yz}a_x - k_{xz}a_y) \\
K_{44} &= \sum (k_{yy}a_z^2 + k_{zz}a_y^2 - 2k_{yz}a_ya_z) \\
K_{45} &= K_{54} = \sum (k_{xz}a_ya_z + k_{yz}a_xa_z - k_{zz}a_xa_y - k_{xy}a_z^2) \\
K_{46} &= K_{64} = \sum (k_{xy}a_ya_z + k_{yz}a_xa_y - k_{yy}a_xa_z - k_{xz}a_y^2) \\
K_{55} &= \sum (k_{xx}a_z^2 + k_{zz}a_x^2 - 2k_{xz}a_xa_z) \\
K_{56} &= K_{65} = \sum (k_{xy}a_xa_z + k_{xz}a_xa_y - k_{xx}a_ya_z - k_{yz}a_x^2) \\
K_{66} &= \sum (k_{xx}a_y^2 + k_{yy}a_x^2 - 2k_{xy}a_xa_y),
\end{aligned}$$

in which k_{xx}, k_{xy}, . . . are given in Eq. 10.4-9.

Eq. 10.4-15 is the general form of equations of motion of the engine. The equations can be simplified by one or more of the following simplifying conditions:

(1) The reference axes X, Y, Z are selected to coincide with the principal axes of the engine; then $I_{xy} = I_{yz} = I_{zx} = 0$. In this case, the mass matrix becomes diagonal.

(2) The engine mounts are so arranged that one or more planes of symmetry exist. That is, motion parallel to the plane of symmetry has no tendency to excite motion perpendicular to it, or rotation about an axis lying in the plane does not excite motion parallel to the

plane. For example, in the first equation of Eqs. 10.4-15, motion in the *XY* plane does not tend to excite motion in the *XZ* or *YZ* plane if

$$K_{13} = \sum k_{xz} = 0 \tag{10.4-20}$$

$$K_{14} = \sum (k_{xz} a_y - k_{xy} a_z) = 0 \tag{10.4-21}$$

$$K_{15} = \sum (k_{xx} a_z - k_{xz} a_x) = 0 \tag{10.4-22}$$

(3) The principal axes p, q, r of each engine mount are orthogonal with respect to the reference axes. Then, in Eqs. 10.4-9, we have

$$k_{xx} = k_p, \quad k_{yy} = k_q, \quad k_{zz} = k_r, \quad k_{xy} = k_{xz} = k_{yz} = 0 \tag{10.4-23}$$

In this case, the supports are then called *orthogonal supports.*

Dynamic Decoupling. If the principal axes of inertia of the engine are parallel to the *X*, *Y*, *Z* axes, respectively, then $I_{xy} = I_{yz} = I_{zx} = 0$. In this case, the mass matrix $[M]$ becomes diagonal, resulting in dynamic decoupling.

Static Decoupling. If the orientations and/or stiffness of the engine mounts can be adjusted such that the stiffness matrix $[K]$ in Eq. 10.4-15 becomes diagonal, then the desired static decoupling can be obtained. In practice, however, this is very difficult to accomplish.

10.5 Mode Shapes and Natural Frequencies

In seeking a harmonic solution $\{q\}$ of Eq. 10.4-15, we assume that $\{q\}$ has the form

$$\{q\} = \{u\}\sin(\omega t + \psi), \tag{10.5-1}$$

where $\{u\}$ is an *eigenvector* or *modal vector*, ω is a natural frequency, and ψ is a constant.

Substituting Eq. 10.5-1 into Eq. 10.4-15 and factoring out the term $\sin(\omega t + \psi)$, we obtain

$$[K]\{u\} = \omega^2 [M]\{u\} \tag{10.5-2}$$

Eq. 10.5-2 is called the *generalized eigenvalue problem*. There are six eigenvalues ω_i^2, and six eigenvectors $\{u\}_i$, representing six rigid body modes of vibration of the engine.

Eq. 10.5-2 can also be transformed to a standard *algebraic eigenvalue problem* or *symmetric eigenvalue problem* (see Section 6.2). Several mathematical

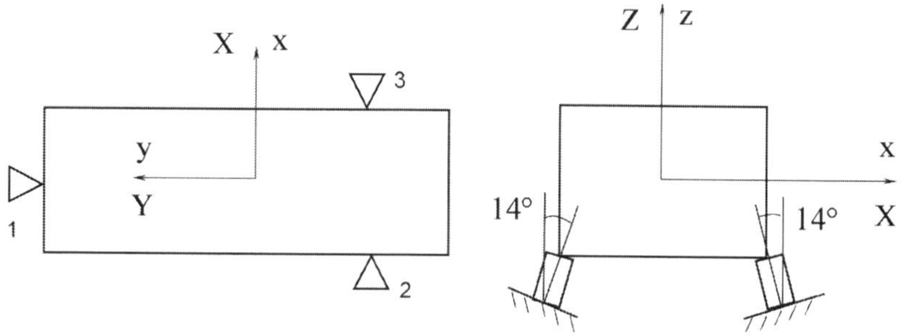

Fig. 10.5-1 An engine mounting system.

software packages can be used to compute the modal vectors and natural frequencies efficiently. For example, the solution of the eigenvalue problem of Eq. 10.5-2 can be obtained with a simple command in Matlab. That is,

```
[V,D]=eig(K,M);
```

where the diagonal elements of D are $\omega_1^2,\cdots,\omega_6^2$, and each column of matrix V contains the corresponding eigenvectors (modal vectors or mode shapes).

Example 10.5-1 An engine is supported by three engine mounts on a vehicle, as shown in Fig. 10.5-1. The coordinates of the mass center of the engine in the global coordinate system are 802.4 mm, 5.5 mm, 616.5 mm. The mass and the principal moments of inertia of the engine are listed in Table 10.5-1. The direction cosines between the principal axes of inertia and the global axes *X*, *Y*, *Z* are listed in Table 10.5-2. The orientation of each engine mount can be defined by a *1-2-3* rotation sequence. The rotation angles and stiffness of each engine mount are listed in Table 10.5-3. The coordinates of each engine mount are listed in Table 10.5-4. Find the natural frequencies and mode shapes of vibration of the engine.

Table 10.5-1 The Mass and Principle Moments of Inertia of the Engine

Mass (kg)	Principle moment of inertia (kg·m^2)		
m	I_1	I_2	I_3
299.1	12.05	25.2	25.68

Table 10.5-2 The Direction Cosines of the Principle Axes of the Engine

	Principle axis 1	Principle axis 2	Principle axis 3
X	0.5736	−0.0184	0.8189
Y	−0.2700	−0.9481	0.1678
Z	0.7734	−0.3174	−0.5488

Table 10.5-3 Orientation Angles and Stiffness of Bushings

	θ_1	θ_2	θ_3	k_p	k_q	k_r
1	0	0	0	100.0	100.0	150.0
2	0	14.0	0	131.1	131.1	524.4
3	0	–14.0	0	131.1	131.1	524.4

Table 10.5-4 Positions of Bushings

	X (mm)	*Y* (mm)	*Z* (mm)
1	790.7	494.1	808.9
2	546.5	–62.3	489.6
3	1087.2	-62.3	418.4

Solution: First, form the stiffness matrix $[K]$ of Eq. 10.4-15. To this end, we need to find the transformation matrix between the principal axes of each engine mount and the global axes X, Y, Z. The orientation of each engine mount in the global coordinate system can be determined as follows: Let the three principal axes of the engine mount be initially aligned with the axes X, Y, Z, respectively. Let the engine mount then be brought to its general orientation relative to the global coordinate system by three successive dextral rotations of the engine mount about its principal axes 1, 2, and 3 through the angles $\theta_1, \theta_2, \theta_3$. Then the transformation matrix between the principal axes of the engine mount and those of the global coordinate system may be expressed as [10.1, 10.2, and 10.5]

$$C = \begin{bmatrix} c_2c_3 & (c_1s_3 + s_1s_2c_3) & (s_1s_3 - c_1s_2c_3) \\ -c_2s_3 & (c_1c_3 - s_1s_2s_3) & (s_1c_3 + c_1s_2s_3) \\ s_2 & -s_1c_2 & c_1c_2 \end{bmatrix} \tag{a}$$

Observe that the matrix C in Eq. a is actually equal to S^T of Eq. 10.3-4. By using Eqs. a and 10.4-9, the stiffness matrix can be calculated from Eq. 10.4-19.

Next, to form the mass matrix $[M]$, we need to find the inertia matrix of the engine in the global coordinate system. This can be accomplished through the expression

$$\begin{bmatrix} I_{xx} & I_{xy} & I_{xz} \\ I_{yx} & I_{yy} & I_{yz} \\ I_{zx} & I_{zy} & I_{zz} \end{bmatrix} = S^T \begin{bmatrix} I_1 & 0 & 0 \\ 0 & I_2 & 0 \\ 0 & 0 & I_3 \end{bmatrix} S, \tag{b}$$

Table 10.5-5 Natural Frequencies and Mode Shapes of the Engine Mounting System

	Frequency	X_c	Y_c	Z_c	α	β	γ
1	5.348	1.000	13.538	0.097	–0.015	0.000	0.003
2	5.487	1.000	–0.119	–0.001	0.000	0.000	0.002
3	6.924	1.000	0.480	–0.054	0.005	–0.001	–0.006
4	7.482	1.000	–1.007	0.127	–0.012	0.000	–0.009
5	9.875	1.000	0.669	163.070	0.022	0.050	0.008
6	11.623	1.000	0.074	–1.061	0.003	0.078	0.011

where S is the direction cosine matrix of Table 10.5-2. Thus, the mass matrix can be formed according to Eq. 10.4-17.

Once the mass and stiffness matrices are obtained, the natural frequencies and mode shapes can be calculated by solving the eigenvalue problem of Eq. 10.5-2. With the Matlab code, we obtain the results as listed in Table 10.5-5.

10.6 Decoupling and Optimization

If the mass and stiffness matrices in Eq. 10.4-15 are diagonal, the equations of motion become six uncoupled (independent) differential equations. Each equation can then be solved independently. In this case, the vibration in one direction does not affect the vibration in other directions. This would be helpful for dynamic analysis and for isolation of vibration. Unfortunately, in practice it is very difficult to design an engine mounting system which is both statically and dynamically decoupled because of packaging constraints on the engine and mount locations, as well as the individual mount stiffness ratio limitations imposed by manufacturability. Therefore, there always exists some degree of coupling for any engine mounting system.

This then raises the question: How should we describe the coupling of an engine mounting system? To answer this question, we consider the kinetic energy of vibration. Specifically, the total kinetic energy of the system is given by

$$T = \frac{1}{2}\omega^2 [q]^T [M][q], \tag{10.6-1}$$

where $\left[q\right]^T = [X \quad Y \quad Z \quad \alpha \quad \beta \quad \gamma]$, and $\left[M\right]$ is the mass matrix. In expanding Eq. 10.6-1, it is convenient to introduce the following parameters:

$$T_x^* = \frac{1}{2}\omega^2 m X^2, \quad T_y^* = \frac{1}{2}\omega^2 m Y^2, \quad T_z^* = \frac{1}{2}\omega^2 m Z^2 \tag{10.6-2a}$$

$$T_\alpha^* = \frac{1}{2}\omega^2\left(I_{xx}\alpha^2 + I_{xy}\alpha\beta + I_{xz}\alpha\gamma\right) \quad (10.6\text{-}2b)$$

$$T_\beta^* = \frac{1}{2}\omega^2\left(I_{xy}\alpha\beta + I_{yy}\beta^2 + I_{yz}\beta\gamma\right) \quad (10.6\text{-}2c)$$

$$T_\gamma^* = \frac{1}{2}\omega^2\left(I_{zx}\gamma\alpha + I_{zy}\gamma\beta + I_{zz}\gamma^2\right), \quad (10.6\text{-}2d)$$

where T_x^*, T_y^*, and T_z^* represent the kinetic energy of translational vibration in the X, Y, and Z directions; and where T_α^*, T_β^*, and T_γ^* represent the kinetic energy of torsional vibration about X, Y, and Z axes. The ratio of these components to the total kinetic energy are defined as

$$T_x = \frac{T_x^*}{T}, \quad T_y = \frac{T_y^*}{T}, \quad T_z = \frac{T_z^*}{T} \quad (10.6\text{-}3)$$

$$T_\alpha = \frac{T_\alpha^*}{T}, \quad T_\beta = \frac{T_\beta^*}{T}, \quad T_\gamma = \frac{T_\gamma^*}{T} \quad (10.6\text{-}4)$$

These ratios are called the *mode coupling indicators* [10.6]. For example, if T_x is equal to one for some natural frequency, this means that this mode is purely translational motion in the X direction; there is no coupling with other modes. The following is a numerical example.

Example 10.6-1 Calculate the mode coupling indicators of the system in Example 10.5-1.

Solution: By using the Matlab code, the mode coupling indicators are calculated and listed in Table 10.6-1. It is seen from Table 10.6-1 that the sixth mode (11.623 Hz) is almost purely torsional vibration (rolling) about the Y-axis ($T_\beta = 0.957$, meaning 95.7% of the kinetic energy is from torsional vibration). The fifth mode (9.875 Hz) is almost purely vertical

Table 10.6-1 The Mode Coupling Indicators of the System of Example 10.5-1

	Frequency	T_x	T_y	T_z	T_α	T_β	T_γ
1	5.348	0.005	0.904	0.000	0.088	0.000	0.003
2	5.487	0.791	0.011	0.000	0.002	0.001	0.195
3	6.924	0.155	0.036	0.000	0.315	0.006	0.488
4	7.486	0.048	0.049	0.001	0.580	0.000	0.322
5	9.875	0.000	0.000	0.994	0.002	0.004	0.000
6	11.623	0.003	0.000	0.004	0.002	0.957	0.034

vibration (bouncing) along the Z-axis ($T_z = 0.994$). The second mode (5.487 Hz) is coupled with 79.1% translational motion along the X-axis and 19.5% torsional vibration about the Z-axis. For engine mounting system design, it is important to decouple the rolling and bouncing modes, that is, to make these two modes independent without coupling with other modes. From Table 10.6-1, this objective has been realized with this mounting system.

(1) Two-Degree-of-Freedom Coupled System

Although in general it is difficult to obtain fully decoupled equations of motion for an engine mounting system, it is possible to design an engine mounting system having two decoupled equations and two sets of coupled equations, which are similar to a two-degree-of-freedom system. This system is called the two-degree-of-freedom coupled system.

Referring to Fig. 10.6-1, assume that the mounts 1 and 2 are installed symmetrically about the YOZ plane, while mounts 3 and 4 are symmetric about the XOZ plane. In this case, we have

$$a_{x_1} = -a_{x_2}, \quad a_{y_1} = a_{y_2} = 0, \quad a_{z_1} = a_{z_2} \tag{10.6-5a}$$

$$a_{x_3} = a_{x_4} = 0, \quad a_{y_3} = -a_{y_4}, \quad a_{z_3} = a_{z_4} \tag{10.6-5b}$$

Next, since the axes of the coordinate system are aligned with those of the engine mounts, the transformation matrix between the principal axes of the engine mounts and the coordinate system is the unit matrix. That is,

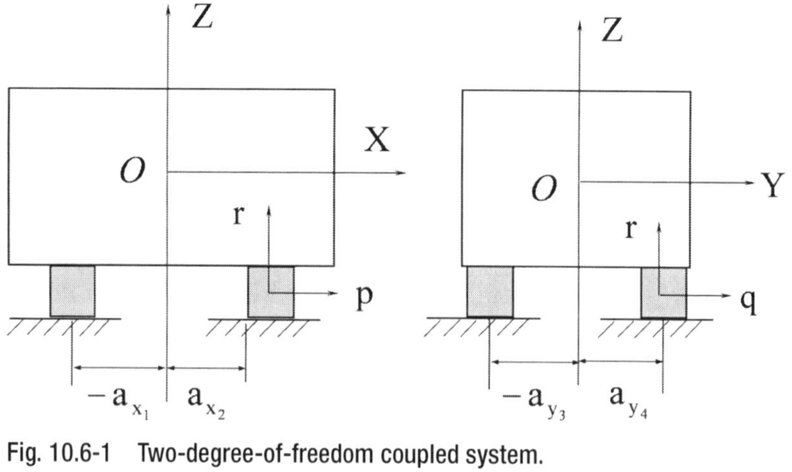

Fig. 10.6-1 Two-degree-of-freedom coupled system.

$$\begin{bmatrix} \lambda_{xp} & \lambda_{yp} & \lambda_{zp} \\ \lambda_{xq} & \lambda_{yq} & \lambda_{zq} \\ \lambda_{xr} & \lambda_{yr} & \lambda_{zr} \end{bmatrix} = \begin{bmatrix} 1 & 0 & 0 \\ 0 & 1 & 0 \\ 0 & 0 & 1 \end{bmatrix} \tag{10.6-6}$$

Thus, from Eq. 10.4-9, we have

$$k_{xx} = k_p, \quad k_{yy} = k_q, \quad k_{zz} = k_r \tag{10.6-7a}$$

$$k_{xy} = k_{yz} = k_{zx} = 0 \tag{10.6-7b}$$

As a result, from Eq. 10.4-19, the nonzero elements of the stiffness matrix are

$$K_{11} = \sum k_p, \quad K_{15} = \sum k_p a_z, \quad K_{22} = \sum k_q \tag{10.6-8a}$$

$$K_{24} = -\sum k_q a_z, \quad K_{33} = \sum k_r, \quad K_{44} = \sum \left(k_q a_z^2 + k_r a_y^2\right) \tag{10.6-8b}$$

$$K_{55} = \sum \left(k_p a_z^2 + k_r a_x^2\right), \quad K_{66} = \sum \left(k_p a_y^2 + k_q a_x^2\right) \tag{10.6-8c}$$

Next, assume that the *X*, *Y*, *Z* axes are parallel to the principal axes of the engine, respectively. In this case, the mass matrix is diagonal. Therefore, the equations of motion of the engine become

$$m\ddot{z}_c + \sum k_r z_c = 0 \tag{10.6-9a}$$

$$I_{zz}\ddot{\gamma} + \sum \left(k_p a_y^2 + k_q a_x^2\right)\gamma = 0 \tag{10.6-9b}$$

$$\begin{bmatrix} m & 0 \\ 0 & I_{yy} \end{bmatrix} \begin{bmatrix} \ddot{x}_c \\ \ddot{\beta} \end{bmatrix} + \begin{bmatrix} \sum k_p & \sum k_p a_z \\ \sum k_p a_z & \sum \left(k_p a_z^2 + k_r a_x^2\right) \end{bmatrix} \begin{bmatrix} x_c \\ \beta \end{bmatrix} = 0 \tag{10.6-9c}$$

$$\begin{bmatrix} m & 0 \\ 0 & I_{xx} \end{bmatrix} \begin{bmatrix} \ddot{y}_c \\ \ddot{\alpha} \end{bmatrix} + \begin{bmatrix} \sum k_q & -\sum k_q a_z \\ -\sum k_q a_z & \sum \left(k_q a_z^2 + k_r a_y^2\right) \end{bmatrix} \begin{bmatrix} y_c \\ \alpha \end{bmatrix} = 0 \tag{10.6-9d}$$

It is seen that Eqs. 10.6-9c are (x_c–β) coupled, while Eqs. 10.6-9d are (y_c–α) coupled. The following is a numerical example illustrating these results.

Example 10-6.2 For the engine mounting system shown in Fig. 10.6-1, determine the mode coupling indicators and mode shapes. Assume the mass of the engine is 299.5 kg, and that the principal moments of inertia are I_x=25 kg·m², I_y=12 kg·m², I_z=28 kg·m². The four mounts are identical and k_p=100 N/mm, k_q=130 N/mm, and k_r=150 N/mm. The coordinates of each mount are listed in Table 10.6-2.

Solution: By using the Matlab code, we obtain the mode coupling indicators and mode shapes as listed in Tables 10.6-3 and 10.6-4, respectively.

It is seen from Table 10.6-3 that the third mode (7.124 Hz) is purely vertical vibration ($T_z=1$). Similarly, the fifth mode (10.2 Hz) is purely torsional vibration about the Z-axis. The second (6.506 Hz) and fourth (8.886 Hz) modes are $Y-\alpha$ coupled, while the first (5.768 Hz) and sixth (12.688 Hz) modes are $X-\beta$ coupled.

Table 10.6-2 Coordinates of Bushing Positions of Example 10.6-2

	a_x (mm)	a_y (mm)	a_z (mm)
1	–500.0	0.0	–50.0
2	500.0	0.0	–50.0
3	0.0	–500.0	–50.0
4	0.0	500.0	–50.0

Table 10.6-3 The Mode Coupling Indicators of the System in Example 10.6-2

	Frequency	T_x	T_y	T_z	T_α	T_β	T_γ
1	5.768	0.996	0.000	0.000	0.000	0.004	0.000
2	6.506	0.000	0.955	0.000	0.045	0.000	0.000
3	7.124	0.000	0.000	1.000	0.000	0.000	0.000
4	8.886	0.000	0.045	0.000	0.995	0.000	0.000
5	10.200	0.000	0.000	0.000	0.000	0.000	1.000
6	12.688	0.004	0.000	0.000	0.000	0.996	0.000

Table 10.6-4 The Mode Shapes of the System in Example 10.6-2

	Frequency	X_c	Y_c	Z_c	α	β	γ
1	5.768	1.000	0.000	0.000	0.000	0.0003	0.000
2	6.506	0.000	1.000	0.000	–7.5E-4	0.000	0.000
3	7.124	0.000	0.000	1.000	0.000	0.000	0.000
4	8.886	0.000	1.000	0.000	0.0159	0.000	0.000
5	10.200	0.000	0.000	0.000	0.000	0.000	1.000
6	12.688	1.000	0.000	0.000	0.000	–0.075	0.000

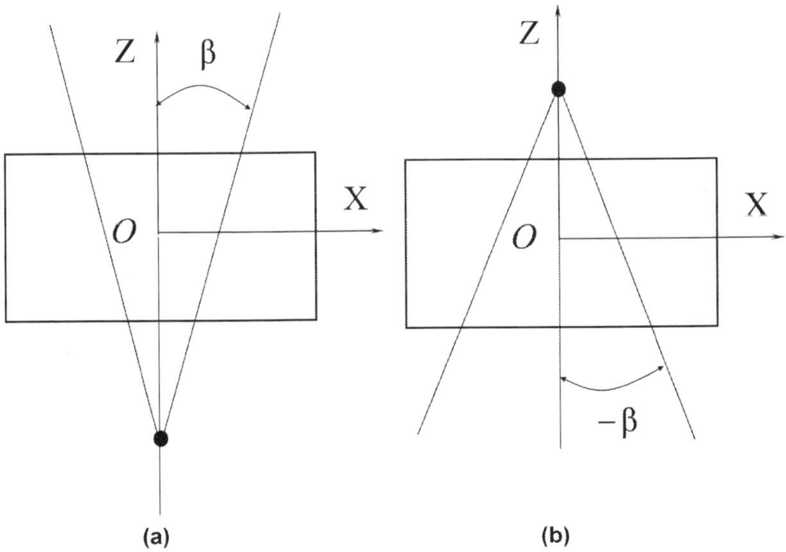

Fig. 10.6-2 The motions of $X-\beta$ coupled modes.

For the $X-\beta$ coupled modes, the vibration involves translational motion along the X-axis and the torsional motion about the Y-axis; the resultant motion can be regarded as rotational motion about a fixed axis that is parallel to the Y-axis. For the first mode (5.768 Hz), since both x_c and β are positive, the fixed axis is below the mass center, as shown in Fig. 10.6-2 (a). Similarly, the fixed axis for the sixth mode (12.688 Hz) is above the mass center, as shown in Fig. 10.6-2 (b).

(2) Three-Degree-of-Freedom Coupled System

In some cases, due to structure limitation, it is impossible to arrange the engine mounts symmetrically with respect to two planes, but it is possible to keep one symmetric plane (see XOZ). In this case, a three-degree-of-freedom coupled system can be obtained.

Referring to Fig. 10.6-3, the engine is supported by four mounts. We thus have

$$a_{x_1} \neq -a_{x_2}, \quad a_{y_1} = a_{y_2} = 0, \quad a_{z_1} = a_{z_2} \tag{10.6-10a}$$

$$a_{x_3} = a_{x_4} = 0, \quad a_{y_3} = -a_{y_4}, \quad a_{z_3} = a_{z_4} \tag{10.6-10b}$$

By using Eq. 10.4-19, the nonzero elements of the stiffness matrix are

$$K_{11} = \sum k_p, \quad K_{15} = \sum k_p a_z, \quad K_{22} = \sum k_q \tag{10.6-11a}$$

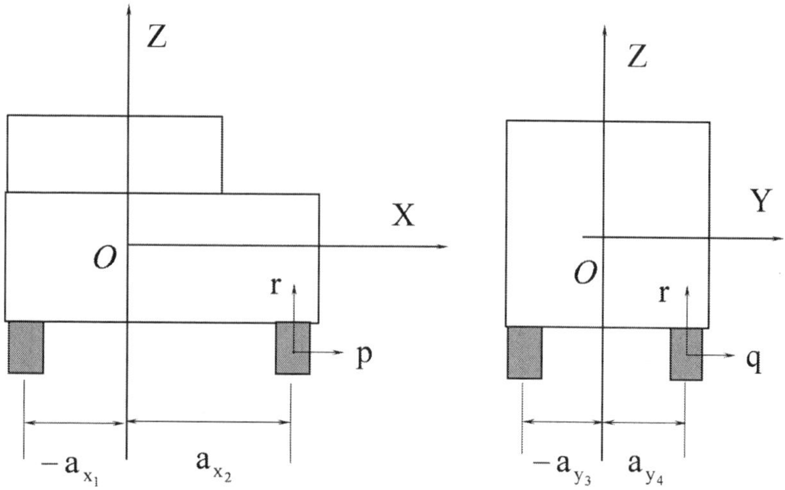

Fig. 10.6-3 Three-degree-of-freedom coupled system.

$$K_{24}=-\sum k_q a_z,\quad K_{26}=\sum k_q a_x,\quad K_{33}=\sum k_r \tag{10.6-11b}$$

$$K_{35}=-\sum k_r a_x,\quad K_{44}=\sum\left(k_q a_z^2+k_r a_y^2\right),\quad K_{46}=-\sum k_q a_x a_z \tag{10.6-11c}$$

$$K_{55}=\sum\left(k_p a_z^2+k_r a_x^2\right),\quad K_{66}=\sum\left(k_p a_y^2+k_q a_x^2\right) \tag{10.6-11d}$$

In this case, we have two set of equations

$$\begin{bmatrix} m & & \\ & m & \\ & & I_y \end{bmatrix}\begin{bmatrix} \ddot{x}_c \\ \ddot{z}_c \\ \ddot{\beta} \end{bmatrix}+\begin{bmatrix} \sum k_p & 0 & \sum k_p a_z \\ 0 & \sum k_r & -\sum k_r a_x \\ \sum k_p a_z & -\sum k_r a_x & \sum\left(k_p a_z^2+k_r a_x^2\right) \end{bmatrix}\begin{bmatrix} x_c \\ z_c \\ \beta \end{bmatrix}=0 \tag{10.6-12a}$$

$$\begin{bmatrix} m & & \\ & I_x & \\ & & I_z \end{bmatrix}\begin{bmatrix} y_c \\ \ddot{\alpha} \\ \ddot{\gamma} \end{bmatrix}+\begin{bmatrix} \sum k_q & -\sum k_q a_z & \sum k_q a_x \\ -\sum k_q a_z & \sum\left(k_q a_z^2+k_r a_y^2\right) & -\sum k_q a_x a_z \\ \sum k_q a_x & -\sum k_q a_x a_z & \sum\left(k_p a_y^2+k_q a_x^2\right) \end{bmatrix}\begin{bmatrix} y_c \\ \alpha \\ \gamma \end{bmatrix}=0$$

(10.6-12b)

It is seen that the equations in the first set are $X-Z-\beta$ coupled. Those in the second set are $Y-\alpha-\gamma$ coupled.

Table 10.6-5 The Mode Coupling Indicators of the System in Example 10.6-3

	Frequency	T_x	T_y	T_z	T_α	T_β	T_γ
1	5.670	0.925	0.000	0.038	0.000	0.037	0.000
2	6.301	0.000	0.905	0.000	0.032	0.000	0.063
3	6.658	0.062	0.000	0.842	0.000	0.096	0.000
4	8.730	0.000	0.001	0.000	0.699	0.000	0.300
5	9.204	0.000	0.095	0.000	0.268	0.000	0.637
6	10.101	0.013	0.000	0.119	0.000	0.867	0.000

Example 10.6-3 Repeat Example 10.6-2 with a_x=−200 mm and leaving the remainder of the data the same. Table 10.6-5 lists the mode coupling indicators.

It is seen from Table 10.6-5 that the modes of 5.670 Hz, 6.658 Hz, and 10.101 Hz are $X-Z-\beta$ coupled, while the modes of 6.301 Hz, 8.730 Hz, and 9.204 Hz are $Y-\alpha-\gamma$ coupled.

(3) Optimization

Most dynamic design principles attempt to design the natural frequencies of the system below and above the excitation frequency range. However, if a system has several resonances within a narrow band, it becomes difficult to accomplish this. Nevertheless, a physically decoupled engine mounting system has a better opportunity of producing fewer resonances over the operating range. The analysts can then concentrate on the vibration that may become resonant, and design effective isolation mounting systems for these vibrations.

In theory, a fully statically decoupled system can be obtained if all the non-diagonal elements of the stiffness matrix are zero. In reality, although it is difficult to realize this condition, it is possible to design an engine mounting system for which the non-diagonal elements of the stiffness matrix become very small, thus reducing the deleterious effects of coupling. This is called optimization design of the engine mounting system. In the following paragraphs, we discuss the underlying theory, and we then discuss numerical methods and software.

Assume that we have m functions $f_1(x), f_2(x), ..., f_m(x)$, where x represents n independent variables $x_1, x_2, ..., x_n$. Let $F(x)$ be

$$F(x)=f_1^2+f_2^2+\cdots+f_m^2 \tag{10.6-13}$$

Our objective is to find the values of these independent variables where $F(x)$ takes on a minimum value. This is called *optimization design*. For an engine mounting system, one can take the following variables as independent variables: the location of the mass center of the engine, the orientation of the engine, the locations of the engine mounts, the orientations of the engine mounts, and their stiffnesses [10.6].

Design engineers are particularly concerned with constrained optimization, where there are a priori limitations on the allowed values of independent variables. With Matlab, this task can be accomplished as follows: First, define the function $F(x)$ as

$$F(x)=\begin{bmatrix} f_1(x) \\ f_2(x) \\ \cdot \\ f_m(x) \end{bmatrix} \tag{10.6.14}$$

Second, define the initial values of the independent variables. Third, define the lower bound $[lb]$, and the upper bound $[ub]$. That is, the desired values of the variables should satisfy the following condition:

$$[lb] < \begin{bmatrix} x_1 \\ x_2 \\ \cdot \\ x_n \end{bmatrix} < [ub] \tag{10.6-15}$$

In Matlab, using the command:

```
[x]=lsqnonlin(fun,x0,lb,ub)
```

one can find the solution for the desired values of the variables where $F(x)$ takes on a minimum value; *fun* is the name of the Matlab m-file defining the function $F(x)$.

We now return to our original problem: to *best* decouple the equations of motion of the engine mounting system. The following procedural outline may be set forth:

(1) Write a Matlab m-file called, for example, *fun.m*, in which the design variables such as the coordinate of the mass center of the engine, the orientation angles of the engine mounts, etc., are denoted by $x(1), \cdots, x(n)$. The 15 non-diagonal elements of the stiffness matrix are expressed as $f_1(x)$, $f_2(x), \cdots, f_{15}(x)$, which are functions of the design variables $x(1), \cdots, x(n)$. Next, define the function $F(x)$, as in Eq. 10.6-14.

(2) Write a Matlab m-file called, for example, *main.m*, which defines the initial values of the variables $x(1), \cdots, x(\text{n})$, and the lower and upper bounds. In this Matlab m-file, the last line is

```
[x]=lsqnonlin(fun,x0,lb,ub)
```

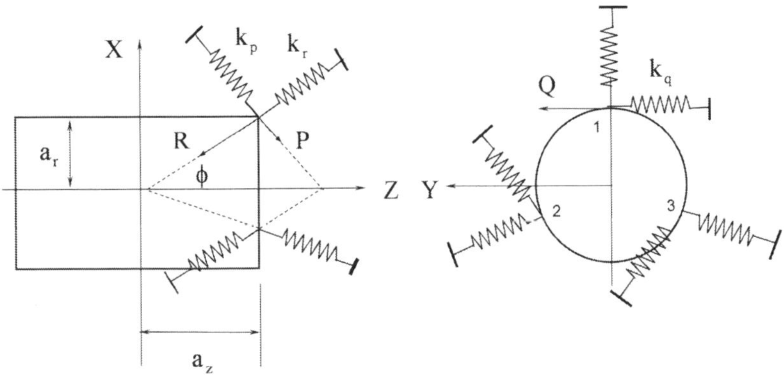

Fig. 10.6-4 A cylinder supported by three mounts.

(3) In Matlab, execute the program *main*.

To demonstrate the validity of this program, let us consider a classical example.

Example 10.6-4 [10.3, pp. 3–28] Fig. 10.6-4 depicts a uniform cylinder supported by three mounts. The mounts are arranged symmetrically about the *Z*-axis; they are attached to one end face of the cylinder at a distance a_r from the *Z*-axis and a distance a_z from the *XY* plane. The mounts are inclined so that their principal axes *R* and *P* intersect, respectively, at two common points on the *Z*-axis. Let the angle between the *Z*-axis and the *R*-axis for each mount be ϕ. Let the angle between the *Z*-axis and the *P*-axis for each be $90^\circ - \phi$. Consequently, the *Q* principal elastic axes are each tangent to the circle of radius a_r, which bounds the end face of the cylinder. We shall show analytically and numerically that complete decoupling of all six rigid body modes of vibration is achieved if the following condition is satisfied:

$$\frac{a_z}{a_r} = \frac{\frac{1}{2}\left[1 - \frac{k_p}{k_r}\right]\sin 2\phi}{\frac{k_q}{k_r} + \frac{k_p}{k_r} + \left[1 - \frac{k_p}{k_r}\right]\sin^2\phi} \tag{a}$$

Solution: It is easy to show that the direction cosines of the three mounts relative to the *X*, *Y*, *Z* axes are as in Tables 10.6-6 to 10.6-8.

By using these direction cosines, from Eq. 10.4-9 we obtain the stiffness coefficients for each mount. That is, for the first mount, we have

$$k_{xx} = k_p \cos^2\phi + k_r \sin^2\phi$$

$$k_{yy} = 0$$

Table 10.6-6 Direction Cosines of Mount 1

	X	***Y***	***Z***
P_1	$-\cos\phi$	0	$-\sin\phi$
Q_1	0	1	0
R_1	$-\sin\phi$	0	$-\cos\phi$

Table 10.6-7 Direction Cosines of Mount 2

	X	***Y***	***Z***
P_2	$\frac{1}{2}\cos\phi$	$\frac{\sqrt{3}}{2}\cos\phi$	$\sin\phi$
Q_2	$-\frac{\sqrt{3}}{2}$	$-\frac{1}{2}$	0
R_2	$\frac{1}{2}\sin\phi$	$-\frac{\sqrt{3}}{2}\sin\phi$	$\cos\phi$

Table 10.6-8 Direction Cosines of Mount 3

	X	***Y***	***Z***
P_3	$\frac{1}{2}\cos\phi$	$\frac{\sqrt{3}}{2}\cos\phi$	$\sin\phi$
Q_3	$-\frac{\sqrt{3}}{2}$	$-\frac{1}{2}$	0
R_3	$\frac{1}{2}\sin\phi$	$\frac{\sqrt{3}}{2}\sin\phi$	$-\cos\phi$

$$k_{zz} = k_p \sin^2\phi + k_r \cos^2\phi$$

$$k_{xy} = k_q$$

$$k_{xz} = -\left(k_p - k_r\right)\sin\phi\cos\phi$$

$$k_{yz} = 0$$

For the second mount, we have

$$k_{xx} = \frac{1}{4}k_p \cos^2\phi + \frac{3}{4}k_q + \frac{1}{4}k_r \sin^2\phi$$

$$k_{yy} = \frac{3}{4}k_p \cos^2\phi + \frac{1}{4}k_q + \frac{3}{4}k_r \sin^2\phi$$

$$k_{zz} = k_p \sin^2 \phi + k_r \cos^2 \phi$$

$$k_{xy} = -\frac{\sqrt{3}}{4} k_p \cos^2 \phi + \frac{\sqrt{3}}{4} k_q - \frac{\sqrt{3}}{4} k_r \sin^2 \phi$$

$$k_{xz} = \frac{1}{2}\left(k_p - k_r\right) \sin\phi \cos\phi$$

$$k_{yz} = -\frac{\sqrt{3}}{2}\left(k_p - k_r\right) \sin\phi \cos\phi$$

For the third mount, we have

$$k_{xx} = \frac{1}{4} k_p \cos^2 \phi + \frac{3}{4} k_q + \frac{1}{4} k_r \sin^2 \phi$$

$$k_{yy} = \frac{3}{4} k_p \cos^2 \phi + \frac{1}{4} k_q + \frac{3}{4} k_r \sin^2 \phi$$

$$k_{zz} = k_p \sin^2 \phi + k_r \cos^2 \phi$$

$$k_{xy} = \frac{\sqrt{3}}{4} k_p \cos^2 \phi - \frac{\sqrt{3}}{4} k_q + \frac{\sqrt{3}}{4} k_r \sin^2 \phi$$

$$k_{xz} = \frac{1}{2}\left(k_p - k_r\right) \sin\phi \cos\phi$$

$$k_{yz} = \frac{\sqrt{3}}{2}\left(k_p - k_r\right) \sin\phi \cos\phi$$

The elements of the stiffness matrix can be calculated using Eq. 10.4-19. As a result, all non-diagonal elements are identical to zero except K_{15}, given by

$$\begin{aligned} K_{15} &= \sum\left(k_{xx} a_z - k_{xz} a_x\right) \\ &= \frac{3}{2} a_z \left(k_p \cos^2 \phi + k_p + k_r \sin^2 \phi\right) + \frac{3}{2} a_r \left(k_p - k_r\right) \sin\phi \cos\phi \end{aligned} \qquad \text{(b)}$$

By letting K_{15} be equal to zero, we obtain Eq. a, which is the condition for complete decoupling modes.

Since complete decoupling is accomplished, the stiffness matrix becomes diagonal. Also, the principal axes of the cylinder are parallel to the axes of the coordinate system. Therefore, the frequencies for the six decoupled modes of the system can be expressed as follows:

The frequencies of translational motions along the axes X, Y, and Z are

$$\omega_x = \sqrt{\frac{K_{11}}{m}}, \quad \omega_y = \sqrt{\frac{K_{22}}{m}}, \quad \omega_z = \sqrt{\frac{K_{33}}{m}} \qquad \text{(c)}$$

The frequencies of rotational motions about the X, Y, and Z axes are

$$\omega_\alpha = \sqrt{\frac{K_{44}}{I_{xx}}}, \quad \omega_\beta = \sqrt{\frac{K_{55}}{I_{yy}}}, \quad \omega_\gamma = \sqrt{\frac{K_{66}}{I_{zz}}} \tag{d}$$

Substituting Eqs. 10.4-19 into Eqs. c and d, we obtain [10.3]

$$\omega_x = \omega_y = \sqrt{\frac{3}{2m}\left(k_p \cos^2\phi + +k_q + k_r \sin^2\phi\right)} \tag{e}$$

$$\omega_z = \sqrt{\frac{3}{m}\left(k_p \sin^2\phi + k_r \cos^2\phi\right)} \tag{f}$$

$$\omega_\alpha = \omega_\beta = \sqrt{\frac{3a_r}{2I_{xx}}\left[k_p \sin\phi\left(a_r \sin\phi + a_z \cos\phi\right) + k_r \cos\phi\left(a_r \cos\phi - a_z \sin\phi\right)\right]} \tag{g}$$

$$\omega_\gamma = a_r \sqrt{\frac{3k_q}{I_{zz}}} \tag{h}$$

We now use the above Matlab code to search these results numerically. We take the stiffness of the mounts and distance a_z as the design parameters. That is, we let $x(1)=k_p$, $x(2)=k_q$, $x(3)=k_r$, $x(4)=a_z$. The following numerical values are used for the simulation: a_r=500 mm, I_{xx}=25 kg·m², I_{yy}=25 kg·m², I_{zz}=12 kg·m², m=200 kg, and $\phi = 30^\circ$.

The results obtained using Matlab are

$$k_p = 50 \text{ N/mm},\ k_q = 100 \text{ N/mm},\ k_r = 150 \text{ N/mm},\ a_z = 123.72 \text{ mm}$$

It is readily seen that these results satisfy Eq. a. Next, the natural frequencies and mode coupling indicators are listed in Table 10.6-9. It is seen that the six rigid body modes are completely decoupled. It is also seen that the frequencies are identical to those obtained by using Eqs. e to h.

Table 10.6-9 Natural Frequencies and Mode Coupling Indicators for Example 10.6-4

	Frequency	T_x	T_y	T_z	T_α	T_β	T_γ
1	5.766	1	0	0	0	0	0
2	5.766	0	1	0	0	0	0
3	6.590	0	0	0	1	0	0
4	6.590	0	0	0	0	1	0
5	6.892	0	0	1	0	0	0
6	12.582	0	0	0	0	0	1

10.7 Torque Roll Axis (TRA)

Consider a rigid body *B* in free space. Suppose an oscillating moment is applied to *B* directed along (or about) a principal axis of inertia of *B*. Then *B* will oscillate about its mass center around an axis parallel to the principal axis of inertia. If, however, the applied oscillating moment on *B* is directed along an axis other than a principal axis of inertia, *B* will oscillate about its mass center around an axis that is neither parallel to a principal axis of inertia, nor parallel to the applied moment axis. This axis of rotation is called the *torque roll axis* (*TRA*).

In an automobile engine, the applied moment axis is along the crankshaft. The crankshaft axis, however, is generally not a principal axis of inertia for the engine. Therefore, if an engine is suspended by very low rate motor mounts, and if it is then subjected to an oscillating crankshaft torque, the engine will oscillate about its *TRA*.

In the following paragraphs, we present a procedure for finding the direction of the *TRA*.

Let *X*, *Y*, *Z* be an engine coordinate system with origin at the engine mass center, and with the *X*-axis parallel to the crankshaft axis. Let the *Y*-axis then be directed to its right when looking along the positive *X*-axis. The positive *Z*-axis is then determined by the right-hand rule. See Fig. 10.7-1.

Let $\mathbf{N}_1$, $\mathbf{N}_2$, $\mathbf{N}_3$ be mutually perpendicular unit vectors parallel to the *X*, *Y*, and *Z* axes, and let $\mathbf{n}_1$, $\mathbf{n}_2$, $\mathbf{n}_3$ be mutually perpendicular unit vectors parallel to the engine's principal axes of inertia. The relation between these two axis systems can be expressed by an orthogonal transformation matrix of direction cosines as

$$\begin{bmatrix} \mathbf{n}_1 \\ \mathbf{n}_2 \\ \mathbf{n}_3 \end{bmatrix} = \begin{bmatrix} \cos\alpha_1 & \cos\beta_1 & \cos\gamma_1 \\ \cos\alpha_2 & \cos\beta_2 & \cos\gamma_2 \\ \cos\alpha_3 & \cos\beta_3 & \cos\gamma_3 \end{bmatrix} \begin{bmatrix} \mathbf{N}_1 \\ \mathbf{N}_2 \\ \mathbf{N}_3 \end{bmatrix}, \qquad (10.7\text{-}1)$$

where α_i, β_i, and γ_i $(i=1,2,3)$ are orientation angles.

Let I_1, I_2, I_3 be the principal moments of inertia of the engine. When an oscillating crankshaft torque M_x is applied to the engine, the equations of motion of the engine can be written as

$$I_1\ddot{\theta}_1 = M_x \cos\alpha_1$$

$$I_2\ddot{\theta}_2 = M_x \cos\alpha_2 \qquad (10.7\text{-}2)$$

$$I_3\ddot{\theta}_3 = M_x \cos\alpha_3$$

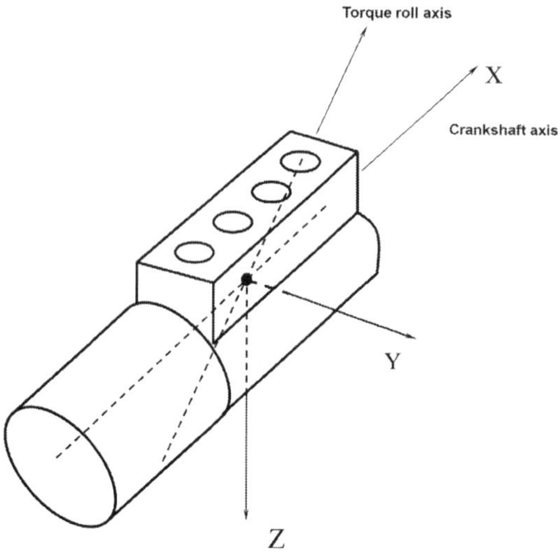

Fig. 10.7-1 Torque roll axis.

The direction of the torque roll axis is parallel to the direction of the resultant angular acceleration, expressed as

$$\ddot{\boldsymbol{\theta}} = \ddot{\theta}_1 \mathbf{n}_1 + \ddot{\theta}_2 \mathbf{n}_2 + \ddot{\theta}_3 \mathbf{n}_3 \tag{10.7-3}$$

From Eq. 10.7-1, we have

$$\mathbf{n}_1 = \cos\alpha_1 \mathbf{N}_1 + \cos\beta_1 \mathbf{N}_2 + \cos\gamma_1 \mathbf{N}_3 \tag{10.7-4a}$$

$$\mathbf{n}_2 = \cos\alpha_2 \mathbf{N}_1 + \cos\beta_2 \mathbf{N}_2 + \cos\gamma_2 \mathbf{N}_3 \tag{10.7-4b}$$

$$\mathbf{n}_3 = \cos\alpha_3 \mathbf{N}_1 + \cos\beta_3 \mathbf{N}_2 + \cos\gamma_3 \mathbf{N}_3 \tag{10.7-4c}$$

Substituting from Eqs. 10.7-4 into Eq. 10.7-3, we obtain

$$\begin{aligned} \ddot{\boldsymbol{\theta}} = & \ddot{\theta}_1 \left[\cos\alpha_1 \mathbf{N}_1 + \cos\beta_1 \mathbf{N}_2 + \cos\gamma_1 \mathbf{N}_3\right] \\ & + \ddot{\theta}_2 \left[\cos\alpha_2 \mathbf{N}_1 + \cos\beta_2 \mathbf{N}_2 + \cos\gamma_2 \mathbf{N}_3\right] \\ & + \ddot{\theta}_3 \left[\cos\alpha_3 \mathbf{N}_1 + \cos\beta_3 \mathbf{N}_2 + \cos\gamma_3 \mathbf{N}_3\right] \end{aligned} \tag{10.7-5}$$

Eq. 10.7-5 may be written as

$$\begin{aligned} \ddot{\boldsymbol{\theta}} = & \ddot{\theta}_1 \left[\cos\alpha_1 + \frac{\ddot{\theta}_2}{\ddot{\theta}_1}\cos\alpha_2 + \frac{\ddot{\theta}_3}{\ddot{\theta}_1}\cos\alpha_3\right] \mathbf{N}_1 \\ & + \ddot{\theta}_1 \left[\cos\beta_1 + \frac{\ddot{\theta}_2}{\ddot{\theta}_1}\cos\beta_2 + \frac{\ddot{\theta}_3}{\ddot{\theta}_1}\cos\beta_3\right] \mathbf{N}_2 \\ & + \ddot{\theta}_1 \left[\cos\gamma_1 + \frac{\ddot{\theta}_2}{\ddot{\theta}_1}\cos\gamma_2 + \frac{\ddot{\theta}_3}{\ddot{\theta}_1}\cos\gamma_3\right] \mathbf{N}_3 \end{aligned} \tag{10.7-6}$$

From Eq. 10.7-2, we have

$$\frac{\ddot{\theta}_2}{\ddot{\theta}_1} = \frac{I_1 \cos\alpha_2}{I_2 \cos\alpha_1}, \qquad \frac{\ddot{\theta}_3}{\ddot{\theta}_1} = \frac{I_1 \cos\alpha_3}{I_3 \cos\alpha_1} \tag{10.7-7}$$

Substituting Eq. 10.7-7 into Eq. 10.7-6, the components of a vector parallel to the torque roll axis with respect to the engine coordinate system may be calculated as

$$m_x = \cos\alpha_1 + \left(\frac{I_1 \cos\alpha_2}{I_2 \cos\alpha_1}\right)\cos\alpha_2 + \left(\frac{I_1 \cos\alpha_3}{I_3 \cos\alpha_1}\right)\cos\alpha_3 \tag{10.7-8a}$$

$$n_y = \cos\beta_1 + \left(\frac{I_1 \cos\alpha_2}{I_2 \cos\alpha_1}\right)\cos\beta_2 + \left(\frac{I_1 \cos\alpha_3}{I_3 \cos\alpha_1}\right)\cos\beta_3 \tag{10.7-8b}$$

$$l_z = \cos\gamma_1 + \left(\frac{I_1 \cos\alpha_2}{I_2 \cos\alpha_1}\right)\cos\gamma_2 + \left(\frac{I_1 \cos\alpha_3}{I_3 \cos\alpha_1}\right)\cos\gamma_3 \tag{10.7-8c}$$

Thus, the direction cosines of the torque axis are given by

$$\cos\alpha = \frac{m_x}{\sqrt{m_x^2 + n_y^2 + l_z^2}} \tag{10.7-9a}$$

$$\cos\beta = \frac{n_y}{\sqrt{m_x^2 + n_y^2 + l_z^2}} \tag{10.7-9b}$$

$$\cos\gamma = \frac{l_z}{\sqrt{m_x^2 + n_y^2 + l_z^2}} \tag{10.7-9c}$$

It is seen from the above results that the torque roll axis of an engine is determined by its moments of inertia and directions of principal axes of inertia.

10.8 Torque Roll Axis (TRA) Mode Decoupling

Mode decoupling strategy has been sought by designers and researchers over the past two decades. In the following paragraphs, we discuss a general mathematical condition for a complete decoupling. We show that decoupling can be obtained in the *TRA* coordinate system where *TRA* is one of the axes [10.7].

First, we introduce a *TRA* coordinate system ($X'\ Y'\ Z'$) as follows: Let the X' direction be parallel to the *TRA* and the origin be located on the mass center of the engine. The axes Y' and Z' may be chosen arbitrarily as long as X', Y', Z' forms an orthogonal right-hand coordinate system.

All equations in this section are defined in the *TRA* coordinate system. However, the superscript used to denote the *TRA* coordinate system is dropped here for brevity. The equations of motion of the engine have the same form as Eqs. 10.4-15, and the matrix elements of the stiffness matrix $[K]$ has the same form as in Eqs. 10.4-19. To achieve a complete decoupling of the *TRA* mode from other modes, the *TRA* direction $\{q\}_{TRA} = [0 \quad 0 \quad 0 \quad 1 \quad 0 \quad 0]^T$ should be one of the natural modes. In other words, the following relationship should be satisfied:

$$[K]\{q\}_{TRA} = \lambda[M]\{q\}_{TRA}, \tag{10.8-1}$$

from which we have

$$K_{14} = 0, \quad K_{24} = 0, \quad K_{34} = 0 \tag{10.8-2}$$

$$K_{44} = \lambda I_{xx}, \quad K_{54} = \lambda I_{xy}, \quad K_{64} = \lambda I_{xz} \tag{10.8-3}$$

Expressed in detail using Eqs. 10.4-19, we have

$$\sum (k_{xz} a_y - k_{xy} a_z) = 0 \tag{10.8-4a}$$

$$\sum (k_{yz} a_y - k_{yy} a_z) = 0 \tag{10.8-4b}$$

$$\sum (k_{zz} a_y - k_{yz} a_z) = 0 \tag{10.8-4c}$$

$$\sum (k_{yy} a_z^2 + k_{zz} a_y^2 - 2k_{yz} a_y a_z) = \lambda I_{xx} \tag{10.8-4d}$$

$$\sum (k_{xz} a_y a_z + k_{yz} a_x a_z - k_{zz} a_x a_y - k_{xy} a_z^2) = \lambda I_{xy} \tag{10.8-4e}$$

$$\sum (k_{xy} a_y a_z + k_{yz} a_x a_y - k_{yy} a_x a_z - k_{xz} a_y^2) = \lambda I_{xz} \tag{10.8-4f}$$

These equations represent the conditions for a complete decoupling of the *TRA* mode. One may vary three stiffness ratios and three orientation angles or even locations while holding other parameters as constants.

Similar results can be obtained if the *TRA* is chosen as the axes Y' or Z'. This is left as an exercise for the reader.

Problems

10.1 The inertia matrix of a rigid body relative to the reference coordinate system *O-xyz* (Fig. P10.1) is

$$[I_{ij}] = \frac{1}{8}\begin{bmatrix} 12 & -2\sqrt{3} & 6 \\ -2\sqrt{3} & 17 & -\sqrt{3} \\ 6 & -\sqrt{3} & 19 \end{bmatrix} (\text{kg}\cdot\text{m}^2),$$

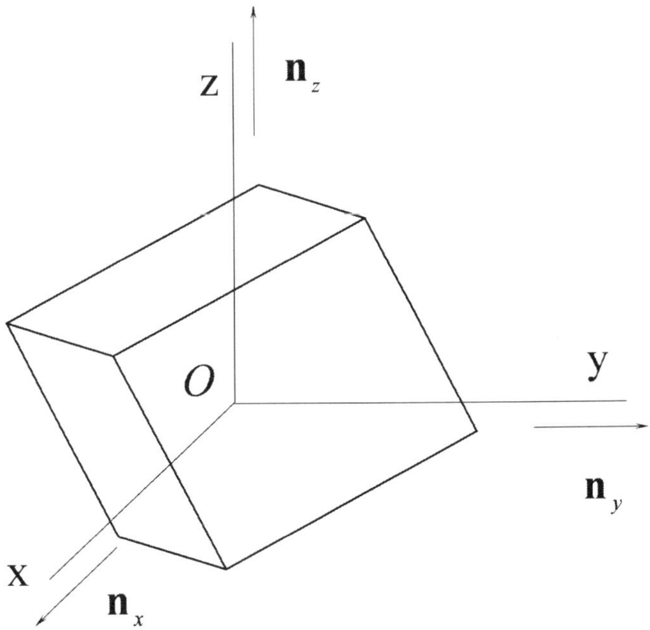

Fig. P10.1

where n_x, n_y, n_z are mutually perpendicular unit vectors parallel to the axes x, y, z, respectively. Find the principal moments of inertia and the corresponding principal directions.

Ans. $I_1 = 1,\ I_2 = 2,\ I_3 = 3\ (\mathrm{kg \cdot m^2})$

$$n_1 = \left(\sqrt{3}/2\right)n_x + (1/4)n_y - \left(\sqrt{3}/4\right)n_z$$

$$n_2 = \left(\sqrt{3}/2\right)n_y + (1/2)n_z$$

$$n_3 = (1/2)n_x - \left(\sqrt{3}/4\right)n_y + (3/4)n_z$$

10.2 Repeat Problem 10.1 using the direction numbers of each principal axis to form a matrix as

$$S = \begin{bmatrix} \sqrt{3}/2 & 0 & 1/2 \\ 1/4 & \sqrt{3}/2 & -\sqrt{3}/4 \\ -\sqrt{3}/4 & 1/2 & 3/4 \end{bmatrix}$$

Show that

$$S^T[I_{ij}]S = \mathrm{diag}(1,2,3)$$

10.3 Assume that the orientation of a coordinate system relative to a reference frame is obtained by a 2-3-1 rotation sequence (through the angles α, β, and γ). Determine the direction cosine matrix between the two coordinate systems.

$$\textbf{Ans. } S = \begin{bmatrix} c_\alpha c_\beta & (-c_\alpha s_\beta c_\gamma + s_\alpha s_\gamma) & (c_\alpha s_\beta s_\gamma + s_\alpha c_\gamma \\ s_\beta & c_\beta c_\gamma & -c_\beta s_\gamma \\ -s_\alpha c_\alpha & (s_\alpha s_\beta c_\gamma + c_\alpha s_\gamma) & (-s_\alpha s_\beta s_\gamma + c_\alpha c_\gamma) \end{bmatrix}$$

10.4 Repeat Problem 10.3 for a 1-3-2 rotation sequences (through the angles α, β, and γ).

$$\textbf{Ans. } S = \begin{bmatrix} c_\beta c_\gamma & -s_\beta & c_\beta s_\gamma \\ (c_\alpha s_\beta c_\gamma + s_\alpha s_\gamma) & c_\alpha c_\beta & (c_\alpha s_\beta s_\gamma - c_\gamma s_\alpha) \\ (s_\alpha s_\beta c_\gamma - c_\alpha s_\gamma) & s_\alpha c_\beta & (s_\alpha s_\beta s_\gamma + c_\alpha c_\gamma) \end{bmatrix}$$

10.5 Show that the determinant of the direction cosine matrix between two coordinate systems is equal to ± 1.

10.6 Show that, for the direction cosine matrix between two coordinate systems, there exists at least one real eigenvalue with value 1.

10.7 From Problem 10.6, show that, for the direction cosine matrix between two coordinate systems, there exists at least one real eigenvector whose coordinate components are the same in both coordinate systems.

10.8 A rigid body is supported by two rubber mounts, as shown in Fig. P10.8. Determine the direction cosines of the mounts relative to the reference axes.

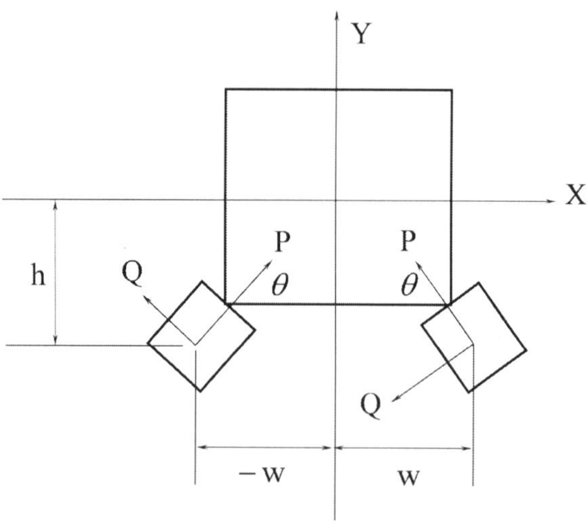

Fig. P10.8

Ans.

$$\begin{Bmatrix} n_p \\ n_q \\ n_r \end{Bmatrix} = \begin{bmatrix} \cos\theta & \sin\theta & 0 \\ -\sin\theta & \cos\theta & 0 \\ 0 & 0 & 1 \end{bmatrix} \begin{Bmatrix} N_x \\ N_y \\ N_z \end{Bmatrix}$$

$$\begin{Bmatrix} n_p \\ n_q \\ n_r \end{Bmatrix} = \begin{bmatrix} -\cos\theta & \sin\theta & 0 \\ -\sin\theta & -\cos\theta & 0 \\ 0 & 0 & 1 \end{bmatrix} \begin{Bmatrix} N_x \\ N_y \\ N_z \end{Bmatrix}$$

10.9 See Problem 10.8. Determine the stiffness of each mount in the reference coordinate system.

Ans. Non-zero terms for the left mount:

$$k_{xx} = k_p\lambda_{xp}^2 + k_q\lambda_{xq}^2 = k_p\cos^2\theta + k_q\sin^2\theta$$

$$k_{yy} = k_p\lambda_{yp}^2 + k_q\lambda_{yq}^2 = k_p\sin^2\theta + k_q\cos^2\theta$$

$$k_{xy} = k_p\lambda_{xp}\lambda_{yp} + k_q\lambda_{xq}\lambda_{yq} = k_p\sin\theta\cos\theta - k_q\sin\theta\cos\theta$$

Non-zero terms for the right mount:

$$k_{xx} = k_p\lambda_{xp}^2 + k_q\lambda_{xq}^2 = k_p\cos^2\theta + k_q\sin^2\theta$$

$$k_{yy} = k_p\lambda_{yp}^2 + k_q\lambda_{yq}^2 = k_p\sin^2\theta + k_q\cos^2\theta$$

$$k_{xy} = k_p\lambda_{xp}\lambda_{yp} + k_q\lambda_{xq}\lambda_{yq} = -k_p\sin\theta\cos\theta + k_q\sin\theta\cos\theta$$

10.10 See again Problem 10.8. Show that if

$$\frac{h}{w} = \frac{\left(1 - \frac{k_q}{k_p}\right)\tan\theta}{1 + \frac{k_q}{k_p}\tan^2\theta},$$

then all the vibration modes are decoupled, and find the natural frequencies for (1) translational motion along the x-axis, (2) translational motion along the y-axis, and (3) rotational motion about the z-axis.

Ans. $\omega_x = \sqrt{\frac{2}{m}\left(k_p\cos^2\theta + k_q\sin^2\theta\right)}$

$$\omega_y = \sqrt{\frac{2}{m}\left(k_p\sin^2\theta + k_q\cos^2\theta\right)}$$

$$\omega_\gamma = \sqrt{\frac{2k_p k_q w^2}{I_{zz}(k_p \cos^2\theta + k_q \sin^2\theta)}}$$

10.11 Step-by-step, calculate the direction cosines shown in Tables 10.6-5 to 10.6-7.

10.12 Step-by-step, calculate the diagonal elements of the stiffness matrix in Example 10.6-4 and the natural frequencies of the complete decoupling modes.

10.13 If we choose the *TRA* as the *Y*-axis, what is the condition for complete decoupling of the *TRA* mode?

Ans. $\sum(k_{xx}a_z - k_{xz}a_x) = 0$

$$\sum(k_{xy}a_z - k_{yz}a_x) = 0$$

$$\sum(k_{xz}a_z - k_{zz}a_x) = 0$$

$$\sum(k_{xz}a_y a_z + k_{yz}a_x a_z - k_{zz}a_x a_y - k_{xy}a_z^2) = \lambda I_{xy}$$

$$\sum(k_{xx}a_z^2 + k_{zz}a_x^2 - 2k_{xz}a_x a_z) = \lambda I_{yy}$$

$$\sum(k_{xy}a_x a_z + k_{xz}a_x a_y - k_{xx}a_y a_z - k_{yz}a_x^2) = \lambda I_{yz}$$

10.14 If we choose the *TRA* as the *Z*-axis, what is the condition for complete decoupling of the *TRA* mode?

Ans. $\sum(k_{xy}a_x - k_{xx}a_y) = 0$

$$\sum(k_{yy}a_x - k_{xy}a_y) = 0$$

$$\sum(k_{yz}a_x - k_{xz}a_y) = 0$$

$$\sum(k_{xy}a_y a_z + k_{yz}a_x a_y - k_{yy}a_x a_z - k_{xz}a_y^2) = \lambda I_{xz}$$

$$\sum(k_{xy}a_x a_z + k_{xz}a_x a_y - k_{xx}a_y a_z - k_{yz}a_x^2) = \lambda I_{yz}$$

$$\sum(k_{xx}a_y^2 + k_{yy}a_x^2 - 2k_{xy}a_x a_y) = \lambda I_{zz}$$

References

10.1. Huston, R. L., *Multibody Dynamics*, Butterworth-Heinemann, Stoneham, MA, 1990.

10.2. Huston, R. L., and Liu, C. Q., *Formulas for Dynamic Analysis*, Marcel Dekker, Inc., New York, 2001.

10.3. Harris, C. M., ed., *Shock and Vibration Handbook*, 3rd ed., McGraw-Hill, New York, 1987.

10.4. Roberson, R. E., and Schwertassek, R., *Dynamics of Multibody Systems*, Springer-Verlag, Berlin, Heidelberg, Germany, 1988.

10.5. Kane, T. R., Likins, P. W., and Levinson, D. A., *Spacecraft Dynamics*, McGraw-Hill, New York, 1983.

10.6. Liu, C. Q., "A Computerized Optimization Method of Engine Mounting Systems," SAE paper 2003-01-1461, SAE International, Warrendale, PA, 2003.

10.7. Jeong, T., and Singh, R., "Analytical Methods of Decoupling the Automotive Engine Torque Roll Axis," *Journal of Sound and Vibration*, Vol. 234, No. 1, 2000, pp. 85–114.

Chapter Eleven

Experimental Modal Analysis

11.1 Introduction

As mechanical systems become increasingly large and complex, as designers need more precise information about the dynamic characteristics of the systems, and as analytical models become less able to accurately describe increasingly complex systems, it has become necessary to look to experimental methods to obtain the needed information about complex system dynamic characteristics.

In recent years, there have been so many advances in measurement technology that it is now possible to experimentally estimate the natural frequencies and the mode shapes of even the most complex mechanical systems.

Experimental modal analysis refers to the process of experimentally determining the modal parameters (natural frequencies, damping ratios, and modal vectors) of a linear time-invariant system [11.1–11.6].

Fig. 11.1-1 provides an illustration of typical vibration analysis processes of a complex structure [11.5]. Initially, there is a theoretical analysis (*theoretical route*) which typically includes three phases: First, the structure's physical characteristics are described in terms of mass, stiffness, and damping matrices. This is called the *spatial model*. Next, using the spatial model, the analyst conducts an analytical modal analysis, yielding natural frequencies, modal damping ratios, and mode shapes. This is called the *modal model*. The third phase is to determine the structure's response under given excitation. The response is dependent not only on the structure's characteristics, but also on the location, amplitude, and direction of the excitation. In this third phase, it is convenient to determine the structure's response to a *standard excitation*. This is called the *response model*. It includes a set of frequency response functions (FRFs).

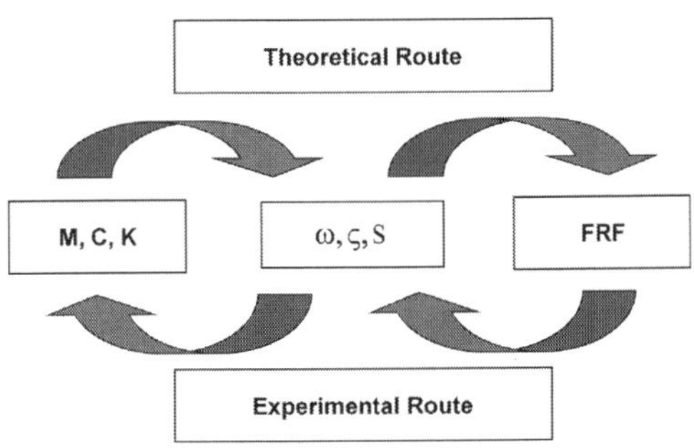

Fig. 11.1-1 Theoretical and experimental routes.

Experimental modal analysis may be viewed as the reverse of a theoretical analysis, as represented by *experimental route* in Fig. 11.1-1. Specifically, from a set of experimentally measured FRFs, the natural frequencies, modal damping ratios, and mode shapes are estimated. Then, from a series of these estimates, the spatial properties are determined. We present details of the analysis in this chapter.

Experimental modal analysis dates back to the 1940s, with the measurement of airplane structure response to excitation. The objective was to understand the flutter phenomenon. At that time, however, transducers and signal processing equipment had very limited capabilities, and thus the entire process was time-consuming and not practical for most applications.

Today, with the advent of digital minicomputers and fast Fourier transforms (FFTs), experimental modal analysis has become an efficient, reliable, and widely used tool in structural vibration analysis.

11.2 Modal Analysis Theory

The study of natural frequencies, damping ratios, and modal vectors in the vibration of linear systems is often called *modal analysis theory*. Modal analysis studies are applied with both lumped-parameter (or discrete) systems and with continuous systems. Modal analysis theory includes the analysis of real (normal) modes as well as complex modes of vibrations. Experimental modal analysis provides a means for relating measured data to the theoretical frequency response.

In Chapter Seven, we discussed the relationship between modal parameters and frequency response functions. It may be helpful to review the principal results. Specifically, in Section 7.6, we saw that for a linear system, with proportional damping, the frequency response matrix $H(\omega)$ may be written as

$$H(\omega)=\sum_{r=1}^{N}\left[\frac{A_r}{j\omega-\lambda_r}+\frac{A_r^*}{j\omega-\lambda_r^*}\right] \tag{11.2-1}$$

and

$$H(\omega)=\sum_{r=1}^{N}\left[\frac{Q_r\mathbf{u}_r\mathbf{u}_r^T}{j\omega-\lambda_r}+\frac{Q_r^*\mathbf{u}_r\mathbf{u}_r^T}{j\omega-\lambda_r^*}\right], \tag{11.2-2}$$

where the notation is

$\lambda_r = \sigma_r + j\omega_r$, system pole

$j=\sqrt{-1}$

σ_r = damping factor

ω_r = damped natural frequency

ω = angular frequency variable

A_r = residue matrix for mode r

$\mathbf{u}_r$ = modal vector for mode r

Q_r = modal scaling factor

$*$ = complex conjugate

N = number of modal frequencies

(See Eqs. 7.6-26 and 7.6-27.)

It may also be helpful to make several observations.

First, Eq. 11.2-2 indicates that each column of the residue matrix is an estimate to within a constant of the modal vector for that pole. Since the residue matrix is symmetric, each row is also an estimate of the modal vector of that pole to within a constant. These statements are correct for each column and each row of the transfer function matrix.

Next, it follows from Eq. 11.2-1 that the element in the position of the p-th row and the q-th column of the FRF matrix can be expressed as

$$h_{pq}(\omega)=\sum_{r=1}^{N}\left[\frac{A_{pqr}}{j\omega-\lambda_r}+\frac{A^*_{pqr}}{j\omega-\lambda^*_r}\right], \qquad (11.2\text{-}3)$$

where A_{pqr} is called the residue, which is the product of the modal deformations at the output point p due to input at point q and a modal scaling factor for mode r. When $p=q$, $h_{pp}(\omega)$ is called the driving-point FRF, meaning that the input and output degrees of freedom are being measured at the same point on the system. When $p\neq q$, $h_{pq}(\omega)$ is called the cross-point FRF. That is, the output is measured at point p, while the input is at point q. From the Maxwell-Betti relations, we have $h_{pq}(\omega)=h_{qp}(\omega)$.

Finally, for the system with non-proportional damping, the modal vectors are complex, and Eq. 11.2-2 has the form (see Eq. 7.6-34)

$$H(\omega)=\sum_{r=1}^{N}\left[\frac{Q_r\mathbf{u}_r\mathbf{u}_r^T}{j\omega-\lambda_r}+\frac{Q^*_r\mathbf{u}^*_r\mathbf{u}^{*T}_r}{j\omega-\lambda^*_r}\right], \qquad (11.2\text{-}4)$$

where $\mathbf{u}^*_r$ is the complex conjugate of modal vector $\mathbf{u}_r$.

Experimental modal analysis is the process of solving Eq. 11.2-4 to determine the modal parameters from the measured FRFs. The equations, however, are nonlinear in terms of the unknown modal parameters. (This is seen by observing the numerators and denominators in Eq. 11.2-4).

To overcome this difficulty, we can attempt to separate the nonlinear problem into two linear problems. That is, we first estimate the values of the $2N$ frequencies and damping factors and then we estimate the $2N$ modal coefficients.

There are several methods for estimating the modal parameters. These include complex exponential, least-squares complex exponential, polyreference time domain, Ibrahim time domain, and eigensystem realization algorithms, among others [11.7–11.15]. D. L. Brown showed that the auto-regressive moving average (ARMA) model provides an excellent general form for use as a starting point for the various parameter estimation methods [11.4]. The ARMA model was historically developed as a set of finite difference equations and was specific to the time domain. However, with the advent of the general use of both Fourier transforms and Laplace transforms, the frequency domain and Laplace domain information may be used in an interchangeable manner. The primary difference between the methods is the order of the ARMA model and the manner in which the data are processed. In the following sections, we will review the various parameter estimation methods.

11.3 The Complex Exponential Algorithm

The complex exponential algorithm (CEA) is also known as the Prony method [11.7]. It is a time domain method. It utilizes a high-order matrix polynomial with scalar coefficients to fit a single free-decay or impulse response measurement. This method is a single input and a single output method.

If the original FRF has been measured, or obtained, in a digital form with equally spaced frequency Δf, then the resulting inverse Fourier transform (impulse response function) will simply be described at a corresponding equally spaced time interval Δt, which is sometimes called the time sampling period.

If we consider the inverse Fourier transform in Eq. 11.2-3, the impulse response function for mode r at output point p due to input at point q can be written as

$$h_{pq}(t)=\sum_{r=1}^{N}\left(A_{pqr}e^{\lambda_r t}+A_{pqr}^{*}e^{\lambda_r^{*} t}\right), \tag{11.3-1}$$

where the notation is:

r = mode number

A_{pqr} = residue for mode r at output point p due to input at point q

A_{pqr}^{*} = complex conjugate of A_{pqr}

$\lambda_r = \sigma_r + j\omega_r$ system pole

λ_r^{*} = complex conjugate of λ_r

σ_r = damping factor for mode r

ω_r = damped natural frequency for mode r

N = number of modes

For the sake of simplicity and brevity, the subscript pq will be omitted in the equations. Let $A^*_{pqr} = A_{pq(N+r)}$, $\lambda^*_r = \lambda_{N+1}$. Then Eq. 11.3-1 can be written as

$$h(t) = \sum_{r=1}^{2N} A_r e^{\lambda_r t} \tag{11.3-2}$$

Let $z_r = e^{\lambda_r \Delta t}$, where Δt is the sampling period. For different sampling times $t_k = k\Delta t$, (k=0,1,2,3,...), we then have [11.4, 11.5]

$$\begin{aligned} h(0) &= h(0\Delta t) = A_1 + A_2 + \cdots + A_{2N} \\ h(1) &= h(1\Delta t) = A_1 z_1 + A_2 z_2 + \cdots + A_{2N} z_{2N} \\ &\vdots \\ h(k) &= h(k\Delta t) = A_1 z_1^k + A_2 z_2^k + \cdots + A_{2N} z_{2N}^k \\ &\vdots \\ h(2N) &= h(2N\Delta t) = A_1 z_1^{2N} + A_2 z_2^{2N} + \cdots + A_{2N} z_{2N}^{2N} \end{aligned} \tag{11.3-3}$$

Multiplying in the first equation of Eqs. 11.3-3 by α_0, in the second by α_1, ..., and adding the results, we obtain

$$\sum_{i=0}^{2N} \alpha_i h(i) = \sum_{j=1}^{2N} A_j \sum_{i=0}^{2N} \alpha_i z_j^i \tag{11.3-4}$$

Or, in matrix form,

$$\sum_{i=0}^{2N} \alpha_i h(i) = \begin{bmatrix} \alpha_0 & \alpha_1 & \cdots & \alpha_{2N} \end{bmatrix} \begin{bmatrix} 1 & 1 & \cdots & 1 \\ z_1 & z_2 & \cdots & z_{2N} \\ z_1^2 & z_2^2 & \cdots & z_{2N}^2 \\ \vdots & \vdots & \vdots & \vdots \\ z_1^{2N} & z_2^{2N} & \cdots & z_{2N}^{2N} \end{bmatrix} \begin{bmatrix} A_1 \\ A_2 \\ \vdots \\ A_{2N} \end{bmatrix} \tag{11.3-5}$$

Observe that the second matrix on the right side of Eq. 11.3-5 has (2N+1) rows and 2N columns. Each column of this matrix can be regarded as a vector in a (2N+1) dimensional space. For a (2N+1) dimensional space, there exists (2N+1) independent vectors which are mutually orthogonal. Thus, we can find the α_i (i=0,1,...,2N), such that the vector $[\alpha_0\ \alpha_1 \ldots \alpha_{2N}]^T$ is orthogonal to the columns of the matrix. Consequently, we have

$$\alpha_0 + \alpha_1 z + \cdots \alpha_{2N} z^{2N} = 0 \tag{11.3-6}$$

By using Eq. 11.3-6 in Eq. 11.3-5, we have another equation

$$\sum_{i=0}^{2N} \alpha_i h(i) = 0 \tag{11.3-7}$$

Without any loss of generality, letting $\alpha_{2N}=1$, we have, from Eqs. 11.3-6 and 11.3-7

$$\alpha_0+\alpha_1 z+\cdots+z^{2N}=0 \tag{11.3-8}$$

and

$$\sum_{i=0}^{2N-1}\alpha_i h(i)=-h(2N) \tag{11.3-9}$$

Eq. 11.3-9 is called the ARMA model, while Eq. 11.3-8 is called the characteristic equation. Once the coefficients α_i (i=0,1,...,2N–1) are found from the ARMA model (Eq. 11.3-9), the 2N characteristic roots can be found from Eq. 11.3-8.

The above analysis does not impose any constraint regarding the starting time. That is, we can take $h(1)$, $h(2)$, . . . as the new $h(0)$, $h(1)$, . . . and establish an ARMA model. Thus, we have

$$\sum_{i=0}^{2N-1}\alpha_i h(i+m)=-h(2N+m) \quad (m=0,1,2,\cdots,2N-1), \tag{11.3-10}$$

where m is the index for the starting time point.

In matrix form, Eq. 11.3-10 can be written as

$$\begin{bmatrix} h(0) & h(1) & \cdots & h(2N-1) \\ h(1) & h(2) & \cdots & h(2N) \\ \vdots & \vdots & \cdots & \vdots \\ h(2N-1) & h(2N) & \cdots & h(4N-2) \end{bmatrix} \begin{bmatrix} \alpha_0 \\ \alpha_1 \\ \vdots \\ \alpha_{2N-1} \end{bmatrix} = -\begin{bmatrix} h(2N) \\ h(2N+1) \\ \vdots \\ h(4N-1) \end{bmatrix} \tag{11.3-11}$$

or simply as

$$[h]\{\alpha\}=-\{\bar{h}\}, \tag{11.3-12}$$

where the matrix $[h]$ is called the Toeplitz matrix [11.4]. Solving Eqs. 11.3-12 for the α_i, we have

$$\{\alpha\}=-[h]^{-1}[\bar{h}] \tag{1.3-13}$$

Substituting the α_i into Eq. 11.3-8, we can find the 2N characteristic roots, and then the natural frequencies and damping ratios. Indeed, we have

$$z_r=e^{\lambda_r \Delta t}=e^{(\sigma_r+j\omega_r)\Delta t}, \tag{11.3-14}$$

from which the damping factor and damped natural frequency are given by

$$\sigma_r=\frac{\ln|z_r|}{\Delta t} \tag{11.3-15}$$

$$\omega_r = \frac{\tan^{-1}(z_r^I / z_r^R)}{\Delta t}, \tag{11.3-16}$$

where z_r^I and z_r^R are the imaginary and real parts of z_r, respectively.

The damping ratio for mode r is given by

$$\varsigma_r = -\frac{\sigma_r}{\sqrt{\sigma_r^2 + \omega_r^2}} \tag{11.3-17}$$

Finally, it is worth noting that the characteristic roots of Eq. 11.3-8 can be found by using software for solving high-order algebraic equations. For example, with Matlab, let p represent the row matrix consisting of the coefficients. Then, with the command `root(p)`, one can find the roots. The second method is to use a companion matrix [11.4]. Specifically, we have the following identities:

$$\begin{aligned} z^0 &= z^0 \\ z^1 &= zz^0 \\ z^2 &= zz^1 \\ &\vdots \\ z^{2N} &= zz^{2N-1} \end{aligned} \tag{11.3-18}$$

Then, Eqs. 11.3-8 and 11.3-18 can be written as

$$\begin{bmatrix} -\alpha_{2N-1} & -\alpha_{2N-2} & -\alpha_{2N-3} & \cdots & -\alpha_0 \\ 1 & 0 & 0 & 0 & 0 \\ 0 & 1 & & & \\ \vdots & \vdots & \vdots & \vdots & \vdots \\ 0 & 0 & 0 & 1 & 0 \end{bmatrix} \begin{bmatrix} z^{2N-1} \\ z^{2N-2} \\ \vdots \\ z^0 \end{bmatrix} = z \begin{bmatrix} z^{2N-1} \\ z^{2N-2} \\ \vdots \\ z^0 \end{bmatrix} \tag{11.3-19}$$

The coefficient matrix on the left side of Eq. 11.3-19 is called the *companion matrix*. Thus, the problem of finding the characteristic roots becomes an eigenvalue problem. As an example, consider the roots of the equation

$$2 - 3z + z^2 = 0 \tag{11.3-20}$$

The corresponding companion matrix is

$$[C] = \begin{bmatrix} 3 & -2 \\ 1 & 0 \end{bmatrix} \tag{11.3-21}$$

Thus, finding the roots of Eq. 11.3-20 is equivalent to finding the eigenvalues, z, of the companion matrix

$$\begin{bmatrix} 3 & -2 \\ 1 & 0 \end{bmatrix} \begin{bmatrix} x_1 \\ x_2 \end{bmatrix} = z \begin{bmatrix} x_1 \\ x_2 \end{bmatrix} \tag{11.3-22}$$

11.4 The Least-Squares Complex Exponential Method

The least-squares complex exponential (LSCE) method was developed directly from the complex exponential algorithm (CEA) with a high-order ARMA model [11.4].

The computational steps of the LSCE are essentially the same as for the CEA. The difference is the utilization of response data. A system of equations is generated such that the number of equations is greater than the number of system poles. A least squares or pseudo-inverse solution is then required to calculate the ARMA coefficients.

The equations for the ARMA model are generated by time shifting. Specifically, letting m=0,1,2,...,$k > 2N$ in Eq. 11.3-10, we have

$$\sum_{i=0}^{2N-1} \alpha_i h(i+m) = -h(2N+m) \quad (m=0,1,2,\cdots,k>2N) \tag{11.4-1}$$

Or, in matrix form,

$$[h]\{\alpha\} = -\{\bar{h}\} \tag{11.4-2}$$

In Eq. 11.4-2, the matrix $[h]$, known as the Toplitz matrix, is not square (the number of rows is larger than that of columns). The ARMA coefficients are then given by

$$\{\alpha\} = -[h]^{+}[\bar{h}], \tag{11.4-3}$$

where $[h]^{+}$ is the pseudoinverse [11.16] of $[h]$. With the coefficients, we can solve for the system poles by the method stated in previous section.

11.5 The Polyreference Time Domain Method

The term *reference* in experimental modal analysis refers to the input (excitation) point. The polyreference time domain (PTD) method is actually a multiple input extension of the LSCE method. This method was developed by Vold in 1982 [11.4, 11.8, 11.9]. The basic concept is to use the whole impulse response function matrix to form an ARMA model. In this way, more information in the measured data is used; thus, the results obtained are more accurate.

Recall that the relationship between the FRF matrix and the residue matrix is

$$H(\omega) = \sum_{r=1}^{N} \left[\frac{A_r}{j\omega - \lambda_r} + \frac{A_r^*}{j\omega - \lambda_r^*} \right] \tag{11.5-1}$$

In experimental modal analysis, the FRF matrix $H(\omega)$ is obtained experimentally. Usually, only one column, or only a few columns, are measured. Thus, the $H(\omega)$ is not square. Let N_o and N_i be the number of rows and columns of $H(\omega)$, respectively. The N_o and N_i are sometimes called the output and input degrees of freedom [11.17], respectively. Note that the residue matrix A_r is of dimension $N_o \times N_i$, and it can be expressed as

$$\left[A_r\right]_{N_o \times N_i} = Q_r \left\{\mathbf{u}_r\right\}_{N_o \times 1} \left\{L_r^T\right\}_{1 \times N_i}, \tag{11.5-2}$$

where

r = mode number

Q_r = modal scaling factor

$\left\{\mathbf{u}_r\right\}$ = modal vector for mode r

$\left\{L_r\right\}$ = modal participation factor column matrix for modal r

Let

$$\left\{\mathbf{u}_r\right\} = \begin{bmatrix} \psi_1 & \psi_2 & \cdots & \psi_{N_o} \end{bmatrix}^T \tag{11.5-3}$$

$$\left\{L_r^T\right\} = \begin{bmatrix} L_1 & L_2 & \cdots & L_{N_i} \end{bmatrix} \tag{11.5-4}$$

The residue matrix can be written as

$$\left[A_r\right] = Q_r \begin{bmatrix} \psi_1 L_1 & \psi_1 L_2 & \cdots & \psi_1 L_{N_i} \\ \psi_2 L_1 & \psi_2 L_2 & \cdots & \psi_2 L_{N_i} \\ \vdots & \vdots & \cdots & \vdots \\ \psi_{N_o} L_1 & \psi_{N_o} L_2 & \cdots & \psi_{N_o} L_{N_i} \end{bmatrix} \tag{11.5-5}$$

Let $[S]$ be the modal matrix composed of all modal vectors, and let $[L]$ be the modal participation matrix composed of all modal participation factors. Then, the frequency response function matrix can be written as

$$[H(\omega)]_{N_o \times N_i} = \left[S\right]_{N_o \times 2N} \left[\frac{Q_r}{j\omega - \lambda_r}\right]_{2N \times 2N} \left[L\right]_{2N \times N_i}^T \tag{11.5-6}$$

Note that the second matrix on the right side of Eq. 11.5-6 is diagonal. The elements on the diagonal are

$$\frac{Q_1}{j\omega - \lambda_1}, \ldots, \frac{Q_N}{j\omega - \lambda_N}, \frac{Q_1^*}{j\omega - \lambda_1^*}, \ldots, \frac{Q_N^*}{j\omega - \lambda_N^*}$$

We are now in a position to derive the polyreference time domain method. Without any loss of generality, in the following, we take $Q_r = 1$. In the time domain, Eq. 11.5-6 becomes

$$[H(t)]_{N_o \times N_i} = \left[S\right]_{N_o \times 2N} \left[e^{\lambda_r t}\right]_{2N \times 2N} \left[L\right]_{2N \times N_i}^T \tag{11.5-7}$$

where

$$\left[e^{\lambda_r t}\right]_{2N\times 2N} = \mathrm{diag}(e^{\lambda_1 t}, e^{\lambda_2 t}, \cdots, e^{\lambda_{2N} t}) \tag{11.5-8}$$

Let $z_r = e^{\lambda_r \Delta t}$, where Δt is the sampling period. For any sampling time $t_k = k\Delta t$ $(k = 0,1,2,\cdots,n)$, from Eq. 11.5-7 we obtain

$$\left[H(k)\right] = \left[S\right]\left[Z\right]^k\left[L\right]^T \tag{11.5-9}$$

Taking the transpose in Eq. 11.5-9, we obtain

$$\left[H^T(k)\right] = \left[L\right]\left[Z\right]^k\left[S\right]^T, \tag{11.5-10}$$

where

$$\left[Z\right] = \mathrm{diag}(z_1, z_2, \cdots, z_{2N}) \tag{11.5-11}$$

Let $[a(i)]_{N_i\times N_i}$ be an arbitrary matrix, and let m be the index of the starting time point $t_m = m\Delta t$. By using Eq. 11.5-10, we can establish the following set of equations [11.4]

$$[a(n)][H^T(m+0)] = [a(n)][L][Z^m][S]^T$$

$$[a(n-1)][H^T(m+1)] = [a(n-1)][L][Z^{m+1}][S]^T$$

$$\vdots$$

$$[a(1)][H^T(m+n-1)] = [a(1)][L][Z^{m+n-1}][S]^T$$

$$[a(0)][H^T(m+n)] = [a(0)][L][Z^{m+n}][S]^T$$

Adding the above equations together, we obtain

$$\sum_{i=0}^{n}[a(i)][H^T(m+n-i)] = \sum_{i=0}^{n}[a(i)][L][Z^{m+n-i}][S]^T \tag{11.5-12}$$

By introducing the following two matrices

$$\left[Q\right]_{N_i\times(n+1)N_i} = \left[[a(n)] \quad \cdots \quad [a(0)]\right] \tag{11.5-13}$$

and

$$[G]_{(n+1)N_i\times 2N} = \begin{bmatrix} [L][Z]^m \\ \vdots \\ [L][Z]^{m+n} \end{bmatrix} \tag{11.5-14}$$

Eq. 11.5-12 can be written in the form

$$\sum_{i=0}^{n}[a(i)][H^T(m+n-i)] = [Q][G][S]^T \tag{11.5-15}$$

When the number of data points n is greater than $2N/N_i$, we see that

$$(n+1)N_i > \left(\frac{2N}{N_i}+1\right)N_i = 2N+N_i \tag{11.5-16}$$

In this case, the $2N$ columns of matrix $[G]$ can be regarded as $2N$ vectors in the $(2N+N_i)$ dimensional space. Thus, there exists N_i vectors, which are orthogonal to the $2N$ vectors. Therefore, it is always possible to find a matrix $[Q]$ satisfying the following equation

$$[Q]_{N_i\times(n+1)N_i}[G]_{(n+1)N_i\times 2N} = [0]_{N_i\times 2N} \tag{11.5-17}$$

Or, equivalently,

$$\sum_{i=0}^{n}[a(i)][L][Z]^{m+n-i} = 0 \tag{11.5-18}$$

Consequently, it follows from Eq. 11.5-15 that

$$\sum_{i=0}^{n}[a(i)][H^T(m+n-i)] = [0] \tag{11.5-19}$$

Letting $m = 0,1,2,\cdots,s$ (that is, taking the time data with different starting points of time), from Eq. 11.5-19, we obtain the ARMA model equations

$$[a(0)][H^T(n)]+[a(1)][H^T(n-1)]+\cdots[a(n)][H^T(0)]=0 \quad (m=0) \tag{11.5-20a}$$

$$[a(0)][H^T(n+1)]+[a(1)][H^T(n)]+\cdots[a(n)][H^T(1)]=0 \quad (m=1) \tag{11.5-20b}$$

$$[a(0)][H^T(n+2)]+[a(1)][H^T(n+1)]+\cdots[a(n)][H^T(2)]=0 \quad (m=2) \tag{11.5-20c}$$

... ...

Without any loss of generality, we can let $[a(0)]=[I]$ (the $N_i \times N_i$ unit matrix). Then, Eqs. 11.5-20 can be written as

$$A\hat{H} = -B, \tag{11.5-21}$$

where A, $\hat{H}$, and B are

$$A = \left[[a(1)] \quad [a(2)] \quad \cdots \quad [a(n)]\right] \tag{11.5-22}$$

$$\hat{H} = \begin{bmatrix} H^T(n-1) & H^T(n) & \cdots & H^T(n+s-1) \\ \vdots & \vdots & \vdots & \vdots \\ H^T(1) & H^T(2) & \vdots & H^T(s) \\ H^T(0) & H^T(1) & \vdots & H^T(s-1) \end{bmatrix} \tag{11.5-23}$$

$$B = \left[H^T(n) \quad H^T(n+1) \quad \cdots \quad H(n+s)\right], \tag{11.5-24}$$

where

n = number of time shifts in column direction

s = number of time shifts in row direction

The matrix $\hat{H}$ defined in Eq. 11.5-23 is called the Hankel matrix [11.4], which is similar to the Toeplitz matrix of CEA. Note that to find all poles of the system, n and s must be sufficiently large.

The coefficient matrix A in Eq. 11.5-21 can be expressed as

$$A = -B\hat{H}^{+}, \tag{11.5-25}$$

where $\hat{H}^{+}$ is the pseudoinverse of $\hat{H}$.

Once the coefficient matrix A is found from Eq. 11.5-25, the characteristic roots can be found from Eq. 11.5-18. In a similar manner as with the complex exponential algorithm, it can be shown that finding the characteristic roots is equivalent to finding the eigenvalues of a companion matrix.

Indeed, since the matrix $[Z]$ is diagonal (see Eq. 11.5-11), from Eq. 11.5-18, we have

$$\sum_{i=1}^{n}[a(i)]z_r^{m+n-i}\{L_r\}=0, \tag{11.5-26}$$

where $\{L_r\}$ is the r-th column of $[L]$. Expanding Eq. 11.5-26, we obtain

$$-[[a(1)]z_r^{m+n-1}+[a(2)]z_r^{m+n-2}+\cdots+[a(n)]z_r^{m}]\{L_r\}=z_r^{m+n}\{L_r\} \tag{11.5-27}$$

Next, it is convenient to define a set of identities as [11.4]

$$\begin{aligned}
&\{V_0\}=z_r^{m}\{L\}\\
&\{V_1\}=z_r^{m+1}\{L\}=z_r\{V_0\}\\
&\{V_2\}=z_r^{m+2}\{L\}=z_r\{V_1\}\\
&\vdots\\
&\{V_{n-1}\}=z_r^{m+n-1}\{L\}=z_r\{V_{n-2}\}
\end{aligned} \tag{11.5-28}$$

Using Eqs. 11.5-27 and 11.5-28, we can establish the eigenvalue equation

$$\begin{bmatrix} -[a(1)] & -[a(2)] & \cdots & -[a(n-1)] & -[a(n)] \\ [I] & [0] & \cdots & [0] & [0] \\ \cdot & \cdot & \cdots & \cdot & \cdot \\ \cdot & \cdot & \cdots & [0] & \cdot \end{bmatrix} \begin{bmatrix} \{V_{n-1}\} \\ \vdots \\ \{V_1\} \\ \{V_0\} \end{bmatrix} = z_r \begin{bmatrix} \{V_{n-1}\} \\ \vdots \\ \{V_1\} \\ \{V_0\} \end{bmatrix} \tag{11.5-29}$$

From Eq. 11.5-29, we can find the characteristic roots z, and, hence, the damped natural frequency and damping factor, as stated before.

Comparing Eqs. 11.3-11 and 11.5-21, we see that the computational steps of the CEA and the PTD methods are similar. The only difference is that the

CEA utilizes one element of the impulse response function matrix to form the Toeplitz matrix by time shifting, while the PTD method utilizes the whole impulse response function matrix to form the Hankel matrix by time shifting. The following paragraphs provide a simple example to illustrate this concept.

Example 11.5-1[1] Fig. 11.5-1 shows two identical balls connected by a massless bar and supported by identical springs. Suppose the *experimentally measured* impulse response function matrix is given by

$$[H(t)]=\begin{bmatrix} h_{11}(t) & h_{12}(t) \\ h_{21}(t) & h_{22}(t) \end{bmatrix}=\begin{bmatrix} \sin\frac{\pi}{2}t & 0 \\ 0 & \cos\frac{\pi}{2}t \end{bmatrix} \tag{a}$$

Use the CEA and the PTD methods to show that the natural frequency of the system is $\pi/2$.

Solution: First, we use the CEA. The basic concept of this method is to utilize an element of the impulse response function matrix to form the ARMA model. By using $h_{11}(t)$ with $\Delta t=1$, Eq. 11.3-11 becomes

$$\begin{bmatrix} h_{11}(0) & h_{11}(1) \\ h_{11}(1) & h_{11}(2) \end{bmatrix}\begin{bmatrix} \alpha_0 \\ \alpha_1 \end{bmatrix}=-\begin{bmatrix} h_{11}(2) \\ h_{11}(3) \end{bmatrix} \tag{b}$$

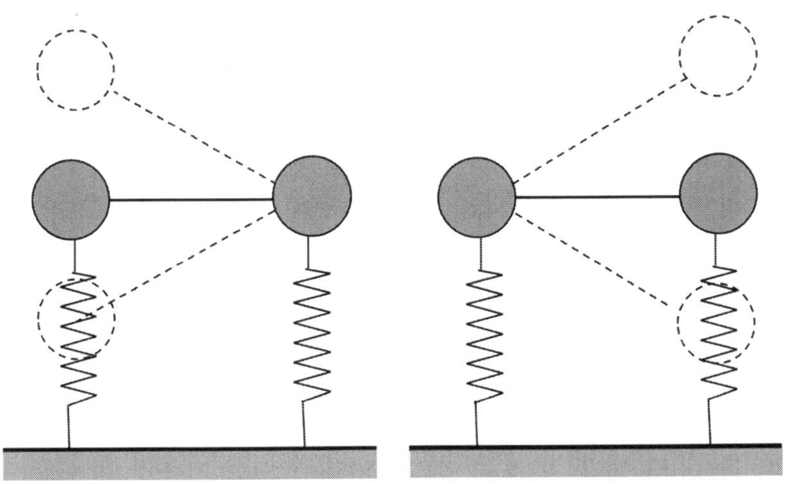

Fig. 11.5-1 Two-degree-of-freedom system.

[1]R. J. Allemang provided a similar example in [11.18].

Substituting the numerical values, we have

$$\begin{bmatrix} 0 & 1 \\ 1 & 0 \end{bmatrix} \begin{bmatrix} \alpha_0 \\ \alpha_1 \end{bmatrix} = - \begin{bmatrix} 0 \\ -1 \end{bmatrix}, \tag{c}$$

from which we obtain

$$\alpha_0 = 1, \quad \alpha_1 = 0$$

Thus, the characteristic equation (Eq. 11.3-8) becomes

$$\alpha_0 z^0 + \alpha_1 z^1 + z^2 = 0$$

or

$$1 + z^2 = 0 \tag{d}$$

The characteristic roots are

$$z = \pm\sqrt{-1} = \pm j \tag{e}$$

Next, from Eqs. 11.3-15 and 11.3-16, the damping factor σ and natural frequency ω are

$$\sigma_r = \frac{\ln|z_r|}{\Delta t} = 0 \tag{f}$$

$$\omega_r = \frac{\tan^{-1}(z_r^I / z_r^R)}{\Delta t} = \frac{\pi}{2} \tag{g}$$

For comparison, we also use the PTD method. The basic concept of this method is to utilize the whole impulse response function matrix to form the ARMA model. As before, taking $\Delta t = 1$, we have

$$H^T(0) = \begin{bmatrix} 0 & 0 \\ 0 & 1 \end{bmatrix}$$

$$H^T(1) = \begin{bmatrix} 1 & 0 \\ 0 & 0 \end{bmatrix}$$

$$H^T(2) = \begin{bmatrix} 0 & 0 \\ 0 & -1 \end{bmatrix}$$

$$H^T(3) = \begin{bmatrix} -1 & 0 \\ 0 & 0 \end{bmatrix}$$

The ARMA model (see Eq. 11.5-21) becomes

$$A\hat{H} = -B, \tag{h}$$

where

$$A=\left[[a(1)] \quad [a(2)]\right]$$

$$\hat{H}=\begin{bmatrix} H^T(1) & H^T(2) \\ H^T(0) & H^T(1) \end{bmatrix}$$

$$B=\left[H^T(2) \quad H^T(3)\right]$$

Substituting the numerical values from Eq. h, we obtain

$$[a(1)]=[0]_{2\times 2}, \quad [a(2)]=[I]_{2\times 2} \tag{i}$$

The corresponding companion matrix can be written as

$$\left[C\right]=\begin{bmatrix} -[a(1)] & -[a(2)] \\ [I] & [0] \end{bmatrix}, \tag{j}$$

of which the eigenvalues are found to be

$$z_{1,2}=\pm j, \quad z_{3,4}=\pm j \tag{k}$$

It is seen that these results are the same as those obtained by using the CEA method (see Eq. e).

11.6 The Ibrahim Time Domain Method

The Ibrahim time domain (ITD) method of parameter estimation [11.10–11.12] uses one column of the impulse response function matrix to form an ARMA model by shifting the time. Thus, this method may be called *the single-reference time domain method*, compared to the PTD method. The optimum test procedure for the ITD method is to simultaneously measure the responses of the test structure at predetermined test points due to a single input. When there are as many response measurements as modes of vibration in the frequency range of interest, then the ARMA model has a matrix with dimension equal to twice the number of modes.

The theory of the method is as follows: Using one column, say, the q-th column of the impulse response function matrix, we can form the following matrices by shifting the time

$$[\bar{h}(0)]_{N_0\times 2n}=\begin{bmatrix} h_{1q}(m+0) & h_{1q}(m+1) & \cdots & h_{1q}(m+2n) \\ h_{2q}(m+0) & h_{2q}(m+1) & \cdots & h_{2q}(m+2n) \\ \cdots & \cdots & \ddots & \cdots \\ h_{N_0q}(m+0) & h_{N_0q}(m+1) & \cdots & h_{N_0q}(m+2n) \end{bmatrix} \tag{11.6-1}$$

$$[\bar{h}(1)]_{N_0\times 2n}=\begin{bmatrix} h_{1q}(m+1) & h_{1q}(m+2) & \cdots & h_{1q}(m+2n+1) \\ h_{2q}(m+1) & h_{2q}(m+2) & \cdots & h_{2q}(m+2n+1) \\ \cdots & \cdots & \ddots & \cdots \\ h_{N_0q}(m+1) & h_{N_0q}(m+2) & \cdots & h_{N_0q}(m+2n+1) \end{bmatrix} \tag{11.6-2}$$

$$[\bar{h}(2)]_{N_0\times 2n}=\begin{bmatrix} h_{1q}(m+2) & h_{1q}(m+3) & \cdots & h_{1q}(m+2n+2) \\ h_{2q}(m+2) & h_{2q}(m+3) & \cdots & h_{2q}(m+2n+2) \\ \cdots & \cdots & \ddots & \cdots \\ h_{N_0q}(m+2) & h_{N_0q}(m+3) & \cdots & h_{N_0q}(m+2n+2) \end{bmatrix}, \tag{11.6-3}$$

where

N_o = the number of output (response)

n = the number of frequencies to be solved for

q = the reference (input) point

m = the index of the starting time point

Following the same procedure as in the PTD method, it can be shown that the ARMA model for the ITD can be written as

$$[A]_{2N_o\times 2N_o}[h(m+0)]_{2N_o\times 2n}=-[I]_{2N_o\times 2N_o}[h(m+1)]_{2N_o\times 2n}, \tag{11.6-4}$$

where

$$[h(m+0)]_{2N_o\times 2n}=\begin{bmatrix}\bar{h}(0)\\ \bar{h}(1)\end{bmatrix}_{2N_o\times 2n} \tag{11.6-5}$$

$$[h(m+1)]_{2N_o\times 2n}=\begin{bmatrix}\bar{h}(1)\\ \bar{h}(2)\end{bmatrix}_{2N_o\times 2n} \tag{11.6-6}$$

The coefficients of the ARMA model in Eq. 11.6-4 can be obtained by the expression

$$[A]_{2N_o\times 2N_o}=-[h(m+1)][h(m+0)]^{+}, \tag{11.6-7}$$

where $[h(m+0)]^{+}$ is the pseudoinverse of $[h(m+0)]$ for the case where $[h(m+0)]$ is not square.

To find the characteristic roots, it is informative to take the Z-transform in Eq. 11.6-4. For present purposes, it suffices to view z^{-1} as a time-delay operator. That is, a shift of a simple period in the time domain is achieved by multiplying by z^{-1} in the Z-domain. Then, Eq. 11.6-4 can be written in delay operator (that is, transformed) form as an eigenvalue problem

$$[A]_{2N_o\times 2N_o}[X(z)]_{2N_o\times 2n}+z\left[I\right]_{2N_o\times 2N_o}[X(z)]_{2N_o\times 2n}=0 \tag{11.6-8}$$

Once the eigenvalues, z, are determined from Eq. 11.6-8, the damping factors and the damped natural frequencies can be computed using the Z-transform relationship

$$z_r=e^{\lambda_r \Delta t}=e^{(\sigma_r+j\omega_r)\Delta t} \tag{11.6-9}$$

$$\sigma_r=\frac{\ln|z_r|}{\Delta t} \tag{11.6-10}$$

$$\omega_r=\frac{\tan^{-1}(z_r^I/z_r^R)}{\Delta t} \tag{11.6-11}$$

Example 11.6-1 Repeat Example 11.5-1 by using the ITD method.

Solution: We take the first column of the impulse response function matrix to form the ARMA model. Letting $m=0$, $n=2$, and $\Delta t=1$, from Eqs. 11.6-1 to 11.6-3, we have

$$[\bar{h}(0)]=\begin{bmatrix} 0 & 1 & 0 & -1 \\ 0 & 0 & 0 & 0 \end{bmatrix} \tag{a}$$

$$[\bar{h}(1)]=\begin{bmatrix} 1 & 0 & -1 & 0 \\ 0 & 0 & 0 & 0 \end{bmatrix} \tag{b}$$

$$[\bar{h}(2)]=\begin{bmatrix} 0 & -1 & 0 & 1 \\ 0 & 0 & 0 & 0 \end{bmatrix} \tag{c}$$

It follows from Eqs. 11.6-5 and 11.6-6 that

$$h(0)=\begin{bmatrix} 0 & 1 & 0 & -1 \\ 0 & 0 & 0 & 0 \\ 1 & 0 & -1 & 0 \\ 0 & 0 & 0 & 0 \end{bmatrix} \tag{d}$$

$$h(1)=\begin{bmatrix} 1 & 0 & -1 & 0 \\ 0 & 0 & 0 & 0 \\ 0 & -1 & 0 & 1 \\ 0 & 0 & 0 & 0 \end{bmatrix} \tag{e}$$

Thus, the ARMA model can be written as

$$[A][h(0)]=-[h(1)] \tag{f}$$

From this, the coefficient matrix $[A]$ is found to be

$$[A]=\begin{bmatrix} 0 & 0 & -1 & 0 \\ 0 & 0 & 0 & 0 \\ 1 & 0 & 0 & 0 \\ 0 & 0 & 0 & 0 \end{bmatrix} \tag{g}$$

The eigenvalues of $[A]$ are found to be $z_{1,2}=\pm j$ and $z_{3,4}=0$. Hence, the damping factor and natural frequency can be computed from Eqs. 11.6-10 and 11.6-11 as

$$\sigma_r=\frac{\ln|z_r|}{\Delta t}=0 \tag{h}$$

$$\omega_r=\frac{\tan^{-1}(z_r^I/z_r^R)}{\Delta t}=\frac{\pi}{2} \tag{i}$$

It is seen that the results are the same as those obtained by using the CEA and PTD methods.

11.7 The Eigensystem Realization Algorithm

The eigensystem realization algorithm (ERA) is a multiple reference, time domain, modal parameter estimation method, developed by Juang and Papa in 1985 [11.4, 11.13, 11.14]. This method is based on the concept of system realization, which is a process of constructing a state space representation from experimental data. The method uses a singular value decomposition (SVD) procedure to condense the data.

The theory of the method is as follows: Without loss of generality, taking the modal scaling factor $Q_r=1$, the impulse response function matrix can be written as

$$[H(t)]_{N_o\times N_i}=\left[S\right]_{N_o\times 2N}\left[e^{\lambda_r t}\right]_{2N\times 2N}\left[L\right]^T_{2N\times N_i}, \tag{11.7-1}$$

where

$$\left[e^{\lambda_r t}\right]_{2N\times 2N}=\operatorname{diag}(e^{\lambda_1 t},e^{\lambda_2 t},\cdots,e^{\lambda_{2N} t})$$

N_o = number of output points

N_i = number of input points

N = number of modes

$\left[S\right]_{N_o\times 2N}$ = modal matrix

$\left[L\right]_{N_i\times 2}$ = modal participation matrix

Define the generalized Hankel matrix as follows:

$$\hat{H}(k)=\begin{bmatrix} [H(k)] & [H(k+1)] & \cdots & [H(k+s-1)] \\ [H(k+1)] & [H(k+2)] & \cdots & [H(k+s)] \\ . & . & \cdots & . \\ . & . & \cdots & . \\ . & . & \cdots & . \\ [H(k+n-1)] & [H(k+n)] & \cdots & [H(k+n+s-2)] \end{bmatrix}_{nN_o \times sN_i}, \qquad (11.7\text{-}2)$$

where

$[H(k)] = [H(k\Delta t)]$

Δt = sampling period

n = number of time shifts in column direction

s = number of time shifts in row direction

It should be noted that both n and s have to be chosen sufficiently large that the rank of the generalized Hankel matrix $[\hat{H}(k)]$ cannot be increased by increasing n and s. The optimum determination of n and s requires further development. A recursive algorithm has been published by Longman and Juang [11.14].

By using Eq. 11.7-1, the generalized Hankel matrices can be expressed as

$$[\hat{H}(k)]=[C][Z]^k[B] \qquad (11.7\text{-}3)$$

$$[\hat{H}(k+1)]=[C][Z]^{k+1}[B], \qquad (11.7\text{-}4)$$

where

$$[C]=\begin{bmatrix} [S] \\ [S][Z] \\ [S][Z]^2 \\ \vdots \\ [S][Z]^{n-1} \end{bmatrix}_{nN_o \times 2N} \qquad (11.7\text{-}5)$$

$$[B]=\begin{bmatrix}[L]^T & [Z][L]^T & \cdots & [Z]^{s-1}[L]^T\end{bmatrix}_{2N \times sN_i} \qquad (11.7\text{-}6)$$

$$[Z]_{2N\times 2N} = \mathrm{diag}(e^{\lambda_1 \Delta t}, \cdots, e^{\lambda_{2N} \Delta t}), \qquad (11.7\text{-}7)$$

where $[C]$ and $[B]$ play a role of observability and controllability matrices, respectively, in control theory.

Since the generalized Hankel matrices have a large size, the transformation method is used to condense the data to match the number of unknowns in the modal parameter estimation algorithm. Mathematically, this can be accomplished by pre-multiplying Eqs. 11.7-3 and 11.7-4 by an $m \times nN_o$ nontrivial matrix, where m is the rank of the generalized Hankel matrix. This transformation matrix, however, is not unique. Indeed, there are a number of ways of finding the transformation matrix—such as QR decomposition, SVD, and the Moore-Penrose pseudoinverse of the generalized Hankel matrix $\hat{H}(k)$.

Consider the SVD of the generalized Hankel matrix $[\hat{H}(k)]$: Let the SVD of the matrix $[\hat{H}(k)]$ be expressed as

$$[\hat{H}(k)] = [U]_{nN_o \times nN_o} [\Sigma]_{nN_o \times sN_i} [V^T]_{sN_i \times sN_i}, \tag{11.7-8}$$

where both $[U]$ and $[V]$ are orthogonal matrices, and where $[\Sigma]$ is a diagonal matrix; the elements on the diagonal are called the singular values. The number of nonzero singular values is equal to the rank of the generalized Hankel matrix $[\hat{H}(k)]$. Let $m = \text{rank}([\hat{H}(k)])$. Let $[\hat{U}]$ be the matrix composed of the first m columns of $[U]$. The matrix $[\hat{U}]$ is of dimension $nN_o \times m$. Specifically, pre-multiplying Eqs. 11.7-3 and 11.7-4 by $[\hat{U}]^T$, we obtain

$$[\hat{U}^T]_{m \times nN_o} [\hat{H}(k)]_{nN_o \times sN_i} = [\hat{U}^T]_{m \times nN_o} [C]_{nN_o \times 2N} [Z]^k_{2N \times 2N} [B]_{2N \times 2N_i} \tag{11.7-9}$$

$$[\hat{U}^T]_{m \times nN_o} [\hat{H}(k+1)]_{nN_o \times sN_i} = [\hat{U}^T]_{m \times nN_o} [C]_{nN_o \times 2N} [Z]^{k+1}_{2N \times 2N} [B]_{2N \times 2N_i} \tag{11.7-10}$$

By multiplying in Eq. 11.7-9 by $[A_1]_{m \times m}$, and in Eq. 11.7-10 by $[A_2]_{m \times m}$, and then adding, we obtain

$$\big[[A_1] \quad [A_2]\big] \begin{bmatrix} [\hat{U}^T][\hat{H}(k)] \\ [\hat{U}^T][\hat{H}(k+1)] \end{bmatrix} = \big[[A_1] \quad [A_2]\big][G]\big[[Z]^k[B]\big], \tag{11.7-11}$$

where

$$[G] = \begin{bmatrix} [\hat{U}^T][C] \\ [\hat{U}^T][C][Z] \end{bmatrix}_{(2nN_o) \times 2N} \tag{11.7-12}$$

When the time shifting number is sufficiently large ($n > N / N_o$), the number of rows of $[G]$ is greater than the number of columns. We can find matrix coefficients such that

$$\big[[A_1] \quad [A_2]\big][G] = [0] \tag{11.7-13}$$

The matrix $[A_2]$ may be chosen as the negative identity matrix of dimension $m \times m$, without loss of generality. Therefore, from Eq. 11.7-11, we obtain the ARMA model

$$[A_1][\hat{U}^T][\hat{H}(k)] = [\hat{U}^T][\hat{H}(k+1)] \tag{11.7-14}$$

Note that from Eq. 11.7-8, we have

$$[\hat{H}(k)] = \begin{bmatrix} [\hat{U}] & [\hat{U}_2] \end{bmatrix} \begin{bmatrix} [\hat{\Sigma}] & [0] \\ [0] & [0] \end{bmatrix} \begin{bmatrix} \hat{V}^T \\ \hat{V}^T \end{bmatrix}, \tag{11.7-15}$$

where each matrix is defined by inspection. Post-multiplying in Eq. 11.7-14 by $[\hat{V}][\hat{\Sigma}]^{-1}$, we can find the coefficient matrix $[A_1]$. That is,

$$[A_1] = [\hat{U}^T][\hat{H}(k+1)][\hat{V}][\hat{\Sigma}]^{-1} \tag{11.7-16}$$

Next, Eq. 11.7-13 can be reduced into a standard eigenvalue problem as follows:

$$[A_1][\psi] = [\psi][Z], \tag{11.7-17}$$

where $[Z]$ is defined as in Eq. 11.7-7, and where $[\psi]$ is

$$[\psi] = [\hat{U}^T][C] \tag{11.7-18}$$

By using standard solution algorithms, we can find the eigenvalues and eigenvectors in Eq. 11.7-17. The damped natural frequency, damping factor, and damping ratio can be calculated as

$$\omega_r = \frac{\tan^{-1}(z_r^I / z_r^R)}{\Delta t}, \text{ (damped natural frequency)} \tag{11.7-19}$$

$$\sigma_r = \frac{\ln|z_r|}{\Delta t}, \text{ (damping factor)} \tag{11.7-20}$$

$$\varsigma_r = -\frac{\omega_r}{\sqrt{\sigma_r^2 + \omega_r^2}}, \text{ (damping ratio)} \tag{11.7-21}$$

The eigenvector matrix found from Eq. 11.7-17 is $[\psi]$, as shown in Eq. 11.7-18. Pre-multiplying in Eq. 11.7-18 by $[\hat{U}]$, we can obtain the matrix $[C]$, as follows:

$$[C] = \begin{bmatrix} [S] \\ [S][Z] \\ [S][Z]^2 \\ \vdots \\ [S][Z]^{n-1} \end{bmatrix} = [\hat{U}][\psi] \tag{11.7-22}$$

Observe that the first $2N \times 2N$ submatrix in $[C]$ is the modal matrix $[S]$, which we have been seeking.

Example 11.7-1 Repeat Example 11.5-1 using the ERA.

Solution: Taking $k=1$, $n=s=2$, and $\Delta t=1$, we have the generalized Hankel matrices

$$[\hat{H}(1)]=\begin{bmatrix} 1 & 0 & 0 & 0 \\ 0 & 0 & 0 & -1 \\ 0 & 0 & -1 & 0 \\ 0 & -1 & 0 & 0 \end{bmatrix}, \quad \text{and} \quad [\hat{H}(2)]=\begin{bmatrix} 0 & 0 & -1 & 0 \\ 0 & -1 & 0 & 0 \\ -1 & 0 & 0 & 0 \\ 0 & 0 & 0 & 1 \end{bmatrix} \tag{a,b}$$

By performing the SVD of $\hat{H}(1)$, we have

$$[\hat{H}(1)]=[U][\Sigma][V]^T, \tag{c}$$

where $[U]$, $[\Sigma]$, and $[V]$ are

$$[U]=\begin{bmatrix} 1 & 0 & 0 & 0 \\ 0 & 0 & 0 & 1 \\ 0 & 0 & 1 & 0 \\ 0 & 1 & 0 & 0 \end{bmatrix}, [\Sigma]=\begin{bmatrix} 1 & 0 & 0 & 0 \\ 0 & 1 & 0 & 0 \\ 0 & 0 & 1 & 0 \\ 0 & 0 & 0 & 1 \end{bmatrix}, [V]=\begin{bmatrix} 1 & 0 & 0 & 0 \\ 0 & -1 & 0 & 0 \\ 0 & 0 & -1 & 0 \\ 0 & 0 & 0 & -1 \end{bmatrix} \tag{d,e,f}$$

It follows from Eq. 11.7-16 that $[A_1]$ is

$$[A_1]=[U^T][\hat{H}(2)][V][\Sigma]^{-1}=\begin{bmatrix} 0 & 0 & 1 & 0 \\ 0 & 0 & 0 & -1 \\ -1 & 0 & 0 & 0 \\ 0 & 1 & 0 & 0 \end{bmatrix} \tag{g}$$

The eigenvalues of $[A_1]$ are $z_{1,2}=\pm j$ and $z_{3,4}=\pm j$. It is seen that the results are the same as those obtained by other methods.

11.8 Estimation of Modal Vectors

In the foregoing sections, we discussed methods for estimating system poles, damping factors, and natural frequencies. In this section, we present a method for estimating the residue matrix and the modal vectors [11.1–11.5].

Once the values of the system poles are known, the substitution of these values into Eqs. 11.2-3 leads to a set of linear equations, enabling the determination of the residue matrices. Specifically, from Eq. 11.2-3, we have the equations

$$\begin{bmatrix} \frac{1}{j\omega_1-\lambda_1} & \frac{1}{j\omega_1-\lambda_2} & \cdots & \frac{1}{j\omega_1-\lambda_{2N}} \\ \frac{1}{j\omega_2-\lambda_1} & \frac{1}{j\omega_2-\lambda_2} & \cdots & \frac{1}{j\omega_2-\lambda_{2N}} \\ \vdots & \vdots & \vdots & \vdots \\ \frac{1}{j\omega_{2N}-\lambda_1} & \frac{1}{j\omega_{2N}-\lambda_2} & \cdots & \frac{1}{j\omega_{2N}-\lambda_{2N}} \end{bmatrix} \begin{bmatrix} A_{pq1} \\ A_{pq2} \\ \vdots \\ A_{pq(2N)} \end{bmatrix} = \begin{bmatrix} h_{pq}(\omega_1) \\ h_{pq}(\omega_2) \\ \vdots \\ h_{pq}(\omega_{2N}) \end{bmatrix}, \quad (11.8\text{-}1)$$

where A_{pqr}, λ_r, ω_r, and N are

A_{pqr} = residue for mode r at output point p due to input at point q

λ_r = system pole for mode r

ω_r = damped natural frequency for mode r

N = number of modes

In Eq. 11.8-1, the $h_{pq}(\omega_1)$, $h_{pq}(\omega_2)$, ..., $h_{pq}(\omega_{2N})$ are known through experimental measurements. Also, the values of the system poles λ_1, λ_2, ..., λ_{2N} are known. Therefore, the residues $A_{pqr}(r=1,2,\cdots,2N)$ can be determined from Eqs. 11.8-1.

By changing the index p in Eqs. 11.8-1, the column q of the residue matrix can be found. This is the modal vector.

The modal vectors obtained from Eqs. 11.8-1 are unscaled. But we can scale them by observing that the frequency response matrix and, consequently, the residue matrices are symmetric. That is, by introducing the modal scaling factor Q_r as

$$Q_r = \frac{1}{2j\omega_r m_r}, \quad (11.8\text{-}2)$$

the residue matrix A_r may be written as

$$A_r = Q_r \begin{bmatrix} \psi_1\psi_1 & \psi_1\psi_2 & \cdots & \psi_1\psi_{N_i} \\ \psi_2\psi_1 & \psi_2\psi_1 & \cdots & \psi_2\psi_{N_i} \\ \vdots & \vdots & & \vdots \\ \psi_{N_o}\psi_1 & \psi_{N_o}\psi_2 & \cdots & \psi_{N_o}\psi_{N_i} \end{bmatrix}_r, \quad (11.8\text{-}3)$$

where m_r, ω_r, N_o, and N_i are

m_r = modal mass for mode r

ω_r = damped natural frequency for mode r

N_o = number of output (response) points

N_i = number of input (reference) points

There are several ways to scale the modal vectors. That is, we can assign the following values [11.3]:

(1) the unit modal mass, $m_r = 1$

(2) the unit modal stiffness, $k_r = m_r \omega_r^2 / (1 - \varsigma_r^2)$

(3) the unit modal vector length, $\sum_{i=1}^{N_o} \psi_{ir}^2 = 1$

(4) the unit scaling factor, $Q_r = 1$

(5) the maximum element in the modal vector is 1

In summary, the elements of the residue matrix, the modal scaling factors, and the modal vectors are related. For example, we have

$$A_{11r} = Q_r \psi_1 \psi_1, \quad A_{12r} = Q_r \psi_2 \psi_1, \dots \tag{11.8-4}$$

Thus, the elements of the modal vector are given by

$$\psi_1 = \sqrt{\frac{A_{11r}}{Q_r}}, \quad \psi_2 = \psi_1 \left(\frac{A_{12r}}{A_{11r}} \right), \dots \tag{11.8-5}$$

11.9 Estimation of Frequency Response Functions

High-quality frequency response functions (FRFs) are critical for experimental modal analysis. In this section, we will discuss the various estimation methods of FRF based on input-output measured data.

The FRF is a frequency domain function expressing the ratio between a response (output) and a reference (input) signal. The position and direction of the measurements are called degrees of freedom (DOFs). An FRF, thus, always depends on two DOFs, that is, the response DOF (numerator) and the reference DOF (denominator). In general, the system FRF is defined as

$$H(\omega) = \frac{S_X(\omega)}{S_F(\omega)}, \tag{11.9-1}$$

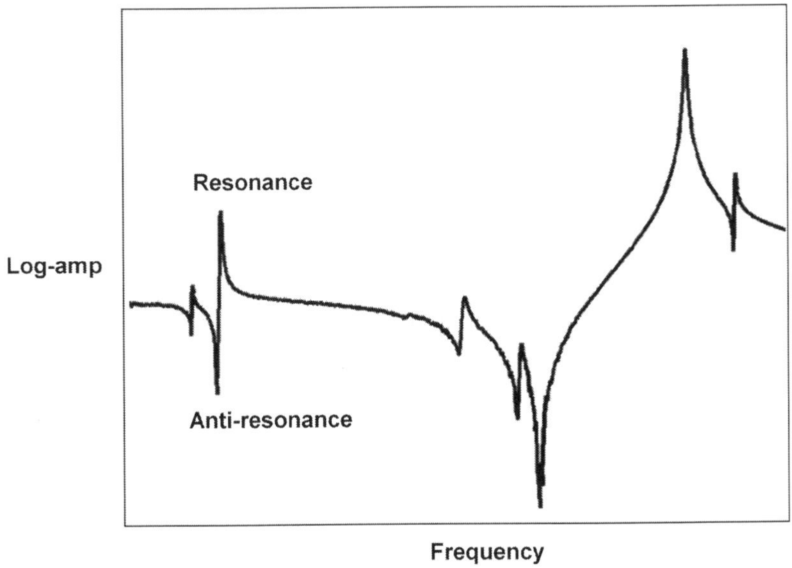

Fig. 11.9-1 Resonance and anti-resonance points in FRF.

where $S_X(\omega)$ is the Fourier transform of the output signal of the system, and $S_F(\omega)$ is the Fourier transform of the input signal to the system.

For experimental modal analysis, the response signal is most commonly the acceleration at response DOF due to a force input at another point. In this case, peaks in the FRF indicate that low input levels generate high response levels (resonance), while minima indicate low response levels, even for high inputs (anti-resonance), as shown in Fig. 11.9-1.

The system described by Eq. 11.9-1 is ideal, where the output is related directly to the input and there is no noise involved. This is not the case in reality, and various estimators are used to estimate the system's FRF from the measured input and output signals. Three estimators are widely used and are summarized as follows [11.3, 11.6, 11.17]:

$\hat{H}_1$ estimator: The measurement noise is assumed in the output.

$\hat{H}_2$ estimator: The measurement noise is assumed in the input.

$\hat{H}_V$ estimator: The measurement noise is assumed in both input and output.

$\hat{H}_1$ estimates the anti-resonance better than resonance. It tends to underestimate the system's FRF. That is, $\hat{H}_1$ is less than the system's true FRF. Upon increasing the number of averaging, $\hat{H}_1$ is approaching the system's true FRF. $\hat{H}_1$ will be exact when there is no noise in the output.

$\hat{H}_2$ estimates resonance better than anti-resonance. It tends to overestimate the system's FRF. That is, $\hat{H}_2$ is larger than the system's true FRF. It is exact when there is no noise on the input. The $\hat{H}_1$ and $\hat{H}_2$ act as lower and upper bounds to the system's true FRF.

$\hat{H}_V$ is never exact, but it gives the best FRF estimate when noise is present on both the input and output.

(1) $\hat{H}_1$ Estimator of FRF

Referring to Fig. 11.9-2, assuming that the measurement noise is in the output, we have, from Eq. 11.9-1,

$$S_X - S_N = HS_F, \tag{11.9-2}$$

where S_N is the Fourier spectrum of the noise signal in the output. Multiplying both sides of Eq. 11.9-2 by S_F^H (the transpose conjugate of S_F), we have

$$G_{XF} - G_{NF} = HG_{FF}, \tag{11.9-3}$$

where G_{XF} is the cross power spectrum between the input and output signals, G_{NF} is the cross power spectrum between the noise and input signals, and G_{FF} is the auto power spectrum of the input.

The $\hat{H}_1$ estimator derives from the condition that the errors are incoherent with the input. That is,

$$G_{NF} = 0 \tag{11.9-4}$$

Thus, from Eq. 11.9-3, the FRF matrix can be written as

$$\hat{H}_1 = G_{XF} G_{FF}^{-1} \tag{11.9-5}$$

There are three important reasons for defining the system FRF in this way. First, this technique measures magnitude and phase since the cross power

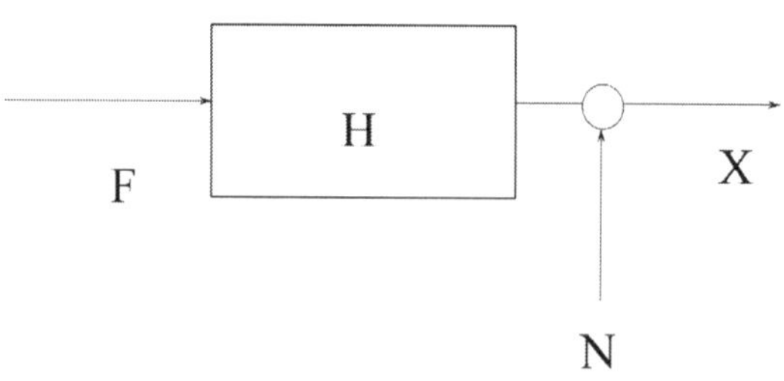

Fig. 11.9-2 Illustration of H_1 estimator.

spectrum contains phase information. Second, this formulation is not limited to sinusoidal waveforms, but may, in fact, be used for any arbitrary waveform that is Fourier transformable. Finally, averaging can be applied to the measurement. This alone is an important consideration, because it will result in large variance if only one measurement is used to estimate the FRF.

Note that since the $\hat{H}_1$ estimator does not account for the noise in the input, the estimated FRF will differ from the true value. This difference may be significant, particularly in the region of resonance, because the force level in this region is very low and because the noise level is of the same magnitude. Thus, the ratio of noise to force is large. Next, to estimate $\hat{H}_1$, it is necessary to find the inverse of G_{FF} (the auto power spectrum of the input). This may cause a numerical problem when G_{FF} is singular. Several reasons can yield singularity of the input auto power spectrum matrix: (1) when some of the input forces are zero, (2) when some of the input forces are correlated (not independent), and (3) when the computational error is large.

(2) $\hat{H}_2$ Estimator of FRF

Referring to Fig. 11.9-3, assuming that the measurement noise is in the input, from Eq. 11.9-1, we have

$$S_X = H(S_F - S_M), \tag{11.9-6}$$

where S_M is the Fourier spectrum of the noise signal in the input. Multiplying both sides of Eq. 11.9-6 by S_X^H (the transpose conjugate of S_X), we have

$$G_{XX} = H(G_{XF} - G_{XM}), \tag{11.9-7}$$

Where, again, G_{XF} is the cross power spectrum between the input and output signals.

The $\hat{H}_2$ technique assumes that the error vectors are incoherent with the output vectors, that is,

$$G_{XM} = 0 \tag{11.9-8}$$

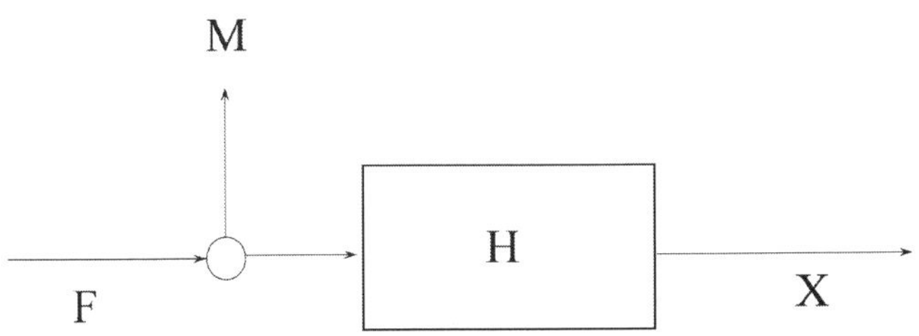

Fig. 11.9-3 Illustration of $\hat{H}_2$ estimator.

Hence, the FRF matrix can be expressed as

$$\hat{H}_2 = G_{XX} G_{XF}^{-1} \tag{11.9-9}$$

Observe that to estimate $\hat{H}_2$, it is necessary to make sure that the inverse of G_{XF} (the cross power spectrum between the input and output signals) exists. If the number of the input is less than the number of the output, G_{XF} is not square. In this case, the pseudoinverse of G_{XF} should be used in Eq. 11.9-9. Next, due to the noise in the input, $\hat{H}_2$ will give a better estimation of the FRF in the region of resonance than $\hat{H}_1$ does. Finally, since the $\hat{H}_2$ estimator does not account for the noise in the output, therefore it will overestimate the system's true FRF.

(3) $\hat{H}_v$ Estimator of FRF

Consider the case where there is uncorrelated noise in both the input and the output signals, as represented in Fig. 11.9-4. In this case, we have

$$(S_X - S_N) = H(S_F - S_M)$$

or

$$S_X - HS_F = S_N - HS_M \tag{11.9-10}$$

Note that in these expressions the H is of dimension $N_o \times N_i$, where N_o is the dimension of the output, and N_i is the dimension of the input. Multiplying both sides of Eq. 10.9-10 by their conjugate transposes, respectively, we obtain

$$(S_X - HS_F)(S_X - HS_F)^H = (S_N - HS_M)(S_N - HS_M)^H \tag{11.9-11}$$

With power spectrum matrices, Eq. 10.9-11 can be written as

$$\begin{bmatrix} -I & H \end{bmatrix} \begin{bmatrix} G_{XX} & G_{XF} \\ G_{FX} & G_{FF} \end{bmatrix} \begin{bmatrix} -I \\ H^H \end{bmatrix} = G_{NN} + HG_{MM}H^H - G_{MN}H^H - HG_{MN}, \tag{11.9-12}$$

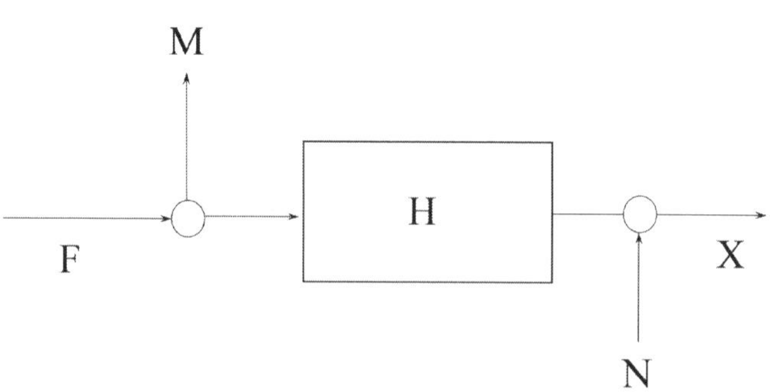

Fig. 11.9-4 Illustration of $\hat{H}_v$ estimator.

where I is the $N_o \times N_o$ unit matrix. Since the input and output noise errors are not correlated, the last two terms in Eq. 11.9-12 are zero. Hence, Eq. 11.9-12 becomes

$$\begin{bmatrix} -I & H \end{bmatrix} \begin{bmatrix} G_{XX} & G_{XF} \\ G_{FX} & G_{FF} \end{bmatrix} \begin{bmatrix} -I \\ H^H \end{bmatrix} = G_{NN} + HG_{MM}H^H, \qquad (11.9\text{-}13)$$

where G_{NN} and G_{MM} are called error matrices. Our purpose is to find the matrix H that will minimize the trace of matrices G_{NN} and G_{MM}. It can be shown [11.17, 11.18] that H (denoted by $\hat{H}_V$) can be found from the eigenvector corresponding to the smallest eigenvalues of the matrix

$$G = \begin{bmatrix} G_{XX} & G_{XF} \\ G_{FX} & G_{FF} \end{bmatrix}_{(N_o+N_i)\times(N_o+N_i)} \qquad (11.9\text{-}14)$$

The $\hat{H}_V$ estimator thus minimizes the global noise contribution in a total least squares sense. The estimator provides the best overall estimate of the FRF. It approximates the $\hat{H}_2$ estimator at resonances and the $\hat{H}_1$ estimator at anti-resonances. However, it requires more computational time than the other two methods.

11.10 Coherence Function

To determine the quality of the FRF, it is not sufficient to know only the relationship between the input and the output. At issue is whether the system's output is totally caused by the input. Noise and/or nonlinear effects of the system can cause large output at various frequencies, thus introducing errors in estimating the FRF. The influence of noise and/or nonlinearities, and thus the degree of noise contamination in the estimated FRF is measured by three types of coherence functions, as follows:

(1) Ordinary Coherence Function

The ordinary coherence function between an input signal F_i and an output signal X_i is defined as [11.17, 11.19]

$$\gamma^2_{oX_iF_i}(\omega) = \frac{\left|\bar{G}_{F_iX_i}(\omega)\right|^2}{\bar{G}_{X_iX_i}(\omega)\cdot\bar{G}_{F_iF_i}(\omega)}, \qquad (11.10\text{-}1)$$

where $\bar{G}_{F_iX_i}$ is the averaged cross power spectrum between the input and the output, $\bar{G}_{X_iX_i}$ is the averaged auto power spectrum of the output, and $\bar{G}_{F_iF_i}$ is the averaged auto power spectrum of the input.

It can be shown analytically that the ordinary coherence function is a frequency-dependent, real value between zero and one (see [11.19], p. 137). A high value (near one) indicates that the output is due almost entirely to the input and, thus, we can be confident in the FRF

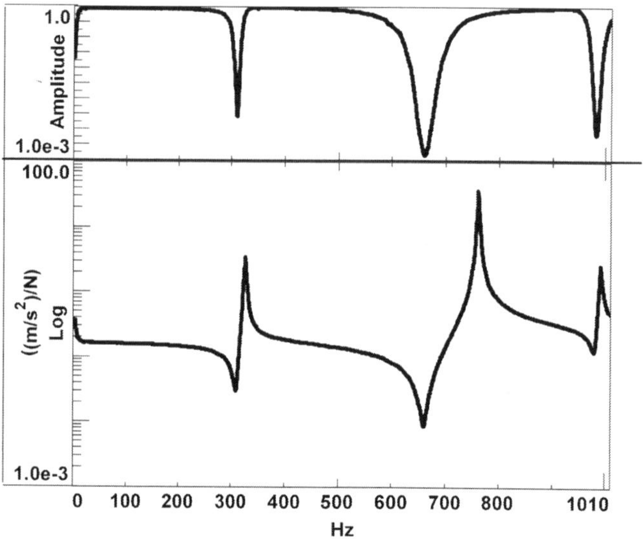

Fig. 11.10-1 A driving point FRF and coherence function.

measurements. A low value (near zero) indicates problems, such as extraneous input signals not being measured, or noise, or nonlinearities, or time delays in the system.

Fig. 11.10-1 shows a typical driving-point FRF (lower chart) and an ordinary coherence function (upper chart) of a crankshaft. It is seen that in the region of resonance, the coherence values are close to one. We can be confident in the FRF measurement, although the coherence values are very small in the region of anti-resonance. Similarly, Fig. 11.10-2 shows a typical cross-point FRF and coherence of the crankshaft. Again, it is seen that the coherence values are close to one in the region of resonance.

(2) Multiple Coherence Function

The multiple coherence function is the coefficient that describes, in the frequency domain, the causal relationship between a signal (an output spectrum) and a set of other signals (the considered input spectra) as a function of frequency. It is the ratio of the energy in an output signal, caused by several input signals, to the total amount of energy in the output signal. It is used to verify the amount of noise in the measurements, as all responses should be related to applied references (inputs).

The multiple coherence function between an output and a set of inputs is calculated from the expression [11.6, 11.19]

$$\gamma^2_{mX_iF}(\omega) = 1 - \frac{S_{nn}(\omega)}{S_{X_iX_i}(\omega)}, \tag{11.10-2}$$

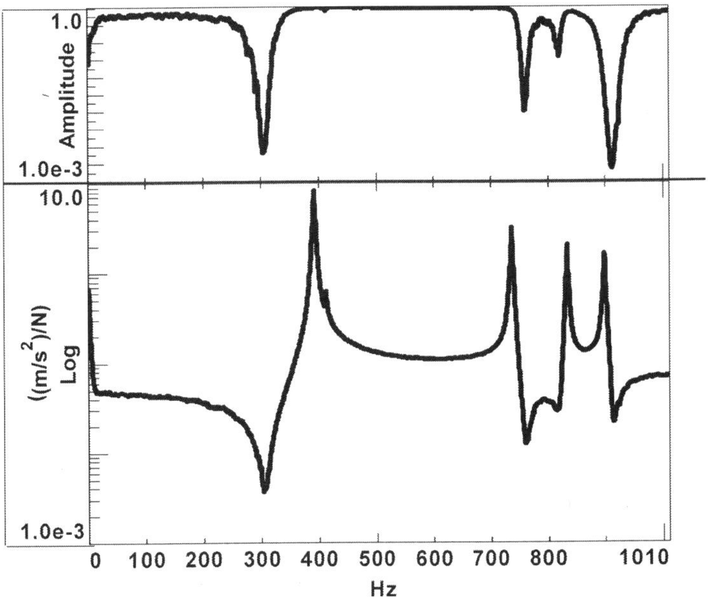

Fig. 11.10-2 A cross point FRF and coherence function.

where $S_{X_iX_i}(\omega)$ is the auto power spectrum of the output signal, and $S_{nn}(\omega)$ is the auto power spectrum of noise in the output signal (in which the contributions of all input spectra have been eliminated).

The value of the multiple coherence is always between zero and one. A high value (near one) indicates that the output is due almost entirely to the inputs. A low value (near zero) indicates that there are problems, such as extraneous input signals not being measured.

(3) Partial Coherence Function

For a multiple input/multiple output system, due to the reaction of the structure or coupling between the inputs, the measured (input and output) signals contain the influence of others. For example, when performing experimental modal analysis for a vehicle, we often use four shakers to excite the system. In this case, it is necessary to determine whether the measured input (force) of one shaker is actually caused by other shakers rather than the true input. The causal relationship between the signals can be measured by the partial coherence function.

The partial coherence function is the ordinary coherence function between conditioned signals. Conditioned signals are those where the causal effects of other signals are removed in a linear least squares sense.

To define the partial coherence, consider the signals X_1, ..., X_i, X_j, ... The partial coherence between X_i and X_j, after eliminating the signals X_1, ...,X_g, is given by (see [11.17, 11.19]

$$\gamma_{pij\times g}^{2}(\omega)=\frac{\left|S_{X_iX_j\times g}(\omega)\right|^2}{S_{X_iX_i\times g}(\omega)\cdot S_{X_jX_j\times g}(\omega)}, \tag{11.10-3}$$

where $S_{X_iX_i\times g}(\omega)$ is the auto power of X_i without the influence of the signals X_1, ..., X_g, $S_{X_jX_j\times g}(\omega)$ is the auto power of signal X_j without the influence of signals X_1, ..., X_g, and $S_{X_iX_j\times g}$ is the cross power between signals X_i and X_j without the influence of signals X_1, ..., X_g.

Each term in Eq. 11.10-3 for calculating the partial coherence can be calculated as follows: After eliminating the influence of signals X_1, ..., X_g from X_i, we have

$$X_{i\times g}=X_i-\sum_{k=1}^{g}\frac{G_{X_iX_k}}{G_{X_iX_i}}X_k, \tag{11.10-4}$$

where $G_{X_iX_i}$ is the auto power spectrum of signal X_i, $G_{X_iX_j}$ is the cross power spectrum between X_i and X_j, and $G_{X_iX_k}/G_{X_iX_i}$ is the transfer function between X_i and X_k. Similarly, after eliminating the influence of signals X_1, ..., X_g from X_j, we have

$$X_{j\times g}=X_j-\sum_{k=1}^{g}\frac{G_{XjX_k}}{G_{X_jX_j}}X_k \tag{11.10-5}$$

Therefore, the terms $S_{X_iX_j\times g}$, $S_{X_iX_i\times g}$, and $S_{X_jX_j\times g}$ in Eq. 11.10-3 and, hence, the partial coherence function can be calculated.

The value of the partial coherence is always between zero and one. A high value (near one) of two inputs (shakers, for examples) indicates that the two signals are correlated after eliminating the influence of other signals. Otherwise they are not correlated.

11.11 Discrete Fourier Transform (DFT) and Fast Fourier Transform (FFT)

Digital computation of a Fourier transform is called the discrete Fourier transform (DFT). It calculates the transform values at discrete points ($m\Delta f$) and performs a numerical integration between fixed limits (N samples). The fast Fourier transform (FFT) algorithm is a dedicated algorithm to calculate the DFT. The FFT algorithm makes use of the following DFT versions [11.20]

$$S_X(m\Delta f) = \Delta t \sum_{n=0}^{N-1} x(n\Delta t)\exp(-j2\pi m\Delta f\, n\Delta t) \quad (m = 0,\cdots,N/2) \qquad (11.11\text{-}1)$$

and

$$x(n\Delta t) = \Delta f \sum_{m=0}^{N/2} S_X(m\Delta f)\exp(j2\pi m\Delta f\, n\Delta t) \quad (n=0,\cdots,N-1) \qquad (11.11\text{-}2)$$

These equations provide the spectrum values $S_X(m\Delta f)$ at the N discrete frequencies $m\Delta f$, and they provide the time series $x(n\Delta t)$ at the N discrete time points $n\Delta t$.

To achieve efficient calculation performance, the FFT algorithm requires that the number of time samples (N) be a power of 2 (such as 2, 4, 8, ..., 512, 1024, 2048).

The DFT of Eq. 11.11-1 is valid only if $x(t)$ is periodic, with period T. The set S_X are the coefficients of the Fourier series expansion of $x(t)$. If $x(t)$ is not periodic, the DFT treats $x(t)$ as though it were periodic and is thus only an estimate of the actual Fourier transform of $x(t)$.

Such a time record of N samples is called a block of data, with N being the block size. As a consequence of Shannon's theorem [11.21], N samples in the time domain convert to $N/2$ spectral (frequency) lines at equally spaced intervals, Δf apart, up to a maximum frequency f_{max}. The time taken to collect the sample block is T. The lowest (minimum) frequency that can be detected, then, is the reciprocal of the time T. The frequency spacing between the spectral lines is, therefore, $1/T$, and the highest (maximum) frequency that can be determined is $(N/2)(1/T)$. The following fundamental relationships apply:

Record length: $$T = \frac{1}{\Delta f} = N\Delta t \qquad (11.11\text{-}3)$$

Block size: $$N = \frac{T}{\Delta t} \qquad (11.11\text{-}4)$$

Maximum frequency: $$f_{max} = \frac{N}{2}\Delta f = \frac{1}{2\Delta t} \qquad (11.11\text{-}5)$$

Sampling frequency: $$f_s = \frac{1}{\Delta t} = 2 f_{max} \qquad (11.11\text{-}6)$$

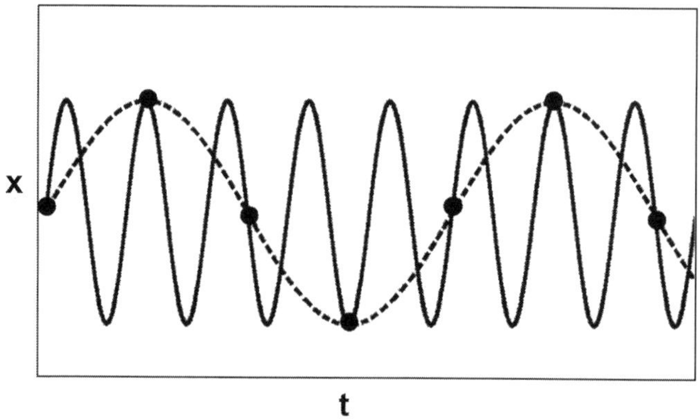

Fig. 11.11-1 Illustration of aliasing error.

Since the waveform is being sampled at discrete intervals and during a finite observation time, we do not have an exact representation. The FFT is susceptible to certain errors, the most important of which is known as *aliasing* and *leakage*.

(1) Aliasing

A sampling frequency that is too low can give rise to the problem of aliasing. Fig. 11.11-1 illustrates how erroneous results can result from aliasing. Here a signal is a sinusoid with a frequency approximately five times the sampling frequency. That is, the sampling frequency is too low (or too slow). If we focus upon the data points sampled, they present themselves as a perfect sinusoid, where in reality that is misleading.

The DFT algorithm of Eq. 11.11-1 cannot detect the error (or the aliasing) if the sampling frequency is too low. The FFT analyzer produces a frequency only one-fifth the actual frequency.

To avoid an aliasing error, we must either use an anti-aliasing filter or increase the sampling frequency. With the latter option, we need a sampling frequency at least twice the highest frequency of interest. This is called the Nyquist criterion. Correspondingly, the Nyquist frequency, designated as f_{max}, is to be less than half the sampling frequency. Higher frequency data must then be removed by analog filtering before digitizing.

(2) Leakage

Leakage is an effect where the energy of a single frequency component of vibration appears to enter and affect adjacent bands [11.20]. The phenomenon is caused by the DFT algorithm instead of the signal itself. Recall that the DFT algorithm is executed digitally using a finite, and thus limited, amount of time data. The algorithm is based on the assumption that data are from a single period of a periodic signal. *Leakage* will therefore

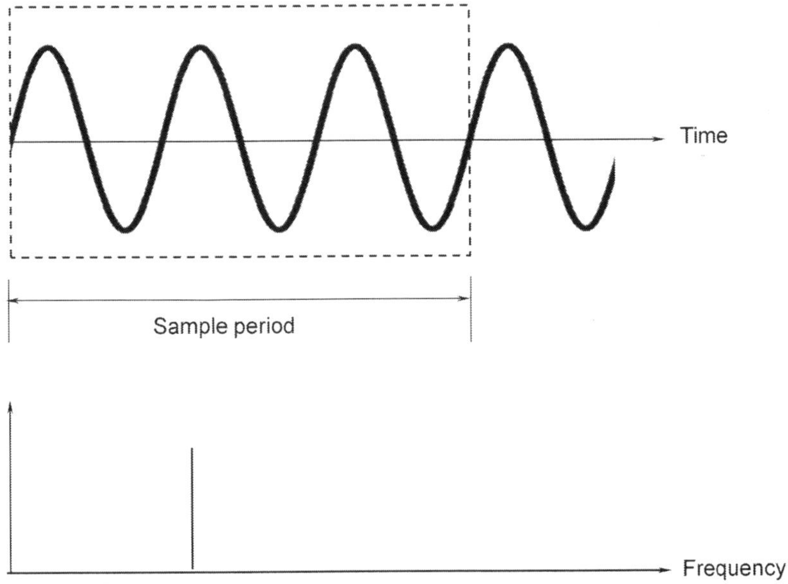

Fig. 11.11-2 If the sample period is an exact integer number of periods of the waveform, there is no leakage error in the frequency domain.

occur if a sampling of data extends beyond or is slightly less than an exact period.

To illustrate this, consider the infinitely long single frequency sinusoid wave, as in Fig. 11.11-2. Observe that the sample length (sample period) is an exact integer multiple of the sinusoid period. In this case, there is no leakage error in the frequency domain. If, however, the sample period does not correspond to an integral multiple of the signal period, as represented in Fig. 11.11-3 (A), the DFT will produce an inaccurate result in the frequency domain. The DFT algorithm embodies the assumption that the sampled data of the time window extends periodically indefinitely, as represented in Fig. 11.11-3 (B). Then, the signal cusp at the ends of the time window will cause a nonzero spectrum, or leakage error, to occur in the frequency domain, as represented in Fig. 11.11-3 (C).

We can reduce leakage in several ways. (1) We can use *burst random* excitation as an input. That is, we can choose the point where the input data are truncated to a location where the signal decays to zero from positive values. (2) We can increase the frequency resolution by using a longer acquisition time. Then a zoom FFT allows the analyst to use all frequency lines within a narrow frequency range to improve the resolution without increasing the block size. (3) We can use a window function to reduce the time data to near zero at the ends of the time window, before using the FFT algorithm.

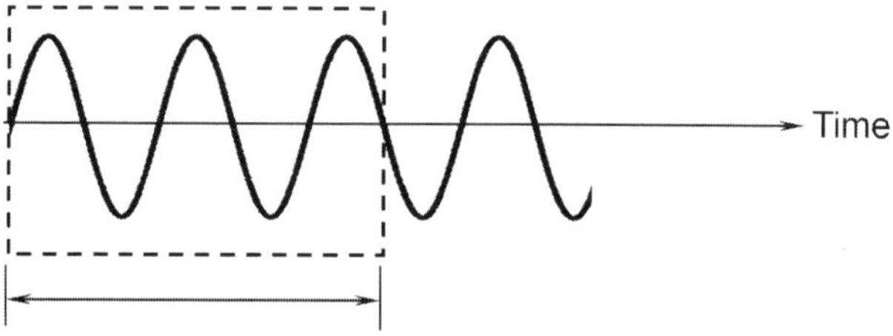

(A)

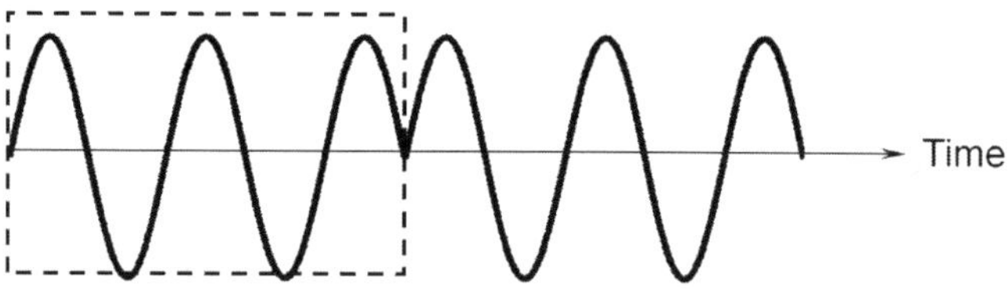

(B)

(C)

Fig. 11.11-3 Illustration of leakage error: (A) discretely sampled wave form, (B) DFT-assumed waveform, (C) frequency spectrum.

11.12 Window Functions

The most widely used method for reducing leakage error is to use a window function. A window function is simply a weighting function applied to the data record before using the FFT. The objective is to minimize the effects of discontinuities occurring when a section of a continuous signal is adjoined to a loop. The choice of a particular window function depends upon both the objective of the analysis and the type of signal being studied.

For impact testing, producing a transient signal, it is generally helpful to use a *very short rectangular* window (also known as a *force window*). A force window is effective in reducing stray noise from a large portion of an input data record.

Also, in impact testing, we can use an *exponential* window with data from the response (output) channels. The exponential window has the effect of reducing the signal to being approximately zero at the end of the record, thus reducing discontinuities. An exponential window is especially useful where there is a burst excitation signal. In this case, the window should be applied to all channels (that is, both forcing and response channels).

A potential problem with exponential windows, however, is that they introduce artificial damping into the measured data. This induced damping needs to be carefully monitored in processing the data.

If we have a signal that is nearly zero at the beginning and end of the sample period, it may be helpful to use a *uniform* (or *rectangular*, or *self-windowing*) function. Signals inherently zero at the ends of the sample period include impulses and transients.

For general purpose analyses of random signals with differing frequency components, it is helpful to use a *Hanning* window, as represented in Fig. 11.12-1.

For a signal with multiple distinguishing tones, with widely different levels, it is often helpful to use a *Kaiser-Bassel* window (see Fig. 11.12-1). But a Kaiser-Bassel window is generally not as effective as a Hanning window for random signals.

Finally, for calibration purposes, we might use a *flat top* window, which is effective in quantifying amplitude, as opposed to frequency resolution.

Fig. 11.12-1 (A) shows the force and exponential windows. The Hanning, Kaiser-Bassel, and flat top windows can be defined as amplitude-modulated cosine waves, as [11.6]

$$w(t)=1-a_1\cos(\omega_0 t)+a_2\cos(2\omega_0 t)-a_3\cos(3\omega_0 t)+a_4\cos(4\omega_0 t)\ (0<t<T)$$

$$w(t)=0\ \text{(elsewhere)}, \tag{11.12-1}$$

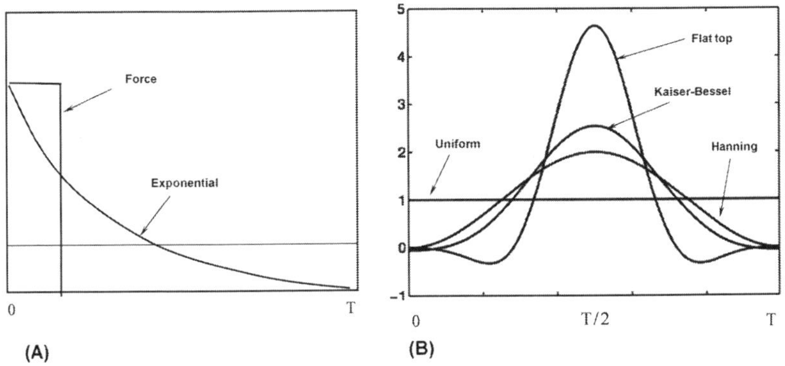

Fig. 11.12-1 Time domain window functions ([11.6], p. 289). (This graph is reproudced with permission of John Wiley & Sons, Inc.)

Table 11.12-1 Window Function Coefficients ([11.6], p. 289) (This table is reproduced with permission of John Wiley & Sons, Inc.)

	a_1	a_2	a_3	a_4
Rectangular	0	0	0	0
Hanning	1	0	0	0
Kaiser-Bessel	1.298	0.244	0.003	0
Flat top	1.933	1.286	0.383	0.032

where ω_0 is the fundamental frequency of the analyzer, and where the coefficients a_i (i=1,2,···,4) are given in Table 11.12-1. By using the expression defined in Eq. 11.12-1, we can obtain these window functions, as shown in Fig. 11.12-1 (B).

Observe that, when we multiply a signal by a window function, we necessarily distort the signal. We can compensate for this distortion by employing numerical correction factors. Specifically, we can scale the resultant signal with either a (1) amplitude correction, or a (2) energy correction. With an amplitude correction, we scale amplitude to its original value. With an energy correction, we adjust the resultant signal energy to provide accurate signal energy for a particular frequency band for broad band analysis. Most software packaging provides the correction factors for each window function.

To illustrate these concepts, consider a sine wave signal, as in Fig. 11.12-2 (A). By multiplying by a Hanning window function, we obtain the windowed sine wave of Fig. 11.12-2 (B). In the frequency domain, the amplitude of the spectrum is only one-half that of the un-windowed signal. To correct this amplitude reduction, we simply multiply by the amplitude correction factor 2 for the Hanning window.

11.13 Averaging

Experimentally measured signals are usually distorted (or *contaminated*) to some extent by random and bias errors. We can reduce random error by averaging a number of measurements. Bias errors, however, are more difficult

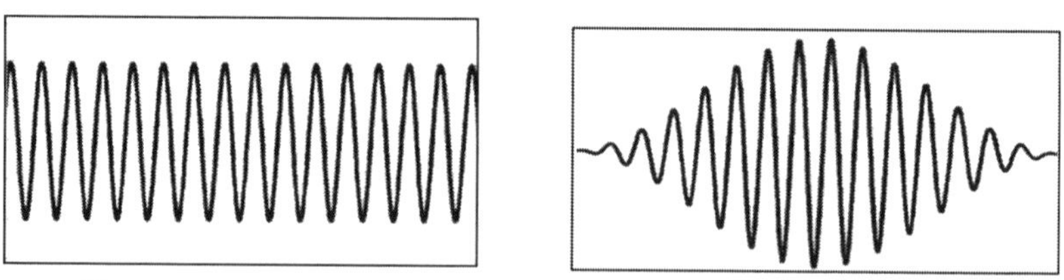

(A) (B)

Fig. 11.12-2 A sine wave (A) and windowed sine wave with Hanning window (B).

to eliminate. These errors typically arise due to nonlinearities, leakage, or mass distortion.

Most software routines employ averaging methods. We discuss two of the most common methods in the following paragraphs (see [11.17]:

(1) Linear Average

With a linear average, all individual measurements have the same influence on the final averaged result. Let n be the number of measurements $x(i)$ $(i=1,2,\cdots,n)$. Then the linear average $\overline{x}$ is simply

$$\overline{x}=\frac{1}{n}\sum_{i=1}^{n}x(i) \tag{11.13-1}$$

(2) Stable Average

With stable averaging, each measurement has the same influence on the final averaged result, but intermediate averaging results are scaled. Specifically, the average is

$$\overline{x}(n)=\left(\frac{n-1}{n}\right)\overline{x}(n-1)+\frac{x(n)}{n} \tag{11.13-2}$$

This is the most commonly used average method for frequency response function measurement with random excitation.

11.14 Overlapping Signal Analysis

As introduced in Section 11.12, the Hanning, Kaiser-Bassel, and flat top window functions taper to zero at their ends. Consequently, a portion of the signal at these ends is diminished or ignored. We can compensate for these losses by using an overlap process [11.3, 11.17].

Fig. 11.14-1 illustrates how a window function (the Hanning window function) ignores portions of the original signal. In the figure, (A) is a graph of the original time signal, (B) to (D) are graphs of Hanning window function weighted time signals, and (E) is the resulting processed data. Approximately 20% of each sample period is ignored in this weighted windowing process.

Fig. 11.14-2 illustrates the concept of overlapping, where, again, (A) is a graph of the original time signal, (B) to (D) are graphs of shifted, overlapping time signals, using again the Hanning window function. [There is a 50% overlap at (C)]. Finally, (E) is the processed data. The result of the overlapping is that portions of the central part of the original signal are no longer ignored.

Overlapping can be used with all window functions except for the rectangular window. It happens, however, that the Hanning window is ideally suited

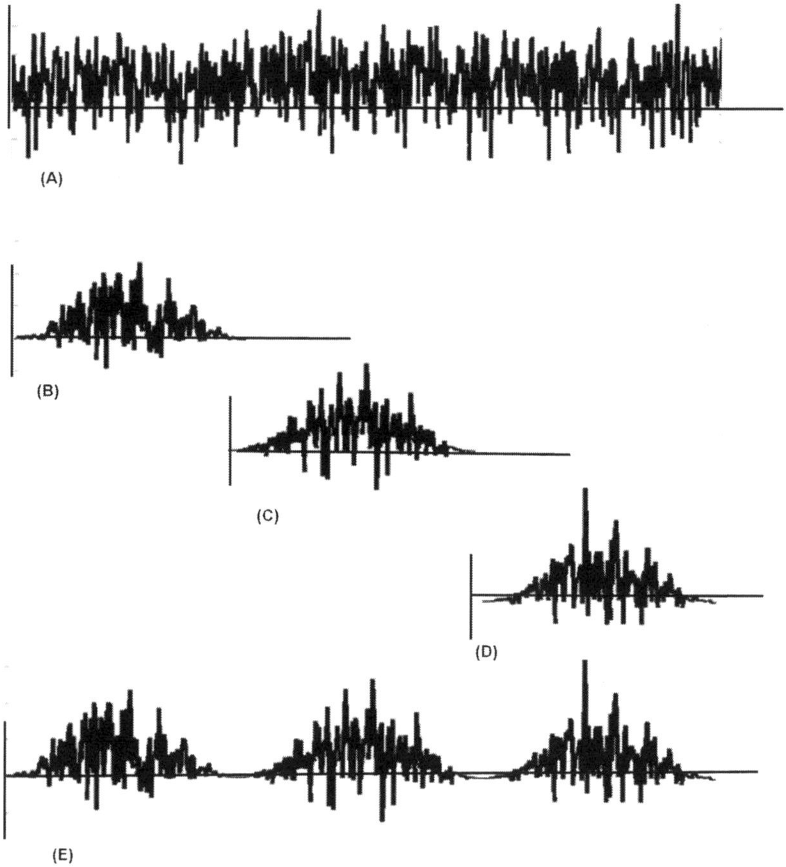

Fig. 11.14-1 No-overlapping process: (A) original signal, (B) to (D) weighted time signals each with a Hanning window, (E) processed data with no overlap [11.3, 11.17].

for overlapping. With the rectangular window, the time data are not being weighted at all. Thus, the overlap processing does not include new information, and therefore it makes no statistical improvement on the estimation procedure.

11.15 Modal Data Acquisition

In this final section of this chapter, we review and discuss measurement techniques. Since the accuracy of an experimental modal analysis is dependent upon accurate and reliable measurements, we need to continually be monitoring the measurement equipment to keep it calibrated, stable, well-maintained, and functioning as designed [11.6]. To do this, we need to have a fundamental understanding of the test equipment itself.

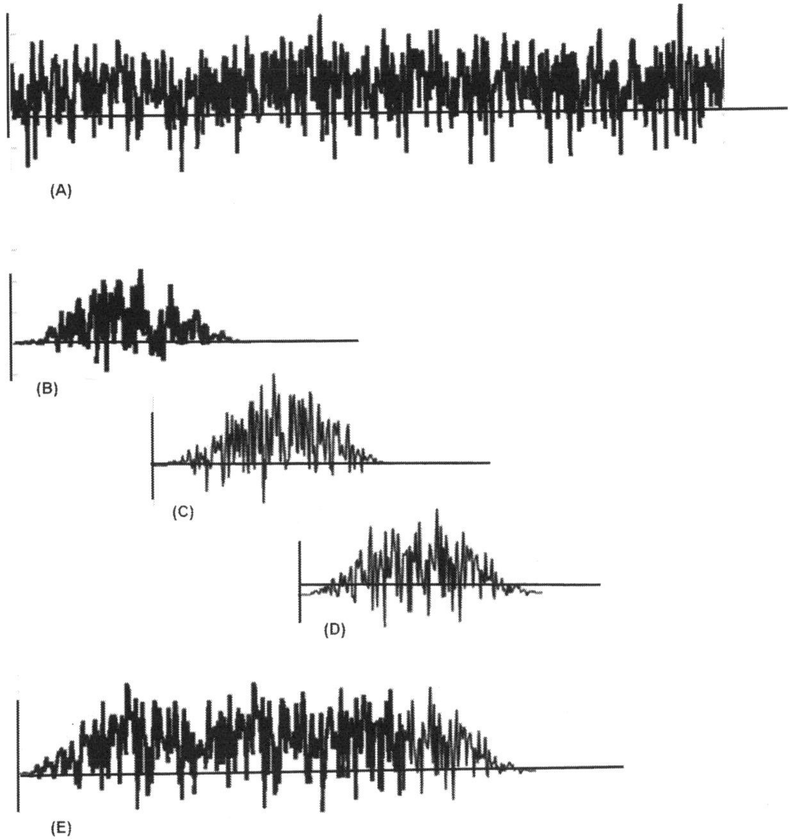

Fig. 11.14-2 Overlapping process: (A) original signal, (B) to (D) three shifted time signals each with a Hanning window, (E) processed data with overlapping [11.3, 11.17].

Typically, the equipment needed in experimental modal analysis includes four major parts:

(1) A test structure

(2) An exciter

(3) A data acquisition system

(4) An analyzer

In the following paragraphs, we discuss each of these parts.

(1) Test Structure

The test structure includes the test object and the support for the test object. There are two general kinds of test object support: (1) free-free, and (2) ground.

Fig. 11.15-1 A crankshaft suspended by light elastic bands representing free-free support condition. (Courtesy of Raven Engineering, Inc.)

With *free-free* support, the test object is freely suspended. That is, none of the points of the test object is fixed or firmly attached to the ground. Thus, the test object has six (6) rigid body degrees of freedom, or rigid body modes of motion. Each of these has frequency zero.

In practice, however, it is not practical, or even feasible, to provide a truly free-free support. This can be accomplished using light elastic bands so that the rigid body modes, while no longer having zero frequencies, will have very low frequency values in comparison with those of the flexible modes. Fig. 11.15-1 shows a crankshaft suspended by light elastic bands representing free-free support condition.

Using computer animation, we can exhibit the rigid body modes, where each point of the test object remains at a fixed distance from each other point of the test object.

As an example, for a specific truck frame mounted on air bags approximating a free-free support, the frequencies for the six rigid body modes were found to be: 2.2, 2.6, 3.0, 3.3, 3.6, and 6.1 Hz. As a contrast, the lowest torsional and bending mode frequencies were 22 and 27.2 Hz, respectively.

The second support, referred to as *ground*, attaches, or fixes, selected points of the test object to the ground (the fixed frame). While we can approximate the free-free support as a soft suspension, it is more difficult to approximate, or model, a ground support since the support is not truly rigid. That is, all supports have some flexibility.

A reasonable procedure is to measure the frequency response of the ground support itself over the anticipated frequency range of the test object. If the natural frequencies of the ground support are considerably higher than those expected of the test object, then we can reasonably assume that the ground is rigid. If this condition is met for all support points, the test structure is considered to be *grounded*.

In summary, the test structure supports are intended to be such that their natural frequencies are either considerably smaller (free-free) or considerably larger (grounded) than the expected frequencies of the test object.

(2) Exciter

A test structure can be excited (brought into vibration) in several ways. The commonly used methods are (1) a hammer, or (2) a shaker (or shakers).

Fig. 11.15-2 provides a sketch of a test configuration (a beam) involving an impact hammer. Fig. 11-15-3 shows a test configuration (an auto body) with an attached shaker. Fig. 11.15-4 shows an experimental setup of a vehicle (body in white) supported by four air bags representing *free-free* support conditions, and where four shakers are attached to the body to provide excitations.

Excitation signals may be classified as either *deterministic*, or *random*. The following signals are commonly used [11.1]:

1) *Slow sweeping sine wave* (periodic deterministic signal)

2) *Periodic chirp* (periodic deterministic signal)

3) *Impulsive* (transient deterministic signal)

4) *Step relaxation* (transient deterministic signal)

5) *Pure random*

6) *Pseudorandom*

7) *Periodic random*

8) *Burst random*

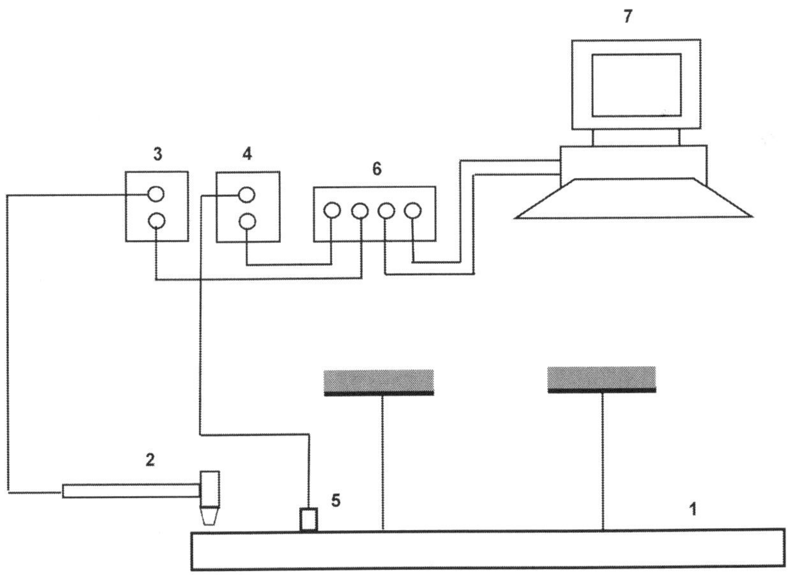

Fig. 11.15-2 Typical test configuration involving impact hammer excitation. 1: test object; 2: impact hammer; 3, 4: signal conditioning; 5: accelerometer; 6: antialiasing filters; 7: Fourier analyzer.

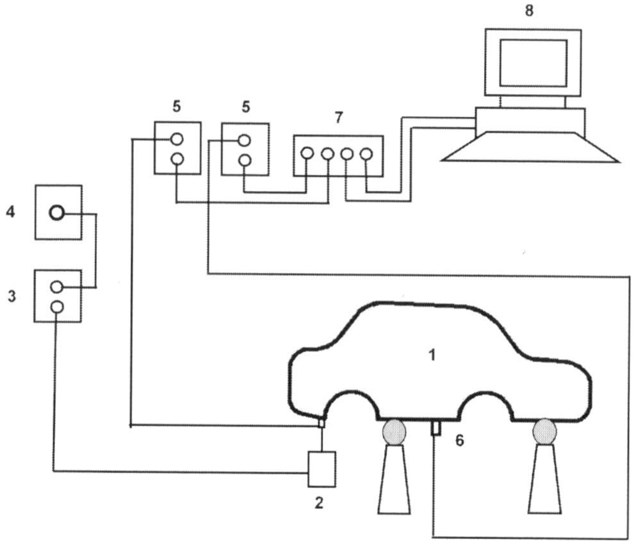

Fig. 11.15-3 Typical test configuration involving shaker excitation. 1: test object; 2: shaker; 3: power amplifier; 4: signal generator; 5: signal conditioning; 6: accelerometer; 7: antialiasing filters; 8: Fourier analyzer.

The selection of input signal depends upon the expected characteristics of the system to be measured and upon the expected utilization of the data. In theory, for a linear system, all input signals should produce the same frequency response functions. Nevertheless, physical systems all have nonlinear properties—at least to some extent.

Fig. 11.15-4 Experimental setup of a vehicle (body in white) supported by four air bags, and where four shakers are attached to the body to provide excitations. (Courtesy of Raven Engineering, Inc.)

Deterministic input signals result in frequency response functions that depend upon the signal type and its amplitude. Thus, a way of quantifying the nonlinearity is to measure frequency response functions using various excitation amplitudes. If the results are significantly different, especially around resonance, it is an indication of important nonlinearities in the test object properties. The excitation amplitude should be large enough to excite a test object in the linear range, but not so high that the test object experiences significant nonlinear vibration.

For a small test object, an impact hammer is usually sufficient to excite the system, but for a large, complex structure, one or more shakers may be needed.

When an instrumented hammer is used to excite a test structure, as represented in Fig. 11.15-1, the procedure is called: *impact testing*. When the test structure is struck by the hammer, the structure receives an impulse having a transient signal lasting for a very short time—only a fraction of a sample period. The width, height, and shape of the impulse determine its spectrum.

Impact signals are popular with analysts, due to the convenience of applying the input and due to the quick measurement time. But impact signals have a poor signal-to-noise ratio and a low root mean square (RMS)-to-peak ratio. Thus, a force window should be applied to the input signal and an exponential window applied to the output signal.

Impact testing can be performed in two ways [11.17]. The first is to measure the response at a single point while the hammer impact is applied at a number of points. This is called the *roving hammer method*. The second method is the opposite: the hammer is applied at a single point and the response is measured at a number of points. This is called the *fixed hammer method*. With both of these methods, it is important to strive to have each impact the same. Also, immediate multiple impacts (or *hammer bounce*) must be avoided since they create difficulties in the signal processing.

Next, in impact testing, the impact hammer and the transducer must be carefully calibrated. A simple method for validating the calibration is to strike a known mass m suspended by a light spring, as represented in Fig. 11.15-5. If the hammer and transducer are properly calibrated, the amplitude of the accelerance (a/F) of the mass should be $1/m$. (The first peak of the accelerance represents the rigid body mode of the mass.)

As noted earlier, for a large complex test structure, we may find it convenient to use a shaker (or shakers) to excite the structure. When using an electromechanical shaker, we need a *source* signal from a signal generator. Since signal generators produce a small signal, we also need a power amplifier to enhance the signal and then to drive the shaker (or shakers).

In the automotive industry, the most commonly used excitation signal is the burst random signal. This is a random transient signal but it is neither a completely transient deterministic signal nor a completely ergodic stationary random signal. Instead, the burst random signal has properties of both transient and random signals. The frequency spectrum of the burst random

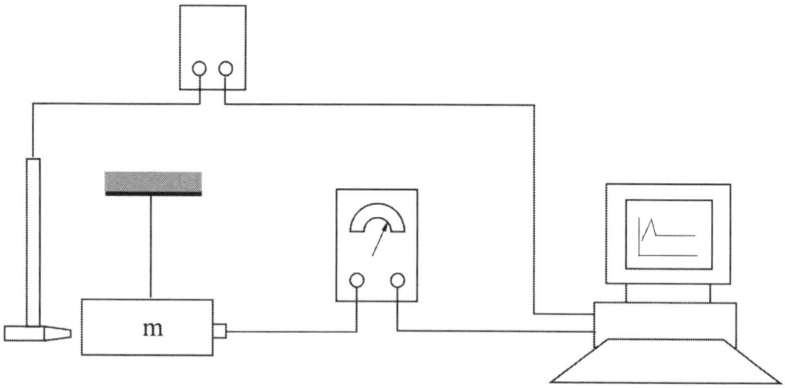

Fig. 11.15-5 Experimental setup for evaluating the calibration of the hammer and the accelerometer. (The amplitude of the accelerance should be equal to $1/m$). 1: a suspended mass; 2: impact hammer; 3: accelerometer; 4, 5: signal conditioning; 6: Fourier analyzer.

signal has a random amplitude and a random phase distribution. It has a nearly uniform energy level throughout its spectrum.

The burst random signal minimizes leakage since the burst random transient history is truncated to zero at both the beginning and the ending points of a sample period. Thus, it is not necessary to use a window function.

Fig. 11.15-6 shows burst random signals used for four shakers in the testing of a truck frame. Observe that the signal is truncated to zero at the end of

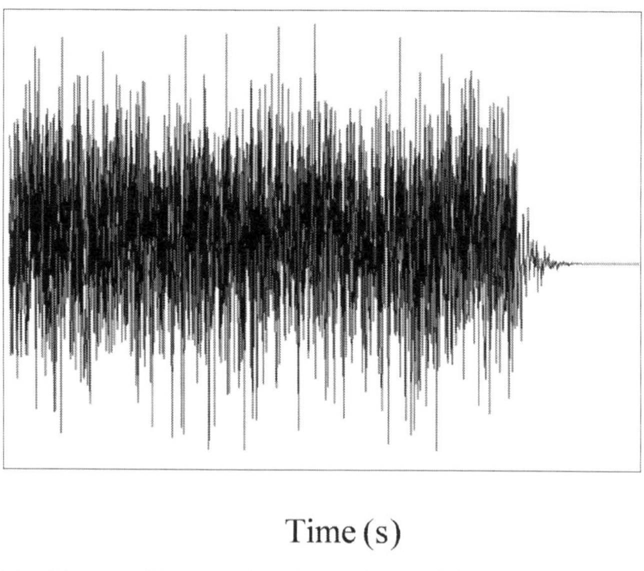

Fig. 11.15-6 Burst random signals for truck frame testing shakers.

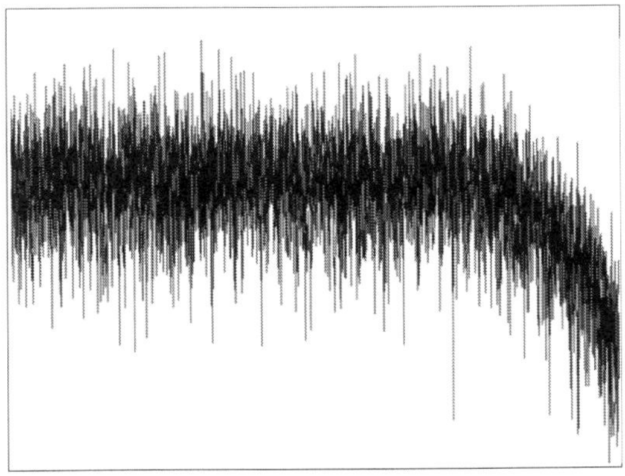

Fig. 11.15-7 Input force spectrum applied by each of the four shakers to the truck frame test structure.

the sample period. Correspondingly, Fig. 11.15-7 shows the input force levels. The force levels of each of the four shakers are the same and they are truncated to zero at the end of the sample period.

(3) Data Acquisition System

Consider again Figs. 11.15-2 and 11.15-3. Observe that the major components for data acquisition are the transducer, the signal conditioning amplifier, and the signal collector, with the antialiasing filter.

The force transducer (also called a *load cell*) measures the excitation force by either a hammer or a shaker. The simplest transducers are piezoelectric devices. The responses (outputs) are measured using accelerometers, directly measuring surface acceleration of the test structure.

In the process of data acquisition, the data are divided into three files, as represented by Fig. 11.15-8. They are [11.17]

(a) *Channel identification*. This file contains the information about transducers used, including the identification number, series number, unit, and calibration value of each transducer.

(b) *Geometry*. This file records the coordinates of each measurement point on the test structure. This information is necessary for modal animation after the mode shapes are estimated. For a large structure, it can be divided into several *components*, and for each component, a local coordinate system can be used if necessary. Fig. 11.15-9 shows an example for an H-frame, divided into three components called *rail1*, *rail2*, and *middle*, respectively.

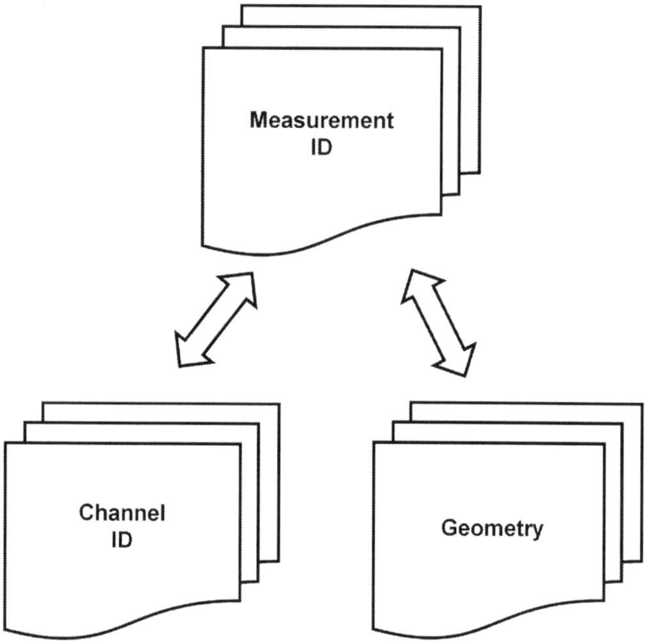

Fig. 11.15-8 Three files involved in data acquisition.

(c) *Measurement point identification.* This file records the information for each measurement point on the test structure, including the component name, point (node) number, and measured direction (x-direction, or y-direction, or z-direction).

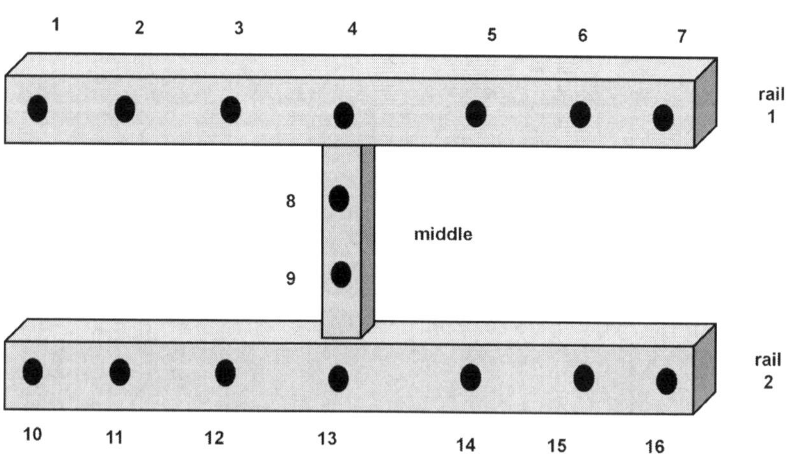

Fig. 11.15-9 Measurement points on an H-frame.

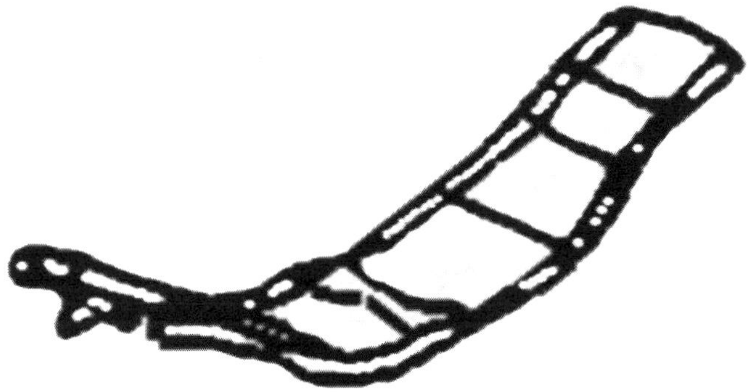

Fig. 11.15-10 Vertical bending mode of a truck frame (27.2 Hz).

(4) Analyzer

The principal function of the analyzer is to perform FFTs of the input and output signals and to then provide the frequency response functions, using either $\hat{H}_1$, or $\hat{H}_2$, or $\hat{H}_V$ estimators. (See Section 11.9.)

Once the frequency response functions are known, modal analysis can be performed using the methods of Sections 11.3 to 11.7.

When the natural frequencies and corresponding modal vectors are obtained, the vibration of each mode can be evaluated as

$$\mathbf{Y}_r = \mathbf{X}_r \sin\omega_r(k\Delta t) \quad (k=0,1,2,\cdots), \tag{11.15-1}$$

where $\mathbf{X}_r$ is the modal vector for mode r, ω_r is the natural frequency for mode r, and Δt is time increment.

Using the result of Eq. 11.15-1 together with the geometry of the test structure, we can exhibit the vibration mode on a computer monitor. This is known as *animation*.

For example, Fig. 11.15-10 shows the vertical bending mode of a truck frame.

11.16 Operational Modal Analysis

In previous sections, we discussed traditional experimental modal analysis techniques. In this method, the modal parameters (natural frequencies, modal damping ratios, and mode shapes) of a system are estimated based on experimentally measured FRFs. In general, this process is conducted in a laboratory. The actual loading conditions to which a structure is subjected, however, often differ considerably from those used in laboratory testing. A vehicle driven on a highway or an engine running in a vehicle are obvious examples. In addition, sometimes we need to consider environmental influences on system behavior (such as pre-stress of the suspension, load-induced stiffening, etc.). Unfortunately, it is difficult to measure FRFs with standard equipment while the system is operating. Therefore, development of a modal analysis method based on operational data is needed. Recently, several new modal parameter estimation techniques have been proposed where modal parameter estimates can be obtained from operational data [11.22–11.27]. This method is called *operational modal analysis* (OMA), or *output-only modal analysis*, since this method uses only output data to estimate modal parameters of a system.

There are several methods for operational modal analysis. They all use time domain auto-correlation and cross-correlation functions, which can be calculated directly from raw time data or be derived from measured auto-power and cross-power by inverse Fourier transform processing. In the following paragraphs, we briefly summarize the procedures of some of these methods. References [11.22–11.27] provide more details.

(1) Stochastic Subspace Identification Method

In the case of unknown inputs acting on a system, a stochastic model can be used to characterize the model. This method uses a state-space discrete-time model, which allows a complete description of both internal and external characteristics of a system. The basic concept is that, given the state of a system at time t, the future response of the system is determined by the input only.

To this end, consider the equation of motion of an n-degree-of-freedom system in the matrix form as

$$M\ddot{q}+C\dot{q}+Kq=f, \tag{11.16-1}$$

where the matrices M, C, and K are the $n\times n$ mass, damping, and stiffness matrices, respectively; q is an $n\times 1$ vector of generalized coordinates; and f is an $n\times 1$ vector of generalized forces.

Eq. 11.16-1 can be reformulated into state-space form by different methods. One commonly used is as follows: Let

$$\{x\}=\begin{Bmatrix} q \\ \dot{q} \end{Bmatrix} \tag{11.16-2}$$

$$f=B\{u\}, \tag{11.16-3}$$

where B is an $n \times s$ matrix, and $\{u\}$ is an $s \times 1$ vector. Next, define matrices $[\bar{A}]$ and $[\bar{B}]$ as

$$[\bar{A}]=\begin{bmatrix} 0 & I \\ -M^{-1}K & -M^{-1}C \end{bmatrix} \tag{11.16-4}$$

$$[\bar{B}]=\begin{bmatrix} 0 \\ M^{-1}B \end{bmatrix} \tag{11.16-5}$$

Then, Eq. 11.16-1 can be transformed into state-space form as

$$\{\dot{x}\}=[\bar{A}]\{x\}+[\bar{B}]\{u\} \tag{11.16-6}$$

Next, let the output or measurement vector be $\{y\}$, representing some or all of the generalized coordinates q. Let $\{y\}$ be an $r \times 1$ vector. It may be expressed as

$$\{y\}=\Theta q, \tag{11.16-7}$$

where Θ is an $r \times n$ matrix. Letting $[\bar{C}]$ be $\begin{bmatrix} \Theta_{r\times n} & 0_{r\times n} \end{bmatrix}$, we obtain the output equation

$$\{y\}=[\bar{C}]\{x\} \tag{11.16-8}$$

Eqs. 11.16-6 and 11.16-8 constitute a state-space continuous-time model of a dynamic system. The triplet $([\bar{A}],[\bar{B}],[\bar{C}])$ is called a state-space realization in control theory.

A number of methods can be used to form a discrete-time approximation of a continuous-time model: Euler's method, the zero-order hold approximation, the bilinear transformation, the impulse invariant approximation, and others (see, for example, [11.28]).

In the stochastic subspace identification method, the unknown input terms are implicitly modeled by the noise terms. Therefore, the following discrete-time stochastic state-space formulation is considered [11.22–11.26]

$$\{x(k+1)\}=[A]\{x(k)\}+\{w(k)\} \tag{11.16-9a}$$

$$\{y(k)\}=[C]\{x(k)\}+\{v(k)\}, \tag{11.16-9b}$$

where $[A]$ and $[C]$ are, respectively, the state-space matrix and the output matrix; $\{w(k)\}$ is the process noise vector, due to disturbances and modeling inaccuracies; $\{v(k)\}$ is the measurement noise, due to sensor inaccuracy; $\{x(k)\}$ is the state vector; and $\{y(k)\}$ is the response of the system. Both $\{w(k)\}$ and $\{v(k)\}$ are assumed to be zero mean, white vector sequences.

The estimation of the matrices $[A]$ and $[C]$ in Eqs. 11.16-9, using only output measurements $\{y(k)\}$, is called the stochastic realization problem [11.23]. This method has been addressed by many researchers in the control and statistics field, and has been successfully adopted in many circumstances [11.24].

Once the state-space matrix $[A]$ and the output matrix $[B]$ are estimated, the natural frequencies and damping ratios of the vibrating system are obtained by performing the eigenvalue decomposition of matrix $[A]$ [11.17, 11.21, 11.29]. That is,

$$[A]=[\Phi][Z][\Phi]^{-1}, \tag{11.16-10}$$

where

$$[\Phi]=\begin{bmatrix} \{\phi\}_1 & \{\phi\}_2 & \cdot & \{\phi\}_n \end{bmatrix} \tag{11.16-11}$$

$$[Z]=\text{diag}(z_1, z_2, \cdots z_n) \tag{11.16-12}$$

Complex eigenvectors and eigenvalues in Eq. 11.16-10 appear in complex conjugate pairs. The discrete eigenvalues z_r on the diagonal of matrix $[Z]$ can be transformed into system poles λ_r

$$z_r = e^{\lambda_r \Delta t}, \quad \lambda_r = \sigma_r + j\omega_r, \tag{11.16-13}$$

where Δt is the sampling period.

The damped natural frequency, damping factor, and damping ratio can be calculated as

$$\omega_r = \frac{\tan^{-1}(z_r^I / z_r^R)}{\Delta t}, \text{ (damped natural frequency)} \tag{11.16-14}$$

$$\sigma_r = \frac{\ln|z_r|}{\Delta t}, \text{ (damping factor)} \tag{11.16-15}$$

and

$$\varsigma_r = -\frac{\omega_r}{\sqrt{\sigma_r^2 + \omega_r^2}}, \text{ (damping ratio)} \tag{11.16-16}$$

Next, the r-th mode shape $\{\psi\}_r$ at the sensor locations is given by

$$\{\psi\}_r = [C]\{\phi\}_r \tag{11.16-17}$$

(2) The Natural Excitation Technique (NExT)

This method applies the LSCE method to auto-correlation and cross-correlation functions between the response signals, yielding global estimates of poles and the mode shapes of the system. This method is also called NExT, standing for natural excitation technique [11.17].

A theory basis of this method is as follows: Under the assumption that the system is excited by stationary white noise, it has been shown that correlation functions between the response signals can be expressed as a sum of decaying sinusoids [11.27]. Each decaying sinusoid has a damped natural frequency and a damping ratio that is identical to that of a corresponding structure mode. Therefore, the traditional modal parameter estimate techniques using impulse response functions as input, such as polyreference, LSCE, EAR, and ITD methods are also appropriate to extract the modal parameters from output-only data measured under operational conditions.

To this end, let us define a sequence of correlation functions

$$[R(k)] = E\left(\left\{y(k+m)\right\}\left\{y(m)\right\}^T\right), \tag{11.16-18}$$

where $E(.)$ denotes the expectation operator; and where $\{y(m)\}$ is a vector containing N_{ref} outputs serving a references. The selection of outputs that function as references have to be chosen in such a way that they contain all of the relevant modal information. Indeed, the selection of output-reference channels is similar to choosing the input-reference locations in a traditional modal test.

In theory, the correlation functions can be decomposed into a sum of decaying sinusoids as

$$[R(k)] = \sum_{r=1}^{N} \{\psi\}_r z_r^k \{L\}_r^T + \{\psi^*\}_r z_r^{*k} \{L^*\}_r^T, \tag{11.16-19}$$

where

$z_r = \exp(\lambda_r \Delta t)$

Δt = sampling period

$*$ = complex conjugate

$\{L\}_r$ = a column vector of N constant multipliers for mode r

r = mode number

$\lambda_r = \sigma_r + j\omega_r$, system pole

λ_r^* = complex conjugate of λ_r

σ_r = damping factor for mode r

ω_r = damped natural frequency for mode r

N = number of modes

Eq. 11.16-19 may be written as

$$[R(k)] = [S][Z]^k[L]^T, \tag{11.16-20}$$

where

$$[S] = \begin{bmatrix} \{\psi\}_1 & \cdots & \{\psi\}_N & \{\psi^*\}_1 & \cdots & \{\psi^*\}_N \end{bmatrix} \tag{11.16-21}$$

$$[L]^T = \begin{bmatrix} \{L\}_1^T & \cdots & \{L\}_N^T & \{L^*\}_1^T & \cdots & \{L^*\}_N^T \end{bmatrix} \tag{11.16-22}$$

$$[Z] = \text{diag}(z_1, ..., z_N, z_1^*, ..., z_N^*) \tag{11.16-23}$$

It is seen that, in its form, Eq. 11.16-20 is identical to Eq. 11.5-9. Therefore, following the same procedure as in Section 11.5, we can obtain the modal parameters.

11.17 Running Modes Analysis

The results of modal analysis are the estimated modal parameters of a given system, which are natural frequencies, modal damping ratios, and mode shapes.

On occasion, a complete modal analysis can be impracticable. It may be that the cost of the test setup is too high, or the test object (for example, a prototype of a vehicle) cannot be made available for the period of time required to perform a detailed modal analysis. In this case, a fast and still useful procedure is to take measurements of the system while it is operating under stationary conditions. This provides a set of measurements, denoted by $X_i(\omega)$, as a function of frequency. The measured quantities $X_i(\omega)$ can be displacements, velocities, or accelerations. It is, however, measured for one particular operating condition, with unknown acting forces or inputs.

If we are interested in a particular phenomenon at a specific frequency, it is helpful to determine what the output levels are at that frequency for each measurement location, called measurement degrees of freedom (DOF). These values can then be assembled in a vector $\{X\}$, having one element for each of measurement DOF.

A better understanding of the phenomenon is to animate the system's wire frame model. This makes it possible to show each motion level at corresponding DOF, in a cyclic manner. Due to the similarity of the animated representation of the vector $\{X\}$ with the mode shape vector, the vector $\{X\}$ is called a running mode, or an operational deflection shape (ODS).

These running modes must be interpreted differently than modal modes. Indeed, running modes reflect only the cycle motion of each DOF under specific operational conditions, and at a specific frequency. Using a displacement/force FRF matrix (see Eq. 11.2-4), the displacement running mode can be expressed as [11.17]

$$X(\omega_p)=\sum_{r=1}^{N}\left[\frac{Q_r\mathbf{u}_r\mathbf{u}_r^T}{j\omega_p-\lambda_r}+\frac{Q_r^*\mathbf{u}_r^*\mathbf{u}_r^{*T}}{j\omega_p-\lambda_r^*}\right]\begin{Bmatrix} F_1(\omega_p) \\ F_2(\omega_p) \\ \cdot \\ F_N(\omega_p) \end{Bmatrix}, \tag{11.17-1}$$

where

$\mathbf{u}_r$ = mode shape vector for mode r

ω_p = particular frequency

$F_k(\omega_p)$ = the k-th force with frequency ω_p

To clarify these concepts, it is helpful to make some observations. First, it is seen from Eq. 11.17-1 that running modes can be identified at any measured frequency ω_p, whereas a modal mode has a fixed frequency determined by the system. Second, running modes depend upon the acting force array. Third, running modes depend upon the FRFs of the system. Finally, running modes will be dominant at structural resonances ($\omega_p \approx \lambda_r$), and also at peaks in the acting force spectra.

Ideally, all response frequency spectra for a running mode analysis would be acquired [11.17] (1) simultaneously, (2) in a short period of time in which the operating conditions of the test object remain the same, and (3) with signals having a high signal-to-noise ratio.

In practice, however, data acquisition systems limit the number of response channels that can be measured simultaneously. For example, for a full vehicle running mode analysis, it may be necessary to measure responses at more than 100 locations, with each location having three degrees of freedom. In this case, it is impractical to take data simultaneously. Therefore, different sets of responses have to be measured at different periods of time.

Since operation conditions may change from one measurement to the next, it is useful to choose a specific response DOF as a reference and then measure all responses relative to this reference. The reference channel for different sets of measured responses should remain the same. This will, we hope, affect all response signals in the same way, and the operating condition change will be averaged out. This procedure also guarantees a fixed phase relationship between the different response signals, using the phase of the reference signal as a reference.

Problems

11.1 The time domain sampled data of an impulsive response for a single-degree-of-freedom system are given by

$$h(0)=1,\ h(1)=0,\ h(2)=-1,\ h(3)=0,\ \Delta t=1$$

Find the natural frequency of the system.

Ans. $\omega=\pi/2$

11.2 Let the system pole λ_r for mode r be written as $\lambda_r=\sigma_r+j\omega_r$. Verify the following results:

$$\text{Modal mass: } \bar{m}_r=\frac{1}{j2Q_r\omega_r},\quad \text{where } Q_r=\frac{1}{j2\bar{m}_r\omega_r}\ \text{(scaling factor)}$$

$$\text{Modal damping: } \bar{c}_r=-2\sigma_r\bar{m}_r$$

$$\text{Modal stiffness: } \bar{k}_r=\bar{m}_r(\sigma_r^2+\omega_r^2)$$

11.3 Repeat Example 7.6-1. Find the modal mass, modal damping, and modal stiffness by using the formulas given in Problem 11.2. Show that the results are the same as those given by S^TMS, S^TCS, S^TKS, where M, C, K are the mass, damping, and stiffness matrices, respectively, and where S is the corresponding modal matrix.

Ans. $Q_1=-j\dfrac{\sqrt{15}}{45},\quad Q_2=-j\dfrac{\sqrt{15}}{135}$

$\bar{m}_1=6,\quad \bar{m}_2=12$

$\bar{c}_1=3,\quad \bar{c}_2=15$

$\bar{k}_1=6,\quad \bar{k}_2=30$

11.4 Repeat Problem 11.3. Choose scaling factors corresponding to unit modal masses.

Ans. $Q_1 = -j\frac{2\sqrt{15}}{15}, \quad Q_2 = -j\frac{4\sqrt{15}}{45}$

$$\bar{m}_1 = 1, \quad \bar{m}_2 = 1$$

$$\bar{c}_1 = 1/2, \quad \bar{c}_2 = 5/4$$

$$\bar{k}_1 = 1, \quad \bar{k}_2 = 5/2$$

$$\bar{S} = \begin{bmatrix} 1/\sqrt{6} & 1/\sqrt{12} \\ 1/\sqrt{6} & -2/\sqrt{12} \end{bmatrix}$$

11.5 For the system shown in Fig. P11.5 (m=10 kg, c= 20 Ns/m, k=16000 N/m), the impulse responses are listed in the following table. Find the poles of the system by using the time domain method. Compare the results with the roots of the equation: $ms^2 + cs + k = 0$.

Ans. $\lambda = -1.000 \pm j39.9875$

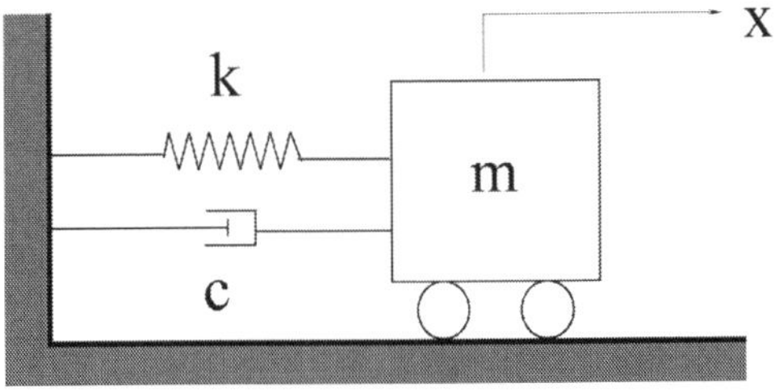

Fig. P11.5

Time(s)	Response (×10^{-2})
0.000	0.0000
0.025	2.0519
0.050	2.1637
0.075	0.3296
0.100	−0.1711

11.6 The Modal Assurance Criterion (MAC) between two modal vectors $\mathbf{u}_1$ and $\mathbf{u}_2$ is defined as [11.1–11.3]

$$MAC_{12} = \frac{\left[(\mathbf{u}_1^*)^T W \mathbf{u}_2\right]^2}{\left[(\mathbf{u}_1^*)^T W \mathbf{u}_1\right]\left[(\mathbf{u}_2^*) W \mathbf{u}_2\right]},$$

where W is the weighting matrix; and $\mathbf{u}_r^*$ is the complex conjugate of $\mathbf{u}_r$. Using the modal vectors of Example 7.6-1 and taking the mass matrix as the weighting matrix, find the MAC values MAC_{11}, MAC_{12}, and MAC_{22}. Explain the physical meaning of the results.

Ans. $MAC_{11} = MAC_{22} = 1, \quad MAC_{12} = 0$

11.7 Repeat Example 7.6-1. Show the validity of the following equation:

$$H(s)=\sum_{r=1}^{n}\frac{\mathbf{u}_r\mathbf{u}_r^T}{\bar{m}_r s^2+\bar{c}_r s+\bar{k}_r},$$

where $\bar{m}_r$, $\bar{c}_r$ and $\bar{k}_r$ are the modal mass, modal damping, and modal stiffness, respectively, and where $\mathbf{u}_r$ is the corresponding modal vector.

11.8 Repeat Example 7.6-1. Find the responses of the system in the time domain.

Ans.

$$\lambda_1=-\frac{1}{4}+j\frac{\sqrt{15}}{4},\quad \lambda_1^*=-\frac{1}{4}-j\frac{\sqrt{15}}{4},$$

$$\lambda_2=-\frac{5}{8}+j\frac{3\sqrt{15}}{8},\quad \lambda_2^*=-\frac{5}{8}-j\frac{3\sqrt{15}}{8}$$

$$h_{11}(t)=-j\frac{\sqrt{15}}{45}\left(e^{\lambda_1 t}-e^{\lambda_1^* t}\right)-j\frac{\sqrt{15}}{135}\left(e^{\lambda_2 t}-e^{\lambda_2^* t}\right)$$

$$h_{12}(t)=-j\frac{\sqrt{15}}{45}\left(e^{\lambda_1 t}-e^{\lambda_1^* t}\right)+j\frac{2\sqrt{15}}{135}\left(e^{\lambda_2 t}-e^{\lambda_2^* t}\right)$$

$$h_{22}(t)=-j\frac{\sqrt{15}}{45}\left(e^{\lambda_1 t}-e^{\lambda_1^* t}\right)-j\frac{4\sqrt{15}}{135}\left(e^{\lambda_2 t}-e^{\lambda_2^* t}\right)$$

11.9 Repeat Example 11.5-1. Suppose the *experimentally measured* impulse response function matrix is given by

$$[H(t)]=\begin{bmatrix} h_{11}(t) & h_{12}(t)\\ h_{21}(t) & h_{22}(t)\end{bmatrix}=\begin{bmatrix} \cos\frac{\pi}{2}t & 0\\ 0 & \cos\frac{\pi}{2}t\end{bmatrix}$$

11.10 Repeat Example 11.6-1. Suppose the *experimentally measured* impulse response function matrix is given by

$$[H(t)]=\begin{bmatrix} h_{11}(t) & h_{12}(t)\\ h_{21}(t) & h_{22}(t)\end{bmatrix}=\begin{bmatrix} \cos\frac{\pi}{2}t & 0\\ 0 & \cos\frac{\pi}{2}t\end{bmatrix}$$

11.11 Repeat Example 11.7-1. Suppose the *experimentally measured* impulse response function matrix is given by

$$[H(t)]=\begin{bmatrix} h_{11}(t) & h_{12}(t) \\ h_{21}(t) & h_{22}(t) \end{bmatrix}=\begin{bmatrix} \cos\frac{\pi}{2}t & 0 \\ 0 & \cos\frac{\pi}{2}t \end{bmatrix}$$

References

11.1. Allemang, R. J., and Brown, D. L., "Experimental Modal Analysis," in *Shock and Vibration Handbook*, C. M. Harris, ed., 3rd ed., McGraw-Hill, New York, 1987.

11.2. Allemang, R. J., "Experimental Modal Analysis," *Proceedings of the Modal Testing and Model Refinement Symposium*, ASME, AMD, Vol. 59, 1983, pp. 1–29.

11.3. Allemang, R. J., *Vibrations: Analytical and Experimental Modal Analysis*, UC-SDRLL-CN-20-263-691, Department of Mechanical, Industrial, and Nuclear Engineering, University of Cincinnati, Cincinnati, OH, 1992.

11.4. Brown, D. L., *System Dynamic Analysis, Parameter Estimation Overview*, Class Notes, Department of Mechanical, Industrial, and Nuclear Engineering, University of Cincinnati, Cincinnati, OH, 1991.

11.5. Ewins, D. J., *Modal Testing: Theory and Practice*, Research Studies Press, Ltd., Letchworth, UK, 1986.

11.6. McConnell, K. G., *Vibration Testing: Theory and Practice*, John Wiley & Sons, New York, 1995.

11.7. Prony, R., "Essai Experimental et Analytique: sur les Lois de la Dilatabilite de Fluides Elastique et sur Celles de la Force Expansive de la Vapeur de l'Alkool, a Differentes Temperatures," *J. de l'Ecole Polytechnique Floreal et Prairial* (Paris), An. III, Vol. 1, 1795, pp. 24–76 (in French).

11.8. Vold, H., et al, "A Multi-Input Modal Estimation Algorithm for Mini-Computers," SAE Paper 820194, SAE International, Warrendale, PA, 1982.

11.9. Vold, H., and Rocklin, T., "The Numerical Implementation of a Multi-Input Modal Estimation Method for Mini-Computers," *Proceedings of the 1st International Modal Analysis Conference*, Orlando, FL, 1982, pp. 542–548.

11.10. Ibrahim, S. R., and Mikulcik, E. C., "The Experimental Determination of Vibration Parameters from Time Responses," *Shock and Vibration Bulletin*, Vol. 46, No. 5, 1976, pp. 187–196.

11.11. Ibrahim, S. R., and Mikulcik, E. C., "A Method for the Direct Identification of Vibration Parameters from the Free Response," *Shock and Vibration Bulletin*, Vol. 47, No. 4, 1977, pp. 183–198.

11.12. Pappa, R. S., "Some Statistical Performance Characteristics of the "ITD" Modal Identification Algorithm," AIAA Paper No. 82-0768, American Institute of Aeronautics & Astronautics, Reston, VA, 1982.

11.13. Juang, J.–N., and Pappa, R. S., "An Eigensystem Realization Algorithm for Parameter Identification and Model Reduction," *AIAA Journal of Guidance, Control, and Dynamics*, Vol. 8, No. 5, 1985, pp. 620–627.

11.14. Juang, J.–N., *Applied System Identification*, Prentice Hall, Englewood Cliffs, NJ, 1994.

11.15. Leuridan, J., Brown, D. L., and Allemang, R. J., "Time Domain Parameter Identification Methods for Linear Modal Analysis: A Unifying Approach," *Journal of Vibration, Acoustics, Stress and Reliability in Design*, Vol. 108, 1986, pp. 1–8.

11.16. Lawson, C. L., and Hanson, E. J., *Solving Least Squares Problems*, Prentice Hall, Englewood Cliffs, NJ, 1974.

11.17. LMS Test Lab 9B, User Manual, Theory Documents, LMS International, Interieuvenlann 68, B-3001, Leuven, Belgium, 2009

11.18. Allemang, R. J., Classroom notes for Vibrations: Analytical and Experimental Modal Analysis, Department of Mechanical, Industrial, and Nuclear Engineering, University of Cincinnati, Cincinnati, Ohio, 1992.

11.19. Bendat, J. S., and Piersol, A. G., *Random Data: Analysis and Measurement Procedures*, 2nd ed., John Wiley & Sons, New York, 1986.

11.20. Randall, R. B., "Vibration Measurement Equipment and Signal Analyzers," in *Shock and Vibration Handbook*, C. M. Harris, ed., 3rd ed., McGraw-Hill, New York, 1987, pp. 13-18 to 13-23.

11.21. Shannon, C. E., "Communication in the Presence of Noise," *Proceedings of the IRE (Institute of Radio Engineers)*, Vol. 37, 1949, pp. 10–21.

11.22. Van Overschee, P., and De Door, B., *Subspace Identification for Linear Systems*, Kluwer Academic Publishers, Dordrecht, The Netherlands, 1996.

11.23. Desai, U. B., Pal, D., and Kirkpatrick, R. D., "A Realization Approach to Stochastic Model Reduction", *International Journal of Control*, Vol. 42, No. 4, 1985, pp. 821–838.

11.24. Hermans, L., and Van der Auweraer, H., "Modal Testing and Analysis of Structures under Operational Conditions: Industrial Applications," *Mechanical Systems and Signal Processing*, Vol. 13, No. 2, 1999, pp. 193–216.

11.25. Brincker, R., Zhang, L., and Andersen, P., "Modal Identification of Output-Only Systems Using Frequency Domain Decomposition," *Smart Material Structure*, Vol. 10, 2001, pp. 441–445.

11.26. Antoni, J., Garribaldi, L., Marchesiello, A., and Sidhamed, M., "New Separation Techniques for Output-Only Modal Analysis," *Shock and Vibration*, Vol. 11, 2004, pp. 227–242.

11.27. James, G. H., III, Carne, T. G., and Laufer, J. P., "The Natural Excitation Technique (NExT) for Modal Parameter Extraction from Operating Structures," *International Journal of Analytical and Experimental Modal Analysis*, Vol. 10, No. 4, 1995, pp. 260–277.

11.28. Burl, J. B., *Linear Optimal Control, H_2 and H_∞ Methods,* Addison Wesley Longman, Inc., New York, 1999.

11.29. Liu, Xiaobo, "A State Space Method for Modal Identification of Mechanical Systems from Time Domain Responses," *Shock and Vibration*, Vol. 12, 2005, pp. 273–282.

Chapter Twelve

Special Topics

12.1 Introduction

In this chapter, we use the methodologies of the preceding chapters to develop specialized methods for solving vibration problems in complex structures. We consider four methodologies:

1) **Substructuring.** Here we divide a complex structure into substructures. We then examine frequency response functions (FRFs) for the somewhat simpler substructures. Having these functions, we deduce vibration characteristics of the overall structure.

2) **Vibration Absorber and Tuning.** Here we consider vibration absorber design and tuning based on baseline FRFs of a structure to intelligently reduce adverse vibration effects.

3) **Hybrid Analyses.** Here we consider procedures of experimental FRFs and finite element analysis (FEA)—known as the HYFEX method.

4) **Sensitivity Analysis.** Here we consider procedures for obtaining eigenvalue and eigenvector changes of a mechanical system as a result of changes in physical parameters of the system.

12.2 Complex Structure Analysis Using the FRFs of Substructures

Most structural systems of practical importance are extremely complex. Consider, for example, an automotive chassis and body. Recently, analysts have used FRFs of portions (substructures) of complex structures to estimate the overall system dynamic characteristics. With the substructures being considerably simpler than the entire structure, it is relatively easy to obtain the FRFs of the substructures.

The substructuring method is attractive not only due to its simplified methodology, but also because it is ideally suited to study individual design changes in the substructures. In addition, the resulting system of equations using substructures is considerably simpler than the equations obtained by studying the system as a whole.

Substructuring methods, however, are not without difficulties. The methods eliminate coupling forces between interfacing substructures by introducing constraint equations for adjoining substructures. These constraint equations generally are not simple. This is especially the case if adjoining substructures (e.g., an auto body and a chassis) are connected with springs and dampers. Moreover, if the coupling force systems are of interest, it is necessary to use back substitution in the constraint equations—an often cumbersome and inefficient procedure.

In 2004, C. Q. Liu and Xiaobo Liu [12.1] presented a new method of substructuring analysis that circumvents the above difficulties. The new method is applicable with software devised for generic systems, and it readily determines the interactive coupling forces. We develop the method in the following paragraphs.

(1) Equations of Motion

Consider a large dynamical structure (or system) S, which is conveniently divided into N substructures, or subsystems. Consider, in turn, a typical subsystem S_k of S. The equations of motion of S_k may be written as

$$\mathbf{x}_k = H_k \mathbf{f}_k \quad (k=1,2,\cdots,N), \tag{12.2-1}$$

where there is no sum on k, but instead k has distinct values: $1,2,\cdots,N$. Also, in Eq. 12.2-1, $\mathbf{x}_k$ is the response vector; $\mathbf{f}_k$ is the force vector, which includes both external forces and coupling forces; and H_k is the frequency response function matrix.

By combining the subsystem equations, we obtain the equations of motion for the entire system S. These equations may be written as

$$\mathbf{x} = H\mathbf{f}, \tag{12.2-2}$$

where $\mathbf{x}$ is the response vector of the total system with dimension $p \times 1$; $\mathbf{f}$ is the total force vector with dimension $q \times 1$, consisting of r coupling forces and $(q - r)$ external forces; and H is the FRF matrix with dimension $p \times q$.

Next, the coupling conditions can be written in a general form as

$$B\mathbf{x} + C\mathbf{f} = 0, \tag{12.2-3}$$

where matrix B has dimension $r \times p$, and matrix C has dimension $r \times q$.

To illustrate these concepts, consider a simple mass-spring system as in Fig. 12.2-1. Let the system have masses m_1 and m_2, and let these masses be connected by a linear spring with stiffness k and a dashpot with damping coefficient c. Let an applied force f_1 be acting on m_1, as shown.

The coupling conditions on m_1 and m_2 are seen to be

$$(k + j\omega c)(x_2 - x_1) = f_{1c} \tag{12.2-4}$$

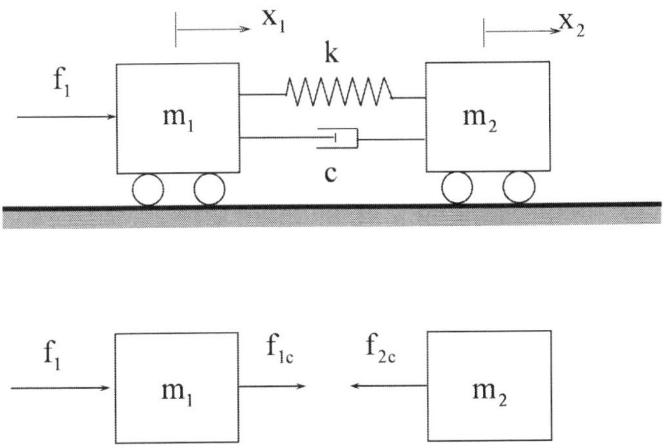

Fig. 12.2-1 A simple mass-spring system [12.1].

and

$$f_{1c} + f_{2c} = 0, \tag{12.2-5}$$

where x_1 and x_2 are the displacements of m_1 and m_2, and f_{1c} and f_{2c} are the coupling forces at the two ends of the connecting spring. Eqs. 12.2-4 and 12.2-5 can be written in the form of Eq. 12.2-3 as

$$\begin{bmatrix} -(k+j\omega c) & k+j\omega c \\ 0 & 0 \end{bmatrix} \begin{bmatrix} x_1 \\ x_2 \end{bmatrix} + \begin{bmatrix} 0 & -1 & 0 \\ 0 & 1 & 1 \end{bmatrix} \begin{bmatrix} f_1 \\ f_{1c} \\ f_{2c} \end{bmatrix} = 0 \tag{12.2-6}$$

By inspection, the matrices B and C are seen to be

$$B = \begin{bmatrix} -(k+j\omega c) & k+j\omega c \\ 0 & 0 \end{bmatrix} \quad \text{and} \quad C = \begin{bmatrix} 0 & -1 & 0 \\ 0 & 1 & 1 \end{bmatrix} \tag{12.2-7}$$

Eqs. 12.2-2 and 12.2-3 form a system of $(p+r)$ equations for the p responses and the r coupling forces. We develop a method for solving these equations in the following subsection.

(2) Solution Method

In theory, we can arrange the elements of the force vector **f** and the multiplying matrix C (see Eq. 12.2-3) so that the last r entries are coupling forces, and, correspondingly, the last r columns of C are linearly independent. Then we may rewrite Eq. 12.2-3 in the form

$$B\mathbf{x}+\begin{bmatrix} C^1_{r\times(q-r)} & C^2_{r\times r}\end{bmatrix}\begin{Bmatrix}\mathbf{f}_a\\ \mathbf{f}_c\end{Bmatrix}=0, \tag{12.2-8}$$

where the subscripts on matrices $C^1_{r\times(q-r)}$ and $C^2_{r\times r}$ designate the corresponding dimensions; $\mathbf{f}_a$ is the applied force vector. The coupling force vector $\mathbf{f}_c$ can then be expressed as

$$\mathbf{f}_c=-\left[C^2_{r\times r}\right]^{-1}\left[B\mathbf{x}+C^1_{r\times(q-r)}\mathbf{f}_a\right] \tag{12.2-9}$$

Hence, by substituting from Eq. 12.2-9 into Eq. 12.2-2, we may solve for the dynamic response vector **x**.

For large systems, however, this approach may be tedious and inefficient. An improved method is as follows: First, by substituting from Eq. 12.2-2 into Eq. 12.2-3, we have

$$\left[BH+C\right]\mathbf{f}=0 \tag{12.2-10}$$

Next, suppose that Φ is an orthogonal complement of $[BH+C]$ such that

$$\left[BH+C\right]\Phi=0, \tag{12.2-11}$$

where Φ has dimension $q\times(q-r)$ with rank $(q-r)$. The force vector **f** can then be expressed as

$$\mathbf{f}=\Phi\boldsymbol{\eta}, \tag{12.2-12}$$

where the vector $\boldsymbol{\eta}$ has $(q-r)$ entries, which may be interpreted as generalized force components. The force vector **f** consists of r coupling forces and $(q-r)$ external forces.

Third, by moving the unknown coupling forces to the right side of Eq. 12.2-12, we have

$$\left[\Phi\left|\begin{matrix}0_{(q-r)\times r}\\ -I_{r\times r}\end{matrix}\right.\right]\begin{Bmatrix}\boldsymbol{\eta}\\ \mathbf{f}_c\end{Bmatrix}=\begin{Bmatrix}\mathbf{f}_a\\ 0\end{Bmatrix}, \tag{12.2-13}$$

where $I_{r\times r}$ is an identity matrix with rank r; $\mathbf{f}_a$, as before, is the applied force vector; and $\mathbf{f}_c$ is the coupling force vector.

Finally, from Eq. 12.2-13, we can solve for $\boldsymbol{\eta}$ and the coupling forces in $\mathbf{f}_c$. Once this is accomplished, the response **x** can be determined as

$$\mathbf{x}=H\Phi\boldsymbol{\eta} \tag{12.2-14}$$

Observe that the key step in this approach is determining the orthogonal complement matrix Φ.

There are several ways of obtaining Φ (see [12.2] to [12.5]). One of the methods is singular value decomposition (SVD). Specifically, let D be $BH+C$; then there are unitary matrices $U_{r\times r}$ and $V_{q\times q}$ such that

$$U^H DV = \begin{bmatrix} \Sigma & 0 \\ 0 & 0 \end{bmatrix}, \tag{12.2-15}$$

where

$$\Sigma = \text{diag}(\sigma_1, \cdots, \sigma_r), \qquad \text{and} \qquad \sigma_1 \geq \sigma_2 \geq \cdots \geq \sigma_r > 0 \tag{12.2-16}$$

The nonzero diagonal elements in the decomposition are called the *singular values* of D. The singular values are unique, but the arrays U and V are not unique.

With selective partitioning of array V, Eq. 12.2-15 can be expressed as

$$D = \begin{bmatrix} U_1 & U_2 \end{bmatrix} \begin{bmatrix} \Sigma & 0 \\ 0 & 0 \end{bmatrix} \begin{bmatrix} V_1^H \\ V_2^H \end{bmatrix} = U_1 \Sigma V_1^H \tag{12.2-17}$$

Array V is unitary. That is, $V_1^H V_2$ is zero. Thus, by post-multiplying in Eq. 12.2-17 by V_2, we have

$$DV_2 = 0 \tag{12.2-18}$$

That is, V_2 is an orthogonal complement of D.

In summary, we can outline the procedure as

Step 1. Perform singular value decomposition of $BH+C$,

Step 2. Form an orthogonal complement matrix Φ (the last $(q-r)$ columns of V),

Step 3. Solve Eq. 12.2-13 for the generalized forces and coupling forces,

Step 4. Solve Eq. 12.2-14 for the responses of the total system.

Example 12.2-1 [12.1] Fig. 12.2-2 shows a system composed of two identical steel plates A and B. Points 1 and 2 of plate A are connected to points 1 and 2 of plate B, respectively, through spring-damper systems. One end of plate A is fixed and one corner point of plate B is fixed. Suppose the frequency response function matrices of the two plates have been made available. The objective is to find the acceleration responses at points 1 and 2 of the two plates due to a unit force f_1 applied vertically at point 1 of plate A.

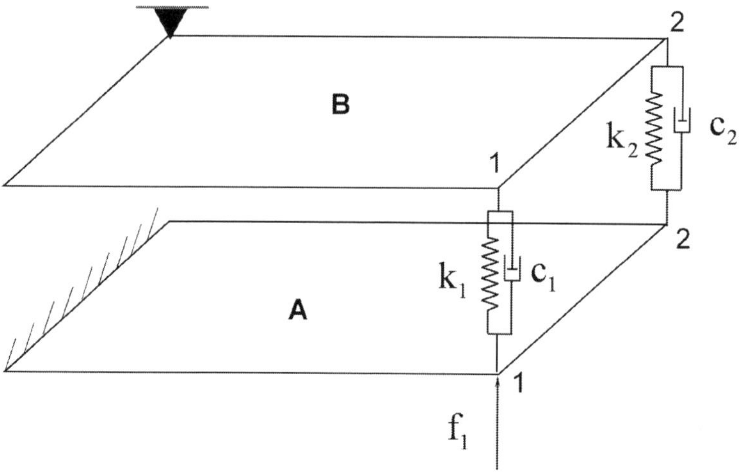

Fig. 12.2-2 A system composed of two plates connected through two spring-damper systems [12.1].

Solution: Eq. 12.2-1, in this case, becomes

$$\begin{bmatrix} a_1^A \\ a_2^A \end{bmatrix} = \begin{bmatrix} H_{11}^A & H_{11}^A & H_{12}^A \\ H_{12}^A & H_{12}^A & H_{22}^A \end{bmatrix} \begin{bmatrix} f_1 \\ f_{1c}^A \\ f_{2c}^A \end{bmatrix} \tag{a}$$

and

$$\begin{bmatrix} a_1^B \\ a_2^B \end{bmatrix} = \begin{bmatrix} H_{11}^B & H_{12}^B \\ H_{12}^B & H_{22}^B \end{bmatrix} \begin{bmatrix} f_{1c}^B \\ f_{2c}^B \end{bmatrix}, \tag{b}$$

where f_1 is the applied force, f_{1c}^A and f_{2c}^A are the coupling forces applied, respectively, at points 1 and 2 of plate A, f_{1c}^B and f_{2c}^B are the coupling forces applied respectively at points 1 and 2 of plate B, H_{ij}^A $(i,j=1,2)$ are the FRFs of plate A, and H_{ij}^B $(i,j=1,2)$ are the FRFs of plate B.

Let the response vector be expressed as

$$\mathbf{x} = \begin{bmatrix} a_1^A & a_2^A & a_1^B & a_2^B \end{bmatrix}^T \tag{c}$$

and let the force vector (including the applied force and the coupling forces) be

$$\mathbf{f} = \begin{bmatrix} f_1 & f_{1c}^A & f_{1c}^B & f_{2c}^A & f_{2c}^B \end{bmatrix}^T \tag{d}$$

Then, the frequency response function matrix H in Eq. 12.2-2 becomes

$$H=\begin{bmatrix} H_{11}^{A} & H_{11}^{A} & 0 & H_{12}^{A} & 0 \\ H_{12}^{A} & H_{12}^{A} & 0 & H_{22}^{A} & 0 \\ 0 & 0 & H_{11}^{B} & 0 & H_{12}^{B} \\ 0 & 0 & H_{12}^{B} & 0 & H_{22}^{B} \end{bmatrix} \tag{e}$$

and the matrices B and C in Eq. 12.2-3 become

$$B=\frac{1}{\omega^{2}}\begin{bmatrix} (k_1+j\omega c_1) & 0 & -(k_1+j\omega c_1) & 0 \\ 0 & (k_2+j\omega c_2) & 0 & -(k_2+j\omega c_2) \\ 0 & 0 & 0 & 0 \\ 0 & 0 & 0 & 0 \end{bmatrix} \tag{f}$$

$$C=\begin{bmatrix} 0 & -1 & 0 & 0 & 0 \\ 0 & 0 & 0 & -1 & 0 \\ 0 & 1 & 1 & 0 & 0 \\ 0 & 0 & 0 & 1 & 1 \end{bmatrix} \tag{g}$$

The solution procedure is as follows: At each frequency ω, perform SVD of $(BH+C)$ such that

$$BH+C=U\begin{bmatrix} \Sigma & 0 \\ 0 & 0 \end{bmatrix}V^{H} \tag{h}$$

Then the matrix Φ, an orthogonal complement of $(BH+C)$, is of dimension 5×1 and is simply the last column of the matrix V. Therefore, Eq. 12.2-13 has the form

$$\begin{bmatrix} V_{15} & 0 & 0 & 0 & 0 \\ V_{25} & -1 & 0 & 0 & 0 \\ V_{35} & 0 & -1 & 0 & 0 \\ V_{45} & 0 & 0 & -1 & 0 \\ V_{55} & 0 & 0 & 0 & -1 \end{bmatrix}\begin{bmatrix} \eta \\ f_{1c}^{A} \\ f_{1c}^{B} \\ f_{2c}^{A} \\ f_{2c}^{B} \end{bmatrix}=\begin{bmatrix} f_1 \\ 0 \\ 0 \\ 0 \\ 0 \end{bmatrix} \tag{i}$$

From Eq. i, we can solve for the generalized force $\boldsymbol{\eta}$ and the coupling forces f_{1c}^{A}, f_{2c}^{A}, f_{1c}^{B}, and f_{2c}^{B}. The responses of the total system can be found by solving Eq. 12.2-14, namely,

$$x=H\Phi\eta \tag{j}$$

To validate this method, we consider a number of cases with various stiffness and damping values of the connecting spring-damper systems. To this end, we use the following typical values for both plates [12.1]: the dimensions are 500.0 mm × 250.0 mm × 10.0 mm, Young's modulus is 250000 N / mm^2, Poison's ratio is 0.3, and the mass density is 7.850 ×10^{-6} kg/mm^2.

Case 1: Moderate stiffness and damping values

First, consider a case in which the spring stiffness values are k_1=195.0 N/mm and k_2=205.0 N/mm, and the damping coefficients are c_1=c_2=0.05 Ns/mm. Let f_1 be a unit force. The responses are the FRFs.

Fig. 12.2-3 shows the FRFs obtained by employing the proposed method as compared to results from a full model finite element analysis. The results are identical.

Case 2: Very large stiffness and damping values

Next, consider the case in which the stiffness and damping values are larger than those in case 1 by three orders of magnitude. That is, k_1=195000.0 N/mm, k_2=205000.0 N/mm, c_1=c_2=50.0 Ns/mm. Fig. 12.2-4

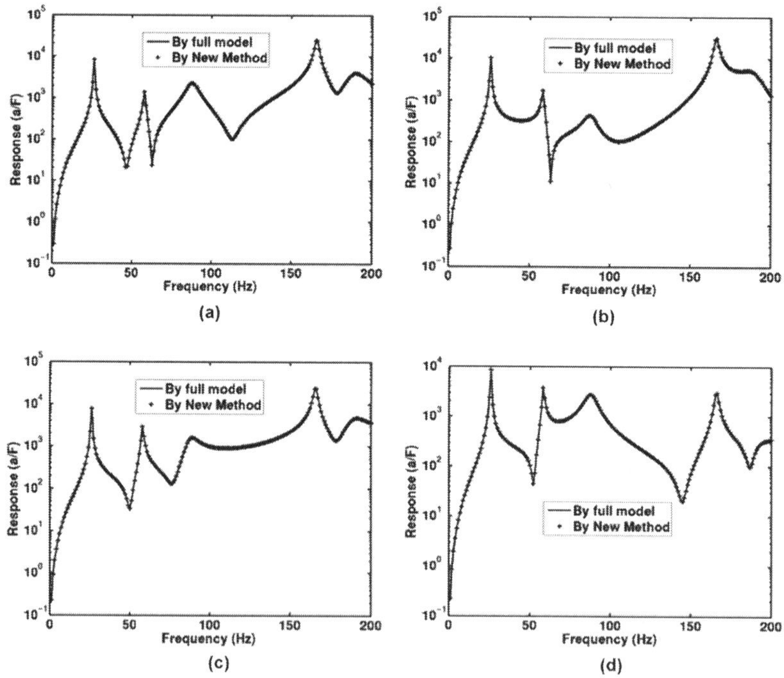

Fig. 12.2-3 Case 1 of example problem. FRFs of the system shown in Fig. 12.2-2: (a) response of point 1 of plate A, (b) response of point 1 of plate B, (c) response of point 2 of plate A, (d) response of point 2 of plate B [12.1].

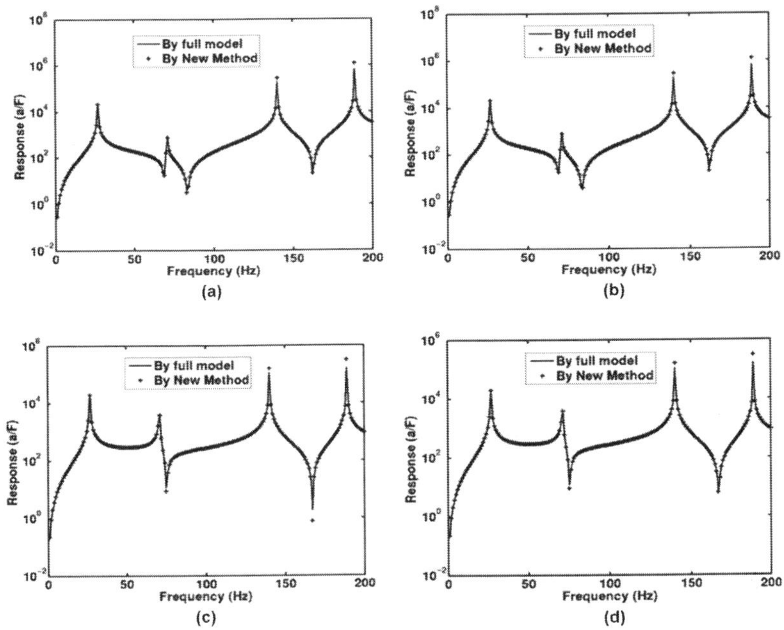

Fig. 12.2-4 Case 2 of example problem. FRFs of the system shown in Fig. 12.2-2: (a) response of point 1 of plate A, (b) response of point 1 of plate B, (c) response of point 2 of plate A, (d) response of point 2 of plate B [12.1].

shows the comparison between the FRFs obtained by the proposed method and by the full model finite element analysis. Again, the results are identical. As expected, the responses of points 1 on the two plates are the same [see Fig. 12.2-4 (a) and (b)]. Since the stiffness values of the connecting spring is extremely large, the two points will move together. The same is true for points 2 of the two plates [see Fig. 12.2-4 (c) and (d)].

12.3 Translational Vibration Absorber

The most effective way to eliminate unwanted vibration is to eliminate the source of the vibration itself. But this often cannot be done. Indeed, any machine with essential moving parts will have inherent vibration that will not be eliminated without removing the moving parts and thus defeating the purpose of the machine.

The alternative to eliminating unwanted vibration is to reduce the harmful effects of the vibration—principally by precision design and manufacturing, by isolation, and by absorption. Precision design, precision manufacture, and isolation are physical and geometrical changes in the system, whereas absorption attempts to control the vibration by absorbing the energy of the vibration. In this section, we discuss translational absorbers, and in the following section, we consider torsional (rotational) absorbers.

A translational vibration absorber (sometimes called a *mass damper*) typically consists of a damped mass-spring device attached to a primary vibrating system with the objective of controlling the vibration amplitude of the system. If, however, the natural frequency and damping of the vibration absorbing device is not properly designed (or *tuned*), the adjoining of the absorbing device may bring the composite system (primary system and absorber) into resonance with the exciting force.

A number of analysts have studied the issues of vibration absorbers [12.6–12.13]. Unfortunately, there are only a few special cases where closed-form solutions for optimal absorbers are known. Moreover, theoretical results are not always applicable with large complex systems. Most practical problems require numerical solutions and these solutions are usually obtained using software developed for generic classes of mechanical systems.

To explore and document such solutions, we present explicit expressions where the FRFs of the composite system are represented in terms of the FRFs of the primary system, and the physical parameters (mass, spring modulus, and damping characteristics) of the adjoining vibration absorber. These expressions are ideally suited for vibration absorber tuning and are generally applicable for complex systems. An additional advantage with these expressions is that alternative absorbers can be evaluated before they are physically adjoined to the vibrating system. In some special cases, the analyses reduce to classical configurations found in the literature. The applicability and validity of this method has been verified by numerical simulations and applications of vibration absorbers in motor vehicles [12.12].

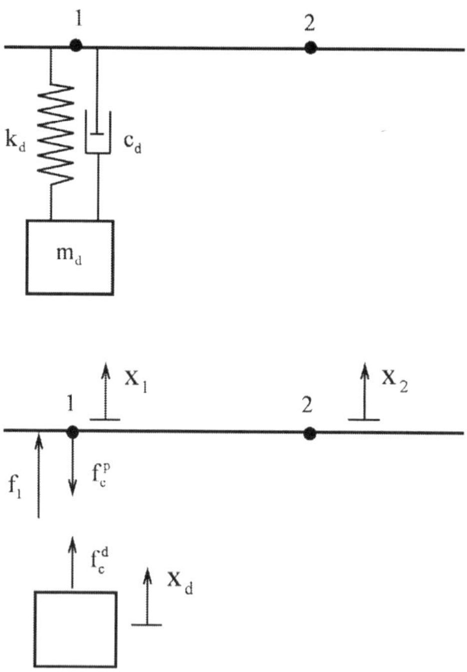

Fig. 12.3-1 A vibration absorber attached to a primary system and a free-body diagram [12.12].

(1) Absorber Tuning Based on the Primary System FRF

Fig. 12.3-1 provides a drawing representing a primary vibrating system together with a vibration absorber [12.12]. The absorber is attached to point 1 of the primary system. Point 2 is some other point (distinct from point 1) of the primary system.

Two questions arise: (1) What is the resulting FRF of the composite system (the primary system with the absorber) at point 1 when the excitation is at point 1? (2) What is the resulting FRF of the composite system at point 2 with the excitation at point 1?

To answer these questions, let the response at point 1 due to a unit excitation force at point 1 be called the *driving-point FRF of point 1*; and let the response at point 1 due to a unit excitation force at point 2 be called the *cross-point FRF between points 1 and 2.*

We can obtain the equation of motion by considering the free-body diagram, as shown in Fig. 12.3-1. From that diagram, we immediately see that the equation of motion may be written as:

$$\begin{Bmatrix} x_1 \\ x_2 \\ x_d \end{Bmatrix} = \begin{bmatrix} H_{11}^x & -H_{11}^x & 0 \\ H_{12}^x & -H_{12}^x & 0 \\ 0 & 0 & -\dfrac{1}{\omega^2 m_d} \end{bmatrix} \begin{Bmatrix} f_1 \\ f_c^p \\ f_c^d \end{Bmatrix}, \tag{12.3-1}$$

where x_1, x_2, and x_d are the responses of point 1, point 2, and the mass of the vibration absorber, respectively; f_1 is the applied force at point 1; f_c^p and f_c^d are the coupling forces applied on the primary system and on the mass of vibration absorber, respectively; H_{11}^x is the driving-point FRF of point 1, H_{12}^x is the cross-point FRF between points 1 and 2 (where the superscript x on H_{11}^x and H_{12}^x designate that the FRFs are measured in terms of displacement over force, or receptance); m_d is the mass of the vibration absorber; and ω is the frequency.

Next, the constraint equations between the primary system and the vibration absorber may be written as

$$\begin{bmatrix} 0 & 0 & 0 \\ k_d + j\omega c_d & 0 & -(k_d + j\omega c_d) \end{bmatrix} \begin{Bmatrix} x_1 \\ x_2 \\ x_d \end{Bmatrix} + \begin{bmatrix} 0 & 1 & -1 \\ 0 & 0 & 1 \end{bmatrix} \begin{Bmatrix} f_1 \\ f_c^p \\ f_c^d \end{Bmatrix} = 0 \tag{12.3-2}$$

Taken together, Eqs. 12.3-1 and 12.3-2 form a system of five equations for five unknowns (three responses and two coupling forces). In Section 12.2, we presented an efficient numerical method for solving such systems of equations. This method, based upon [12.1], is also easy to implement. Since the present analysis involves only two subsystems, however, we can obtain explicit expressions for the responses at points 1 and 2. This will provide a basis for vibration absorber tuning, making the process of tuning considerably simpler.

To this end, let us eliminate f_c^d between the third of Eqs. 12.3-1 and the second of Eqs. 12.3-2. This yields

$$x_d = \frac{k_d + j\omega c_d}{k_d - m_d\omega^2 + j\omega c_d} x_1 \quad (12.3\text{-}3)$$

Next, eliminating f_c^d between the third of Eqs. 12.3-1 and the first of Eqs. 12.3-2 leads to

$$f_c^p = -m_d\omega^2 x_d \quad (12.3\text{-}4)$$

Next, it follows from the first of Eqs. 12.3-1 that

$$x_1 = H_{11}^x f_1 - H_{11}^x f_c^p \quad (12.3\text{-}5)$$

Substituting from Eq. 12.3-4 into Eq. 12.3-5 and using Eq. 12.3-3 yields

$$x_1 = H_{11}^x f_1 + H_{11}^x \frac{(k_d + j\omega c_d) m_d\omega^2}{k_d - m_d\omega^2 + j\omega c_d} x_1 \quad (12.3\text{-}6)$$

Dividing both sides of Eq. 12.3-6 by f_1 and solving for x_1/f_1 yields

$$\frac{x_1}{f_1} = \frac{H_{11}^x}{1 - \dfrac{(k + j\omega c_d) m_d\omega^2}{k_d - m_d\omega^2 + j\omega c_d} H_{11}^x} \quad (12.3\text{-}7)$$

To simplify Eq. 12.3-7, it is helpful to introduce an equivalent mass m_{eq} of the vibration absorber. That is, let m_{eq} be

$$m_{eq} = \frac{(k_d + j\omega c_d) m_d}{k_d - m_d\omega^2 + j\omega c_d} \quad (12.3\text{-}8)$$

or, in terms of nondimensional quantities,

$$m_{eq} = \frac{1 + 2\varsigma_d \beta_d j}{(1 - \beta_d^2)^2 + 2\varsigma_d \beta_d j} m_d \quad (12.3\text{-}9)$$

where

$\beta_d = \dfrac{\omega}{\omega_d}$, the ratio of the forcing frequency to the natural frequency of the mass damper

$\omega_d = \sqrt{\dfrac{k_d}{m_d}}$, the natural frequency of the mass damper

$$\varsigma_d = \frac{c_d}{2\sqrt{k_d m_d}}, \text{ the damping ratio of the mass damper}$$

Observe that x_1/f_1 is simply the resulting driving-point FRF (receptance) of the system with the vibration absorber attached at point 1. Let x_1/f_1 be denoted by $\hat{H}_{11}^x$. Then, by substituting from Eq. 12.3-8 into Eq. 12.3-7, we have

$$\hat{H}_{11}^x \equiv \frac{x_1}{f_1} = \frac{H_{11}^x}{1 - m_{eq}\omega^2 H_{11}^x} \tag{12.3-10}$$

Eq. 12.3-10 provides the answer to the first question raised at the beginning of this section: that is, when a vibration absorber is attached to a vibrating system, the resulting driving-point FRF (receptance) $\hat{H}_{11}^x$, at the attachment point, can be expressed in terms of the equivalent mass of the absorber and the driving point FRF: H_{11}^x, of the original system, at the attachment point of the vibration absorber.

The resulting FRF can also be expressed in terms of accelerance (acceleration over force). To this end, if we multiply both sides of Eq. 12.3-10 by $-\omega^2$ and note that $-\omega^2 x_1 = a_1$ and $-\omega^2 H_{11}^x = H_{11}^a$, we can write

$$\hat{H}_{11}^a \equiv \frac{a_1}{f_1} = \frac{H_{11}^a}{1 + m_{eq} H_{11}^a} \tag{12.3-11}$$

Consider now the second question: What is the cross-point FRF between points 1 and 2 when a vibration absorber is attached to point 1? It follows from the second equation of Eq. 12.3-1 that

$$x_2 = H_{12}^x f_1 - H_{12}^x f_c^p \tag{12.3-12}$$

Substituting Eq. 12.3-4 into Eq. 12.3-12 and using Eqs. 12.3-3 and 12.3-8 yields

$$x_2 = H_{12}^x f_1 + H_{12}^x \omega^2 m_{eq} x_1 \tag{12.3-13}$$

Dividing both sides of Eq. 12.3-13 by f_1 and substituting Eq. 12.3-10 into the result, we obtain the following:

$$\hat{H}_{12}^x \equiv \frac{x_2}{f_1} = \frac{H_{12}^x}{1 - m_{eq}\omega^2 H_{11}^x} \tag{12.3-14}$$

Similarly, the FRF can be expressed in terms of accelerance. That is,

$$\hat{H}_{12}^a \equiv \frac{a_2}{f_1} = \frac{H_{12}^a}{1 + m_{eq} H_{11}^a} \tag{12.3-15}$$

Eqs. 12.3-14 and 12.2-15 indicate that the resulting cross-point FRF between points 1 and 2 of the composite system when a vibration absorber is attached to point 1 can be expressed in terms of the baseline (*original*) driving-point FRF of point 1, the baseline cross-point FRF between points 1 and 2, and the equivalent mass of the vibration absorber.

Eqs. 12.3-11 and 12.3-15 provide a basis for vibration absorber tuning for complex systems. Specifically, the baseline FRFs of the primary system can be obtained by either analytical or experimental methods. Then, a number of alternative vibration absorbers can be evaluated. That is, by changing the physical parameters (mass, spring modulus, and damping characteristics) of the vibration absorber, the new resulting FRFs of the composite system can be calculated from Eqs. 12.3-11 and 11.3-15. Then, by comparing these FRFs, a *best* vibration absorber can be found. *Best* vibration absorber means that the FRF peak of the original (baseline) system is replaced by two side peaks with approximately the same amplitudes. Also, the side peaks are to be as low as possible. (See [12.12] for more details.)

It is important to observe that this tuning process is not physically on a test structure. Instead, the tuning is the result of computer simulations. In this way, we can avoid the empirical time and expenses.

(2) Special Cases: A Single-Degree-of-Freedom System

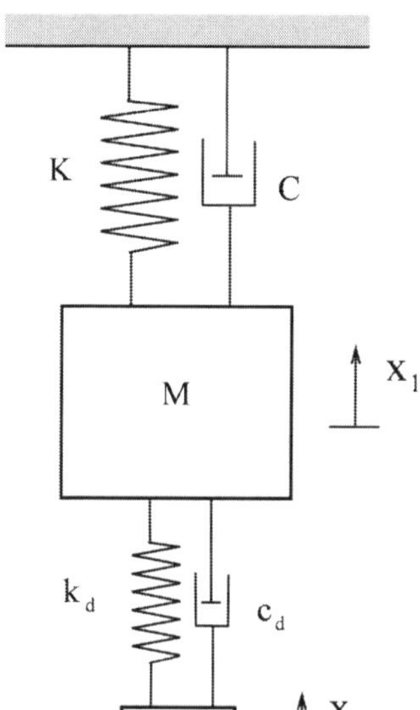

Fig. 12.3-2 A single-degree-of-freedom system with a vibration absorber [12.12].

As mentioned earlier, these results are applicable to general complex mechanical systems. They are, thus, applicable to a single-degree-of-freedom system as well. Fig. 12.3-2 shows a single-degree-of-freedom system with a vibration absorber. It can be shown that classical results recorded in the literature [12.9, 12.13] are simply special cases of the equations provided in Eqs. 12.3-11 and 12.3-15.

To see this, consider first that the baseline FRF of the single-degree-of-freedom system is simply

$$H_{11}^{x} = \frac{1}{K - M\omega^2 + j\omega C} \tag{12.3-16}$$

By substituting from Eq. 12.3-16 into Eq. 12.3-10, and by introducing the new parameters as

$$\beta = \frac{\omega}{\sqrt{\frac{K}{M}}} = \frac{\omega}{\omega_o}$$

$$\alpha=\frac{\omega_d}{\omega_o}=\frac{\beta}{\beta_d}$$

$$\varsigma_o=\frac{C}{2\sqrt{KM}}$$

$$\mu=\frac{m_d}{M}$$

and

$$\varsigma_d=\frac{c_d}{2\sqrt{k_d m_d}},$$

we see that the amplitude of the FRF of the combined system, without regard to phase, can be written as

$$\left|\frac{x_1}{f_1}\right|=\frac{1}{K}\sqrt{\frac{A^2+B^2}{C^2+D^2}}, \tag{12.3-17}$$

where

$$A=\alpha^2-\beta^2$$

$$B=2\varsigma_d\alpha\beta$$

$$C=(\alpha^2-\beta^2)(1-\beta^2)-\mu\alpha^2\beta^2-4\alpha\beta^2\varsigma_d\varsigma_o$$

and

$$D=2\varsigma_d\alpha\beta(1-\alpha^2-\mu\beta)+2\varsigma_o\beta(\alpha^2-\beta^2)$$

If the mass ratio μ is zero (no vibration absorber), then Eq. 12.3-17 reduces to the FRF of the original single-degree-of-freedom system. That is,

$$\left|\frac{x_1}{f_1}\right|=\frac{1}{K}\sqrt{\frac{1}{(1-\beta^2)^2+(2\varsigma_o\beta)^2}} \tag{12.3-18}$$

If the primary system has no damping, that is, ς_o is zero, then Eq. 12.3-17 becomes

$$\left|\frac{x_1}{f_1}\right|=\frac{1}{K}\left\{\frac{(\alpha^2-\beta^2)^2+(2\varsigma_d\alpha\beta)^2}{[(\alpha^2-\beta^2)(1-\beta^2)-\mu\alpha^2\beta^2]^2+(2\varsigma_d\alpha\beta)^2(1-\beta^2-\mu\beta^2)^2}\right\}^{\frac{1}{2}} \tag{12.3-19}$$

Eq. 12.19 is exactly the same as that recorded in [12.9].

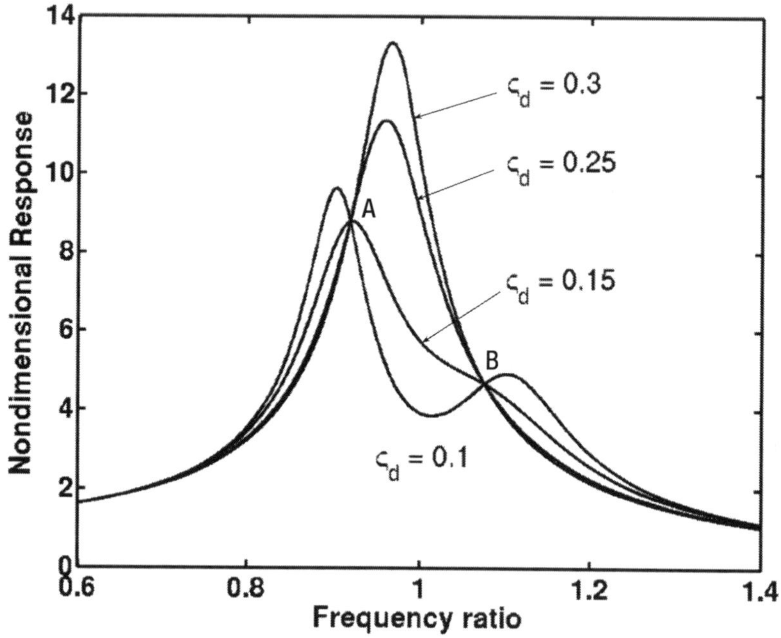

Fig. 12.3-3 Magnification factor for $\mu = 0.05$, $\alpha = 1$.

Frequency Tuning. From Eq. 12.3-19, the magnification factor can be written as

$$\left|\frac{x_1}{f_1/K}\right| = \left\{\frac{(\alpha^2-\beta^2)^2+(2\varsigma_d\alpha\beta)^2}{[(\alpha^2-\beta^2)(1-\beta^2)-\mu\alpha^2\beta^2]^2+(2\varsigma_d\alpha\beta)^2(1-\beta^2-\mu\beta^2)^2}\right\}^{\frac{1}{2}} \tag{12.3-20}$$

Fig. 12.3-3 shows four curves with different damping ratios. Note that all curves pass through two points A and B on the graph, independent of the damping ratio, ς_d, of the damper. These points are called fixed points. Their locations are obtained by studying the structure of the magnification factor in Eq. 12.3-20. If the ratio of the coefficient of ς_d^2 to the term independent of ς_d is the same in both the numerator and the denominator of Eq. 12.3-20, then the amplitude is independent of the value of ς_d. This leads to [12.9]

$$\frac{(2\alpha\beta)^2}{(\alpha^2-\beta^2)^2} = \frac{(2\alpha\beta)^2(1-\beta^2-\beta^2\mu)^2}{[(\alpha^2-\beta^2)(1-\beta^2)-\alpha^2\beta^2\mu]^2} \tag{12.3-21}$$

or

$$\frac{2\alpha\beta}{\alpha^2-\beta^2} = \pm\frac{(2\alpha\beta)(1-\beta^2-\beta^2\mu)}{(\alpha^2-\beta^2)(1-\beta^2)-\alpha^2\beta^2\mu} \tag{12.3-22}$$

Thus, we have

$$2\alpha\beta = 0 \tag{12.3-23a}$$

$$\frac{1}{\alpha^2-\beta^2} = +\frac{(1-\beta^2-\beta^2\mu)}{(\alpha^2-\beta^2)(1-\beta^2)-\alpha^2\beta^2\mu} \tag{12.3-23b}$$

$$\frac{1}{\alpha^2-\beta^2} = -\frac{(1-\beta^2-\beta^2\mu)}{(\alpha^2-\beta^2)(1-\beta^2)-\alpha^2\beta^2\mu} \tag{12.3-23c}$$

The first two solutions are trivial ($\beta = 0$). The third yields the equation

$$\left(1+\mu/2\right)\beta^4 - \left(1+\alpha^2+\mu\alpha^2\right)\beta^2 + \alpha^2 = 0 \tag{12.3-24}$$

Let the two roots of this equation be β_1^2 and β_2^2, corresponding to each fixed point in Fig. 12.3-3. In an equation having unity for the coefficient of its highest power, the sum of the roots is equal to the coefficient of the second term with the sign changed. Hence, we have

$$\beta_1^2 + \beta_2^2 = \frac{1+\alpha^2+\mu\alpha^2}{1+\mu/2} \tag{12.3-25}$$

Next, the amplitude of motion at each fixed point may be obtained by substituting each value of β_1 and β_2. Since the amplitude is independent of the damping ratio ς_d, we can let $\varsigma_d = \infty$ in Eq. 12.3-20 for a simple calculation. Thus, we obtain

$$\left|\frac{x_1}{f_1/K}\right| = \sqrt{\frac{1}{(1-\beta^2-\mu\beta^2)^2}} = \pm\frac{1}{1-\beta^2-\mu\beta^2}, \tag{12.3-26}$$

where the plus sign applies to β_1 below the resonance frequency ($\beta = 1$), and the minus sign applies to β_2 above the resonance frequency. The optimal damping parameters are determined by specifying equal amplitudes of the magnification factor at the fixed points. This leads to

$$\frac{1}{1-\beta_1^2-\beta_1^2\mu} = -\frac{1}{1-\beta_2^2-\beta_2^2\mu}, \tag{12.3-27}$$

from which we have

$$\beta_1^2 + \beta_2^2 = \frac{2}{1+\mu} \tag{12.3-28}$$

Eq. 12.3-28 provides the condition that the fixed points have the same amplitude of motion. From Eqs. 12.3-25 and 12.3-28, the optimum tuning is obtained [12.9, 12.10]:

$$\alpha_{\text{opt}} = \frac{1}{1+\mu}, \tag{12.3-29}$$

where $\alpha = \omega_d / \omega_o$. Thus, the tuning frequency of the vibration absorber is given by

$$\omega_d = \frac{\omega_o}{1+\mu} \tag{12.3-30}$$

Optimal Damping Ratio. If the effect of the damping is considered, it is possible to select a value of the damping parameter ς_d that will make the fixed points nearly equal to the points of greatest amplitude of the motion. First, we find the values of β at fixed points: This is accomplished by solving Eq. 12.3-24 using the value of α as determined by Eq. 12.3-29. This yields

$$\beta^4 - \frac{2\beta^2}{1+\mu} + \frac{2}{(2+\mu)(1+\mu)^2} = 0 \tag{12.3-31}$$

Solving for β to obtain the abscissas at the fixed points leads to

$$\beta_1^2 = \frac{1}{1+\mu}\left(1 - \sqrt{\frac{\mu}{2+\mu}}\right), \quad \beta_2^2 = \frac{1}{1+\mu}\left(1 + \sqrt{\frac{\mu}{2+\mu}}\right) \tag{12.3-32}$$

The value of the magnification factor at the fixed points is found by substituting from Eq. 12.3-32 into Eq. 12.3-26:

$$\left|\frac{x_1}{f_1/K}\right|_{\text{fixed points}} = \sqrt{1+\frac{2}{\mu}} \tag{12.3-33}$$

A simple and direct way to determine a suitable value of the damping ratio ς_d is to select a frequency located centrally in the interval between the fixed points, and then select ς_d such that the amplitude of the motion is equal at these three frequencies. A convenient value for β^2 lying between the two fixed points A and B is defined by the arithmetic mean of β_1^2 and β_2^2. From Eq. 12.3-28, this leads to

$$\beta_L^2 = \frac{1}{2}\left(\beta_1^2 + \beta_2^2\right) = \frac{1}{1+\mu} \tag{12.3-34}$$

The frequency corresponding to the frequency ratio β_L of Eq. 12.3-34 is the natural frequency of the composite system when the damping ratio is infinite. It is called the *locked frequency*. Substituting α from Eq. 12.3-29 and β from Eq. 12.3-34 into Eq. 12.3-20, we obtain

$$\left|\frac{x_0}{f_1/K}\right|^2 = \frac{\mu^2 + 4\varsigma_d^2(1+\mu)}{\mu^2} \tag{12.3-35}$$

By equating the amplitude of the magnification factor at β_L to that at β_1 and β_2 given by Eq. 12.3-33, the optimal damping ratio is determined as [12.9]

$$\varsigma_{\text{opt}} = \sqrt{\frac{\mu}{2(1+\mu)}} \tag{12.3-36}$$

It is possible to find the value of ς_d that gives maximum magnification factor at the fixed point *A*, and also to find the value of ς_d that gives maximum magnification factor at the point *B*. The average of the two values thus obtained indicates optimal damping ratio [12.11]

$$\varsigma_{opt} = \sqrt{\frac{3\mu}{8(1+\mu)^3}} \tag{12.3-37}$$

Fig. 12.3-4 illustrates the magnification factor for mass ratio $\mu = 0.05$ for three different values of damping parameters. In the figure, curve 1 corresponds to the value of damping given by Eq. 12.3-36, while curve 2 corresponds to the value of damping given by Eq. 12.3-37, and curve 3 corresponds to the value of damping ratio $\varsigma_d = 0.25$.

Next, for constant tuning ($\alpha = \omega_d / \omega_o = 1$), it can be shown that the optimal damping ratio is given by [12.11]

$$\varsigma_{opt} = \frac{\mu(3+\mu)[1+\sqrt{\mu/(2+\mu)}]}{8(1+\mu)} \tag{12.3-38}$$

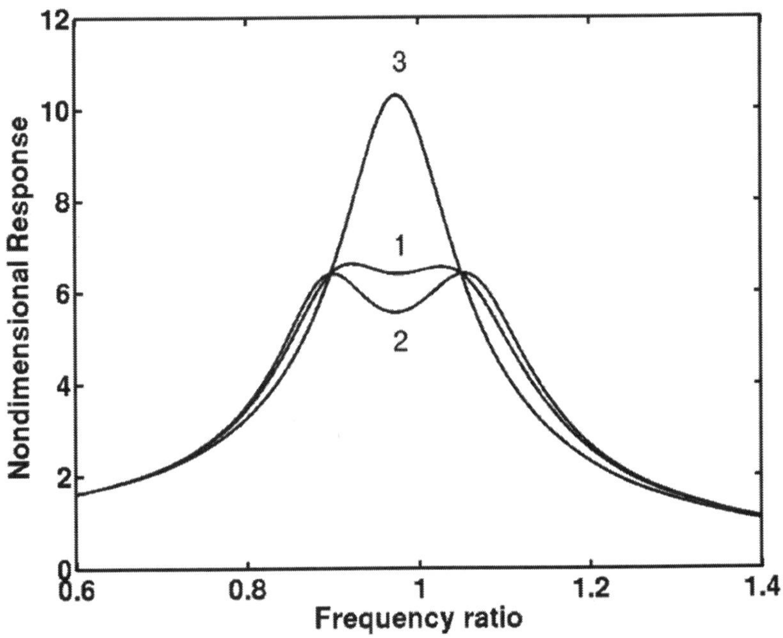

Fig. 12.3-4 Magnification factor for $\mu = 0.05$: Curve 1 for damping ratio given by Eq. 12.3-36, curve 2 for damping ratio given by Eq. 12.3-37, and curve 3 for damping ratio $\varsigma = 0.25$.

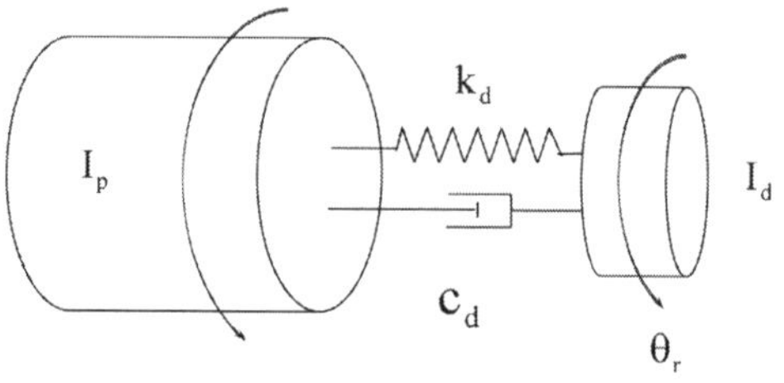

Fig. 12.4-1 A torsional vibration absorber attached to a vibrating system. (Reprinted from SAE paper 2009-01-2152, [12.14]).

12.4 Torsional Vibration Absorber

Fig. 12.4-1 provides a schematic representation of a torsional vibration absorber [12.14]. It consists of a rotor, with axial moment of inertia I_d, attached to a vibrating system, with axial moment of inertia I_p. The attachment consists of a torsional damper with damping coefficient c_d together with a linear torsional spring with stiffness k_d.

As with the translational vibration absorber, if the torsional vibration absorber is not properly tuned, the composite system (the original system and the attached absorber) may have a resonance frequency within the frequency range of an exciting moment. Therefore, it is necessary to evaluate the dynamic properties of the combined system.

As with the translational vibration absorber, two questions arise: First, what is the resulting driving-point FRF of the composite system at the attachment point? Second, what is the resulting cross-point FRF of the composite system between the attachment point and other points of interest?

To answer these questions, we can conduct an analysis similar to that of the previous section. That is, we can develop expressions analogous to Eqs. 12.3-11 and 12.3-15.

To this end, let I_e be the equivalent axial moment of inertia of the absorber defined as

$$I_e = \frac{(k_d + j\omega c_d)I_d}{k_d - \omega^2 I_d + j\omega c_d} \qquad (12.4\text{-}1)$$

Then the resulting driving-point FRF, $\hat{H}_{11}^{\alpha}$ (angular acceleration divided by moment), of the combined system is given by

$$\hat{H}_{11}^{\alpha} = \frac{H_{11}^{\alpha}}{1 + I_e H_{11}^{\alpha}}, \qquad (12.4\text{-}2)$$

where H_{11}^{α} is the baseline driving-point FRF. The resulting cross-point FRF, $\hat{H}_{12}^{\alpha}$ (angular acceleration divided by moment), of the combined system between points 1 and 2 is given by

$$\hat{H}_{12}^{\alpha} = \frac{H_{12}^{\alpha}}{1 + I_e H_{11}^{\alpha}}, \qquad (12.4\text{-}3)$$

where H_{12}^{α} is the baseline cross-point FRF.

The derivation of Eqs. 12.4-2 and 12.4-3 is left to the reader as an exercise.

12.5 Experimental Measurement of the Torsional FRF

In tuning a torsional vibration absorber, we need to know the torsional FRF of the subject system—that is, of the torsionally vibrating system we are attempting to control. In this section, we describe a method for experimentally measuring the torsional FRF—the so-called *roving hammer test.*

Fig. 12.5-1 provides a schematic representation for the procedure [12.14]. Two small, relatively light blocks (aluminum) are mounted on the subject system at points 1 and 2. Accelerometers are then attached to the blocks. In the ensuing analysis, we can reasonably neglect the masses of the small blocks since they are relatively light.

Next, we strike the block at point 1, as indicated in Fig. 12.5-1, and the accelerances are measured at points 1 and 2. Let a_1/F_1 and a_2/F_1 be the driving-point FRF at point 1 and the cross-point FRF between point 1 and 2, respectively. Let D (= $2R$) be the distance between the accelerometers. Then the torsional FRF (receptance) can be expressed as [12.14]

$$\frac{\theta}{M}=\frac{180}{\pi}\frac{1}{R}\tan^{-1}\left[\frac{(a_1/F_1)-(a_2/F_1)}{-(2\pi f)^2 D}\right]\approx-\frac{180}{\pi}\left[\frac{(a_1/F_1)-(a_2/F_1)}{2\pi^2 D^2 f^2}\right], \quad (12.5\text{-}1)$$

where the torsional angle θ is expressed in degrees; M is the applied moment; and, as before, f is the frequency. Thus, we have the torsional accelerance [12.14]

$$\frac{\ddot{\theta}}{M}\approx\frac{360}{\pi}\left[\frac{(a_1/F_1)-(a_2/F_1)}{D^2}\right] \quad (12.5\text{-}2)$$

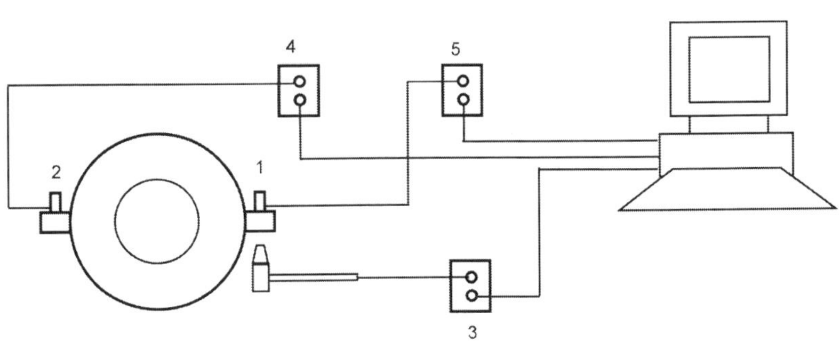

Fig. 12.5-1 Experiment setup for the measurement of torsional FRF. 1 and 2: accelerometers; 3: ICP sensor power unit; 4 and 5: charge amplifiers. (Reprinted from SAE paper 2009-01-2152, [12.14].)

12.6 HYFEX Method

Based upon the pioneering work of Klosterman and Lemon [12.15], many researchers have attempted to use experimentally measured FRFs to represent portions of a total system dynamic analysis. They then use finite element analysis for the remaining portions of the dynamic analysis. The overall procedure is called the Hybrid FEA-Experimental FRF method, or, simply, the *HYFEX* method [12.16].

The HYFEX method allows substructures of a dynamic system described by FRF matrices to be coupled via a FEA model. The coupled model can then be studied using modal frequency response analysis.

If, however, in, say, an auto body analysis, a chassis model does not include bushing models, we cannot directly connect the FRFs of the auto body to the chassis model for performing a total system dynamic analysis. In other cases, where the chassis model includes the bushings, and the bushing dynamic rates are modeled as constant stiffnesses rather than frequency-dependent stiffnesses, the direct use of the HYFEX will yield unsatisfactory results. In this latter regard, it is usually advantageous to represent the bushings by experimentally obtained FRFs instead of detailed finite element models. This is due to the time currently needed to build the finite element model.

In this section, we discuss how the FRFs of an auto body and the frequency-dependent stiffness data of the bushings can be combined through a mathematical formulation to better represent the dynamic characteristics of the automobile.

Consider Fig. 12.6-1 where A and B represent substructures of a vibrating system. For example, we can think of A and B as a chassis and an auto body, respectively. Let A and B be connected rigidly.

For substructure A, we express the equation of motion in the form

$$\begin{bmatrix} Z_{rr}^{A} & Z_{rc}^{A} \\ Z_{cr}^{A} & Z_{cc}^{A} \end{bmatrix} \begin{Bmatrix} X_{r}^{A} \\ X_{c}^{A} \end{Bmatrix} = \begin{Bmatrix} F_{r}^{A} \\ F_{c}^{A} \end{Bmatrix}, \tag{12.6-1}$$

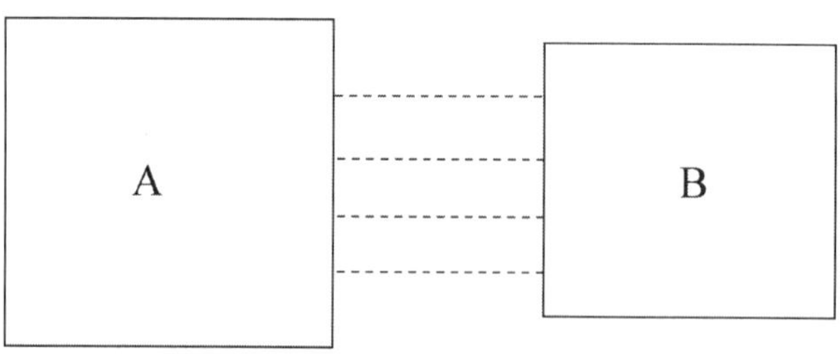

Fig. 12.6-1 Substructures *A* and *B* rigidly connected.

where *F*s are force components, the *X*s are displacement components, and the *Z*s are impedance matrices. The subscript *c* represents the coupling points and the subscript *r* represents the other response points.

For the coupling points of substructure *B*, we have

$$\begin{bmatrix} 0 & 0 \\ 0 & Z_{cc}^B \end{bmatrix} \begin{Bmatrix} X_r^A \\ X_c^B \end{Bmatrix} = \begin{Bmatrix} 0 \\ F_c^B \end{Bmatrix} \tag{12.6-2}$$

The boundary conditions can be written as

$$\{X_c^A\} = \{X_c^B\}, \ \{F_c^A\} + \{F_c^B\} = \{0\} \tag{12.6-3}$$

Hence, from Eqs. 12.6-1 and 12.6-2, we have

$$\begin{bmatrix} Z_{rr}^A & Z_{rc}^A \\ Z_{cr}^A & Z_{cc}^A + Z_{cc}^B \end{bmatrix} \begin{Bmatrix} X_r^A \\ X_c^A \end{Bmatrix} = \begin{Bmatrix} F_r^A \\ 0 \end{Bmatrix} \tag{12.6-4}$$

Eq. 12.6-4 provides a basis for the HYFEX method. Specifically, the FRFs of the substructure *B* at the coupling points can be experimentally determined. The inversion of the FRFs is the impedance matrix Z_{cc}^B. Then the response of the total system can be determined by solving Eq. 12.6-4.

Next, consider the structure represented in Fig. 12.6-2, where the substructures *A* and *B* are connected by several dampers (in the figure, only one damper is shown). In this case, Eq. 12.6-4 is not applicable. But an equation for determining the response of the total system can be derived as follows: For substructure *A*, Eq. 12.6-1 is still valid. For the dampers *D*, we have

$$\{X_c^A\} - \{X_c^B\} = [H^D]\{F_c^B\}$$

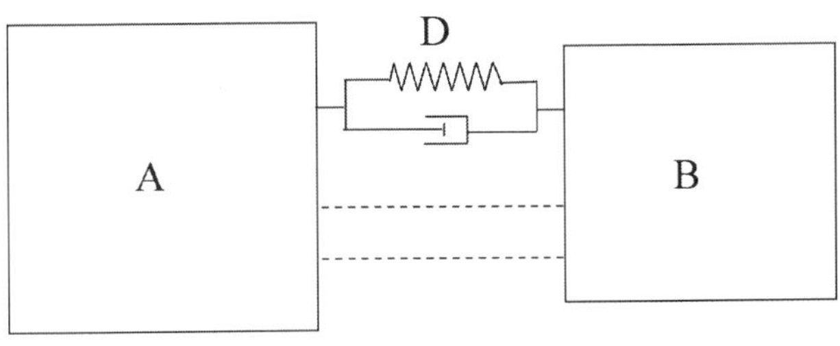

Fig. 12.6-2 Substructures *A* and *B* connected by dampers.

or

$$\{X_c^A\}=\{X_c^B\}+[H^D]\{F_c^B\}$$

or

$$\{X_c^A\}=[H_{cc}^B+H^D]\{F_c^B\} \tag{12.6-5}$$

Hence, we have

$$[H_{cc}^B+H^D]^{-1}\{X_c^A\}=\{F_c^B\}=-\{F_c^A\} \tag{12.6-6}$$

Combining Eqs. 12.6-1 and 12.6-6, we obtain

$$\begin{bmatrix} Z_{rr}^A & Z_{rc}^A \\ Z_{cr}^A & Z_{cc}^A+[H_{cc}^B+H^D]^{-1} \end{bmatrix}\begin{Bmatrix} X_r^A \\ X_c^A \end{Bmatrix}=\begin{Bmatrix} F_r^A \\ 0 \end{Bmatrix} \tag{12.6-7}$$

Eq. 12.6-7 forms a basis for the dynamic analysis of the total system. Specifically, the procedure can be outlined as follows:

1) Determine the stiffness matrix of the dampers (force over displacement, or *F*/*X*) experimentally.
2) Invert the stiffness matrix of the dampers to obtain the frequency response functions, H^D.
3) Add the FRFs of the body and the dampers together to obtain $H_{cc}^B+H^D$.
4) Invert the combined FRFs matrix $H_{cc}^B+H^D$.
5) Perform the actual coupled analysis based on Eq. 12.6-7.

Once the solutions of Eq. 12.6-7 are obtained, the interface force components at the coupling points can be determined from Eq. 12.6-6. That is,

$$\{F_c^B\}=[H_{cc}^B+H^D]^{-1}\{X_c^A\} \tag{12.6-8}$$

These interface force components can be used to recover the responses at the auxiliary (structure and acoustic) response points with the following equation:

$$\{X_r^B\}=[H_{cr}^B]\{F_c^B\}, \tag{12.6-9}$$

where the H_{cr}^{B} are the transfer function matrices between the auxiliary response points and the coupling points within the subsystem B.

Example 12.6-1 To illustrate the HYFEX method, consider the simple system shown in Fig. 12.6-3 (a). The system consists of two masses connected by springs and viscous dampers. Fig. 12.6-3 (b) shows the system divided into two subsystems *A* and *B* together with a connecting spring and damper *D*.

Let the masses, stiffnesses, and damping coefficients have values:

m_1=4 kg, m_2=2 kg, k_1=4 N/m, k_2=k_3=2 N/m, c_1=2 Ns/m, c_2=c_3=1 Ns/m

The FRFs of *D* and of subsystem *B* can be evaluated separately. That is,

$$H^D = \frac{1}{k_2 + j\omega c_2} = \frac{1}{2 + j\omega} \tag{a}$$

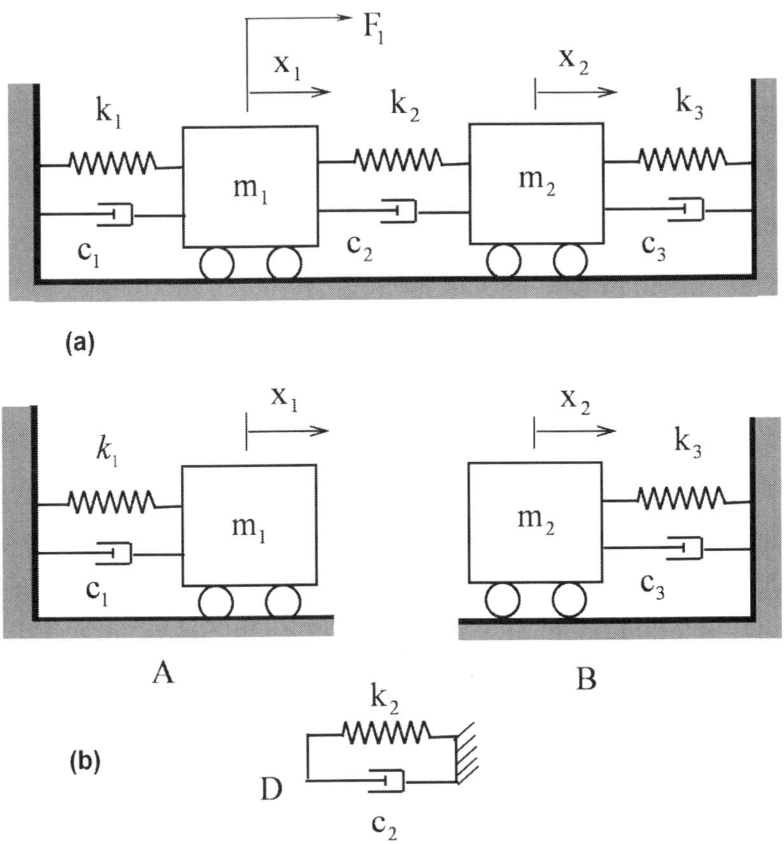

Fig. 12.6-3 A two-degree-of-freedom system and its division into subsystems.

$$H_{cc}^{B} = \frac{1}{k_3 - m_2\omega^2 + j\omega c_3} = \frac{1}{2 - 2\omega^2 + j\omega} \tag{b}$$

Next, the impedance of the subsystem A is

$$Z_{cc}^{A} = k_1 - m_1\omega^2 + j\omega c_1 = 4 - 4\omega^2 + j2\omega \tag{c}$$

Eq. 12.6-7 then becomes

$$\left[Z_{cc}^{A} + \left(H_{cc}^{B} + H^{D}\right)^{-1}\right]X_1 = F_1 \tag{d}$$

The FRF of subsystem A at the connecting point is given by

$$\frac{X_1}{F_1} = \frac{1}{\left[Z_{cc}^{A} + \left(H_{cc}^{B} + H^{D}\right)^{-1}\right]} \tag{e}$$

Substituting from Eqs. a to c into Eq. e, we obtain

$$h_{11} = \frac{X_1}{F_1} = \frac{4 - 2\omega^2 + j2\omega}{\Delta}, \tag{f}$$

where Δ is

$$\Delta = 8\omega^4 - j14\omega^3 - j33\omega^2 + j20\omega + 20 \tag{g}$$

This result can be verified by direct calculation as follows: The mass, stiffness, and damping matrix of the system are given by

$$M = \begin{bmatrix} 4 & 0 \\ 0 & 2 \end{bmatrix}, \quad K = \begin{bmatrix} 6 & -2 \\ -2 & 4 \end{bmatrix}, \quad C = \begin{bmatrix} 3 & -1 \\ -1 & 2 \end{bmatrix}$$

The dynamic matrix of the system is

$$B(\omega) = K - M\omega^2 + j\omega C = \begin{bmatrix} 6 - 4\omega^2 + j3\omega & -2 - j\omega \\ -2 - j\omega & 4 - 2\omega^2 + j2\omega \end{bmatrix} \tag{h}$$

The FRF matrix of the system is then

$$H = \begin{bmatrix} h_{11} & h_{12} \\ h_{21} & h_{22} \end{bmatrix} = \frac{adj(B)}{\left|K - M\omega^2 + j\omega C\right|} \tag{i}$$

From Eq. i, we obtain

$$h_{11} = \frac{X_1}{F_1} = \frac{4 - 2\omega^2 + j2\omega}{\Delta}, \tag{j}$$

where Δ is as defined in Eq. g.

As expected, it follows from Eqs. f and j that the results from the two methods are identical.

References [12.16] and [12.17] document the successful application of the HYFEX method in studying vehicle axle noise. Fig. 12.6-4 shows a flow chart of the HYFEX method.

In the analysis of Liu et al. (see [12.16, 12.17]), the FRFs (a/F and p/F) of the body at the connecting points with suspension were experimentally measured by using a test setup, as represented in Fig. 12.6-5.

The bushing stiffnesses were measured using a servo-hydraulic rate test machine. The frequency response functions (a/F) of the bushings can be obtained by inverting the stiffness and then differentiating the results twice.

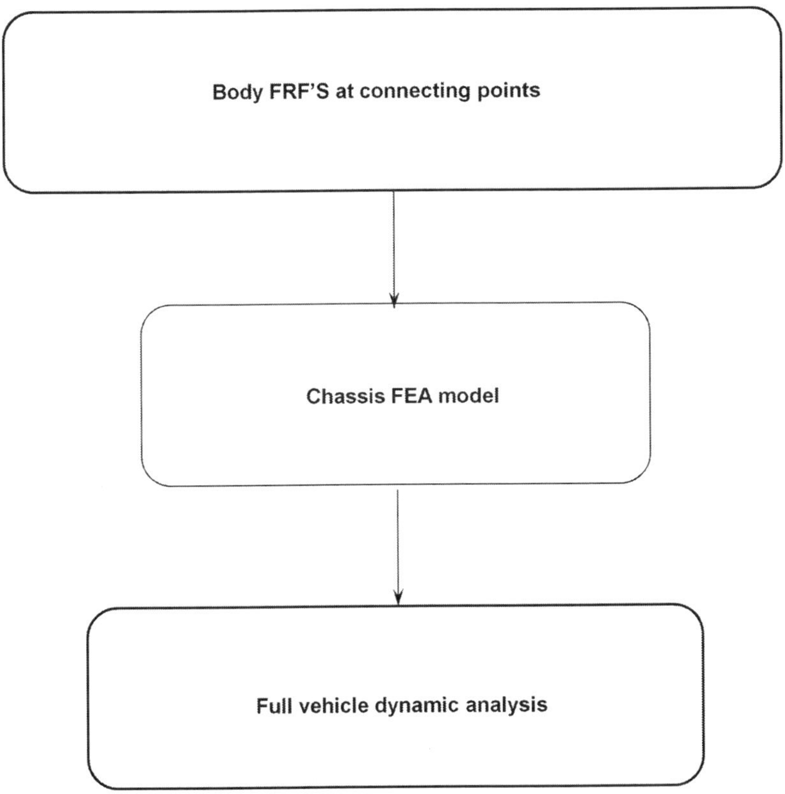

Fig. 12.6-4 Application of HYFEX method to vehicle dynamic analysis.

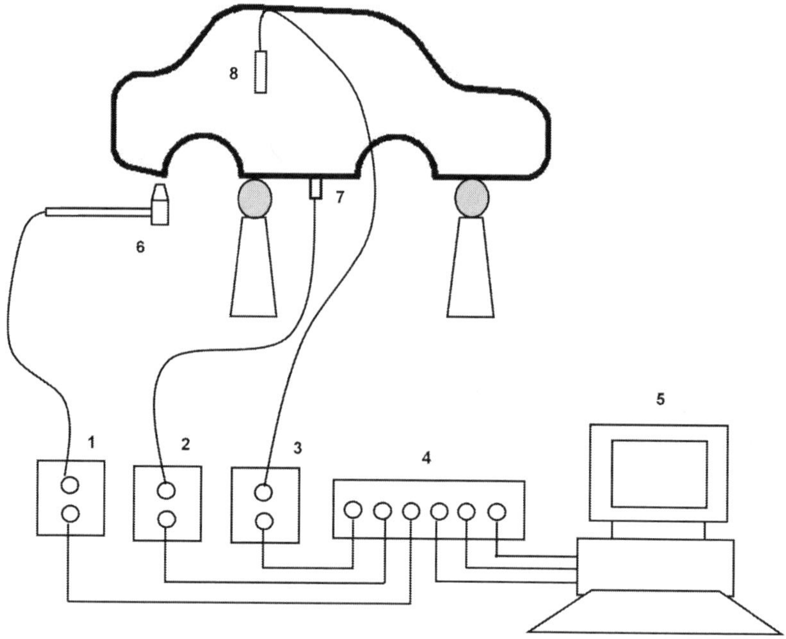

Fig. 12.6-5 Experimental impact test setup for measuring auto body FRFs. 1, 2, 3: signal conditioning; 4: antialiasing filter; 5: FFT analyzer; 6: impact hammer; 7: accelerometer; 8: microphone.

The calculation steps are as follows:

1) Use Eq. 12.6-7 to calculate the responses, $\begin{bmatrix} X_r^A & X_c^A \end{bmatrix}^{-1}$, of the connecting points of the chassis.

2) Use Eq. 12.6-8 to calculate the coupling forces, $\left\{F_c^B\right\}$, acting on the connecting points of the vehicle body.

3) Use Eq. 12.6-9 to calculate the responses, $\left\{X_r^B\right\}=\left[H_{cr}^B\right]\left\{F_c^B\right\}$, at the auxiliary response points, where $\left[H_{cr}^B\right]$ is the transfer function between the auxiliary response points and the connecting points within subsystem *B*. For example, the sound pressure level at the driver's right ear position can be calculated by the transfer function *p*/*F* multiplying the coupling force.

From the above calculations, the partial contribution of each connecting point to the sound pressure level at the driver's ear position can be obtained.

The direction lines to each connecting point are called *paths*. Of greatest interest are those paths for the dominant axle noise. To find these dominant paths, it is helpful to inspect a three-dimensional display of path sound pressure level as a function of source position and of frequency. The major noise paths are ready identified (see [12.16] for more details).

12.7 Sensitivity Analysis

Sensitivity analysis is the study of system response changes as a result of changes in physical parameters of the system. Recent work in sensitivity analysis has been intense, and it is continuing. Interest in sensitivity analysis dates back to the work of Fox and Kapoor [12.18] in 1968. It spans engineering disciplines ranging from automatic control theory to the study of large physical systems. Haltka and Adelman [12.19] provide a survey of developments in sensitivity analysis up through 1986.

To determine the effect of a design change on a system's dynamic performance, it is helpful to know the way eigenvectors and eigenvalues change with respect to the design variables. To this end, analysts have developed a number of methods for computing eigenvector and eigenvalue derivatives. Sutter et al. [12.20] provide a comparison of the numerical performance of these methods.

Methods for evaluating eigenvector and eigenvalue derivatives include finite-difference method, modal method [12.18], a modified modal method [12.20–12.23], and Nelson's method [12.24].

The finite-difference method approximates the derivatives using a difference formula. This requires knowing the eigenvector and eigenvalue at a nominal point and at least one perturbed design point. The finite-difference method is sensitive to round-off and truncation errors associated with the step size.

The modal method approximates the eigenvector derivatives as a linear combination of the mode shapes. This can be computationally expensive for large systems with many eigenvectors. Indeed, determining all the eigenvectors of a large system is a daunting task.

As a result of this difficulty with the modal method, analysts were led to develop a modified modal method requiring fewer modes to determine the derivatives. This is accomplished by including an additional term in the linear combination of the eigenvectors.

In contrast to these methods, Nelson's method is an exact procedure. Nelson's method expresses the eigenvector derivative in terms of a particular solution and a complementary solution of the system governing equations. Finding a particular solution, however, typically requires system matrix factorization. As the system size becomes large, the cost of system matrix factorization proportionally increases.

Liu et al. [12.25] developed a method requiring only the eigenvector and associated eigenvalue for the eigenvector derivative being sought. In Liu's method, using QR decomposition, a condensed coefficient matrix is automatically obtained. This matrix is upper triangular so that the determination of the eigenvector derivative can be efficiently computed by back-substitution with minor extra computational cost.

(1) Eigenvalue Derivatives

Consider the basic second-order ordinary differential equation for vibration of a mechanical system:

$$M\ddot{\mathbf{x}}+C\dot{\mathbf{x}}+K\mathbf{x}=\mathbf{f}(\mathrm{t}), \tag{12.7-1}$$

where M, C, and K are the system's mass, damping, and stiffness matrices of order $N\times N$. By using the Laplace transform, we obtain the algebraic expressions

$$B(s)\mathbf{X}=\mathbf{F}(s), \quad B(s)=Ms^2+Cs+K \tag{12.7-2}$$

or

$$H(s)\mathbf{F}=\mathbf{X}, \quad H(s)=[Ms^2+Cs+K]^{-1}, \tag{12.7-3}$$

where, as before, $\mathbf{X}$ and $\mathbf{F}$ are the Laplace transforms of $\mathbf{x}$ and $\mathbf{f}$. Thus, we have

$$HB=I, \tag{12.7-4}$$

where I is the $N\times N$ unit matrix.

By taking the partial derivative in Eq. 12.7-4 with respect to a design parameter p, we have

$$\frac{\partial H}{\partial p}B=-H\frac{\partial B}{\partial p} \tag{12.7-5}$$

or

$$\frac{\partial H}{\partial p}=-H\frac{\partial B}{\partial p}H \tag{12.7-6}$$

The transfer matrix $H(s)$ in Eq. 12.7-6 can be expressed as a summation of residue matrices in partial fractions as (see Eq. 7.6-17)

$$H=\sum_{r=1}^{2N}\frac{A_r}{s-\lambda_r}, \tag{12.7-7}$$

where λ_r $(r=1,2,\cdots 2N)$ are the system poles. Taking the partial derivative in Eq. 12.7-7 with respect to the design parameter p leads to

$$\frac{\partial H}{\partial p}=\sum_{r=1}^{2N}\frac{\frac{\partial A_r}{\partial p}(s-\lambda_r)+A_r\frac{\partial \lambda_r}{\partial p}}{(s-\lambda_r)^2} \tag{12.7-8}$$

Substituting from Eqs. 12.7-7 and 12.7-8 into Eq. 12.7-6, we obtain

$$\sum_{r=1}^{2N} \frac{\frac{\partial A_r}{\partial p}(s-\lambda_r)+A_r\frac{\partial \lambda_r}{\partial p}}{(s-\lambda_r)^2} = -\left[\sum_{r=1}^{2N}\frac{A_r}{s-\lambda_r}\right]\frac{\partial B}{\partial p}\left[\sum_{r=1}^{2N}\frac{A_r}{s-\lambda_r}\right] \tag{12.7-9}$$

Multiplying both sides of Eq. 12.7-9 by $(s-\lambda_r)^2$, and then letting $s=\lambda_r$, we obtain

$$A_r\frac{\partial \lambda_r}{\partial p} = -A_r\frac{\partial B}{\partial p}A_r \tag{12.7-10}$$

The residue matrix A_r can be expressed as (see Eq. 7.6-24)

$$A_r = Q_r\mathbf{u}_r\mathbf{u}_r^T \quad (r=1,2,\cdots,2N), \tag{12.7-11}$$

where Q_r is a scaling factor, and $\mathbf{u}_r$ is the eigenvector associated with the eigenvalue λ_r.

Substituting from Eq. 12.7-11 into Eq. 12.7-10 and using the orthogonality property of the eigenvectors, the derivative of the eigenvalue becomes

$$\frac{\partial \lambda_r}{\partial p} = -Q_r\mathbf{u}_r^T\left[\left.\frac{\partial B(s)}{\partial p}\right|_{s=\lambda_r}\right]\mathbf{u}_r, \tag{12.7-12}$$

where

$$\left.\frac{\partial B(s)}{\partial p}\right|_{s=\lambda_r} = \lambda_r^2\frac{\partial M}{\partial p}+\lambda_r\frac{\partial C}{\partial p}+\frac{\partial K}{\partial p} \tag{12.7-13}$$

(2) Eigenvector Derivatives

Let $\mathbf{u}_r$ be an eigenvector of the system for mode r associated with the eigenvalue λ_r. Then we have

$$B(\lambda_r)\mathbf{u}_r = 0 \quad (r=1,2,\cdots,N), \tag{12.7-14}$$

where

$$B(\lambda_r) = \lambda_r^2 M + \lambda_r C + K \tag{12.7-15}$$

Differentiating in Eq. 12.7-14 with respect to the design parameter p, we have

$$B(\lambda_r)\frac{\partial \mathbf{u}_r}{\partial p} = -\frac{\partial B(\lambda_r)}{\partial p}\mathbf{u}_r \quad (r=1,2,\cdots,N) \tag{12.7-16}$$

It should be noted that the matrix $B(\lambda_r)$ on the left side of Eq. 12.7-16 has rank $(N-1)$. This means that the equations cannot be solved uniquely for the eigenvector derivatives. That is, the equations are redundant and the determinant of $B(\lambda_r)$ is zero.

There are several methods available to treat systems with a singular (zero determinant) multiplying matrix: The modal method approximates the eigenvector derivative as a linear combination of eigenvectors. This method can be computationally expensive—especially if a large number of modes are needed to accurately represent the mode shape derivative.

The modified modal method was developed to reduce the number of modes needed to represent the derivative by including an additional term in the linear combination of the system mode shapes.

Finally, Nelson [12.24] proposed an exact analytical method, and Liu et al. [12.25] provided two exact analytical methods for determining eigenvector derivatives. We review these two methods in the following paragraphs.

Formulation (1)

First, we find the QR decomposition of matrix $B(\lambda_r)$as

$$B(\lambda_r) = Q\begin{bmatrix} R \\ 0 \end{bmatrix}, \tag{12.7-17}$$

where Q is an $N \times N$ orthogonal matrix, R is an $(N-1)\times N$ upper triangle matrix, and 0 is $1\times N$ row matrix. Let U_1 be the matrix composed of the first $N-1$ columns of Q, and let U_2 be the matrix composed of the remaining columns of Q. Then, from Eq. 12.7-17, we have

$$U_1^T B(\lambda_r) = R \tag{12.7-18}$$

and

$$U_2^T B(\lambda_r) = 0 \tag{12.7-19}$$

Eq. 12.7-18 shows that the matrix $U_1^T B$ is of rank $(N-1)$. Multiplying in Eq. 12.7-16 by U_1^T, we obtain $(N-1)$ independent equations

$$R\frac{\partial \mathbf{u}_r}{\partial p} = -U_1^T \frac{\partial \lambda_r}{\partial p}\mathbf{u}_r \tag{12.7-20}$$

To determine uniquely each element of the eigenvector derivative, it is necessary to consider the norm for the eigenvector in the form

$$\mathbf{u}_r^T M \mathbf{u}_r = 1, \tag{12.7-21}$$

where M is a symmetric nonsingular matrix. Taking the partial derivative in Eq. 12.7-21 with respect to the design parameter p, we have

$$2\mathbf{u}_r^T M \frac{\partial \mathbf{u}_r}{\partial p} + \mathbf{u}_r^T \frac{\partial M}{\partial p} \mathbf{u}_r = 0 \tag{12.7-22}$$

Eqs. 12.7-20 and 12.7-22 can be used to solve for the eigenvector derivative. It should be noted that the norm condition of Eq. 12.7-21 must be used in both Eqs. 12.7-20 and 12.7-22. The eigenvector derivative can be evaluated via

$$A \frac{\partial \mathbf{u}_r}{\partial p} = F_r, \tag{12.7-23}$$

where

$$A = \begin{bmatrix} R \\ 2\mathbf{u}_r^T M \end{bmatrix} \qquad \text{and} \qquad F_r = -\begin{bmatrix} U_1^T \dfrac{\partial B(\lambda_r)}{\partial p} \mathbf{u}_r \\ \mathbf{u}_r^T \dfrac{\partial M}{\partial p} \mathbf{u}_r \end{bmatrix} \tag{12.7-24}$$

It should be noted that $\dfrac{\partial B(\lambda_r)}{\partial p}$ may be expressed as

$$\frac{\partial B(\lambda_r)}{\partial p} = \lambda_r^2 \frac{\partial M}{\partial p} + \lambda_r \frac{\partial C}{\partial p} + \frac{\partial K}{\partial p} + \frac{\partial \lambda_r}{\partial p}(2\lambda_r M + C) \tag{12.7-25}$$

or, by using Eq. 12.7-12, as

$$\begin{aligned} \frac{\partial B(\lambda_r)}{\partial p} = & -Q_r \mathbf{u}_r^T \left[\lambda_r^2 \frac{\partial M}{\partial p} + \lambda_r \frac{\partial C}{\partial p} + \frac{\partial K}{\partial p} \right] + \mathbf{u}_r (2\lambda_r M + C) \\ & + \lambda_r^2 \frac{\partial M}{\partial p} + \lambda_r \frac{\partial C}{\partial p} + \frac{\partial K}{\partial p} \end{aligned} \tag{12.7-26}$$

It is helpful to note that matrix A in Eq. 12.7-23 has the form

$$A = \begin{bmatrix} * & * & * & * & . & . & . & * & * \\ & * & * & * & . & . & . & * & * \\ & & * & * & . & . & . & * & * \\ & & & * & . & . & . & * & * \\ & & & & . & . & . & . & . \\ & & & & & & & * & * \\ * & * & * & * & * & * & * & * & * \end{bmatrix} \tag{12.7-27}$$

Hence, its conversion to an upper triangle form needs to take place only for its last row. Therefore, determination of the eigenvector derivative requires only a back-substitution procedure.

Formulation (2)

The solution to Eq. 12.7-16 may be expressed in terms of a particular solution $\mathbf{V}_r$ and a complementary solution $c_r\mathbf{u}_r$, where c_r is an undetermined coefficient [12.24]. Thus, we have

$$\frac{\partial \mathbf{u}_r}{\partial p} = \mathbf{V}_r + c_r\mathbf{u}_r \tag{12.7-28}$$

Since $\mathbf{V}_r$ is a particular solution, we can, without loss of generality, let the last element of $\mathbf{V}_r$ be zero. Note also that $\mathbf{u}_r$ is an eigenvector; then $B(\lambda_r)\mathbf{u}_r = 0$. Hence, substitution from Eq. 12.7-28 into Eq. 12.7-16 gives

$$B(\lambda_r)\begin{Bmatrix} v_1 \\ v_2 \\ \cdot \\ v_{N-1} \\ 0 \end{Bmatrix} = -\frac{\partial B(\lambda_r)}{\partial p}\mathbf{u}_r \equiv b_r \tag{12.7-29}$$

Multiplying both sides of Eq. 12.7-29 by U_1^T, and in view of Eq. 12.7-18, we obtain

$$R\begin{Bmatrix} v_1 \\ v_2 \\ \cdot \\ v_{N-1} \\ 0 \end{Bmatrix} = U_1^T b_r \tag{12.7-30}$$

or equivalently

$$R_1\begin{Bmatrix} v_1 \\ v_2 \\ \cdot \\ v_{N-1} \end{Bmatrix} = U_1^T b_r, \tag{12.7-31}$$

where R_1 is the matrix composed of the first $(N-1)$ columns of matrix R.

It should be noted that matrix R_1 is a strict upper triangle matrix of the form

$$R_1 = \begin{bmatrix} * & * & * & * & . & . & * \\ & * & * & * & . & . & * \\ & & * & * & . & . & * \\ & & & * & . & . & . \\ & & & & & & * \end{bmatrix} \tag{12.7-32}$$

Thus, the first $(N-1)$ elements of $\mathbf{V}_r$ can be calculated efficiently by a back-substitution procedure. (Note: the last element of $\mathbf{V}_r$ is assigned to be zero).

To determine c_r, it is necessary to consider the norm in the form given by Eq. 12.7-21. Taking the partial derivative in Eq. 12.7-21 with respect to design parameter p, we have

$$2\mathbf{u}_r^T M \frac{\partial \mathbf{u}_r}{\partial p} + u_r^T \frac{\partial M}{\partial p} \mathbf{u}_r = 0 \tag{12.7-33}$$

Substituting from Eq. 12.7-28 into Eq. 12.7-33, we obtain

$$c_r = -\mathbf{u}_r^T M V_r - \frac{1}{2} \mathbf{u}_r^T \frac{\partial M}{\partial p} \mathbf{u}_r \tag{12.7-34}$$

Thus, the calculation for the eigenvector derivative is accomplished.

Example 12.7-1 Consider the idealized lumped parameter system shown in Fig. 12.7-1, where the mass, stiffness, and damping data are as follows:

m_1=0.9 kg, m_2=m_3=m_4=2 kg

k_1=k_2=k_3=k_4=k_5= 60 N/m

c_1=c_2=c_3=c_4=c_5= 0.2 Ns/m

(1) Find the eigenvalues and eigenvectors. (2) Plot FRFs to show that near the third resonance, the FRF shifts much faster with increasing k_4. (3) Find the derivatives of the eigenvalues and eigenvectors. (4) Let the stiffness k_4 be increased by a 0.5 increment ($\Delta k_4 = 0.5$); find the new eigenvectors and show that

$$\{\mathbf{u}_i\}_{\text{new}} \approx \{\mathbf{u}_i\}_o + \left\{\frac{\partial \mathbf{u}}{\partial k_4}\right\} \Delta k_4 \tag{a}$$

Solution: The eigenvalues and eigenvectors can be found by using the standard procedure as described in Chapter Six. The eigenvalue derivatives are calculated by Eq. 12.7-12 and listed in Table 12.7-1.

Table 12.7-1 shows that the magnitude of the derivative of λ_3 with respect to k_4 is the largest. This means that λ_3 is more sensitive to k_4 than any of

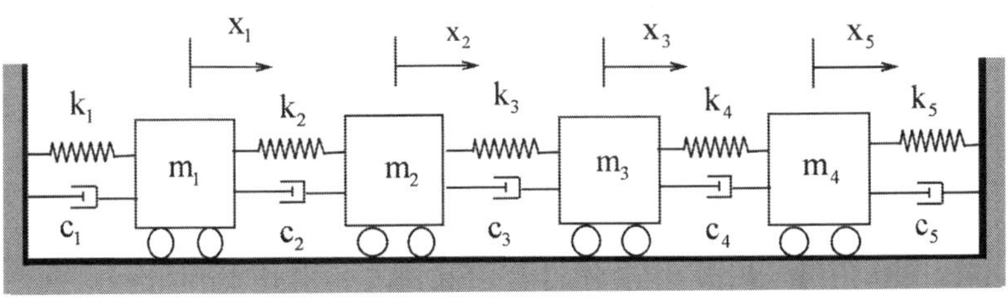

Fig. 12.7-1 Four-degree-of-freedom system.

Table 12.7-1 Eigenvalues and Their Derivatives with Respect to k_4

	λ_r	$\partial\lambda_r/\partial k_4$
1	$-0.0205 + j3.5049$	$j0.4019 \times 10^{-2}$
2	$-0.0817 + j6.9993$	$j0.6839 \times 10^{-2}$
3	$-0.1592 + j9.7721$	$j4.3752 \times 10^{-2}$
4	$-0.2609 + j12.5086$	$j0.0839 \times 10^{-2}$

the others. This fact is clearly demonstrated in Fig. 12.7-2: near the third resonance frequency, the FRF shifts much faster with increasing k_4. The eigenvector derivatives are calculated with Eq. 12.7-23, and the results are listed in Table 12.7-2. To verify the results, let the stiffness k_4 be increased by 0.5 units. The eigenvector of the *original* system and the *new* system are found as follows:

$$\{\mathbf{u}_1\}_o = \begin{Bmatrix} 0.2395 \\ 0.4379 \\ 0.4521 \\ 0.2842 \end{Bmatrix} \quad \{\mathbf{u}_2\}_o = \begin{Bmatrix} 0.3343 \\ 0.4229 \\ -0.1792 \\ -0.4886 \end{Bmatrix}$$

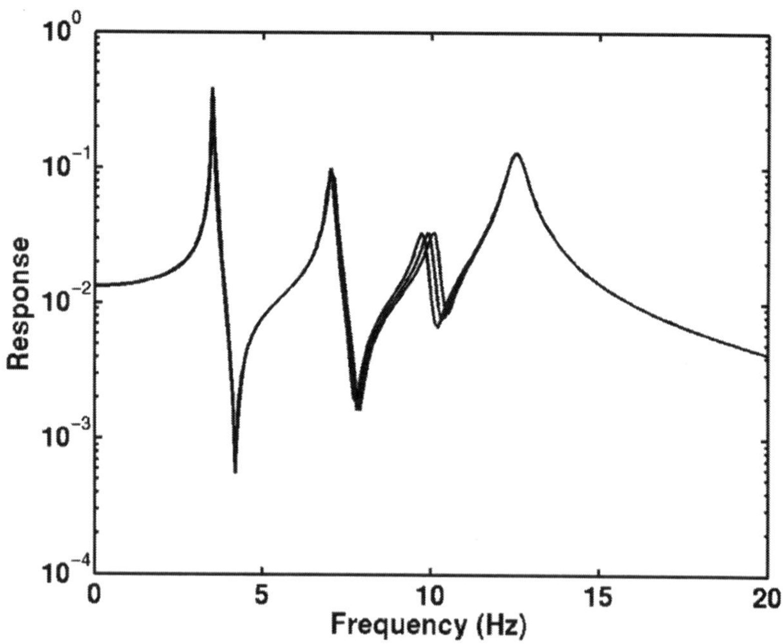

Fig. 12.7-2 The FRF H_{11}. Near the third resonance frequency, the FRF shifts faster with increasing k_4.

Table 12.7-2 Eigenvector Derivatives (×10[5])

$\partial u_1/\partial k_4$	$\partial u_2/\partial k_4$	$\partial u_3/\partial k_4$	$\partial u_4/\partial k_4$
−6.9230 + *j*0.0809	179.8091 + *j*4.1963	−47.6038 + *j*1.5515	−47.8247 + *j*1.9958
−22.6878 + *j*0.2651	179.4687 + *j*4.1884	−411.4725 + *j*13.4103	−12.3794 + *j*0.5166
−69.9857 + *j*0.8177	−248.9163 + *j*5.8091	50.1502 − *j*1.6344	110.1003 + *j*4.5947
148.6419 + *j*1.7367	301.9777 − *j*7.0474	239.8609 − *j*7.8173	−108.4268 + *j*4.5248

$$\{\mathbf{u}_3\}_o = \begin{Bmatrix} 0.2999 \\ 0.1701 \\ -0.5013 \\ 0.4234 \end{Bmatrix} \quad \{\mathbf{u}_4\}_o = \begin{Bmatrix} 0.9231 \\ -0.3212 \\ 0.1105 \\ -0.034 \end{Bmatrix}$$

$$\{\mathbf{u}_1\}_{\text{new}} = \begin{Bmatrix} 0.2394 + j0.04 \times 10^{-5} \\ 0.4347 + j0.13 \times 10^{-5} \\ 0.4517 + j0.4 \times 10^{-5} \\ 0.2897 - j0.9 \times 10^{-5} \end{Bmatrix} \quad \{\mathbf{u}_2\}_{\text{new}} = \begin{Bmatrix} 0.3352 - j2.1 \times 10^{-5} \\ 0.4238 - j2.0 \times 10^{-5} \\ -0.1804 + j2.9 \times 10^{-5} \\ -0.4871 - j3.5 \times 10^{-5} \end{Bmatrix}$$

$$\{\mathbf{u}_3\}_{\text{new}} = \begin{Bmatrix} 0.2997 + j0.7 \times 10^{-5} \\ 0.1681 + j6.7 \times 10^{-5} \\ -0.5011 - j0.8 \times 10^{-5} \\ 0.4246 - j3.8 \times 10^{-5} \end{Bmatrix} \quad \{\mathbf{u}_4\}_{\text{new}} = \begin{Bmatrix} 0.9228 + j1.0 \times 10^{-5} \\ -0.3213 + j0.3 \times 10^{-5} \\ 0.1111 - j2.3 \times 10^{-5} \\ -0.0349 + j0.23 \times 10^{-5} \end{Bmatrix}$$

From these results and from Table 12.7-2, we see that Eq. a is reasonably satisfied.

References

12.1. Liu, C. Q., and Liu, X., "A New Method for Analysis of Complex Structures Based on FRF's of Substructures," *Shock and Vibration*, Vol. 11, No. 1, 2004, pp. 1–7.

12.2. Walton, W. C., Jr., and Steeves, E. C., "A New Matrix Theorem and Its Application for Establishing Independent Coordinates for Complex Dynamical Systems with Constraints," NASA Technical Report TR R-326, NASA Headquarters, Washington, DC, 1969.

12.3. Singh, R. P., and Likins, P. W., "Singular Value Decomposition for Constrained Dynamical Systems," *Journal of Applied Mechanics*, Vol. 52, No. 4, 1985, pp. 943–948.

12.4. Huston, R. L., *Multibody Dynamics*, Butterworth-Heinmann, Boston, MA, 1990.

12.5. Huston, R. L., and Liu, C. Q., *Formulas for Dynamic Analysis*, Marcel Dekker, New York, 2001.

12.6. Thomson, W. T., *Theory of Vibration with Applications*, Prentice Hall, Inc., Upper Saddle River, NJ, 1972.

12.7. Tse, F. S., Morse, I. E., and Hinkle, R. T., *Mechanical Vibration, Theory and Applications*, Allyn and Bacon, Inc., Boston, 1979.

12.8. Den Hartog, J. P., *Mechanical Vibrations*, 4th ed., McGraw-Hill, New York, 1956.

12.9. Everett Reed, F., "Dynamic Vibration Absorbers and Auxiliary Mass Dampers," in *Shock and Vibration Handbook*, C. M. Harris, ed., 3rd ed., McGraw-Hill, New York, 1987.

12.10. Varburton, G. B., "Optimal Absorber Parameters for Various Combinations of Response and Excitation Parameters," *Earthquake Engineering and Structural Dynamics*, Vol. 10, 1982, pp. 381–401.

12.11. Brock, J. E, "A Note on the Damped Vibration Absorber," *Journal of Applied Mechanics*, Vol. 13, 1946, p. A-284.

12.12. Liu, C. Q., and Chang, C. C., "A Method for Vibration Absorber Tuning Based on Baseline Frequency Response Functions," *Proceedings of the Institute of Mechanical Engineering, Part D: Journal of Automobile Engineering*, Vol. 21, No. 9, 2007, pp. 1071–1078.

12.13. Krenk, Steen, "Frequency Analysis of the Tuned Mass Damper," *Journal of Applied Mechanics*, Vol. 72, 2005, pp. 936–942.

12.14. Liu, C. Q., "A Method for Torsional Damper Tuning Based on Baseline Frequency Response Functions," SAE paper 2009-01-2152, SAE International, Warrendale, PA, 2009.

12.15. Klosterman, A. L., and Lemon, J. R., "Dynamic Design Analysis via the Building Block Approach," *Sound and Vibration Bulletin*, Vol. 42, 1972, pp. 97–104.

12.16. Liu, C. Q., and Mir, Hamid, "A Hybrid Method for Vehicle Axle Noise Simulation with Experimental Validation," SAE paper 2003-01-1707, SAE International, Warrendale, PA, 2003.

12.17. Liu, C. Q., "Combination of an Improved FRF-Based Substructure Synthesis and Power Flow Method with Application to Vehicle Axle Noise Analysis," *Shock and Vibration*, Vol. 15, 2008, pp. 51–60.

12.18. Fox , R. L., and Kapoor, M. P., "Rates of Change of Eigenvalues and Eigenvectors," *AIAA Journal*, Vol. 6, 1968, pp. 2426–2429.

12.19. Haftka, R. T., and Adelman, R. H., "Sensitivity Analysis of Discrete Structure Systems," *AIAA Journal,* Vol. 24, No. 5, 1986, pp. 823–832.

12.20. Sutter, T. R., et al, "A Comparison of Several Methods for the Calculation of Vibration Mode Shape Derivatives," *AIAA Journal,* Vol. 26, 1988, pp. 1506–1511.

12.21. Taylor, D. R., and Kane, T. R., "Multiparameter Quadratic Eigenvalue Problems," *Journal of Applied Mechanics,* Vol. 42, 1975, pp. 478–483.

12.22. Plaut, R. H., and Huseyin, K., "Derivatives of Eigenvalues and Eigenvectors in Non-Self-Adjoint Systems," *AIAA Journal,* Vol. 11, 1973, pp. 250–251.

12.23. Wang, B. P., "Improved Approximate Method for Computing Eigenvector Derivatives in Structure Dynamics," *AIAA Journal,* Vol. 29, No. 6, 1991, pp. 1018–1020.

12.24. Nelson, R. B., "Simplified Calculation of Eigenvector Derivatives," *AIAA Journal,* Vol. 14, 1976, pp. 1201–1205.

12.25. Liu, C. Q., Liu, X. B., and Chang, C. C., "On Sensitivity Analysis of Discrete Structure Systems," *Computer and Structure,* Vol. 56, No. 1, 1995, pp. 141–145.

Appendix A:

Systems of Units/Conversion Factors

	To convert from:	To:	Multiply by:
Length	centimeter (cm)	meter (m)	10^{-2}
		inch (in.)	0.3937
		foot (ft)	3.281×10^{-2}
	meter (m)	kilometers (km)	10^{-3}
		inch (in.)	39.37
		foot (ft)	3.281
	kilometer (km)	foot (ft)	3.281×10^{-3}
		mile (mi)	0.6214
	inch (in.)	centimeter (cm)	2.54
		meter (m)	2.54×10^{-2}
		foot (ft)	1/12
	foot (ft)	centimeter (cm)	30.48
		meter (m)	0.3048
		kilometer (km)	3.048×10^{-4}
		inch (in.)	12
		mile (mi)	1.894×10^{-4}
Mass	kilogram (kg)	pound-mass (lb · m)	2.2046
		slug	6.8466×10^{-2}
	pound-mass (lb · m)	kilogram (kg)	0.4536
		slug	3.1056×10^{-2}
	slug	kilogram (kg)	14.593
		pound-mass (lb · m)	32.2
Force	Newton (N)	pound-force (lb · f)	0.2248
	Dyne (dyn)	Newton (N)	10^{-5}
	pound-force (lb · f)	Newton (N)	4.448

	To convert from:	To:	Multiply by:
Velocity	meters/second (m/s)	kilometers/hour (km/h)	3.6
		miles/hour (mph)	2.237
	kilometers/hour (km/h)	miles/hour (mph)	0.6214
		feet/second (ft/s)	0.9114
	foot/second (ft/s)	miles/hour (mph)	0.6818
		meters/second (m/s)	0.3048
	mile/hour (mph)	kilometers/hour (km/h)	1.609
		inches/second (in./s)	17.6004
Torque	Newton-meter (N · m)	foot-pound (ft · lb)	0.7376
	foot-pound (ft · lb)	Newton-meter (N · m)	1.356
Moment of inertia	kg · m^2	slug · ft^2	0.7376
		lb · ft^2	23.75
		lb · in. · s^2	8.851
Unbalance	gram-millimeter (g · mm)	ounce-inch (oz · in.)	1.3887×10^{-3}
	ounce-inch (oz · in.)	gram-millimiters (g · mm)	720.09

Appendix B:

Vehicle Speed, Engine rpm, and Prop-shaft rpm

In the study of vehicle vibration, it is helpful to know the relationship between the vehicle speed, engine rpm, and prop-shaft rpm.

To this end, we introduce the following notations:

N_e = engine rpm (revolutions per minute) (rev/min)

N_p = prop-shaft rpm (rev/min)

N_t = tire rpm (rev/min)

f_e = engine rotation frequency (Hz)

f_p = prop-shaft rotation frequency (Hz)

f_t = tire rotation frequency (Hz)

n = axle ratio = N_p / N_t

OD = overdrive ratio = N_e / N_p

R_t = tire effective radius (in.)

V = vehicle speed (mph)

We then have

$$f_e = \frac{N_e}{60} \text{ (Hz)}$$

$$f_p = \frac{N_p}{60} = \frac{f_e}{OD} \text{ (Hz)}$$

$$f_t = \frac{N_t}{60} = \frac{f_p}{n} = \frac{f_e}{n(OD)} \text{ (Hz)}$$

$$V = 2\pi f_t R_t = \frac{2\pi f_p R_t}{n} = \frac{2\pi f_e R_t}{n(OD)} \text{ (in./s)}$$

or

$$V = 0.357 f_t R_t = 5.95 \times 10^{-3} \frac{N_p R_t}{n} \text{ (mph)}$$

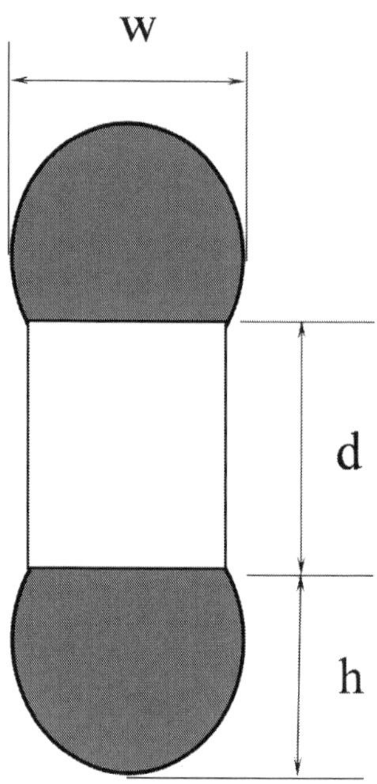

Fig. A.B-1 Tire profile.

$$V = 0.357\frac{f_e R_t}{n(OD)} = 5.95\times10^{-3}\frac{N_e R_t}{n(OD)} \text{ (mph)}$$

The tire effective radius may be determined as follows: referring to Fig. A.B-1, the tread width (w), aspect ratio (100 h/w), and rim diameter (d) of a tire are usually shown on the side of the tire in the form: P*xxx*/*yy*R*zz*, where

$$\text{P} = \text{passenger vehicle tire}$$

$$xxx = \text{tread width (mm)} = w$$

$$yy = \text{aspect ratio} = \frac{h}{w}\times100$$

$$\text{R} = \text{radial tire}$$

$$zz = \text{rim diameter (in.)} = d$$

Under standard tire pressure, the tire effective radius is given by

$$R_t = \frac{d}{2} + \left(\frac{h}{w}\right)\left(\frac{w}{25.4}\right) = \frac{zz}{2} + \left(\frac{yy}{100}\right)\left(\frac{xxx}{25.4}\right). \text{ (in.)}$$

Table B-1 Vehicle Speeds of a Jeep with Tire: P255/55R18

Tire type	P*xxx*/*yy*R*zz*	P255/55R18	
Tire radius (in.)		R_t=*zz*/2+*yy*/100**xxx*/25.4	14.52
Axle ratio			3.7
Prop rpm	Vehicle speed (mph)		
1000	23.4		
1200	28.0		
1400	32.7		
1600	37.4		
1800	42.0		
2000	46.7		
2200	51.4		
2400	56.0		
2600	60.7		
2800	65.4		
3000	70.1		
3200	74.7		

As an example, consider a Jeep with tire: P255/55R18. The tire effective radius is then

$$R_t = \frac{18}{2} + \left(\frac{55}{100}\right)\left(\frac{255}{25.4}\right) = 14.52 \text{ (in.)}$$

Table B-1 lists the vehicle speed as a function of prop-shaft rpm, where the axle ratio for this vehicle is 3.7.

Appendix C:

Transformation Matrices[1]

As discussed in Chapter Ten, in engine mounting system design it is necessary to know the transformation matrices between the local (the engine) coordinate system and the global (the vehicle) coordinate system. Transformation matrices between two coordinate systems play an important role in modal analysis and system vibration analysis.

Consider a body B moving in a reference frame R as in Fig. A.C-1. Let $\hat{B}$ be a reference frame fixed in B. The axes of $\hat{B}$ are called *body-fixed axes*. The axes of R are called *space-fixed axes*. Let $\mathbf{N}_i$ and $\mathbf{n}_i$ (i=1,2,3) be mutually perpendicular unit vectors parallel to the axes of R and $\hat{B}$. The transformation matrices between R and $\hat{B}$ may be defined in terms of orientation angles as follows: Let the axes of $\hat{B}$ be initially aligned with those of R. Let $\hat{B}$ then be

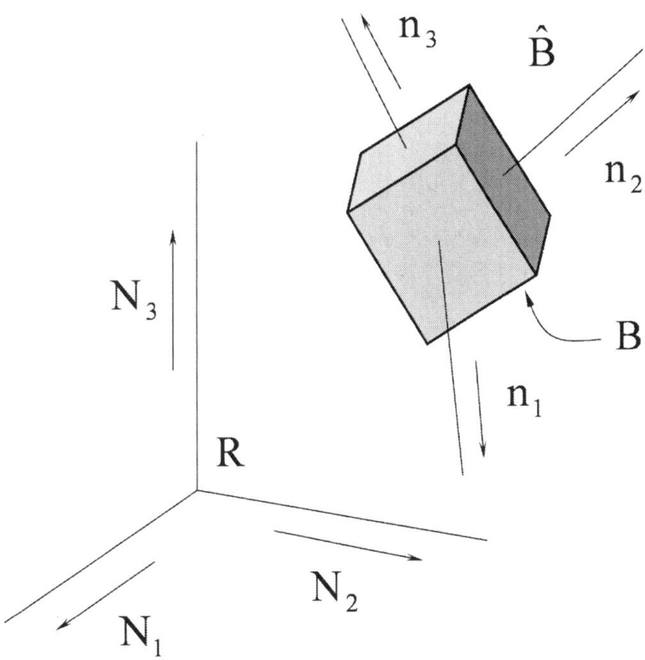

Fig. A.C-1 A body *B*, a reference frame *R*, and unit vector sets $\mathbf{n}_i$ and $\mathbf{N}_i$ (i=1,2,3) fixed in *B* and *R*.

[1]See [10.2, 10.5].

brought to its general orientation relative to R by three dextral successive rotations either about body-fixed axes or about space-fixed axes.

The transformation matrices between R and $\hat{B}$ are dependent on the rotation sequence. In the following, we list them for 24 rotation sequences. Specifically, for *body-fixed axis rotation* the transformation matrix is denoted by S; that is, $\mathbf{N} = S\mathbf{n}$ and $\mathbf{n} = S^T\mathbf{N}$, where $\mathbf{N}$ is the column array of unit vectors $\mathbf{N}_i$ (i=1,2,3), and where $\mathbf{n}$ is the column array of unit vectors $\mathbf{n}_i$ (i=1,2,3). Similarly, for *space-fixed axis rotation* the transformation matrix is denoted by P; that is, $\mathbf{N} = P\mathbf{n}$ and $\mathbf{n} = P^T\mathbf{N}$. Note further that s_i and c_i represent $\sin\theta_i$ and $\cos\theta_i$ (i=1,2,3). Finally, the matrices S_i and P_i (i=1,2,3) are defined as

$$S_1(\theta_i) = P_1(\theta_i) = \begin{bmatrix} 1 & 0 & 0 \\ 0 & c_i & -s_i \\ 0 & s_i & c_i \end{bmatrix}$$

$$S_2(\theta_i) = P_2(\theta_i) = \begin{bmatrix} c_i & 0 & s_i \\ 0 & 1 & 0 \\ -s_i & 0 & c_i \end{bmatrix}$$

and

$$S_3(\theta_i) = P_3(\theta_i) = \begin{bmatrix} c_i & -s_i & 0 \\ s_i & c_i & 0 \\ 0 & 0 & 1 \end{bmatrix}$$

Transformation Matrices for Body-Fixed Axis Rotation Sequence:

1. 1-2-3 rotation:

$$S = S_1(\theta_1)S_2(\theta_2)S_3(\theta_3) = \begin{bmatrix} c_2c_3 & -c_2s_3 & s_2 \\ (s_1s_2c_3 + c_1s_3) & (-s_1s_2s_3 + c_1c_3) & -s_1c_2 \\ (-c_1s_2c_3 + s_1s_3) & (c_1s_2s_3 + s_1c_3) & c_1c_2 \end{bmatrix}$$

2. 2-3-1 rotation:

$$S = S_2(\theta_1)S_3(\theta_2)S_1(\theta_3) = \begin{bmatrix} c_1c_2 & (-c_1s_2c_3 + s_1s_3) & (c_1s_2s_3 + s_1c_3) \\ s_2 & c_2c_3 & -c_2s_3 \\ -s_1c_2 & (s_1s_2c_3 + c_1s_3) & (-s_1s_2s_3 + c_1c_3) \end{bmatrix}$$

3. 3-1-2 rotation:

$$S = S_3(\theta_1)S_1(\theta_2)S_2(\theta_3) = \begin{bmatrix} (-s_1s_2s_3 + c_1c_3) & -s_1c_2 & (s_1s_2c_3 + c_1s_3) \\ (c_1s_2s_2 + s_1c_3) & c_1c_2 & (-c_1s_2c_3 + s_1s_3) \\ -c_2s_3 & s_2 & c_2c_3 \end{bmatrix}$$

4. 1-3-2 rotation:

$$S = S_1(\theta_1)S_3(\theta_2)S_2(\theta_3) = \begin{bmatrix} c_2c_3 & -s_2 & c_2s_3 \\ (c_1s_2c_3 + s_1s_3) & c_1c_2 & (c_1s_2s_3 - s_1c_3) \\ (s_1s_2c_3 - c_1s_3) & s_1c_2 & (s_1s_2s_3 + c_1c_3) \end{bmatrix}$$

5. 2-1-3 rotation:

$$S = S_2(\theta_1)S_1(\theta_2)S_3(\theta_3) = \begin{bmatrix} (s_1s_2s_3 + c_1c_3) & (s_1s_2c_3 - c_1s_3) & s_1c_2 \\ c_2s_3 & c_2c_3 & -s_2 \\ (c_1s_2s_3 - s_1c_3) & (c_1s_2c_3 + s_1s_3) & c_1c_2 \end{bmatrix}$$

6. 3-2-1 rotation:

$$S = S_3(\theta_1)S_2(\theta_2)S_1(\theta_3) = \begin{bmatrix} c_1c_2 & (c_1s_2s_3 - s_1c_3) & (c_1s_2c_3 + s_1s_3) \\ s_1c_2 & (s_1s_2s_3 + c_1c_3) & (s_1s_2c_3 - c_1s_3) \\ -s_2 & c_2s_3 & c_2c_3 \end{bmatrix}$$

7. 1-2-1 rotation:

$$S = S_1(\theta_1)S_2(\theta_2)S_1(\theta_3) = \begin{bmatrix} c_2 & s_2s_3 & s_2c_3 \\ s_1s_2 & (-s_1c_2s_3 + c_1c_3) & (-s_1c_2c_3 - c_1s_3) \\ -c_1s_2 & (c_1c_2s_3 + s_1c_3) & (c_1c_2c_3 - s_1s_3) \end{bmatrix}$$

8. 1-3-1 rotation:

$$S = S_1(\theta_1)S_3(\theta_2)S_1(\theta_3) = \begin{bmatrix} c_2 & -s_2c_3 & s_2s_3 \\ c_1s_2 & (c_1c_2c_3 - s_1s_3) & (-c_1c_2s_3 - s_1c_3) \\ s_1s_2 & (s_1c_2c_3 + c_1s_3) & (-s_1c_2s_3 + c_1c_3) \end{bmatrix}$$

9. 2-1-2 rotation:

$$S = S_2(\theta_1)S_1(\theta_2)S_2(\theta_3) = \begin{bmatrix} (-s_1c_2s_3 + c_1c_3) & s_1s_2 & (s_1c_2c_3 + c_1s_3) \\ s_2s_3 & c_2 & -s_2c_3 \\ (-c_1c_2s_3 - s_1c_3) & c_1s_2 & (c_1c_2c_3 - s_1s_3) \end{bmatrix}$$

10. 2-3-2 rotation:

$$S = S_2(\theta_1)S_3(\theta_2)S_2(\theta_3) = \begin{bmatrix} (c_1c_2c_3 - s_1s_3) & -c_1s_2 & (c_1c_2s_3 + s_1c_3) \\ s_2c_3 & c_2 & s_2s_3 \\ (-s_1c_2c_3 - c_1s_3) & s_1s_2 & (-s_1c_2s_3 + c_1c_3) \end{bmatrix}$$

11. 3-1-3 rotation (Euler angles):

$$S = S_3(\theta_1)S_1(\theta_2)S_3(\theta_3) = \begin{bmatrix} (-s_1c_2s_3 + c_1c_3) & (-s_1c_2c_3 - c_1s_3) & s_1s_2 \\ (c_1c_2s_3 + s_1c_3) & (c_1c_2c_3 - s_1s_3) & -c_1s_2 \\ s_2s_3 & s_2c_3 & c_2 \end{bmatrix}$$

12. 3-2-3 rotation:

$$S = S_3(\theta_1)S_2(\theta_2)S_3(\theta_3) = \begin{bmatrix} (c_1c_2c_3 - s_1s_3) & (-c_1c_2s_3 - s_1c_3) & c_1s_2 \\ (s_1c_2c_3 + c_1s_3) & (-s_1c_2s_3 + c_1c_3) & s_1s_2 \\ -s_2c_3 & s_2s_3 & c_2 \end{bmatrix}$$

Transformation Matrices for Space-Fixed Axis Rotation Sequence:

1. 1-2-3 rotation:

$$P = P_3(\theta_3)P_2(\theta_2)P_1(\theta_1) = \begin{bmatrix} c_2c_3 & (s_1s_2c_3 - c_1s_3) & (c_1s_2c_3 + s_1s_3) \\ c_2s_3 & (s_1s_2s_3 + c_1c_3) & (c_1s_2s_3 - s_1c_3) \\ -s_2 & s_1c_2 & c_1c_2 \end{bmatrix}$$

2. 2-3-1 rotation:

$$P = P_1(\theta_3)P_3(\theta_2)P_2(\theta_1) = \begin{bmatrix} c_1c_2 & -s_2 & s_1c_2 \\ (c_1s_2c_3 + s_1s_3) & c_2c_3 & (s_1s_2c_3 - c_1s_3) \\ (c_1s_2s_3 - s_1c_3) & c_2s_3 & (s_1s_2s_3 + c_1c_3) \end{bmatrix}$$

3. 3-1-2 rotation:

$$P = P_2(\theta_3)P_1(\theta_2)P_3(\theta_1) = \begin{bmatrix} (s_1s_2s_3 + c_1c_3) & (c_1s_2s_3 - s_1c_3) & c_2s_3 \\ s_1c_2 & c_1c_2 & -s_2 \\ (s_1s_2c_3 - c_1s_3) & (c_1s_2c_3 + s_1s_3) & c_2c_3 \end{bmatrix}$$

4. 1-3-2 rotation:

$$P = P_2(\theta_3)P_3(\theta_2)P_1(\theta_1) = \begin{bmatrix} c_2c_3 & (-c_1s_2c_3 + s_1s_3) & (s_1s_2c_3 + c_1s_3) \\ s_2 & c_1c_2 & -s_1c_2 \\ -c_2s_3 & (c_1s_2s_3 + s_1c_3) & (-s_1s_2s_3 + c_1c_3) \end{bmatrix}$$

5. 2-1-3 rotation:

$$P = P_3(\theta_3)P_1(\theta_2)P_2(\theta_1) = \begin{bmatrix} (-s_1s_2s_3 + c_1c_3) & -c_2s_3 & (c_1s_2s_3 + s_1c_3) \\ (s_1s_2c_3 + c_1s_3) & c_2c_3 & (-c_1s_2c_3 + s_1s_3) \\ -s_1c_2 & s_2 & c_1c_2 \end{bmatrix}$$

6. 3-2-1 rotation:

$$P = P_1(\theta_3)P_2(\theta_2)P_3(\theta_1) = \begin{bmatrix} c_1c_2 & -s_1c_2 & s_2 \\ (c_1s_2s_3 + s_1c_3) & (-s_1s_2s_3 + c_1c_3) & -c_2s_3 \\ (-c_1s_2c_3 + s_1s_3) & (s_1s_2c_3 + c_1s_3) & c_2c_3 \end{bmatrix}$$

7. 1-2-1 rotation:

$$P = P_1(\theta_3)P_2(\theta_2)P_1(\theta_1) = \begin{bmatrix} c_2 & s_1s_2 & c_1s_2 \\ s_2s_3 & (-s_1c_2s_3 + c_1c_3) & (-c_1c_2s_3 - s_1c_3) \\ -s_2c_3 & (s_1c_2c_3 + c_1s_3) & (c_1c_2c_3 - s_1s_3) \end{bmatrix}$$

8. 1-3-1 rotation:

$$P = P_1(\theta_3)P_3(\theta_2)P_1(\theta_1) = \begin{bmatrix} c_2 & -c_1s_2 & s_1s_2 \\ s_2c_3 & (c_1c_2c_3 - s_1s_3) & (-s_1c_2c_3 - c_1s_3) \\ s_2s_3 & (c_1c_2s_3 + s_1c_3) & (-s_1c_2s_3 + c_1c_3) \end{bmatrix}$$

9. 2-1-2 rotation:

$$P = P_2(\theta_3)P_1(\theta_2)P_2(\theta_1) = \begin{bmatrix} (-s_1c_2s_3 + c_1c_3) & s_2s_3 & (c_1c_2s_3 + s_1c_3) \\ s_1s_2 & c_2 & -c_1s_2 \\ (-s_1c_2c_3 - c_1s_3) & s_2c_3 & (c_1c_2c_3 - s_1s_3) \end{bmatrix}$$

10. 2-3-2 rotation:

$$P = P_2(\theta_3)P_3(\theta_2)P_2(\theta_1) = \begin{bmatrix} (c_1c_2c_3 - s_1s_3) & -s_2c_3 & (s_1c_2c_3 + c_1s_3) \\ c_1s_2 & c_2 & s_1s_2 \\ (-c_1c_2s_3 - s_1c_3) & s_2s_3 & (-s_1c_2s_3 + c_1c_3) \end{bmatrix}$$

11. 3-1-3 rotation:

$$P = P_3(\theta_3)P_1(\theta_2)P_3(\theta_1) = \begin{bmatrix} (-s_1c_2s_3 + c_1c_3) & (-c_1c_2s_3 - s_1c_3) & s_2s_3 \\ (s_1c_2c_3 + c_1s_3) & (c_1c_2c_3 - s_1s_3) & -s_2c_3 \\ s_1s_2 & c_1s_2 & c_2 \end{bmatrix}$$

12. 3-2-3 rotation:

$$P = P_3(\theta_3)P_2(\theta_2)P_3(\theta_1) = \begin{bmatrix} (c_1c_2c_3 - s_1s_3) & (-s_1c_2c_3 - c_1s_3) & s_2c_3 \\ (c_1c_2s_3 + s_1c_3) & (-s_1c_2s_3 + c_1c_3) & s_2s_3 \\ -c_1s_2 & s_1s_2 & c_2 \end{bmatrix}$$

Notation

Chapter One

L	= length, m
g	= acceleration of gravity, m/s^2
θ	= angle, rad
$x(t)$	= displacement, m
A	= amplitude of vibration, also a constant of integration
ω	= angular frequency, rad/s
f	= $\omega/2\pi$, frequency, Hz
T	= $1/f$, period of vibration, s
ω_n	= angular natural frequency, rad/s
ϕ	= phase angle, rad
x_0	= an initial displacement, m
$\dot{x}_0$	= initial velocity, m/s
ω_0	= fundamental frequency, rad/s
a_n, b_n	= Fourier coefficients
A_n	= $\sqrt{a_n^2 + b_n^2}$, Fourier frequency spectrum
$h(t)$	= impulse response function

Chapter Two

$F_1, F_2, \ldots$	= force vectors, N
$\sum \boldsymbol{F}$	= sum of all the forces acting on a body, N
$\boldsymbol{a}$	= acceleration vector, m/s^2
$\boldsymbol{T}_1, \boldsymbol{T}_2, \ldots$	= torque vectors acting on a body, Nm
$\sum \boldsymbol{T}$	= sum of all the torques acting on a body, Nm
$\boldsymbol{\alpha}$	= angular acceleration vector of a body, rad/s^2
g	= acceleration of gravity, m/s^2
R, r	= radius, m
L, l	= length, m

$\frac{d}{dt}()$	= time derivative of ( )
j	= $\sqrt{-1}$, the imaginary unit
x, y, z	= displacements in the x, y, and z directions, m
$\ddot{x}, \ddot{y}, \ddot{z}$	= acceleration components, m/s^2
θ	= angular displacement, rad
$\ddot{\theta}$	= angular acceleration component, rad/s^2
K_θ	= torsional stiffness of a shaft, Nm/rad
G	= shear modulus, N/m^2
E	= Young's modulus, N/m^2
J	= second moment of area when used with E, m^4
J_p, I	= polar moment of inertia of a shaft or disk, kg·m^2
T	= kinetic energy, also period of vibration, Nm, s
U	= potential energy, Nm
T_{max}	= maximum kinetic energy, Nm
U_{max}	= maximum potential energy, Nm
m_s	= mass of a spring, kg
m_{ef}	=effective mass of a beam, kg
δ	= static deflection of a spring, also logarithmic decrement, m
k	= spring constant (stiffness), N/m
k_e	= equivalent spring constant (stiffness), N/m
c	= viscous damping coefficient, Ns/m
m	= mass of a particle, kg
c_r	= $2\sqrt{km}$, critical viscous damping coefficient, Ns/m
ς	= c/c_r damping ratio
λ	= constant used in solving differential equation
A	= amplitude of vibration, also a constant of integration
B_1, B_2, C	= integration constants
ϕ	= phase angle, rad

x_0 = an initial displacement, m

$\dot{x}_0$ = initial velocity, m/s

ω = angular frequency, rad/s

ω_n = undamped angular natural frequency, rad/s

ω_d = $\omega_n\sqrt{1-\varsigma^2}$, damped angular natural frequency, rad/s

T_d = $2\pi/\omega_d$, damped period of vibration, s

Chapter Three

M, m = mass, kg

k = spring constant (stiffness), N/m

c = viscous damping coefficient, Ns/m

c_{eq} = equivalent viscous damping coefficient, Ns/m

j = $\sqrt{-1}$, the imaginary unit

δ = static deflection of a spring due to a attached mass, m

F_0 = amplitude of a harmonic force, N

$F(t)$ = an external force, N

ω = excitation frequency, rad/s

ω_n = undamped angular natural frequency, rad/s

ω_d = $\omega_n\sqrt{1-\varsigma^2}$, damped angular natural frequency, rad/s

ω_r = $\omega_n\sqrt{1-2\varsigma^2}$, resonant frequency, rad/s

ς = $c/(2\sqrt{km})$, damping ratio

h = F_0/m, m/s²

A = amplitude of vibration

ϕ = phase angle of free vibration, rad

ψ = phase angle of forced vibration with respect to the excitation force, rad

x_p = the particular solution

X = amplitude of the particular solution

r = ω/ω_n, frequency ratio

MF	= magnification factor
MF_{max}	$\approx 1/2\varsigma$, the maximum value of magnification factor
X_0	$= F_0/k$, m
A_1, A_2	= integration constants
T_{beat}	$= 2\pi/\Delta\omega$, beat period, s
e	= eccentricity of an unbalance mass (not to be confused with exponential)
τ	= dummy time variable of integration
F_T	= magnitude of transmitted force, N
TR	= transmissibility
W_d	= energy lost per cycle, Nm
η	$= 2\varsigma$, structural damping lost factor
a_n, b_n	= Fourier coefficients

Chapter Four

m	= mass, kg
r	= radius, m
x, y	= displacements, m
ω	= angular velocity, rad/s
e	= eccentricity of the mass center of a disk, also piston pin offset (not to be confused with exponential), m
k	= bending stiffness of a shaft, N/m
c	= viscous damping coefficient, Ns/m
ω_n	= undamped natural frequency, rad/s
ς	$= c/(2\sqrt{km})$, damping ratio
h	$= e\omega^2$
ψ	= phase angle, rad
OS	= deflection of shaft center, m
x_s, y_s	= coordinates of shaft center, m

j	$= \sqrt{-1}$, the imaginary unit
$\mathbf{Z}$	= complex amplitude or phasor of vibration of a shaft
$\mathbf{U}$	= unbalance vector, kg·m
$\mathbf{S}(\omega)$	= influence coefficient (sensitivity)
$\mathbf{S}_{ij}$	= elements of influence coefficient matrix
$\mathbf{Z}_t$	= complex amplitude of vibration of shaft when a trial mass added
$\mathbf{U}_t$	= new unbalance vector after a trial mass is added
$\mathbf{Z}_1, \mathbf{Z}_2$	= vibration phasor at planes 1 and 2, respectively
$\mathbf{Z}_{11}, \mathbf{Z}_{21}$	= vibration phasor at planes 1 and 2 after a trial mass added at plane 1
$\mathbf{Z}_{12}, \mathbf{Z}_{22}$	= vibration phasor at planes 1 and 2 after a trial mass added at plane 2
L	= length, m
s	= distance of origin of coordinate system to the mass center of crank
m	= mass of crank and crankpin, kg
m_1	= rotating mass of an engine, kg
m_2	= reciprocating mass of an engine, kg
F_x, F_y	= inertia force components of an engine, N
ϕ	$= \omega t$, rotating angle of crankshaft, rad
γ	= angle between axes of two cylinders, deg
M_x, M_y	= moments of inertia forces about the axes x and y, respectively, kg·m^2
$\mathbf{N}_x, \mathbf{N}_y$	= unit vectors parallel to global axes X and Y, respectively
$\mathbf{n}_x, \mathbf{n}_y$	= unit vectors parallel to the axes x and y fixed in crank, respectively
a, b	= distances between two cylinders of engine, m
$W_{1,4}$	= total ring weights on crankpins 1 and 4, respectively, kg
$W_{2,3}$	= total ring weights on crankpins 2 and 3, respectively, kg

Chapter Five

x_1, x_2	= displacements, m
$\mathbf{x}$	= displacement column vector, m
m_1, m_2	= masses, or generalized masses, kg
c_1, c_2	= damping coefficients, or generalized damping coefficients, Ns/m
k_1, k_2	= spring constants, or generalized stiffness, N/s
X_1, X_2	= amplitude of vibration
Δ	= determinant of the coefficients
ω	= angular frequency, rad/s
ω_i	= angular frequency for mode i
λ	= ω^2, characteristic root, eigenvalue
$\mathbf{u}_1, \mathbf{u}_2$	= modal vectors or mode shapes
$\hat{\mathbf{u}}$	= normalized modal vector
M	= mass matrix
K	= stiffness matrix
Z	= system impedance matrix
C	= cofactor of a matrix
$adj(Z)$	= adjoint matrix of Z
C^T	= transpose of C
Z^{-1}	= inverse of Z
S	= modal matrix
$\overline{S}$	= normalized modal matrix
L	= length (not to be confused with Lagrangian function), m
a	= length, m
θ	= angle, rad
g	= acceleration of gravity, m/s^2
ψ	= phase angle, rad
$\lambda_0, \lambda_1, \lambda_2$	= eigenvalues

C_i	= constants
A, B	= integration constants
ψ_1, ψ_2	= integration constants, phase angles
J_1, J_2	= inertia moments of mass, kg·m^2
k_t	= torsional stiffness, N·m/rad
e	= distance (not to be confused with exponential), m
y_i	= new displacement variable, m
$\mathbf{f}$	= excitation force column vector, N
f_i	$= \mathbf{u}_i^T \mathbf{f}$
F_1, F_2	= amplitudes of excitation forces, N
$\mathbf{h}$	$= \overline{S}^T \mathbf{f}$
$\mathbf{x}_0$	$= \mathbf{x}(0)$, initial displacement column vector, m
$\dot{\mathbf{x}}_0$	$= \dot{\mathbf{x}}(0)$, initial velocity column vector, m/s
ς_i	$= c_i/(2\sqrt{k_i m_i})$, the damping ratio for mode i
α, β	= constants $(2\varsigma_i\omega_i = \alpha + \beta\omega_i^2)$
I	= the unit matrix
q_i	= the i-th generalized coordinate
δW	= virtual work, Nm
δr_i	= virtual displacement of the i-th position vector, m
δq_i	= virtual displacement of the i-th generalized coordinate
Q_i	= the i-th generalized force
T	= kinetic energy, Nm
U	= potential energy, Nm
L	$= T-U$, Lagrangian function, Nm
$\frac{d}{dt}(\ )$	= time derivative of ( )

Chapter Six

M	= mass matrix
K	= stiffness matrix

C	= damping matrix
λ	= ω^2, eigenvalue
C_i	= constants
ϕ_i	= phase angle, rad
M^{-1}	= the inverse of the matrix M
j	= $\sqrt{-1}$, the imaginary unit
$\mathbf{x}$	= displacement column vector
$\mathbf{u}$	= eigenvector
L	= lower triangular matrix from Cholesky decomposition
L^{-1}	= the inverse of the matrix L
$M^{1/2}$	= the matrix square root of the matrix M
$\mathbf{q}$	= generalized coordinate column vector
$\mathbf{q}^*$	= complex conjugate of $\mathbf{q}$
I	= the unit matrix
$\tilde{K}$	$= L^{-1}K(L^T)^{-1}$
$\tilde{C}$	$= L^{-1}M(L^T)^{-1}$
$\mathbf{v}_1, \mathbf{v}_2, \ldots$	= modal vectors
V	$= [\mathbf{v}_1 \cdots \mathbf{v}_n]$
$\mathbf{y}_1$	$= \mathbf{x}$
$\mathbf{y}_2$	$= \dot{\mathbf{x}}$
$\mathbf{y}$	$= [\mathbf{y}_1 \quad \mathbf{y}_2]^T = [\mathbf{x} \quad \dot{\mathbf{x}}]^T$
$\mathbf{z}$	= nonzero vector of constants
A	$= \begin{bmatrix} 0 & I \\ -M^{-1}K & -M^{-1}C \end{bmatrix}$, for $\dot{\mathbf{y}} = A\mathbf{y}$
A	$= \begin{bmatrix} C & M \\ M & 0 \end{bmatrix}$, for $A\dot{\mathbf{y}} + B\mathbf{y} = 0$
B	$= \begin{bmatrix} K & 0 \\ 0 & -M \end{bmatrix}$, for $A\dot{\mathbf{y}} + B\mathbf{y} = 0$
$\mathbf{b}$	$= \begin{bmatrix} 0 \\ M^{-1}\mathbf{f} \end{bmatrix}$

λ_i	= eigenvalue
λ_i^*	= complex conjugate of λ_i
ω_i	= angular natural frequency for mode i
λ_i	= damping ratio for mode i
Re( )	= real part of ( )
Im( )	= imaginary part of ( )
$\mathbf{u}_i$	= modal vector for mode i
$\mathbf{u}_i^*$	= complex conjugate of $\mathbf{u}_i$
$\mathbf{z}_i$	$= \begin{bmatrix} \mathbf{u}_i \\ \lambda_i \mathbf{u}_i \end{bmatrix}$
$\mathbf{z}_i^*$	$= \begin{bmatrix} \mathbf{u}_i^* \\ \lambda_i \mathbf{u}_i^* \end{bmatrix}$
a_i	$= \mathbf{z}_i^T A \mathbf{z}_i$
b_i	$= \mathbf{z}_i^T B \mathbf{z}_i$
P	$= [\mathbf{z}_1 \ \cdots \ \mathbf{z}_n \ \mathbf{z}_1^* \ \cdots \ \mathbf{z}_n^*]$
Λ	$= \text{diag}(\lambda_1, \cdots, \lambda_n)$
Λ^*	$= \text{diag}(\lambda_1^*, \cdots, \lambda_n^*)$
A_{rs}	= the elements of the residual matrix
$\mathbf{f}$	= force vector
$\mathbf{x}_0$	= initial displacement column vector
$\dot{\mathbf{x}}_0$	= initial velocity column vector
μ	= constant

Chapter Seven

$x(t)$	= a function of time
s	$= \sigma + j\omega$
$L[x(t)]$	$= X(s)$, the Laplace transform of $x(t)$
$L^{-1}[x(s)]$	= the inverse Laplace transform of $X(s)$
j	$= \sqrt{-1}$, the imaginary unit

m	= mass, kg
k	= spring constant (stiffness), N/m
c	= viscous damping coefficient, Ns/m
λ	= characteristic root
ς	$= c/(2\sqrt{km})$ damping ratio
ω_n	= undamped angular natural frequency, rad/s
ω_d	= damped angular natural frequency, rad/s
$h(t)$	= impulse response function
$H(s)$	= transfer function
H_d	$= x/F$, receptance transfer function
H_v	$= v/F$, mobility transfer function
H_a	$= a/F$, accelerance transfer function
Re( )	= real part of ( )
Im( )	= imaginary of ( )
$\delta(t)$	= the Dirac delta function
H_d^R	= real part of H_d
H_d^I	= imaginary part of H_d
r	$= \omega/\omega_n$, frequency ratio
ψ	= phase angle, rad
λ_n	$= \alpha_n + j\beta_n$, complex poles
λ_n^*	= complex conjugate of λ_n
A	= complex residue
A*	= complex conjugate of A
$a(s)$, $b(s)$	= polynomials
a_n, b_n	= coefficients of polynomials
M	= mass matrix
C	= damping matrix
K	= stiffness matrix

$Z(s)$	$= Ms^2 + Cs + K$, system impedance matrix
$[Z(s)]^{-1}$	= inverse of $Z(s)$
$F(s)$	= Laplace transform of force function $f(t)$
S	= modal matrix
m_r	= modal mass for mode r
c_r	= modal damping for mode r
k_r	= modal stiffness for mode r
$\mathbf{u}_r$	= modal vector for mode r
Q_r	= modal scaling factor for mode r
A_r	= residue matrix for mode r
A_r^*	= complex conjugate of A_r
A_{pqr}	= matrix element in the (p,q)-position of A_r
FRF	= frequency response function
h_{pq}	= the element in the (p,q)-position of transfer function matrix H

Chapter Eight

δ_{ij}	= deflection at point i owing to a unit force applied at point j, m
$f_1, f_2, \ldots$	= forces applied at stations 1, 2, …, N
$x_1, x_2, \ldots$	= deflections at stations 1, 2, …, m
$[\delta]$	= flexibility matrix
$[k]$	$= [\delta]^{-1}$, stiffness matrix
W	= work, Nm
L	= length, m
E	= Young's modulus, N/m^2
J	= second moment of area when used with E, m^4
y	= deflection, m
M_1, M_2	= concentrated masses, kg
M	= mass matrix

m_{ij}	= elements of mass matrix M
D	= $[\delta]M$, dynamic matrix
m	= mass, kg
k	= spring constant (stiffness), N/m
W_1, W_2	= weights, N
R_1, R_2, R	= reaction forces, N
T_{max}	= maximum kinetic energy, Nm
U_{max}	= maximum potential energy, Nm
$C_1, C_2, \ldots$	= constants
S_1, S_2	= sweeping matrices
m_n	= mass at station n, kg
F_n^L, F_n^R	= forces applied on m_n at left and right sides, respectively
Z_n	= state vector at station n
$H_1, H_2, \ldots$	= transfer matrices
θ	= angular displacement, rad
T	= torque, N·m
J_n	= polar moment of inertia of the n-th disk on a shaft, kg.m^2
M_n^L, M_n^R	= moments applied on mass m_n from the left and right sides, respectively, Nm
V_n^L, V_n^R	= shear forces applied on m_n from the left and right sides, respectively, N
Ω	= angular velocity, rad/s

Chapter Nine

T	= tension in strings, N
ρ	= mass/length, kg/m; also mass/volume, kg/m^3
θ	= angular displacement, rad
$u(x)$	= deflection, m
$\phi(x)$	= mode shape function
$q(t)$	= generalized coordinate

A = cross sectional area, m^2; also a constant

L = length, m

G = shear modulus, N/m^2

E = Young's modulus, N/m^2

J = second moment of area when used with E, m^4

J_p = polar moment of inertia of cross sectional area when used with G, m^4

T = kinetic energy, Nm

U = potential energy, Nm

T_{max} = maximum kinetic energy, Nm

U_{max} = maximum potential energy, Nm

c = $\sqrt{T/\rho}$, ρ = mass/length, for lateral vibration of strings, kg/m

c = $\sqrt{AE/\rho}$, ρ = mass/volume, for longitudinal vibration of rods, kg/m^3

c = $\sqrt{G/\rho}$, ρ = mass/volume, for torsional vibration of rods, kg/m^3

a = $\sqrt{EJ/\rho}$, ρ = mass/length, for lateral vibration of beams, kg/m

γ = mass/volume, kg/m^3

M = bending moment, Nm

V = shear force, N

m_{ij} = generalized mass

k_{ij} = generalized stiffness

Q_n = generalized force for generalized coordinate q_n

ω_n = natural frequency, rad/s

ω_c = natural frequency corrected for the effect of rotary inertia and shear deformation, rad/s

Chapter Ten

I_d = moment of inertia of a disk about its center line, kg·m^2

T_d, T = period of torsional vibration, s

I_L	= moment of inertia of an engine about suspension rod, kg·m^2
$\vec{\lambda}$	= unit vector parallel to suspension rod
H_x, H_y, H_z	= heights of ends of a test reference frame relative to a plane, m
H_o	= height of origin of the test reference frame relative to a plane, m
α, β, γ	= angle displacements, rad, deg
θ, ϕ, ψ	= Euler angles, rad, deg
$[I]$	= inertia matrix
I_{ij}	= elements of $[I]$
$\{x\}$	= eigenvector
λ	= eigenvalue
$\mathbf{n}_x, \mathbf{n}_y, \mathbf{n}_z$	= unit vectors
$\hat{\mathbf{n}}_x, \hat{\mathbf{n}}_y, \hat{\mathbf{n}}_z$	= unit vectors
$\mathbf{N}_x, \mathbf{N}_y, \mathbf{N}_z$	= unit vectors
$\hat{\mathbf{N}}_x, \hat{\mathbf{N}}_y, \hat{\mathbf{N}}_z$	= unit vectors
$\lambda_{Xx}, \lambda_{Xy}, \ldots$	= direction cosines
S	= transfer matrix
$q_1, \cdots, q_6$	= auxiliary variables
x_c, y_c, z_c	= coordinates of the mass center of an engine, m
m	= mass of the engine, kg
F_x, F_y, F_z	= reaction forces of a bushing due to translational motion of engine, N
F_{RX}, F_{RY}, F_{ZR}	= reaction forces of a bushing due to torsional motion of engine, N
M_X, M_Y, M_Z	= reaction moments of a bushing due to translation motion of engine, Nm
M_{RX}, M_{RY}, M_{RZ}	= reaction moments of a bushing due to torsional motion of engine, Nm
k_p, k_q, k_r	= stiffness of a bushing in principal directions, N/m

$k_{xx}, k_{xy}, \ldots$ = stiffness of a bushing in global coordinate system, N/m

a_x, a_y, a_z = coordinates of a bushing in global coordinate system, m

$[M]$ = mass matrix

$[K]$ = stiffness matrix

K_{ij} = elements of $[K]$

$\{u\}$ = modal vector

ω = frequency, rad/s

T = total kinetic energy, Nm

T_x^*, T_y^*, T_z^* = translational kinetic energy, Nm

$T_\alpha^*, T_\beta^*, T_\gamma^*$ = rotational kinetic energy

T_x, T_y, T_z = $T_x^*/T, T_y^*/T, T_z^*/T$, translational mode coupling indicators

$T_\alpha, T_\beta, T_\gamma$ = $T_\alpha^*/T, T_\beta^*/T, T_\gamma^*/T$, torsional mode coupling indicators

$[lb]$ = lower bound of design variables

$[ub]$ = upper bound of design variables

Chapter Eleven

M = mass matrix

K = stiffness matrix

Z = system impedance matrix

C = cofactor of a matrix

ω = angular frequency variable, rad/s

FRF = frequency response function

ς = damping ratio

λ_r = $\sigma_r + j\omega_r$, system pole

j = $\sqrt{-1}$

σ_r = damping factor

ω_r = damped natural frequency

A_r = residue matrix for mode r

$\mathbf{u}_r$ = modal vector for mode r

Q_r	= modal scaling factor
*	= complex conjugate
N	= number of modal frequencies (also sample data block size)
s	= Laplace variable (also the number of time shifts in FRF matrix)
$Z(s)$	= system impedance
r	= mode number
A_{pqr}	= residue for mode r at output point p due to input at point q
A^*_{pqr}	= complex conjugate of A_{pqr}
$h(t)$	= impulse response function
$\{L_r\}$	= modal participation factor column matrix for modal r
Δt	= sampling period, s
$\hat{H}_1$	= FRF estimator; the noise is assumed in the output
$\hat{H}_2$	= FRF estimator; the noise is assumed in the input
$\hat{H}_V$	= FRF estimator; the noise is assumed in both input and output
$\gamma^2_{oX_iF_i}(\omega)$	= ordinary coherence function
$\gamma^2_{mX_iF}(\omega)$	= multiple coherence function
$\gamma^2_{pijxg}(\omega)$	= partial coherence function
T	= $N\Delta t$, record length, s
$f_{\max}$	= maximum frequency, Hz
f_s	= $2f_{\max}$, sampling frequency, Hz
CEA	= complex exponential algorithm
LSCE	= least square complex exponential method
PTD	= polyreference time domain method
ITD	= Ibrahim time domain method
ERA	= eigensystem realization algorithm

Chapter Twelve

$\mathbf{x}_k$	= response vector of subsystem k, m
H_k	= FRF matrix of subsystem k
$\mathbf{f}_k$	= force vector applied on subsystem k, N
$\mathbf{x}$	= response vector of total system, m
$\mathbf{f}$	= force vector applied on total system, N
H	= FRF matrix of the total system
B, C	= coefficient matrix used in constraint equation
Φ	= orthogonal complement matrix of $[BH + C]$
η	= generalized force component
σ_r	= singular values of $[BH + C]$
Σ	= diag(σ_1, σ_2, …)
m_d	= mass of vibration absorber, kg
c_d	= damping coefficient of vibration absorber, Ns/m
k_d	= stiffness of vibration absorber, N/s
f_d	= natural frequency of damper, Hz
f_c^d	= coupling force applied on damper, N
f_c^p	= coupling force applied on primary system, N
ς_d	= damping ratio of vibration absorber
α_{opt}	= optimal frequency ratio for damper tuning
ς_{opt}	= optimal damping ratio for damper tuning
Z_{rr}^A	= impedance matrix of subsystem A at points other than coupling points.
Z_{cc}^A	= impedance matrix of subsystem A at coupling points
X_r^A	= response of subsystem A at points other than coupling points
X_c^A	= response of subsystem A at coupling points
F_c^A	= force vector applied on coupling points of subsystem A, N

F_r^A	= force applied on points other than coupling points of subsystem A, N
H_{11}^a, H_{12}^a	= baseline accelerance matrix of a primary system
$\hat{H}_{11}^a, \hat{H}_{12}^a$	= new accelerance matrix of the primary system with a mass damper
H_{11}^x, H_{12}^x	= baseline receptance matrix of a primary system
$\hat{H}_{11}^x, \hat{H}_{12}^x$	= resulting receptance matrix of the primary system with a mass damper

Index

Note: page numbers followed by *f* indicate a figure; page numbers followed by a *t* indicate a table.

F

G

H

I

K

L

M

N

V

About the Authors

Ronald L. Huston is professor of mechanics in mechanical engineering at the University of Cincinnati. He also holds the titles distinguished research professor and Herman Schneider professor.

Dr. Huston has received a B.S. in mechanical engineering (1959), an M.S. in engineering mechanics (1961), and a Ph.D. in engineering mechanics (1962), all from the University of Pennsylvania in Philadelphia.

Dr. Huston has been a member of the faculty at the University of Cincinnati since 1962. During this time he has held a number of administrative positions, including head, Department of Engineering Analysis; interim head of chemical and materials engineering; director of the Institute of Applied Interdisciplinary Research; and acting senior vice president and provost. He has also served as a secondary faculty member in the Department of Biomedical Engineering and as adjunct professor of orthopedic surgery research.

From 1979 to 1980, on special leave from the university, Dr. Huston served as division director of civil and mechanical engineering at the National Science Foundation. In 1978, he was visiting professor of applied mechanics at Stanford University.

Dr. Huston's research interests are primarily in dynamics, with applications in multibody dynamics, biomechanics, and automotive engineering.

Dr. Huston has written more than 150 journal articles and 150 conference papers, as well as 75 book reviews and 7 books. He has served as a technical editor of *Applied Mechanics Reviews*, as associate editor of the *Journal of Applied Mechanics*, and as book review editor of the *International Journal of Industrial Engineering*.

C. Q. Liu is a Senior Technical Specialist at the Chrysler Corporation LLC, Auburn Hills, Michigan, USA. He is the recipient of the Chrysler Corporation Recognition Award (1999). He has over 30 years of experience as a noise, vibration, and harshness (NVH) specialist and 13 years of university teaching and research experience. He is the author/coauthor of four books and over 40 journal articles.

Dr. Liu has extensive analysis and experimental skills in NVH of vehicle and powertrain systems. He is sought after by peers as a consultant on hybrid finite element analysis (FEA) and the experimental (HYFEX) method for vehicle axle noise analysis, FRF-based substructure synthesis, experimental modal analysis, transfer path analysis (TPA), driveline sensitivity analysis, balance technologies, engine mounting system optimization design, vehicle instrumentation, data acquisition, digital signal processing, and engineering interpretation of results.

Dr. Liu received a B.S. degree in Engineering Mechanics (1969) from Tsinghua University, Beijing, China; an M.S. degree in Applied Mechanics (1982) from Chongqing University, Chongqing, China; and a Ph.D. degree in Mechanical Engineering (1991) from the University of Cincinnati, Ohio, USA.